Extreme Value Theory with Applications to Natural Hazards

Nicolas Bousquet · Pietro Bernardara
Editors

Extreme Value Theory with Applications to Natural Hazards

From Statistical Theory to Industrial Practice

Springer

Editors
Nicolas Bousquet
EDF Lab, Saclay-Chatou, and
Sorbonne Université
Paris, France

Pietro Bernardara
EDF Research and Development
London, UK

ISBN 978-3-030-74941-5 ISBN 978-3-030-74942-2 (eBook)
https://doi.org/10.1007/978-3-030-74942-2

Translation from the French language edition: *Événements naturels extrêmes : théorie statistique et mitigation du risque* by Nicolas Bousquet and Pietro Bernardara, © Lavoisier 2018. Published by Edition Lavoisier. All Rights Reserved.

This Springer imprint is published by the registered company Springer Nature Switzerland AG
The registered company address is: Gewerbestrasse 11, 6330 Cham, Switzerland

Acknowledgements

This book could not have been written without the help of many colleagues of EDF R&D and engineering units, our academic collaborators, students, Ph.D. students, postdoctoral fellows, friends, and family. We would like to thank them for their contributions, advice, and support:

Didier Dacunha-Castelle (University of Paris XI), Ambre Diet (ENSAI), Mathieu Couplet (EDF), Antonin Couturier (ENSAI), Sandrine Dyèvre (EDF), Stéphane Fortier (EDF), Anne-Laure Fougères (Camille Jordan Institute), Jean-Philippe Fontes (EDF), Roberto Frau (EDF), Marie Gallois (EDF), Stéphane Girard (INRIA), François Gourand (Météo-France), Nicole Goutal (EDF), Kaniav Kamary (AgroParisTech), Merlin Keller (EDF), Bertrand Iooss (EDF), Tiphaine Le Morvan (EDF), Cécile Mercadier (Camille Jordan Institute), Vincent Moutoussamy (EDF), Eric Parent (AgroParisTech), Alberto Pasanisi (EDF), Hélène Pesme (EDF), Florent Pezet (ENSAI), Anne-Laure Popelin (EDF), Fabrizio Ruggeri (CMR IMATI Milan). Thanks also to Amandine Marrel (CEA), whose post-review advices helped us to clarify the text.

The EDF authors would like to thank Emmanuel Garnier, CNRS Research Director and specialist in climate history, for having kindly produced a major scientific contribution opening this book, and introduced its authors to the richness and variety of this disciplinary field, as well as Hervé Boll, Deputy Director of Programmes at EDF R&D, and Jean-Paul Chabard, Scientific Director at EDF R&D, for their encouragement during the writing.

Contents

Contributors

Marc Andreewsky EDF R&D, Chatou, France

Pietro Bernardara EDF R&D, Chatou, France

Nicolas Bousquet EDF R&D, Chatou, France

Anne Dutfoy EDF R&D, Palaiseau, France

Emmanuel Garnier Research Professor CNRS, University of Besançon, Besançon, France

Thi-Thu-Huong Hoang EDF R&D, Palaiseau, France

Emmanuel Paquet EDF-DTG, Grenoble, France

Sylvie Parey EDF R&D, Palaiseau, France

Nicolas Roche EDF R&D, Palaiseau, France

Alain Sibler EDF R&D, Palaiseau, France

Jérôme Weiss EDF R&D, Palaiseau, France

Notations

The following notations are defined at their first occurrence within the text. They will be reused without reminded afterward. Very generally, random variables will be denoted using capital letters, while their realizations will be denoted with usual ones. Vectors and matrices will be written in bold type, unlike scalars.

Probability Distributions

Centered reduced Gaussian	$\mathcal{N}(0, 1)$
Generalized Pareto	$\mathcal{G}_{PD}(u, \sigma, \xi)$
Exponential	$\mathcal{E}(\lambda)$
Gaussian	$\mathcal{N}(\mu, \sigma^2)$
Generalized extreme values	$\mathcal{G}_{EV}(\mu, \sigma, \xi)$
Bernoulli	$\mathcal{B}_e(p)$
Binomial	$\mathcal{B}(N, p)$
Poisson	$\mathcal{P}(\lambda)$

Abbreviations

iid	Independent and identically distributed
rv	Random variable
MLE	Maximum likelihood estimator
LSE	Least square estimator
$\mathbf{X}$	Multidimensional random vector of study (in dimension d)
X	Unidimensional random variable of study
$\mathbb{P}(.)$	Usual probability measure (Lebesgue-dominated)
$\Omega_{\mathbf{X}}$	Canonic notation for the probabilized space of $\mathbf{X}$
$\mathcal{B}(A)$	Borelian $\sigma-$algebra over a space A
$P(A)$	Set of partitions of A

$\mathbf{1}_{\{x\in A\}}$	Indicator function
$\varnothing$	Void or null set
Z	Random variable of interest
$\mathbf{x}_i = (x_{1,i}, \ldots, x_{d,i})$	realization of $\mathbf{X}$
$F_{\mathbf{X}}$	Cumulative distribution function of $\mathbf{X}$
$f_{\mathbf{X}}$	Probability density function of $\mathbf{X}$
$\boldsymbol{\theta}$	Generic notation for a parameter vector of the distribution of $\mathbf{X}$
Θ	Space of parameter vector $\boldsymbol{\theta}$
$F_{\mathbf{X}}(. \mid \boldsymbol{\theta})$	Cumulative distribution function of $\mathbf{X}$, parametrized by $\boldsymbol{\theta}$
$f_{\mathbf{X}}(. \mid \boldsymbol{\theta})$	Probability density function of $\mathbf{X}$, parametrized by $\boldsymbol{\theta}$
$L(\mathbf{x}_1, \ldots, \mathbf{x}_n \mid \boldsymbol{\theta})$	Statistical likelihood of observations $(\mathbf{x}_1, \ldots, \mathbf{x}_n)$ conditional to $\boldsymbol{\theta}$
$\pi(\boldsymbol{\theta})$	Bayesian prior density function of $\boldsymbol{\theta}$
$\pi(\boldsymbol{\theta} \mid \mathbf{x}_1, \ldots, \mathbf{x}_n)$	Bayesian posterior density function given a sample $(\mathbf{x}_1, \ldots, \mathbf{x}_n)$
$\Pi(\boldsymbol{\theta})$	Prior cumulative distribution function
$\Pi(\boldsymbol{\theta} \mid \mathbf{x}_1, \ldots, \mathbf{x}_n)$	Posterior cumulative distribution function
$\mathbf{a}_n, \mathbf{b}_n$	Sequences of vectors related in the statement of limit theorems (e.g., Fisher–Tippett–Gnedenko)
N_u	Random number of exceedances of a threshold u
N_{Cu}	Random number of clusters obtained using the threshold u
$\chi(\chi^{\text{sup}}, \chi^{\text{inf}})$	Tail dependence coefficients (sup / inf)
$\bar{\chi}$	Extremal dependence coefficient under asymptotic independence
$\mathcal{D}(F)$	Attraction domain of a cumulative distribution function F
ℓ	Stable tail dependence function
$C_{\boldsymbol{\theta}}$ or C_F	Copula of parameter or linked to a multivariate distribution F
η	Extremal index (or coefficient)
ρ	Generic notation for a correlation coefficient
$\boldsymbol{\rho}$	Generic notation for a correlation matrix
sign(x)	Sign of x
Supp(f)	Support of density function f
$\mathbb{E}_X[.]$	Expectated value (expectation) of X (X can be removed when non-ambiguous)
$\mathbb{V}_X[.]$	Variance of X
$\mathbb{C}ov_{\mathbf{X}}[.]$	Covariance matrix of $\mathbf{X}$
$\mathbb{M}_k[.]$	Moment of order k
$\mathbb{M}_{k,r,s}[.]$	Weighted moment of orders (k, r, s)
x_p	Return level associated to a probability p
p	Probability of exceedance (e.g., annual) $\mathbb{P}(X > x_p)$
T	Time period of interest (e.g., $T = 1/p$)
$\mathbb{R}$	Set of real numbers
$\mathbb{N}$	Set of integer numbers

z_α Quantile of order α of the centered reduced Gaussian distribution

$\phi(.)$ Density function of the centered reduced Gaussian distribution

L^2 Space of square-integrable functions

$\mathcal{C}$ Generic notation for a functional regularity class

$< ., . >$ Canonic scalar product

A^T Transpose of A

$\mathrm{tr}(A)$ Trace of A

$\mathrm{diag}(A)$ Diagonal vector of A

$| A |$ Determinant of A

$\nabla(X)$ Gradient of X

$X_1 \vee X_2$ Component-wise maximum

$\xrightarrow{\mathcal{L}}$ In-law convergence

$\xrightarrow{\mathbb{P}}$ Convergence in probability

$\xrightarrow{a.s.}$ Almost sure convergence

log Neperian logarithm

exp(.) Exponential function

Maps

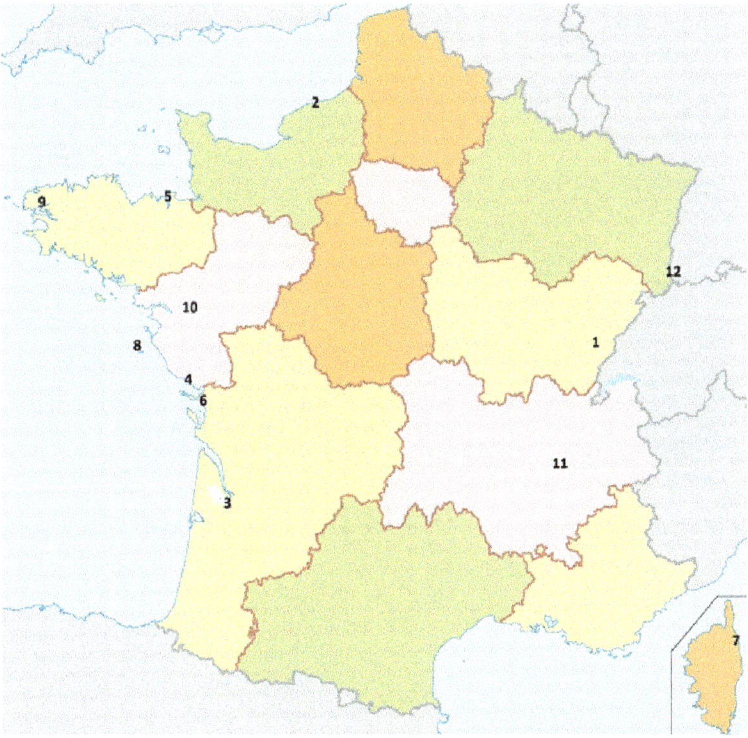

Location of the main French sites studied in this book: Besançon (1), Gravelines (2), Bordeaux (3), La Faute-sur-Mer (4), Saint-Servan (5), La Rochelle (6), Penta-di-Casinca (7), Ile d'Yeu (8), Brest (9), Nantes (10), Grenoble (11), Kembs (12)

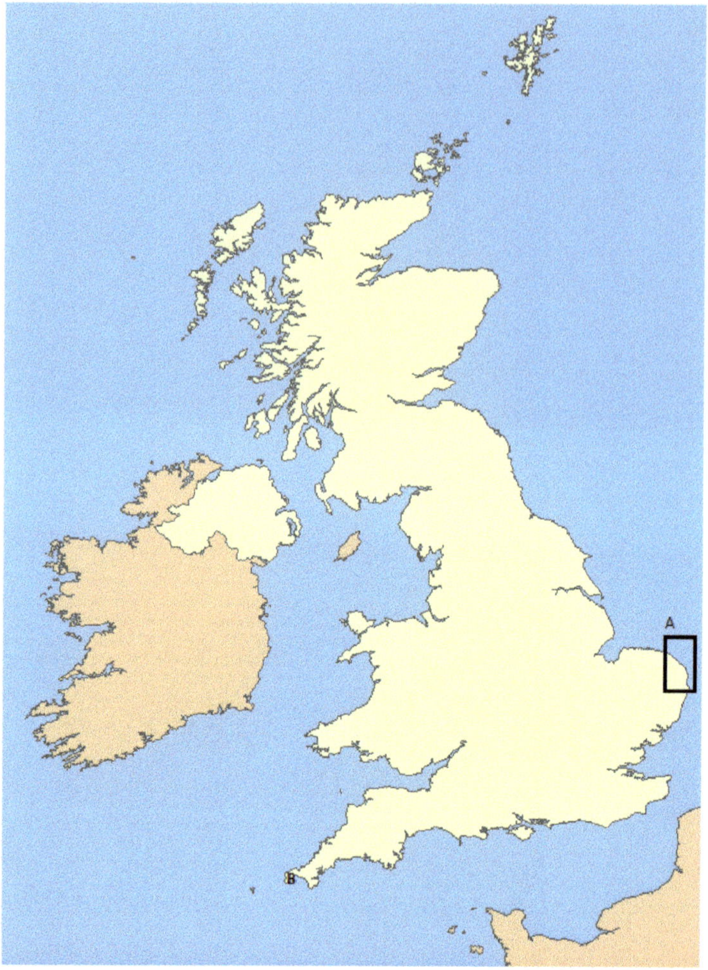

Location of the main British sites studied in this book: North Norfolk Coast (A), Newlyn (B)

Chapter 1
Extreme Events and History: For a Better Consideration of Natural Hazards

Emmanuel Garnier

Abstract This preface aims to remind us of the fundamental importance, for our modern but fragile societies, of the difficult study of historical data on natural hazards. Forgetting is a regular feature of our decisions, and the preservation of the memory of natural disasters depends heavily on our ability to quantify their effects. The author, a climate historian, recalls here some fundamental aspects of this discipline and its necessary connection to statistical approaches to risk analysis.

At a first glance it may seem surprising, if not debatable, to ask a historian to introduce a work entitled *Extreme Value Theory with Applications to Natural Hazards* written by seasoned and experienced engineers. However, the book's contributors have indeed judged it sound to bestow me with the great honor of writing the introduction to this educational collective work.

De facto, this book is based on two major observations. First, the obvious scarcity of textbooks on extreme value statistics, and directly related to this, the difficulties encountered by students and engineers in improving their expertise in this strategically important domain. The second observation—*deo gratias!*—legitimizing to a certain extent the presence of a historian, is an acknowledgement of the fragility of statistical methods when forced to confront unpredictable events, like for instance a beaver dam in Quebec which made a mockery of patiently developed hydrological models and led to a disastrous flood.

However, by diving deeply into the archival veins of ancient societies, the historian injects a plethora of "unpredictable events" that can potentially disrupt models, a reality that explains the sustained interest in the world of reinsurance for this type of information [316].

Is more proof needed to convince you? In 2008, as I was conducting research in a renowned French climatology laboratory, I shared with my colleagues and other

E. Garnier (✉)
Research Professor CNRS,
University of Besançon, Besançon, France
e-mail: egarnier.cea-cnrs@orange.fr

N. Bousquet and P. Bernardara (eds.), *Extreme Value Theory with Applications to Natural Hazards*, https://doi.org/10.1007/978-3-030-74942-2_1

disaster management specialists the historical risk of a volcanic eruption affecting the European continent. Deeply disconcerted, my interlocutors politely helped me understand that this was extremely unlikely and that such an idea probably came from an over-interpretation of historical sources which were, "by definition, unreliable". I evoked the terrible precedent of the Icelandic volcano Laki in the summer of 1783, which killed tens of thousands of people, but they were still not convinced, merely 2 years later the Eyjafjöll volcano erupted, with well-known significant economic and social consequences [52, 309].

1.1 A Neglected Gold Mine: Historical Archives

To study extreme events of the past, historians use direct and indirect data concurrently. The former refers to archival sources, which describe, if one is interested in the climate, the weather conditions on a given date. Though this information comes mainly from instrumental measurements—like meteorological readings in personal diaries or by scientists—it also contains precise descriptions of extreme weather events. In contrast, the latter represents the influence of meteorology and geology on the natural components of the biosphere and the hydrosphere. More specifically, it provides indirect information on floods, droughts, storms, volcanoes, earthquakes, log-jams, and vegetation.

Numerous and varied phenological sources allow us to study the influence of climate variation on annual or periodic phenomena related to vegetation (germination, flowering, and dates at which fruit ripen) and the animal world (birdsong, migration, nest-making, etc.). Such records, existing from the thirteenth century on in France, haphazardly jumble things like official grain, grape, and other fruit (apples, olives, etc.) harvest dates. Although these are certainly not forgotten by paleoclimatologists, their use remains open to discussion. Their users are frequently criticized for not taking into account a number of fundamental anthropic parameters such as cultural changes and geopolitical and epidemiological contexts.

It is at levels of uncertainty like this that the historian—an archival specialist *par excellence*—intervenes to "separate" time series data from its social imprint. For instance, the contextualization of the grape harvest banns of the city of Besançon (cf. Figure 0.1) between 1525 and 1850 proved that military events and plague epidemics played a decisive role in the grape harvest dates [318]. Also, we should not forget that the use of phenology to reconstruct climates of the past is limited by seasonality: primacy is given to seasons in which plants grow best.

An inexhaustible source of Clio's disciples, administrative archives concurrently offer two major advantages: chronological continuity (generally from the fifteenth century) and homogeneity of information. Town archives contain first-hand sources on municipal debates and city events in which catastrophes are omnipresent. Any extreme event had an impact on infrastructure (bridges, mills, canals, etc.) and the fragile socio-economic balance of a town. It therefore provoked debate, as well as decision-making, on the part of elected representatives. Moreover, urban sources solve the problem of seasonal representativeness. Indeed, unlike the countryside, the off-season does not exist in towns, and elected representatives discuss summer droughts just as much as winter ice-jams.

Engaged in a centralization phase, the Princely States were endowed with new bureaucratic tools as early as the seventeenth century. The process is particularly evident in France, where the monarchy created specialized archive-producing administrations. The windthrow register (trees uprooted by wind), maintained since the 1560s by the forestry administration, as well as the reports written by the Admiralty of the Royal Navy along the coast, describing extreme maritime events (storms, submersions, tsunamis, etc.), are two important examples of such specialized archives. The first modern meteorological observations in Europe begin to be collected at this time, with the founding of the *Observatoire de Paris* (1669) and its sister institution, the *Académie royale des sciences*— and a century later the *Société royale de médecine* in France (1778), the *Royal Society* of London, and the *Societas Meteorologica Palatina* in Mannheim [308].

Additional sources such as personal diaries and family ledgers can also be used. These too mention catastrophic events that have occurred. More surprisingly perhaps, archives of the Catholic and Protestant churches around the world provide a very wide range of extraordinary documents. These include references to rain and other (wished-for) weather processions, depending on the climate of the time, which

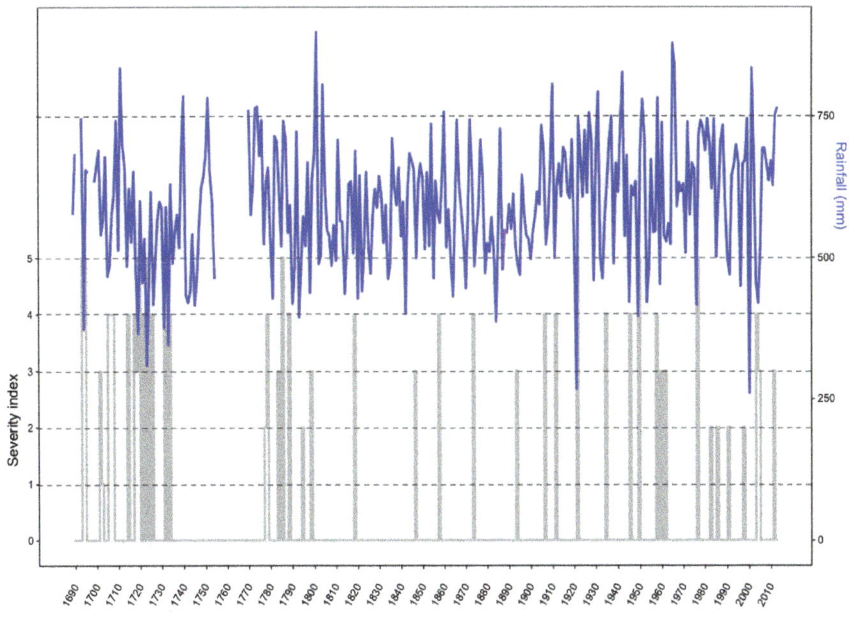

Fig. 1.1 Comparison of droughts in France (right-hand side y-axis: quantified according to the *Historical Severity Drought Scale* (HSDS)) with the annual rainfall in Paris (left-hand side y-axis: in mm) in the period 1689–2013. *Source* Garnier [316]

provide information on severe events like droughts, floods, and extreme cold. Another very fruitful track is parish registers in which the priest or pastor recorded deaths. These make it possible to apprehend the demographic effect of a crisis or the impact of a storm on a church steeple. Purely Catholic, the *ex-voto* translates into visual form past catastrophes and the social perception of them.

More than mere commemorative paintings, the point of these at a time of strong religious practice was to recall persisting dangers (maritime, volcanic, and climatic) to the community [613]. Such ecclesiastical sources can inform natural hazard researchers about often unexpected parts of the world. Would it surprise you that, for example, it is now quite possible (if the financial means are there) to reconstruct a history of natural disasters in China from the archives of French missionaries? In its highly bureaucratic way, the Catholic church generated thousands of reports between the 1600s and the mid-twentieth century. Well-informed in the sciences, priests carried out systematic meteorological surveys and recorded all the disasters (tsunamis, submersions, earthquakes, etc.) that they had witnessed (Fig. 1.1).

1.2 Turning Words into Figures

In a step-by-step way, the process consists of tracking in archives—as any historian does—non-instrumentally based information appearing in the form of direct or indirect data, then "converting" it into a measure of climate variation (temperature, precipitation, and barometric pressure) during the selected period. This is where the most daring scientific challenge lies and extreme care is required, where it is necessary to transcribe this predominantly word-based information into actual temporal data series.

This process can only be implemented starting with the creation of a database in which all information brought together is organized and placed in chronological order so as to offer an overall view of the results, and furthermore, facilitate statistical analyses. Nevertheless, it is important to note that to be useful, these data must be collected at relevant geographic scales (by region, for instance). Otherwise, biases related to climatic and topographic variability are likely to lead to aberrant time series. To date, around 100 000 text-based and instrumentally based data points have been collected. Far from being definitive, this database will grow due to progress and projects in the years to come.

One item of data here corresponds to one historical reference but not necessarily one data point, as the same event can easily involve two or more phenomena. For instance, a violent storm may be accompanied by heavy precipitation. From left to right in a given row of the database presented in Table 1.2, we have the event, the precise source, the estimated reliability of the source, the place, the year, the exact date when indicated, and lastly the text itself, in order that any researcher can see in detail the original source when needing to retrieve accurate information or reinterpret the content.

Table 1.1 Severity categories and associated level of damage for historical floods before the arrival of measurement instruments

5	Very severe event in terms of geographic magnitude (local, regional, national), financial and societal cost (shortages, riots, large numbers of deaths)
4	Significant regional damage: hydraulic infrastructure, bridges, farms, cattle, harvests, communication lines
3	Significant damage but limited to a few localities or a city: costs, shortages, drowned cattle, human deaths
2	Light and local damage: a few villages, agricultural land, wetlands
1	Event mentioned in sources, limited and localized damage
–1	Mentioned in historical sources but nothing more

After this preliminary step, it remains to "derive" values from these archives by transforming the basic data into meteorological severity indices (temperature and precipitation) or seismic measures (the EMS[1] scale), for instance, which make it possible to interweave time-marked text sources with instrument data [313]. To do this, the descriptive content of written archives can be used to establish an intensity scale for catastrophic events of the past, provided that historians, engineers, modelers, and/or insurers work well as a team. Taking the example of storm surge, it is easy to compare an event of the past with its modern-day counterpart by looking at the damage caused to infrastructure and society, such as the destruction of dunes, dykes, flooded areas, and even loss of life [317] (Table 1.1). In this way, the intensities indexation system can lead to drawing regular curves over extremely long periods of time, meaning that comparisons can now be considered as statistically relevant (Fig. 1.2).

1.3 Some Extreme Hazards in the Light of History

In practice, historical documents are riddled with "monstrosities" and "disturbances", expressions used in the past to designate what we now call extreme hazards or natural disasters. And for good reason: calling an event catastrophic, both now and in the past, implies that humans were or have been impacted. On the climate front, our decision-makers, media, and even a part of the scientific community are constantly asserting that we will experience an upsurge and intensification of natural disasters in the future. Of all the uncertainties and interrogations raised by this debate, a major question emerges: can recent extreme events be studied with a view to comparing them to historical events, in order to more accurately assess social, economic, political, and cultural consequences? To the joy of historians, the extreme nature of these types of events has left a trace across historic sources, like coins falling one by one from a

[1] European macroseismic scale, which measures the strength and severity of earthquakes.

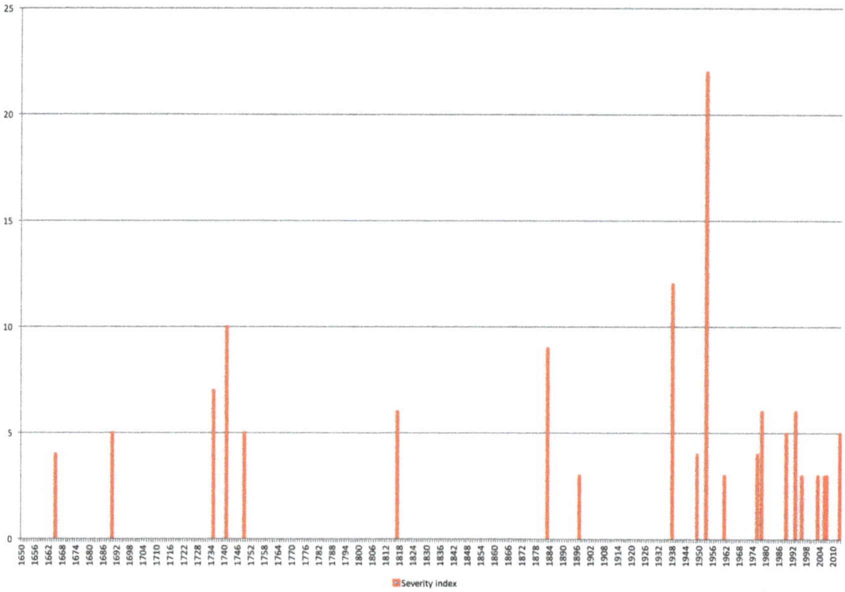

Fig. 1.2 History of storm surge impact on the North Norfolk coast (the United Kingdom, cf. Fig. 0.2). The impact score (y-axis) is based on many types of damage, such as flooding of buildings, loss of life, crops, and livestock, flooded areas, damage to infrastructure, and financial cost. *Source* [317]

pocket with a hole in it, enough even to reconstruct an extremely robust history of the worst natural events of the past five centuries.

1.3.1 Continental Storms

Based on archival analysis, the verdict of history refutes the comments of those who, in the aftermath of the huge European storm of December 1999, insisted on its exceptional wind strength. During France's *Ancien Régime*, for example, it would have been more appropriate to speak of the "storm of the century", as the elements tend to be unleashed with regularity due to climatic conjunctures at given times. The five hundred years (1500–2000) represented in Fig. 1.3 thus reveal a chronology of contradictory peaks and troughs in the number of severe events recorded. Over the period as a whole, 22 events between storm (force 10) and hurricane (force 12) strength affected northern Europe, with a return period of about 15 years, with 80% of them occurring in the months of January, February, and March. However, we should be careful not to draw any general conclusions from this, because the apparent return periods do not take into account geographical and temporal variability.

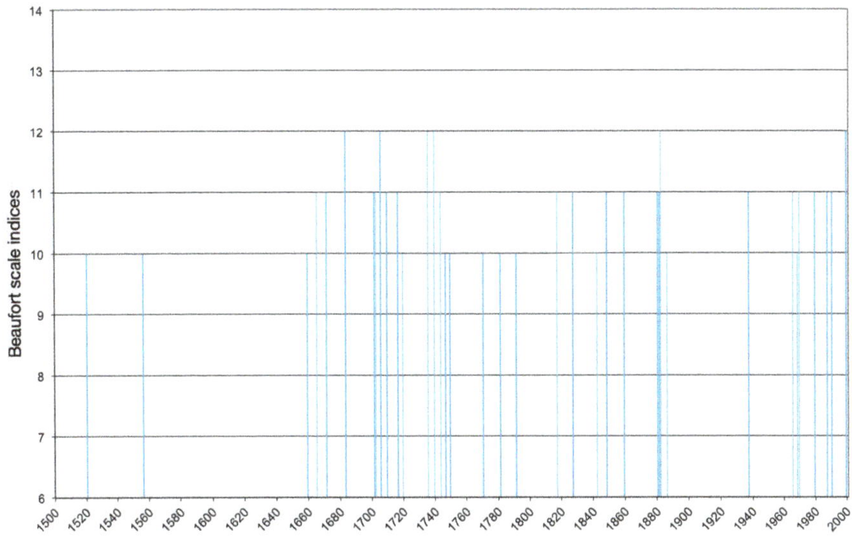

Fig. 1.3 Severe storms (force $\geq$ 10) in Europe from 1500 to 2000. *Source* Garnier [316]

Nevertheless, certain periods appear more critical than others; for instance, records of the forestry administration indicate windthrow years clustered around 1580–1640, 1710–1770, and 1870–1920. If we consider only the most severe events, France was touched by 15 storms, whose force was between 10 and 12, during this period. The early modern period (sixteenth and seventeenth centuries) is a poor sibling—though gaps in the archives are probably numerous—compared to the last century of the *Ancien Régime* (before 1789), where eight disasters were recorded, five in the short period 1720–1760. This increase in wind speed was not entirely a matter of chance, however. It suffices to cross-reference it with the global climate context of the time to see that it developed during a relative warming period accompanied by strong low-pressure activity.

Subsequently, the meteorological situation continued to improve despite a few returning cold snaps, induced by the last counter-offensives of the Little Ice Age in the 1780s. However, the end of the eighteenth century was not exempt from severe storms, as evidenced by the events of January 17, 1784, which devastated the Atlantic coast. The national repercussions were such that the prestigious *Société royale de médecine* carried out an investigation in the days following the event. It resulted in the drafting of a comprehensive report on meteorological conditions and damage suffered at the time. It must be said that on this day, Aeolus had a particularly heavy hand, causing the sinking of some 30 ships and the death of around 50 people [310].

In the following century, in a Europe engaged in the industrial revolution, France had an impressive sequence of violent storms and hurricanes that ravaged both the north and the south in the years 1842, 1869, 1872, 1876, and 1894. Nevertheless, major natural catastrophes were in fact more numerous in the eighteenth century (9 events) than in the twentieth (7 events), even when taking into account the violent Lothar and Martin windstorms of 1999.

1.3.2 Droughts

Drought has always marked rural and urban societies and is regularly mentioned in historical sources. For the record, note that the general term "drought" covers several different things, though in its most frequent usage, it describes a lack of rainfall.

In this sense of the word, and being unable to work with variables giving the deviation from mean or normal rainfall—non-existent before the middle of the eighteenth century—the usual approach consists in relying on quantitative mentions found in sources, like duration, frequency, and geographic extent. This is acceptable given that today, both for the World Meteorological Organization and *Météo-France*, drought severity is characterized by the number of days without rain. Archives have, however, the immense advantage of not normally introducing ambiguity, typically specifying both the beginning and end dates (most often religious) of dry periods, as well as numerous mentions like "great hurry to mill because of the low water" and indications that "fish are dying in the rivers because of the drought". It would be wrong though to exaggerate by suggesting that droughts mentioned in the archives correspond to a total absence of precipitation. Such droughts—uncovered by historians—simply refer to long and extremely dry episodes that severely affected societies and economies [312].

In the case of France, 68 droughts were recorded between 1500 and 2014. These are quite unevenly distributed, chronologically speaking, and correspond to a wide range of levels of severity (Fig. 1.4).

The disparities are even more striking when these events are looked at over 50-year periods. Three main trends can be seen since the sixteenth century. The first is between 1500 and 1700, characterized by a sustained number of droughts with on average one every 8 years. A big change occurs in the eighteenth century, which sees a dramatic increase in the number of droughts, 21 in all, with an impressive peak visible in the first half of the century (14 droughts). During this time, a drought occurred on average once every 3.5 years, which is not even close to being matched over the rest of the 500 years studied. Never before 1700 or after 1800 was this kind of frequency observed. This string of dry episodes is particularly well documented in archives and in the proceedings of the *Académie royale des sciences* in Paris. Indeed, the scientists of the *Observatoire de Paris* were particularly concerned about this long dry and hot period. For this reason, a large number of instrument data and observations are readily available for today's historians [308].

The nineteenth and twentieth centuries are characterized by more frequent droughts, even though the Little Ice Age faded away from the 1850s on. Nevertheless, the twentieth century has witnessed a real upsurge, though without it being possible to observe a turning point from 1950 onwards. Remember, however, that this

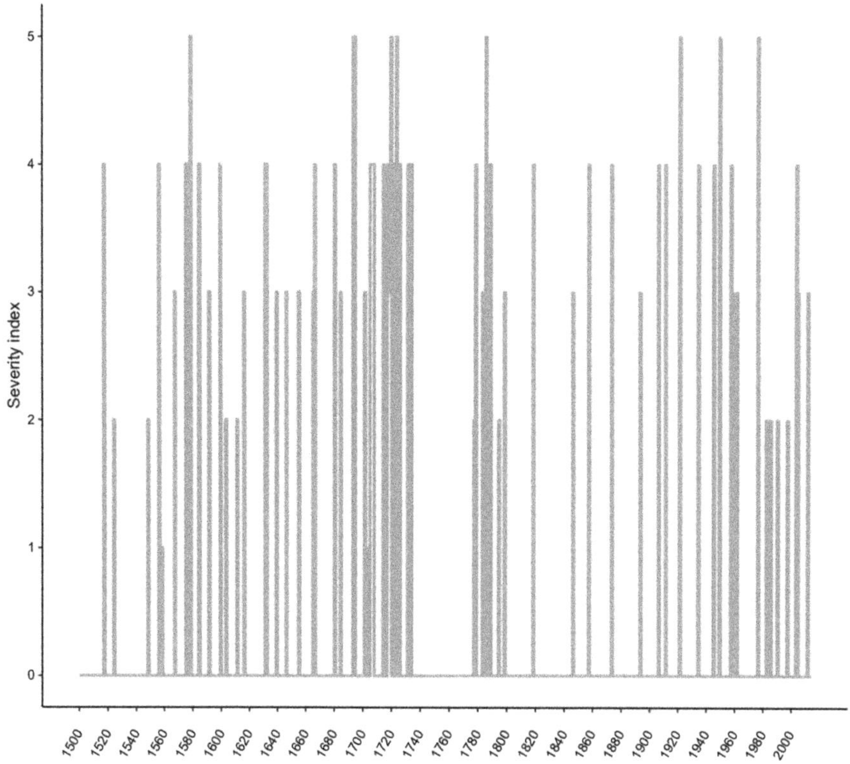

Fig. 1.4 Chronology and severity of droughts in France from 1500 to 2014 according to the Historical Severity Drought Scale. *Source* Garnier [316]

increase is notable only in the context of a 200-year comparison. The level remains much below the records of the eighteenth century.

The available British records allow us to identify 42 droughts of varying severity on the Historical Severity Drought Scale (HSDS). The chronology thus reconstructed (Fig. 1.5) shows very large differences from one century to the next, both in terms of frequency and severity.

First, the distribution of extreme events by half-century shows the great arid phases of Great Britain's history since the 1500s. There is a clear break between the period before 1700 and all that comes after. In the former, only 11 events occurred, whereas there were 31 in the latter. This low number before 1700 has two possible explanations. The first is simply gaps in the archives, while the second—probably the more relevant—is linked to the climate context of the period. Indeed, the sixteenth and seventeenth centuries corresponded in Europe to a very severe phase of the Little Ice Age, with many more humid and cool seasons, especially between 1540–1640 and 1683–1693.

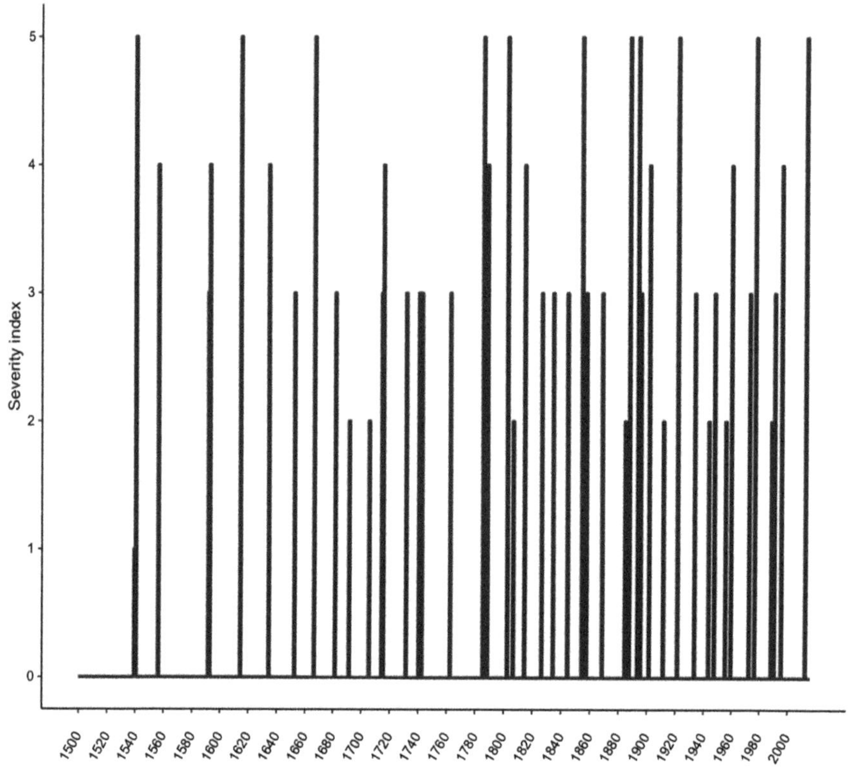

Fig. 1.5 Chronology and severity of droughts in Great Britain from 1500 to 2014 according to the Historical Severity Drought Scale. *Source* Garnier [316]

After the 1700s, droughts appear to be much more frequent. The first period corresponds to the years 1700–1750 during which the European continent underwent a new set of climatic conditions. From 1705 until the end of the 1730s, the climate was warmer and dryer, although 1709 still remains today the year of the "great frost" in Europe. Moreover, several droughts like those of 1714, 1715, and 1740 began during very cold and dry winters. The second break occurs after 1800 with an increased frequency of droughts, most probably attributable to the beginning of global warming during 1830–1850. Henceforth, every 50-year period suffered six to seven drought events. We note in passing that there has been no increase in this since the 1950s.

Lastly, monthly precipitation data from Oxford, collected since the eighteenth century, can be brought together with the HSDS drought assessment (Fig. 1.6). There are good matches between the historical events recorded in the archives and periods of rainfall deficit. Nevertheless, droughts of severity 5 do not consistently correspond to the lowest precipitation. For instance, the extremely severe droughts of 1785 and 1976 do not coincide with the lowest precipitation levels in Oxford. Reasons for this

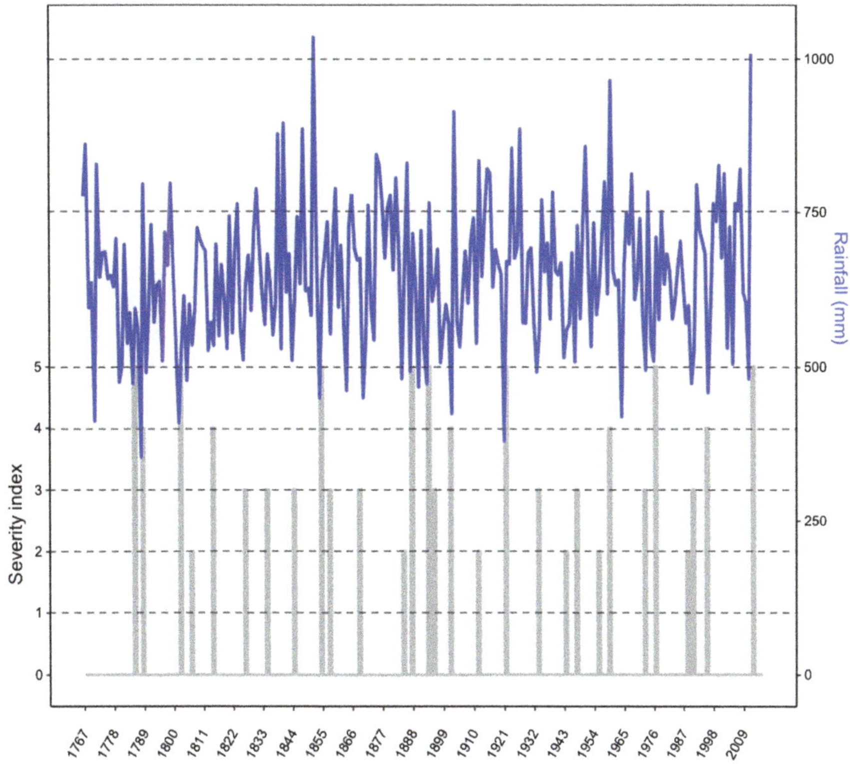

Fig. 1.6 Plot of British droughts (measured on the HSDS scale) superimposed on precipitation in Oxford (right-hand side *y*-axis: in mm) between 1767 and 2012. *Source* Garnier [316]

may include local rainfall conditions which may be different from the rest of the country, and other meteorological factors—like wind—that may have exacerbated droughts [314].

1.3.3 Storm Surges

On February 28, 2010, at around 2 am, Cyclone Xynthia struck France's Atlantic coast, mainly between the south of Brittany and Bordeaux, causing human and material losses concentrated in Vendée and Charente-Maritime. Dykes, dunes, and other structures were overcome, resulting in the flooding of more than 50,000 hectares and around 50 deaths. Among the reasons for the high human and material cost, the failure to take historical experience into account—in particular, the North Sea disaster of 1953—played a major role, thus increasing the exposure of coastal societies to such events (Table 1.2).

Table 1.2 Extract from a French historical database (Gravelines, North Sea). *Source* Garnier (2018)

Hazard type	Source name	Quality index of the source	Exact location	County (Province)	Date	Description
Sea surge	Arch. Mun. Saint-Omer, BB 34	4	Saint-Omer	Pas-de-Calais (Nord-Pas-de-Calais)	12/011613	*“Le 12 de janvier, les vents et les pluies continuelles augmentant de plus en plus les inondations causent partout des pertes et des dommages considerables et principalement dans la plus part de tous nos biens de Gravelines, Bourbourg, pays de Langles, & dont toutes les terres n’étoient qu’une mer. Depuis 1606 (plutôt 1609) on ne vit point ni pareilles tempêtes ni de telles inondations. Mgr l’évêque de St Omer ordonna des prières publiques avec litanies pour implorer la divine misericorde”*
Flood	Arch. Mun. Saint-Omer, BB 34	4	Saint-Omer	Pas-de-Calais (Nord-Pas-de-Calais)	08/031610	*“Les eaux du ciel se sont tellement “repandues dans le pais quelles l’ont tout a fait inondées et empeché la naviguation et la chatelnie de bourbourg a fait fermer des ecluses le vingt neuf decembre, ainsy point encore de naviguation a esperer pour le suppliant”*
Frozen sea	Arch. Mun. Dunkerque, AncDK2, folio 67	4	Dunkerque	Nord (Nord-Pas-de-Calais) .	14/011697	*“Bourgm(est)re et eschevins de la ville et territoire de Dunkerque ordonnent aux gens de loy de Zudcoot (Zuydcoote), Leffrenouck (Leffrinckoucke), branche de Coudekerke (Coudekerque), Teteghem, Uxem et Ghiuel (Ghyvelde) de continuer a faire rompre pressament la glace dans le canal de Furnes avec la mesme quantite dhommes qu’ils ont fait pendant la derniere gelee”*
Earth-quake	Arch. Mun. Dunkerque, AncDK87	4	Dunkerque	Nord (Nord-Pas-de-Calais)	18/051773	“Le trois décembre dernier un petit mouvement qui toucha le petit pont de Coppenhaguefort a fracassé et emporté sept fenêtres a un de mes carosses qui luy a coûté considérablement sans y comprendre l’intérêt...”

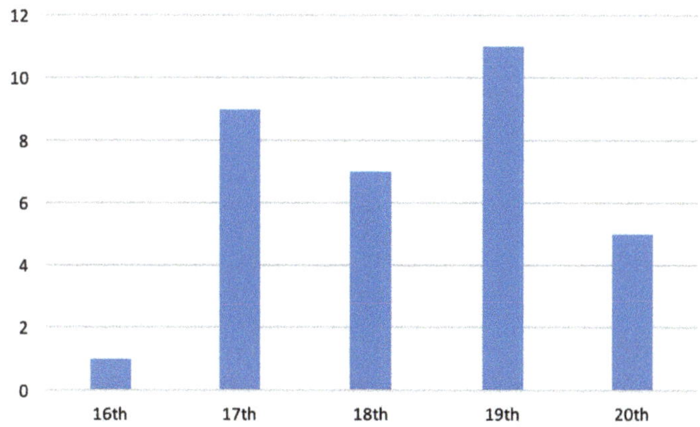

Fig. 1.7 Distribution of storm surges by century near Gravelines (North Sea, cf. Fig. 0.1). *Source* Garnier (2018)

A historical approach refutes the idea that a catastrophe of the February 2010 magnitude was totally unexpected. Indeed, between 1550 and 2010, 117 such events are recorded in the archives for the French coastline. Among them, 30 hit the Atlantic coast exclusively. More interesting are the average return times calculated from these time series. Once again, these are relatively homogeneous—between 14 and 19 years—with an estimated period of 15 years on the Atlantic coast [317, 375].

Focusing on the Atlantic coast alone, the chronology shows wide disparities across 500 years. Contrary to what one might think, the last 50 years have not seen an upsurge in this kind of extreme event. The most catastrophic century is the eighteenth, with nine storm surge events, whereas the twentieth century totaled only five (see, for example, Fig. 1.7). It also seems that the apogee of the little ice age in the seventeenth century resulted in a lower frequency of such extreme events (three only). During the last 100 years, six storm surge events struck France's Atlantic regions, but notably, the great majority occurred between 1924 and 1957.

Receiving the full force of Xynthia, the town of La Faute-sur-Mer did not discover the dangers of the sea only in 2010. At least seven submersions have affected the town since 1711, an average return time of 42 years (Fig. 1.8). More astonishing is the fact that the municipality experienced three storm surges in the first half of the twentieth century (1928, 1937, and 1941), rather recent events, which once again seem to have been totally ignored when developing the tourist potential of the town [315].

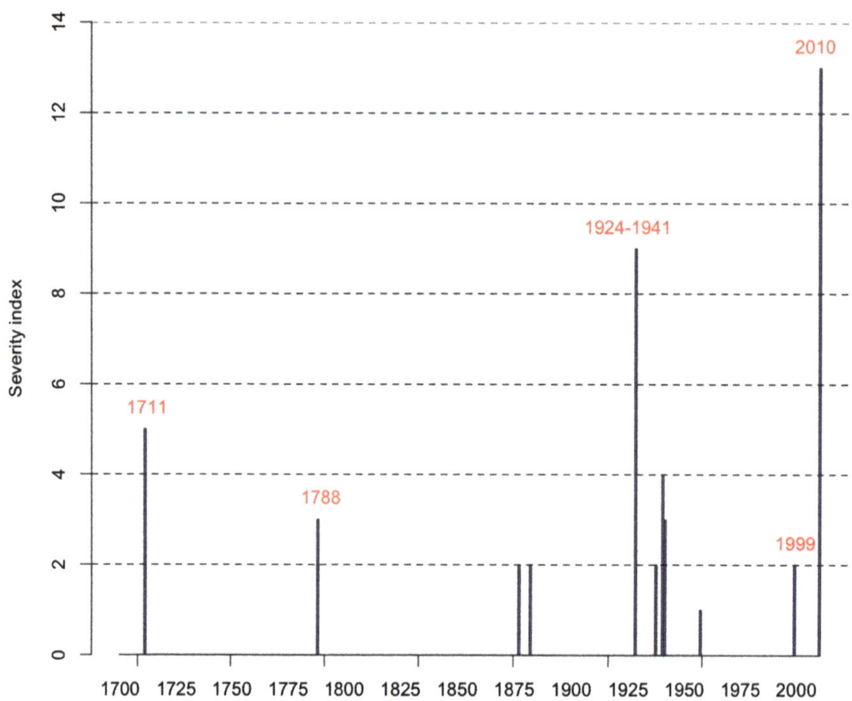

Fig. 1.8 Storm surge events at La Faute-sur-Mer between 1700 and 2010 in terms of impact score. This score is based on many types of damage, such as flooding of buildings, loss of life, crops, and livestock, flooded areas, damage to infrastructure, and financial cost. *Source* [317]

1.4 Lessons of History

The paradigm of civilization as presented by the philosopher Montesquieu in the eighteenth century postulates that the more a society evolves, the more it can protect itself from the consequences of natural disasters. The archives examined here tend to prove the contrary, underlining the capacity of pre-industrial societies to preserve and transmit the memory of natural catastrophes, and beyond that, draw lessons from experience to improve resilience, using visual means like *ex-votos*, more resilient landscape features such as hedgerows, and simply by not constructing directly on the coastline, among others. More trivially, a simple question needs to be asked: can historical examples be used to strengthen the capacity of contemporary society to protect itself against extreme events?

Though the experience of the past should not be idealized, it must be taken into account and used to design prevention strategies based on adaptation ideas developed by our predecessors. For them, the risk was not a fatality, but rather a state of expectation leading to anticipation of possible crises. They therefore kept alive the memory of past catastrophes and tried to be ready for ongoing threats. Indeed, communities, from medieval times to the first half of the twentieth century, developed flexible and effective alert/prevention systems, enabling people to take refuge, either upstairs in their houses or in safe areas, known from the past to be protected from flooding. Unfortunately, such teaching has been lost due to a break in the continuity of passed-on memories, demographics (rural exodus, the attraction of the seaside), and loss of technical know-how in the post-World War II context. While empirical, these ways of the elders illustrate the passing on of memories, based on the ability of a group to acquire and develop, over time, knowledge of its own survival legacy.

Besançon, July 9, 2017,

Emmanuel Garnier

CNRS Senior Scientist, University of Bourgogne Franche-Comté, Besançon, France
Honorary Senior Member of the French University Institute (2010–2015)
Visiting Professor at the Prevention Research Institute
(University of Kyoto, Japan)
Visiting Professor and French Government Overseas Fellow
(Cambridge University, Churchill College, UK; 2012–2013)
Visiting Professor at Geneva University (2014)

Chapter 2
Introduction

Nicolas Bousquet and Pietro Bernardara

Abstract This introduction recalls the considerable socio-economic challenges associated with extreme natural hazards. The possibilities of statistical quantification of past hazards and extrapolation, offered by extreme value theory, make it an essential tool to improve risk mitigation. The objectives of this book, answering the questions of engineers, decision-makers, researchers, professors, and students, are briefly presented.

2.1 Major Economic and Societal Challenges

Understanding natural hazards and the industrial risks they lead to is a topic of great importance to society. With all its human, economic, and ecological dimensions, society has repeatedly found itself vulnerable to "extreme" natural events. Floods, storms, heavy rains, and the like, have the ability to significantly weaken or even destroy civil engineering works and industrial sites. High or extremely low temperatures have the potential to induce heavy rain [59] or drastically reduce water resources at production sites where machines need to be cooled [482, 764]. Industrial accidents provoked by extreme natural events can in turn disrupt the habitats and ecosystems of a whole region. As an instance, strong concerns have recently been expressed about the robustness of aging American dams to extreme precipitation, the destruction of which would have colossal socio-economic and environmental costs [478]. The potentially cumulative effect of more than one extreme event is especially feared, as it may lead to even worse consequences. For instance, a storm might cut off access to a site that needs to be protected from a second event linked to the first, such as flooding caused by heavy rain.

The quantification of extreme natural hazards, in terms of *probability of occurrence* and *associated costs*, became a topic of major interest in the mid-twentieth century. The catastrophic flood of February 1953, simultaneously in the Netherlands,

N. Bousquet (✉) · P. Bernardara
EDF R&D, 6 quai Watier, 78401 Chatou, France
e-mail: nicolas.bousquet@edf.fr

N. Bousquet and P. Bernardara (eds.), *Extreme Value Theory with Applications to Natural Hazards*, https://doi.org/10.1007/978-3-030-74942-2_2

England, and Belgium (over 1 800 dead [451]) alerted the scientific and political worlds to the importance of being able to estimate the risk associated with observed extreme natural events and even to be able to extrapolate risk assessment to different or more extreme scenarios. From the last years of the twentieth century, extensive media coverage of such events, like the mudslides in Venezuela in 1999 (over 30 000 deaths [175]) and the storms Lothar and Martin which hit Europe in December 1999 (and partially flooded the Blayais nuclear power plant in Gironde), has again brought to our attention the increased vulnerability of our complex societies to environmental risks [111, 477]. Only 20 days after the storm Xynthia, Iceland's Eyafjallajökull volcano erupted, significantly changing the perception of the risk of volcanic eruption in Europe, at least for the aviation industry. After the Earth Summit in Rio in 1992, at last, this subject was finally given a more important place at the table, when it was realized that consequences of climate change could include an increase—in severity and frequency—of certain types of extreme natural events [111, 520, 765, 788]. Though more recent works are more conservative on the subject and suggest that the influence of climate change is yet to be proved for cyclones, hail, and torrential rain (among others), more frequent and unprecedented heatwaves [554], faster winds [799], and rising sea levels are already upon us [18].[1] Recent major civil and industrial emergencies caused by extreme natural events—-sometimes cumulative—such as the Fukushima nuclear accident in March 2011 involving an earthquake followed by a tsunami—only strengthen the need for methods for quantifying extreme events, that can effectively take into account their inherent complexity.

Generally, risk mitigation, that is, the set of measures taken ex ante to mitigate the effect of cumulative extreme events, is of great interest to authorities [9, 96]. Large industries—including energy companies such as EDF—more than share these concerns. Indeed, it is essential for them—a legal obligation even—to ensure the robustness of their facilities against natural hazards, preserve production, and minimize societal consequences due to a partial or total loss of them. Such mitigation is generally carried out by designing site protection projects and by setting up warning systems and crisis management processes. These call for studies consisting of:

(i) understanding and characterizing physical phenomena that generate environmental phenomena;
(ii) characterizing the intensity of possible extreme events.

It is clear however that, depending on the issues and the vulnerability of the sites to be protected, we may wish to completely protect them from these extreme events, or instead accept a certain level of risk (with zero risk equivalent to infinite cost). The level of risk we decide to accept is a choice which society, regulatory agencies, and project leaders can then use to optimize protective structure designs and mobilize resources. This *interpretive* dimension to the risk, which may involve important philosophical and societal concerns [70, 379, 541], is common to many domains

[1] Extreme events happen anyway, whether influenced by climate change or not. Moreover, uncertainty about the effects of climate change is even higher for extreme events than for changes for regular, average events.

that make use of formal methods for quantifying risk. It is particularly true for actuarial studies [124, 644, 666] and financial studies [249, 733], where extreme value statistics are frequently used. We point readers with a strong interest in this subject toward to the works [108, 109] for more details.

In its most standard accepted form, particularly in industry, an industrial risk*[2] is generally defined as the product of the probability of occurrence of a hazard and a measure of the cost it would induce [6]. Quantifying the vulnerability of a site involves various types of cost (human, goods, etc.), while the probability measure is a generic diagnostic tool that helps us compare diverse scenarios.

Rather than probabilities, in practice we often speak of the *frequency of occurrence* (a probability typically representing the meaning of a limiting frequency [551]). For other reasons, given in Sect. 3.1, it is often appropriate to define a phenomenon[3] as truly *random* and try to characterize it using probability distributions and processes, themselves quantified using statistical tools. Indeed, it is quite useful to be able to associate a frequency of occurrence (for example, annual) to the intensity of a natural phenomenon. Obviously, the more extreme the phenomenon, the rarer it will be, and vice versa. A more practical way to express the rarity of an event is to use the concept of *return period*. This, and the frequency of occurrence, have been borrowed from statisticians by engineers and modified to serve their needs. Good practice, methodology manuals, and regulatory frameworks have therefore integrated these concepts.

It logically follows that there is a need for statistical estimation methods and tools for analyzing the frequency of occurrence of rare events. This link between the statistical characterization of natural phenomena and the design of civil and industrial engineering works is long-established [260]. New perspectives are opening up today: the contribution of new contextual data, sometimes in large numbers (*Big Data* paradigm), the complexity of situations where risks accumulate, and taking into account trends, some of which being linked to climate change, mean constantly improving these methods and tools and updating the regulatory frameworks (Fig. 2.1).

2.2 A Statistical Theory of Extreme Values

Historically the use of probability and statistics for the characterization of extreme natural events took off in the first half of the twentieth century, thanks to the theoretical work of Fréchet [291], Fisher and Tippett [283] and Gumbel [364]. They showed that, under certain assumptions, an extreme data sample (non-stationary over a time period or values exceeding a threshold) can be seen as a set of samples from a group of probability distributions called *extreme value* distributions, which can be statistically estimated from an observed sample.

---

[2] The asterisk indicates a term found in the glossary at its first occurrence in the text.

[3] Or at least the nature of the mechanism producing the actual effects of the phenomenon.

Fig. 2.1 Flooding of the Seine in Paris, June 2, 2016

This theory had its first real practical use following the events of 1953 mentioned above. Since then, the application of statistical distributions from extreme value theory to samples of observations of environmental variables collected at sites of interest has become common methodological practice. An important ingredient in the decision support process, it allows a scientific characterization of random natural hazards. Accepted by authorities, scientists and engineers worldwide, it is now the go-to methodology in certain regulatory frameworks [1].

Today, thanks to the availability of large regional and spatial databases, the capacity of numerical models to simulate extreme events, and advances in multivariate spatial statistics methods, the relevance of a simple application of an extreme value distribution to a sample of observations for a given site, has become more limited. However, new approaches are invariably constructed from results that originated in extreme value theory. This, therefore, represents a body of theory and practical techniques that engineers responsible for characterizing random natural hazards need to know.

The use of this theory is of course not without critics. Working with a statistical distribution, necessarily estimated from a small number of observations, to extrapolate beyond them even, poses a major epistemological problem [107]. From "how many observations" to what "extreme intensity" can we be reasonably confident in our extrapolation, even when accompanied by measures of uncertainty? Do we even have the "right" (is it even science?) to conduct such an analysis?[4] Proof of these legitimate doubts is the existence of other approaches to extreme risk modeling, generally founded on *deterministic* principles, e.g., the *probable maximum precipitation* calculation, widespread in North America [803], or *modeling* the physics of the phenomenon using numerical methods (see Sect. 3.2.2). A quite comprehensive list of alternative approaches such as these can be found in [9].

Nevertheless, this book will provide theoretical and practical answers to the above-mentioned doubts. From an epistemological perspective, we agree with the viewpoint of Stuart Coles [173], who asserts that extreme value theory provides probably the only practical solution to the need for extrapolation based on rational, objective, and technical arguments:

By definition, extreme values are scarce, meaning that estimates are often required for levels of a process that are much greater than have already been observed. This implies an extrapolation from observed levels to unobserved levels, and extreme value theory provides a class of models to enable such extrapolation. In lieu of an empirical or physical basis, asymptotic argument is used to generate the extreme value models. It is easy to be cynical about this strategy, arguing that the extrapolation of models to unseen levels requires a leap of faith, even if the models have an underlying asymptotic rationale. There is no simple defense against this criticism, except to say that applications demand extrapolation, and that it is better to use techniques that have a rationale of some sort. This argument is well understood and, notwithstanding objections to the general principle of extrapolation, there are no serious competitor models to those provided by extreme value theory ([173], p. vii).

Extreme value theory provides a strict and precise framework, in order to be applicable. In particular, it is important to note that phenomena must be studied using continuous variables, and by defining orderable variables, i.e., quantities for which one can define an ordering for comparing and ranking purposes [107]. Furthermore, one is not allowed to provide a precise point estimate of an extreme value return level or very low probability without also providing one or several measures of uncertainty.

[4] In particular, questioning the actual relevance of the concept of probability in the extreme value setting is a legitimate line of enquiry [107].

2.3 Objectives of the Book

2.3.1 A Need for a Cross-Disciplinary Approach to Extreme Values

A brief literature or internet search will quickly confirm to anyone the current popularity of extreme value theory. However, most of this content remains stuck in the world of academia [65, 173, 261], essentially developed by mathematicians, with a clear lack of critical analysis of its use in modeling and estimating real-world phenomena with real-world data. From a practical point of view, probability estimation methods for rare events have progressed in parallel in several scientific disciplines (finance [485], economics [47], hydrology [429], meteorology and climatology [23], maritime hydraulics [394], etc.), partly in cohesion with extreme value theory, partly by breaking away to answer needs related to each subject's specific requirements [145]. This lack of results transfer from one domain to the next may lead, on the one hand, to consolidation of the use of certain methods in certain domains, whose theoretical backing continues to be refined over time. On the other hand, potentially useful and important cross-disciplinary research directions may unfortunately fall by the wayside.

In response, researchers and engineers from EDF R&D have for several years come together in a research group on extreme value statistics, focusing on defining a good practice framework at this subject's scientific and practical interface. This book is the direct result of thoughts and discussions emanating from the research group. We aim to provide an in-depth look at the common methodological heart of theoretical and applied subjects, via a critical analysis of methods used and results obtained. It is clear is that extreme value probability estimation studies for various disciplines broadly share the same underlying framework with, of course, specific steps related to the study in question. This cross-disciplinary point of view is a specific focus of the first part of the book, which presents the "classical" theory of extreme value statistics.

2.3.2 Which Random Events?

Natural hazards that may impact human activities and infrastructures include a wide range of environmental phenomena. Here are some examples:

- in *meteorology*: hot and cold air phenomena, rain, extreme wind events and tornados, snow, hail, lightning;
- in *hydrology*: flooding or baseflow* events, high and low water temperatures;
- certain *maritime phenomena*: abnormally high and low sea levels, extreme tides due to meteorological processes, astronomical tides, waves (chop* and swell*), high and low water temperatures;

- in *geology* and *seismotectonics*: earthquakes, tsunamis, volcanic eruptions and landslides;
- *space weather*: solar storms and meteorites outbreak.

This list is inspired by regulations in the nuclear industry (e.g., [12, 15, 17, 19]), a particularly proactive and attentive actor in this area. However, the list is by no means exhaustive and only serves to give the reader an idea of phenomena that can be characterized using methods presented in the book. The applicability of tools based on this theory often demands simplification, with hazards represented by a limited number of variables that together vary with time. Such constraints do not prevent the theory being used to characterize many other natural phenomena, such as large wildfires [35, 724].

2.3.3 Answering to New Risk Perceptions

The increasing complexity of risk perception requires the development of new theoretical and practical aspects of extreme value statistics. These fall into four main categories, mostly treated in the second part of the book, and include recent research results, often the product of the work of EDF's engineers and researchers.

1. The *cumulative effect of several hazards*, the importance of which we mentioned earlier.
2. Taking *trends* into account: If there is a trend, can it increase the impact of events today considered extreme? How can we, in this case, modify mitigation strategies to take this into account over relevant time intervals?
3. *Taking advantage of spatial data and other information sources*, notably for refining results from other studies and producing new ones. Demands for increasing geographic and temporal precision can limit the number of directly applicable data samples obtained from the past. In parallel, detection and measuring instruments are becoming increasingly important and increasingly precise. The refinement of old results and the production of new risk quantification methods thus requires the construction and deployment of a whole arsenal of modern techniques, based on the use of spatio-temporal data from the neighborhood of the site in question, and/or any specific site expertise available. The goal is to use these varied sources of information in the best way possible to confront potentially high levels of uncertainty that could result from these statistical predictions.
4. Lastly, the use of *stochastic and numerical models* is an alternative approach that may prove crucial when too few extreme data values are available to ensure the adequacy of the previous approaches. This strategy involves seeking to "mimic" the natural phenomenon at the source of the hazard, be it through the use of empirical–statistical models, based on all available data (not only extreme values), or using mechanistic models of the physics involved [767]. Simulation techniques can then be used to replicate the phenomenon a huge number of times, thus "detecting" extreme situations; the probability of occurrence can then be quantified.

2.4 Key Messages and Warnings for the Reader

This book is aimed at a broad category of users of extreme statistics: engineers, researchers, professors, students. It aims to bring them, through a transversal state of the art at diérents professions and thanks to the practical experience of the authors, to adopt the practice and legitimate questioning of an intelligent customer faced with a concrete problem.

The "quest" proposed to the reader, throughout this book, is indeed mainly oriented to provide answers to an engineer, confronted with basic questions of the type: *What is the intensity of a hazard associated with a given return period?* Starting from theoretical building blocks and the characteristics of observed samples, the book turns to questions about the temporal homogeneity of the phenomenon being looked at, then shows how to attempt to gradually improve the statistical significance of a sample by integrating various information sources. Questions of *inference* (the explanatory approach) then start to mingle with those of *forecasting* and *extrapolation* (the predictive approach). The last section of the book, devoted to a comprehensive look at examples from industry, will help readers put elements of theoretical and applied critical analysis from earlier parts of the book into practice.

Ultimately, we think that this approach promotes learning, and raises—as we go along—legitimate questions about the appropriateness of each method, thus allowing us to develop a critical eye for all tools issued from extreme value theory.

Warning

The methods and tools described in this book are useful resources for the intended audience of engineers and researchers. An engineer has to be a critical analyst of methods sprouting from the general theory of extreme values, which are constantly evolving. The search for and understanding of data requires a delicate statistical dance, built on a necessary dialogue between engineers and experts. Similarly, selection and then application of any statistical model that attempts to be consistent with observed natural hazard data, cannot occur without ongoing critical analysis of the model's merits, especially when testing and validating it. Note that the aim of the book is not to give a comprehensive inventory of all existing extreme value methods in the literature; numerous references to highly specialized research results are given in the text, and curious readers are encouraged to follow these to their source. Instead, our underlying goal is to provide a general methodological framework, constantly accompanied by useful practical tips.

Therefore, the book should not be looked at as scientific doctrine, to be followed to the letter. The actual design of protective structures—at EDF and elsewhere—is in reality conducted using several approaches to risk modeling, some of which will be specific to the geographic site and hazard being considered. Extreme value theory is a possible ingredient in this process, its advantage being its general and non-specific nature. For this reason, it is often called upon to play a complementary role, or used for verification purposes.

Part I
Standard Extreme Value Theory

Chapter 3
Probabilistic Modeling and Statistical Quantification of Natural Hazards

Pietro Bernardara and Nicolas Bousquet

Statistics are no substitute for judgment

Henry Clay

Abstract This chapter presents the main useful concepts of the statistical approach to extreme values, such as return periods or levels, as well as a general methodology for conducting a statistical study. It also introduces alternative approaches, statistical or not, that can be valuable in dealing with extreme natural hazards.

3.1 The Rationale for a Statistical Approach to Extreme Values

3.1.1 Practical Relevance and Scope

3.1.1.1 Considering Natural Phenomena to Be Random

The statistical approach to which this book is devoted is based on the fundamental assumption that extreme phenomena belong to a family of events possessing a certain stochastic regularity, and that observations $x_1, \ldots, x_n, \ldots$ exist of these events.[1] A swollen river can be interpreted as an extreme water flow, significantly different from the usual one. However, the extreme phenomenon will still be characterized by the same underlying mechanisms that lead to the normal phenomenon (rainfall

[1] We will use the generic word *data* in the rest of the book.

P. Bernardara · N. Bousquet (✉)
EDF R&D, 6 quai Watier, 78401 Chatou, France
e-mail: nicolas.bousquet@edf.fr

N. Bousquet and P. Bernardara (eds.), *Extreme Value Theory with Applications to Natural Hazards*, https://doi.org/10.1007/978-3-030-74942-2_3

upstream, glacial melting, etc.). Statistical quantification then appears naturally from the following hypothesis.

Hypothesis 3.1 *The data $x_1, \ldots, x_n$ can be considered to be observed values of a random variable X.*

Extreme values can then be defined as maxima of samples of observed values of X (*block maxima* approach, or MAXB), or as observed values that exceed a certain threshold (*Peaks Over Threshold* approach, or POT). More broadly in statistics, extreme values are extreme quantiles x_p of X, characterized by probabilities of not exceeding such values, $F = \mathbb{P}(X < x_p)$, that are very close to 1. The word *significantly*, expressed above, has a precise meaning proposed by the classical statistics of Fisher [282] and Pearson [774], and recently deepened by Johnson [417].

From an engineer's point of view, this hypothesis gives a practical way to model the phenomenon of interest, and therefore to answer questions of risk quantification. This consensual definition of randomness corresponds well with the nature of observations and predictions of typically considered phenomena (see Sect. 3.1.2 for details). Probability theory, whose framework is introduced in Sect. 4, is the best-adapted mathematical environment for handling *random* events [442] (Sect. 1.2: "The Relation to Experimental Data") [623, 700]. The term *natural hazard* will be used extensively in this book to refer to such phenomena, both in terms of observations and predictions pertaining to them.

3.1.1.2 Two Fundamental Theorems

The two sampling approaches introduced above (MAXB and POT) are amenable to powerful results from extreme value theory, itself essentially built on two fundamental theorems: the Fisher-Tippett [283] and Pickands [615] theorems. These two theoretical results lead to two statistical modeling methods for extreme values.

- The Fisher-Tippett theorem states that the probability distribution of a sample of maxima or minima drawn from various samples of independent and identically distributed (iid) variables converges to a family of distributions called *generalized extreme value* (GEV) distributions, under standard conditions [173].
- Under similar conditions, Pickands' theorem states that the probability distribution of a sample of values above (or below) a fixed extreme threshold u (excess sample) converges to a family of distributions called *generalized Pareto distributions* (GPD).

Techniques for sampling observations (MAXB and POT), and using the corresponding statistical distributions (GEV and GPD) studied in the literature and applied to real-world situations, have been chosen and developed based on these two theorems. Taking a simplistic point of view, an analogy can be made between them and the central limit theorem (see Sect. 4.2.2), which establishes the convergence of the mean of a sample of random variables to a normal distribution. In particular, the first theorem above, seen from a simplistic point of view, establishes the convergence of the maxima of samples of random variables to the GEV family.

Remark 3.1 Important properties also derive from these theorems, which reinforce their universality: the shape of the distribution tail is consistent between the two approaches, and the law of maxima obeys a fundamental principle known as max-stability: the maximum of a sample of maxima follows a law belonging to the same GEV family (see Definition 6.5 for a formalization of this notion). The dual principle is the stability of the GPD family when the threshold is modified for a higher value (see Proposition 6.3 and [588]), a property also called *heredity*.

3.1.1.3 Frequency of Occurrence and Return Levels

For phenomena described using time series, it is common usage to quantify the occurrence of an extreme hazard in terms of *annual exceedance probability* [260], denoted by p. This is defined as the probability that a given value of the relevant variable X, often denoted by x_p, is exceeded in a given one-year period. The value x_p is called the **return level**.

The annual frequency of occurrence is a *classical statistical estimator* of p (Sect. 4.2.2). For example, we can associate a river flow of 300 m^3/s with the value $p = 0.01$ if we estimate that the probability of observing a flow of $X > 300$ m^3/s over a one-year period is equal to 0.01.

In the more general context of sampling according to a different frequency, the relationship between $F(x_p)$ and the annual exceedance probability p is more complex. The occurrences of exceedances are then considered independent and follow a homogeneous Poisson process (see Sect. 4.1.2.1 and Definition 4.2), and the average number λ of events per year present in the sample must be taken into account. It comes to $p = 1 - F^{\lambda}(x_p)$, and consequently $T = 1/(1 - F^{\lambda}(x_p))$. In practice, the latter formula is often approximated by $T = 1/\lambda(1 - F(x_p))$ [667].

3.1.1.4 Return Periods

The *return period* concept, more intuitive for engineers and decision-makers (and even the general public) and commonly used in regulation, was popularized by Gumbel [364] and is defined as follows. If the probability p of observing a flow of $X > 300$ m^3/s in a one-year period is equal to 0.01, it follows that over a 1 000-year period, we can expect to see, on average, ten events where such a flow occurs. Ten times in 1 000 years means on average once every 100 years. This is what we call the return period, which is given by the simple formula

$$T = 1/p.$$

Nevertheless, this concept must be used with caution. The formulation "on average once every 100 years" has often led neophytes in the field to believe that the extreme phenomenon in question should occur periodically every 100 years, which is of course false [661]. A more formal explanation of the return period concept is offered in Proposition 6.5.

3.1.1.5 Typical Return Levels

In practice, the dimensioning of industrial installations requires the assessment of extremely rare events, occurring on average once every 100, 1 000, or even 10 000 years. The hazard is thus labeled (especially in hydrology) *centennial*, *millennial*, or *decamillennial*. Checking for the respect of regulations therefore requires estimating annual exceedance probabilities typically between 10^{-2} and 10^{-4}.

However, available data series for natural phenomena tend to go back 20–40 years only (see Sect. 5.2.2 for more details). It is thus frequent to try to estimate return levels that have never been observed, meaning attempts to *extrapolate* the available statistical information. In such cases, it becomes impossible to provide a "classical" validation of the modeling in which we compare real data with the model's predictions [29]. For example, to validate a millennial rainfall model using observed data, we would require access to a sequence of several millennial observations, not to mention the fact that the validation would be weakened by intrinsic variability affecting this observed sample. The actual limits up to which extrapolation remains credible are still a subject of debate today [174, 484, 645]. From a methodological standpoint, our positioning is stated in the introduction of the book: extreme value theory is a useful and objective ingredient, but validating the acceptability of results remains the task of expert decision-makers.

3.1.2 Phenomenological Relevance

Most natural hazards are characterized by permanent uncertainty as to their future state [234]. For example, it is impossible to exactly predict tomorrow's temperature at a given time and place, as it is a measurement of the thermal agitation of particles, i.e., a measurement of energy exchange at the heart of a thermodynamic system subject to an indeterminacy principle. Furthermore, the measurement necessarily involves a spatial and temporal approximation. Any sequence of temperature measurements or predictions thus has the following property: the "missing" information on the size of the error cannot be set to zero; there always remains an incompressible margin of uncertainty that has, by definition, the properties of a random phenomenon [150, 212, 240]. This is also characteristic of phenomena such as rain and waves.

The physics of water flow, on the other hand, might seem at first glance to be deterministic: if all parameters are known, the phenomenon's behavior is entirely predictable, and a probabilistic framework would seem misplaced. However, this does not take into account our ignorance as to the exact parameter values, nor whether they indeed entirely characterize the situation. Are we really capable of completely characterizing a natural phenomenon by a set of parameters and structural equations? If so, are we capable of absolute precision, in terms of what we know about them? *Epistemic* uncertainty, due to missing knowledge, characterizes such problems. It is typically distinguished from random uncertainty, by considering that it can be reduced by *adding knowledge*, whereas random uncertainty cannot [216, 386, 671]. However, the notion of knowledge addition is conditional on the nature of the tools used to represent this knowledge and translate it into information. In contemporary scientific practice, the addition of knowledge is most often achieved by improving *models* [109, 767]. These necessarily suffer from a form of error that is linked to the deterministic and discretized way they are computationally implemented (see, for example, [390] and [240], Sect. 4.2). In other words, when characterizing complex phenomena, there always remains an irreducible error whose features, again, are those of a random phenomenon.[2]

Besides the practical relevance discussed in Sect. 3.1.1, these considerations offer an epistemological justification for the use of probability distributions (Sect. 4.1.2) and processes (Sect. 4.1.5) to represent natural phenomena in a simplified way, as well as the use of statistical techniques to estimate these distributions and processes and make predictions and extrapolations.

3.1.3 How Should We Conduct a Standard Study?

The work undertaken to characterize good practice for extreme value statistics studies within EDF revealed six fundamental steps that should be followed. These can be summarized by the following keypoints:

1. **Defining variables and probability levels.** Here, we try to define the problem to be addressed. Which rare event do we wish to model? Which variable z can be used to model it? (the *variable of interest.*) Is this choice dictated by regulations? Is it accessible to measurement or simulation ($x = z$), or do we need to study another variable $x \neq z$? (a *study variable.*) What is the probability level to estimate? Does it depend on current regulations, and should the evolution of these be expected? Is this level scientifically acceptable, as discussed in Introduction (Sect. 1.2)? If several hazards are simultaneous, then what do we want to estimate? (because here, the very meaning of probability level needs to be defined.) Similarly, if there

[2] However, many other theories of uncertainty can usefully, and sometimes more appropriately, be proposed to represent epistemic uncertainty, as possibility theory [231] and fuzzy logic [195].

is a trend in the phenomenon, the return level concept is no longer pertinent; what then should we quantify?

Example 3.1 (*Eurocodes*)

The study variable x (temperature, precipitation, etc.) is sometimes similar to the variable of interest z defined by regulations or norms; this is the case, for example, in the Eurocodes, a set of European norms which standardize methods of calculation, and are used to verify the stability and dimensioning of buildings and civil engineering structures. See http://normalisation.afnor.org/thematiques/eurocodes for more details.

Since it is the decision variable related to a project, we may need to work with a study variable x that is connected to the variable of interest by a relationship such as

$$Z = f(x, d),$$

where d is a vector of deterministic parameters, and the function f may be a deterministic or stochastic physical model [660], or a probabilistic model of connected or combined hazards.

Example 3.2 (*Dimensioning of dikes*)

In a design study for dikes protecting a riverside power station, the variable of interest or *project variable* is the height (level) of the river. Nevertheless, the extreme value statistics study may be conducted on the *river flow*. The river flow is more conducive to modeling than the height [10], and for this reason, this approach is also recommended in regulations [1]. The mapping between the study variable (river flow) and the variable of interest (river height) is then performed using a calibration curve or hydraulic modeling (see Example 4.7). Another transformation of the variable has to be made if hydraulic structures at the site can influence the flow (known as the *natural flow*). In this case, it is necessary to reconstruct the sequence of natural flows that would be observed if no hydraulic structures were present. This procedure is known as *flow naturalization* [10].

Example 3.3 (*Sticky snow*)

Sticky snow is a type of snow which can lead to accretion* on electric power cables. The variable of interest is the value of the load on a cable (which could lead to rupture), which can be indirectly determined from meteorological data [214].

2. **Data collection.** This is a crucial step in any statistical study, and particularly for extreme value statistics, where available data is rare. What data is available for the study? Is it directly related to quantifying the hazard? Is the information it contains reliable? Sufficient?
2. **Sampling extreme values.** Extreme value theory allows for two main types of sampling: *block maxima (periods)* (MAXB approach) or *threshold exceeding* (POT approach). Which is the most pertinent with respect to the amount and type of data available? When two or more hazards are related, how to adapt the sampling?
3. **Modeling and statistical estimation.** Observed data is then modeled using a statistical distribution—either of type GEV or GPD—which is appropriate to the extreme data in question and coherent with the choice of sampling made above.
4. **Testing and visual checks.** The chosen distribution can be partly validated using statistical tests and plots.
5. **Calculation of results and validation.** In this final step, estimation of rare probabilities and return levels (or functions of interest specific to the case of joint hazards or the existence of a trend) are carried out using the chosen distribution, and results are interpreted according to know-how and expertise in each domain.

Extreme value sampling, estimation, testing, and graphical verification are classical steps when applying extreme value theory, and for each of which it is possible to draw on a wide range of methods. On the contrary, in the definition of variables and probability levels (1), in data collection (2), and in the calculation of results and validation (6), the need for specific expertise may arise. Within these three stages (1–2–6), it is therefore particularly important to take a critical look at the strict relevance of extreme value theory. It is in these three stages in particular that the practical experience of EDF's engineers and researchers bring a certain originality to this book.

3.2 Alternative Approaches

3.2.1 Alternative Statistical Approaches

The classical hypotheses of the extreme value theory are hard to be fully verified for numerous natural hazards, so an engineer's work consists in getting as close as possible to them. Because there remains the possibility of error in the historical and current use of these hypotheses in each domain, exemptions to the strict application of extreme value statistical modeling exist and other distributions are regularly used by engineers instead of extreme distributions (exponential, lognormal [452], generalized exponential [491], or mixtures of simple laws [587]), which can be estimated from other types of sampling. In this book, deviations from the theory will be clearly signaled as such. These historical and even current practices may also be the result of adaptations of the theory of extremes. The Gumbel distribution, for example, which is a special case of the GEV distribution, is still widely used by hydrologists (see Remark 6.5).

3.2.2 Stochastic Scenario Approaches

Other approaches, which may be termed *stochastic*, differ from statistical approaches because they seek to model the behavior of the phenomenon giving rise to the extreme hazard. These approaches are considered further in Chap. 10. This modeling strategy may be based on an empirical examination of *all* observations of the phenomenon (not just extreme ones) and a choice of a random process, possibly *stationary* (Sect. 4.1.5). Such approaches make it possible to re-simulate the empirical behavior of the phenomenon, possibly with the help of covariates, and naturally produce extreme events.

A more physics-based approach can also be developed using deterministic mechanistic models, implemented as code. EDF has developed, for example, tools for simulating the physics of the transformation of rainfall into flow rates (such as the MORDOR code for hydrology [306]), and from flow rates to water height (the MASCARET [351] and TELEMAC [301] hydraulics codes).

These deterministic tools can then be associated with Monte Carlo stochastic simulations, as in EDF's SCHADEX method [586], for simulating long series of "artificial" observations whose statistical properties can be analyzed.

Chapter 4
Fundamental Concepts of Probability and Statistics

Nicolas Bousquet

Abstract This chapter recalls the fundamental concepts and results of probability theory and statistical theory. These are essential to understand and use the tools of extreme value theory in an appropriate way. From the notions of randomness, probability distributions to random processes, from classical estimators to regression models, this chapter aims to facilitate the technical understanding of the rest of this book.

4.1 Main Concepts and Notions

Probabilistic modeling of a hazard X rests on the *random variable* character given to X, evolving in a d-dimensional sampling set Ω. Since $\Omega \neq \varnothing$, subsets of values $\mathscr{A} \subset \Omega$ that X can take are non-empty, and come with a certain stability: the countable union of a number of $\mathscr{A}_i$ is still in Ω, as is the complement of any subset $\mathscr{A}$.

These fundamental properties allow us to "pave over" (*measure*) the set Ω so that any observation (occurrence) of an *event* $A \in \mathscr{A}$ can be associated with a numerical value $\mathbb{P}(A)$. The set of these values is contained in the interval [0, 1], and we have that

$$\mathbb{P}(\Omega) = 1.$$

$\mathbb{P}$ is therefore called a *probability measure*.

In probability theory, the triplet $(\Omega, \mathscr{A}, \mathbb{P})$ is called a *probability space*, where the set Ω is the *sample space*, and $\mathscr{A}$ is a σ*-algebra*. In general, $\mathscr{A}$ is chosen to be the set of subsets of Ω which are Lebesgue measurable (see Sect. 4.1.1), but it is usually not defined explicitly in applied problems.

N. Bousquet (✉)
EDF R&D, 6 quai Watier, 78401 Chatou, France
e-mail: nicolas.bousquet@edf.fr

N. Bousquet and P. Bernardara (eds.), *Extreme Value Theory with Applications to Natural Hazards*, https://doi.org/10.1007/978-3-030-74942-2_4

4.1.1 One-Dimensional Problems

Consider to begin with the case where $d = 1$. If Ω is a *discrete* set (for example, $\Omega = \{1, 2, 3, \ldots, \})$ or a *categorical* one (for example, $\Omega = \{A_1, A_2, \ldots, \})$, where each A_i defines a class of values of X that are disjoint from all other A_j, and more generally if Ω is *countable*, the probability distribution is said to be discrete, and is determined by the probabilistic *mass function*

$$f(x) = \mathbb{P}(X = x)$$

for each $x \in \Omega$. However, the great majority of random variables encountered in this book will be *continuous*. In particular, the values of X (wind speed, temperature, flow rate, etc.) are continuous, which is an essential condition for applying extreme value theory, and Ω is generally a continuous subset of $I\!R^d$, even if measuring devices are limited to a certain accuracy in practice. This accuracy does not play a role in the construction of the probabilistic model but instead, in this, of the statistical model, which encompasses the probabilistic model by establishing a direct link with noisy observations (see Sect. 4.1.6). In practice, the models are identical if the noise affecting the data can be considered negligible.

In the continuous case, i.e., when Ω is no longer countable, the probability distribution can be defined via the *cumulative distribution function*

$$F_X(x) = \mathbb{P}(X \le x)$$

for each value $x \in \Omega$. In order to satisfy the probability axioms [442], this function must be increasing, and for $d = 1$,

$$\lim_{x \to x_{\inf}} F_X(x) = 0$$
$$\lim_{x \to x_{\sup}} F_X(x) = 1,$$

where $(x_{\inf}, x_{\sup})$ are the upper and lower bounds (perhaps infinite) of Ω. The multidimensional case ($d > 1$) is examined in Sect. 4.1.4. Returning to $d = 1$, an approximate "equivalent" to the discrete probability $f(x)$ in the continuous case is given by the probability that X is between $x - a$ and $x + b$ (with $a, b \ge 0$):

$$\mathbb{P}(x - a \le X \le x + b) = F_X(x + b) - F_X(x - a).$$

This property pushes us to define, in cases where F_X is differentiable, the derivative of F_X as the limiting case $a = b = \varepsilon \to 0$,

$$f_X(x) = \frac{dF_X}{dx}(x),$$

which is called the *probability density* of X, and therefore satisfies

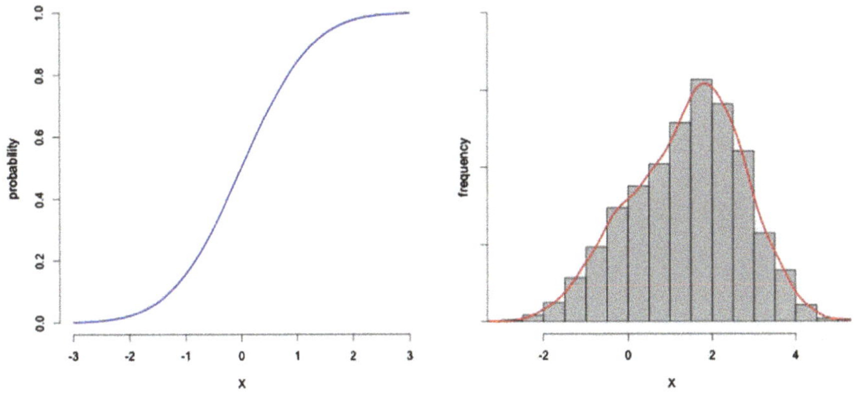

Fig. 4.1 Left: Example of a cumulative distribution function. Right: Frequency histogram of X and the corresponding probability density (curve)

$$F(x) = \int_{-\infty}^{x} f_X(u)\, du$$

and

$$\mathbb{P}(x - a \leq X \leq x + b) = \int_{x-a}^{x+b} f_X(u)\, du.$$

Necessarily, $\int_{\Omega} f_X(u)\, du = 1$. Thus, any continuous probability distribution in dimension $d = 1$ (though this is also true for $d > 1$) can be written interchangeably (if differentiable[1]) in terms of its cumulative distribution function or its density (Fig. 4.1).

Informally, f_X can be seen as the limit of a histogram giving the frequencies of possible values of X in smaller and smaller classes (Fig. 4.1). More formally, the cumulative distribution function and probability density need to be seen as tools useful for taking a *measure* of the distribution of X with respect to that of the space Ω.

Suppose, for example, that $\Omega = I_1 \times I_2 \times \ldots \times I_d$, where each I_k is an interval of $\mathbb{R}$ (closed, open, and half-open). It is therefore a parallelepiped containing all observable values of X. This solid (or space) can be characterized by several measures, such as, for example, its volume. The Lebesgue measure [468], written as μ_L, was constructed to be a reference measure that allows the description of these types of spaces in a universal and uniform way. Like for volumes, it has a finite value if Ω is compact. The density f_X defines another measure on Ω which specifies the shape of the distribution of X and allows us to distinguish it from uniformity. One should therefore interpret it as a *relative* measure with respect to the Lebesgue one (we also say: *dominated* by the Lebesgue measure). For readers interested in a detailed introduction to measure theory, we suggest the book [106] (for an engineer's approach) and [459] (for a more mathematical one).

[1] And more generally, differentiable in $d > 1$ dimensions.

Information contained in probability distributions can often be summarized using what are called the kth-order *moments* M_k, $k \in \mathbb{N}$, defined as the means of the variables X^k:

$$M_k = \mathbb{E}[X^k] = \int_\Omega x^k f_X(x)\, dx.$$

If these exist for $k = 1$ and $k = 2$, we can define the *expectation* $\mathbb{E}[X]$, and *variance*

$$\mathbb{V}[X] = \int_\Omega (x - \mathbb{E}[X])^2 f_X(x)\, dx.$$

The expectation gives us a measure of the mean of X with respect to its distribution f_X, while $\mathbb{V}[X]$ is a measure of the variation in (or dispersion of) f_X. The *standard deviation* of X is defined by

$$\sigma_X = \sqrt{\mathbb{V}[X]}.$$

Alternatively, the *coefficient of variation*

$$\mathrm{CV}[X] = \frac{\mathbb{E}[X]}{\sigma_X}$$

provides another *relative* measure of the variability in or dispersion of f_X (and is especially used by engineers). Lastly, we will say that a variable X is standardized if it has been centered (by subtracting its expectation), and divided by its standard deviation:

$$X' = \frac{X - \mathbb{E}[X]}{\sigma_X}, \tag{4.1}$$

thus transformed into a variable X' with mean 0 and variance 1.

4.1.2 Families of Parametric Models

Let us now recall several fundamental probabilistic and statistical models which will appear often in more complex situations in this book. In our treatment, we are going to focus on *parametric* models, i.e., ones that can be entirely characterized by a finite set of parameters.

The first reason for this choice is connected to the framework of our studies. Indeed, the extreme value behavior of random samples follows, under certain theoretical conditions, parametric distributions (see Chap. 6). The same is true for the behavior of statistical estimators (Sect. 4.2.2), which obey (parametric) laws of large numbers.

This underlying argument is consolidated by the following observation: when interested in extreme-valued events, the number of available data may be very low.

Attempting to describe the mechanism creating such events using a random process characterized by an infinite (or merely large) number of parameters (i.e., more than the number of observations) is rather impractical, since most of the parameters will have to remain unknown, or have several possible values. The final model would be of little relevance.

In the rest of the book, in its most general form, ψ will indicate this vector of parameters, living in a finite-dimensional space. The conditioning by ψ on the random mechanism producing events will be made clear in the notation for densities and cumulative distribution functions: $f(x) = f(x|\psi)$ and $F(x) = F(x|\psi)$.

4.1.2.1 Distributions

In the discrete case, we may be interested in the occurrence of an event of the type $Z > z_0$, where Z is, for example, the maximum monthly water level, and z_0 the height of a protective dike. Suppose that we have a sample of indicator values $(\delta_1, \ldots, \delta_n) \in \{0, 1\}^n$ equal to 1 if the event occurs, and 0 otherwise. Under the hypothesis that the δ_i are independent and each corresponds to a submersion test occurring with the same probability p, and denoting $X_n = \sum_{i=1}^n \delta_i$ as the total number of "successes" from n trials, the probability (mass) distribution of X_n is

$$f(x) = \binom{n}{x} p^x (1-p)^{n-x}$$

for $x \in \Omega = \{0, 1, 2, \ldots, n\}$, where

$$\binom{n}{x} = \frac{n!}{x!(n-x)!}.$$

The random variable X_n is then said to follow the *binomial distribution* $\mathcal{B}(n, p)$ (Fig. 4.2). In the case of a risk-quantification study, we would aim to estimate the overflow probability p from the observed statistic x_n.

The variable X_n defined above, known as a *count* variable, can be generalized with the intention of estimating the occurrence of random events during a fixed time period (e.g., a year). If we suppose that an event happens with a mean frequency of $\lambda > 0$ in this time period, then the probability that it happens exactly $X_n = x \in \Omega = \{0, 1, 2, \ldots, \infty\}$ times is

$$f(x) = \frac{\lambda^x}{x!} \exp(-\lambda),$$

which is none other than the probability mass function of the *Poisson distribution* with mean λ (Fig. 4.2). This distribution plays an important role notably in establishing statistical distributions associated with historical observations, since it allows us to model the arrival of a number of events between two time points (e.g., several dozen years apart) that have not been directly observed. The technical connection between the binomial and Poisson distributions can be stated as in the following lemma:

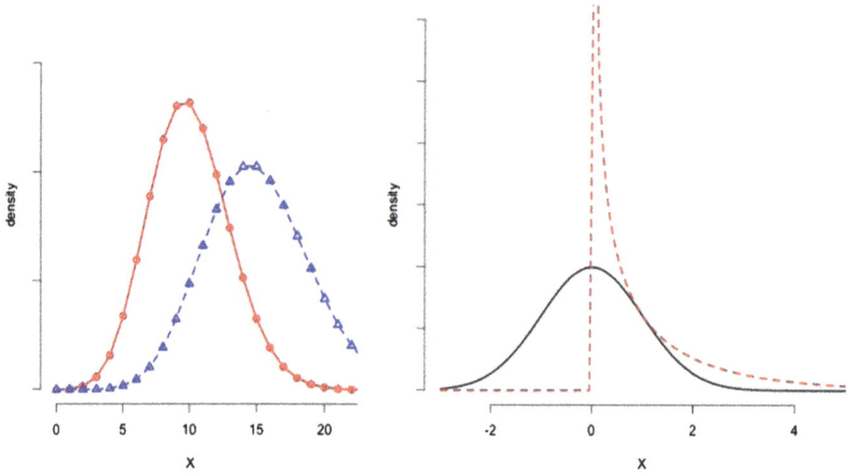

Fig. 4.2 Top: mass function of the discrete binomial distribution $\mathcal{B}_n(100, 0.1)$ (circles) and the Poisson distribution $\mathcal{P}(15)$ (triangles). Down: continuous probability distribution of the standardized normal distribution $\mathcal{N}(0, 1)$ (solid line) and a χ^2_1 distribution (dashed line)

Lemma 4.1 *If X_n follows a binomial distribution $\mathcal{B}(n, p)$ with $p \ll 1$, then it can be approximated by a Poisson distribution with mean np as $n \to \infty$.*

Recall furthermore—in the continuous case—the fundamental importance of the *normal distribution* $X \sim \mathcal{N}(\mu, \sigma^2)$ which has, when $d = 1$ and $\Omega = I\!R$, mean μ and variance σ^2:

$$f_X(x) = \frac{1}{\sqrt{2\pi}\sigma} \exp\left(-\frac{1}{2\sigma^2}(x-\mu)^2\right).$$

It can be used to model a great number of phenomena, in particular the distribution of the mean of a random sample (law of large numbers). The variable $(X - \mu)/\sigma$ follows what is called a *standardized* normal distribution $\mathcal{N}(0, 1)$ (Fig. 4.2). Usually, we denote by $\phi(.)$ and $\Phi(.)$ the density and cumulative distribution function, respectively, of this standardized variable. The convergence of statistics to normal distributions (see Sect. 4.2) is a classical result in the literature (central limit theorem).

4.1.3 Statistical Testing

The general approach of statistical testing consists in rejecting or not rejecting (without necessarily accepting) a so-called *null* statistical hypothesis H_0 given a set of data $\mathbf{x_n}$. For example, in a parametric framework this assumption can correspond to the specific choice of a value $\psi = \psi_0$ in the same family $f(x|\psi)$ or a domain $\psi \in \Psi_0$. Defining a test is defining a statistic

$$R_n = R(X_1, \ldots, X_n)$$

which is a random variable, whose distribution $\mathcal{F}_{R_n}$ is known (at least asymptotically, i.e., when $n \to \infty$) when the H_0 hypothesis is true, and independent of the hypothesis value (ex: independent of ψ). More precisely, in a parametric framework when ψ is tested, $\mathcal{F}_{R_n}$ must not depend on ψ, and the variable R_n is called *pivotal*. When R_n is defined independently of ψ, this statistic is called *ancillary*.

The localization of the observed statistic $r_n = r(x_1, \ldots, x_n)$ with respect to $\mathcal{F}_{R_n}$ Fisher (1926; [282]) as the probability p_{r_n} (called *p-value*) of observing an event more "extreme" (smaller or larger) than r_n. The lower this probability, the further away the event r_n is from the highest density R_n values, and the less likely H_0 is (remember that the p-value is not the probability that H_0 is true). In other words, if H_0 is false, then r_n should be an extreme value of $\mathcal{F}_{R_n}$.

The current approach to testing, known as the Neyman–Pearson approach (1928; [474]), requires setting a threshold of significance $\alpha \ll 1$ defining extremality and comparing the quantile $q_{1-\alpha}$ of $\mathcal{F}_{R_n}$ with p_{r_n}; if $p_{r_n} < q_{1-\alpha}$, event r_n is even less likely than α, and H_0 must be rejected. Otherwise, this assumption is plausible (but not necessarily validated). The common practice in all experimental sciences, again, is to set $\alpha = 5\%$ or $\alpha = 1\%$, but these arbitrary thresholds are increasingly criticized [258, 565], and it is currently recommended [73, 417] to conduct several tests and test very low thresholds (e.g., $\alpha \in [1‰, 5‰]$).

In many cases, R_n is chosen positive, so that the p-value $p_{r_n} = \mathbb{P}(R_n > r_n)$ can be simply defined. This is the case with the Kolmogorov–Smirnov test, which is widely used to test the validity of the exponential law on waiting times between two cluster maxima (see Sect. 6.5.2.1).

Example 4.4 (*Kolmogorov–Smirnov test* [723])

With the classical empirical estimator (cf. Sect. 4.2.2) $x \mapsto \hat{F}_n(x)$ of the cumulative distribution function F of a unidimensional iid sample $x_1, \ldots, x_n$, defined by

$$\hat{F}_n(x) = \frac{1}{n}\sum_{i=1}^{n} \mathbf{1}\{x_i \leq x\},$$

and given a candidate F_0 for F, it is required to test the hypothesis $H_0 : F = F_0$. The test statistic is defined by

$$R_n = \sqrt{n} \sup_{x \in I\!R} \left\| \hat{F}_n(x) - F_0(x) \right\|.$$

Under H_0 and for a large n, R_n approximately follows the Kolmogorov distribution, with cumulative distribution function

$$F_{KS}(x) = 1 - 2\sum_{k=1}^{\infty}(-1)^{k+1}\exp\left(-2k^2x^2\right) \quad \text{for } x \in I\!R^+,$$

which is generally tabulated within traditional software tools.

For an important class, of so-called χ^2 (Chi-squared) tests, R_n is designed to follow a χ^2 *distribution with* $q \geq 1$ *degrees of freedom*

$$R_n \sim \chi_q^2$$

whose density function is displayed on Fig. 4.2 for $q = 1$. χ^2 distributions are intrinsically linked to normal distributions through a quadratic relation. For instance, the sum of the squares of n independent $\mathcal{N}(0, 1)$ random variables follows a χ_n^2 distribution with n degrees of freedom. The quantiles of the latter are provided in practice by specific tables or algorithms.

Test power. Different procedures testing the same property are not necessarily all equally relevant. One way to compare them is by their statistical *power*, i.e., their respective probability of rejecting the null hypothesis H_0 when it is indeed false. When we use statistical tests, we always want to have high power, or at least the highest possible. Statistical power is defined as

$$1 - \beta,$$

where β is the rate of *type II* error: the probability of incorrectly accepting H_0. This rate is equivalent to the *false positive rate* in detection procedures. A classical example of the most powerful test for H_0: $\mathbb{P} = P_0$ against H_1: $\mathbb{P} = P_1$, where P_0 is a submodel of P_1 (which keeps certain parameters of P_1 fixed), is the *maximum likelihood test* (Neyman–Pearson's theorem). It is also called the LRT test (*likelihood ratio test*).

Example 4.5 (*Adequacy* χ^2 *test (discrete case)* [784])

Let $\mathbf{x_n} = (x_1, \ldots, x_n)$ be a sample of iid realizations of X, taking values in $\{1, \ldots, M\}$. It is required to test the null hypothesis H_0 according to which the probabilities that X takes the value 1 to M are, respectively, $p_1, \ldots, p_M$ with $\sum_{k=1}^{M} p_k = 1$. Denote

$$\hat{p}_k = \frac{1}{n}\sum_{j=1}^{n}\delta_{\{x_j=k\}}$$

where $\delta_{\{x_j=k\}} = 1$ if $x_j = k$ and 0 else. Then

$$R_n = \sqrt{n \sum_{k=1}^{M} \frac{(\hat{p}_k - p_k)^2}{p_k}} \tag{4.2}$$

follows, under H_0, the χ^2_{M-1} distribution.

Theorem 4.1 (LRT test (likelihood ratio test)) *Let $\mathbf{X_n} = (X_1, \ldots, X_n)$ be a sample of iid random variables with common distribution $\mathbb{P}$ with density function f. It is required to test H_0: $\mathbb{P} = P_0$ versus H_1: $\mathbb{P} = P_1$. Denote by $L_i(\mathbf{X_n}) = \prod_{k=1}^{n} f_i(X_k)$ the maximized statistical likelihood under Hypothesis H_i, with $\in \{0, 1\}$ (see Sect. 4.2.2.3 for a detailed definition of the likelihood and its maximization). Consider*

$$R_n = 2 \log \frac{L_1(\mathbf{X_n})}{L_0(\mathbf{X_n})}.$$

Then, if P_0 designates a parametric model, with parameter ψ such that $\psi \in \Psi_0$, and P_1 is specified by $\psi \notin \Psi_0$, then R_n asymptotically follows a mixture of Dirac measures and χ^2 distributions whose degrees of freedom are equal to or lower than the number of constraints q imposed by the null hypothesis.

Many details on the mechanisms, specifications, and warnings on the interpretation of statistical tests (parametric, nonparametric, compliance, adequacy, homogeneity, independence, association tests, etc.) are provided in [207] and [356]. The specific case of LRT tests is particularly detailed in [349], Chap. 21. Applied to the specific case of extreme models, the reader interested in a general review may consult article [555].

Example 4.6 (*LRT test*)

In the specific case where Ψ_0 is in the strict interior of Ψ, then

$$R_n \overset{n \to \infty}{\sim} \chi^2_q. \tag{4.3}$$

Consider a normal distribution $\mathcal{N}(\mu, \sigma)$ with $\psi = (\mu, \sigma) \in I\!R \times I\!R^+_*$. It is required to test $H_0 : \mu = 0$ versus $H_1 : \mu \neq 0$. A single constraint differentiates the two hypotheses, and $0 \in I\!R$. Then $q = 1$ and the result (4.3) applies. If it is required to test $H_0 : \mu = 0$ versus $H_1 : \mu > 0$, the domain Ψ is then restricted to $I\!R^+ \times I\!R^+_*$, and (4.3) must be replaced by

$$R_n \overset{n \to \infty}{\sim} \frac{1}{2}\delta_0 + \frac{1}{2}\chi^2_1.$$

4.1.4 Multidimensional Cases

The study of joint hazards (Chap. 9) requires a generalization of the main concepts and ideas presented in Sect. 4.1.1. Let $\mathbf{X} = (x_1, \ldots, x_d)^T$ be the vector of hazards under consideration. The *joint cumulative distribution function* is defined by

$$F_X(\mathbf{x}) = \mathbb{P}\left(X_1 \leq x_1, \ldots, X_d \leq x_d\right),$$

where $\mathbf{x} = (x_1, \ldots, x_d)$. If the X_i are continuous random variables and F_X is differentiable, the joint probability density is given by

$$f_X(\mathbf{x}) = \frac{\partial^d F_X}{\partial x_1 \ldots \partial x_d}(\mathbf{x}).$$

Thus, for any set $\mathscr{A} \subset \Omega \subset I\!R^d$,

$$\mathbb{P}\left(\mathbf{X} \in \mathscr{A}\right) = \int_{\mathscr{A}} f_X(\mathbf{u})\, d\mathbf{u}.$$

In particular, if $\Omega = I\!R^d$,

$$F_X(\mathbf{x}) = \int_{-\infty}^{x_1} \cdots \int_{-\infty}^{x_d} f_X(\mathbf{u})\, du_1 \ldots du_d.$$

Each *marginal density* characterizing X_i independent of the other variables can be obtained by integrating them out. If $\Omega = \bigotimes_{i=1}^d \Omega_i$, then

$$f_{X_i}(x_i) = \iint_{\bigotimes\limits_{j \neq i} \Omega_j} f_X(u_1, \ldots, u_{i-1}, x_i, u_{i+1}, \ldots, u_d)\, du_1 \ldots du_d.$$

The notion of *covariance* helps us to summarize the pairwise dependencies between the X_i:

$$\mathbb{C}ov(X_i, X_j) = \int_{\Omega_i} \int_{\Omega_j} \left(x_i - \mathbb{E}[X_i]\right)\left(x_j - \mathbb{E}[X_j]\right) f_{X_i, X_j}(x_i, x_j)\, dx_i dx_j,$$

where $\mathbb{E}[X_i]$ is the marginal expectation of X_i and f_{X_i, X_j} the bivariate joint density of X_i and X_j, defined as the marginal:

$$f_{X_i, X_j}(x_i, x_j) = \int_{\bigotimes\limits_{k \neq i, j} \Omega_k} f_X(\ldots, u_{i-1}, x_i, \ldots, u_{j-1}, x_j, \ldots, u_d)\, du_1 \ldots du_d.$$

Covariance generalizes the notion of variance: $\mathbb{C}ov(X_i, X_i) = \mathbb{V}[X_i]$ (variance of the marginal distribution of X_i). In practice, a multivariate distribution is often summarized in terms of its vector of expectations $\mathbb{E}[\mathbf{X}] = (\mathbb{E}[X_1], \ldots, \mathbb{E}[X_d])^T$ and variance–covariance matrix

$$\Sigma = (\mathbb{C}ov(X_i, X_j))_{i,j},$$

or *correlation matrix* $\Sigma' = (\rho_{i,j})_{i,j}$, defined by

$$\rho_{i,j} = \frac{\mathbb{C}ov(X_i, X_j)}{\sqrt{\mathbb{V}[X_i]\mathbb{V}[X_j]}}.$$

Each $\rho_{i,j}$ is between -1 and 1, giving information on *linear* dependency between variables X_i and X_j. Nonetheless, this summary is in general rather incomplete. For example, if there is independence between X_i and X_j, then $\mathbb{C}ov(X_i, X_j) = 0$, but the reverse is not necessarily true. The correlation matrix Σ' completely characterizes the dependency structure only in very specific cases, notably when $\mathbf{X}$ is a Gaussian vector, and in general does not provide an entirely relevant measure of dependency. It is therefore important to fight the well-established practice of exaggerating this indicator's importance [791]. In Chap. 9, we describe what is meant by *exhaustive information on the dependency structure*, and provide more useful tools for dealing with multivariate distributions in extreme value theory.

4.1.5 Random Processes and Stationarity

The probability distributions found in this book are special cases of discrete-time[2] *random processes* (or stochastic processes), which describe the general behavior of sequences of random variables $X_1, \ldots, X_n$. These variables do not have to be independent and identically distributed (iid). The distribution f_{X_i} for each X_i can be different for each i. However, the X_i may not be independent of each other while still having similar distributions. In such cases, the process is said to be *stationary*.

Definition 4.1 A random process $X_1, \ldots, X_n$ is stationary if for any set of integers $\{k_1, \ldots, k_s\}$ and integer m, the joint probability distributions of $(X_{k_1}, \ldots, X_{k_s})$ and $(X_{k_1+m}, \ldots, X_{k_s+m})$ are identical.

This definition allows us, in general, to characterize environmental time series data more appropriately than with hypotheses, thus better taking into account a certain inertia often found in natural phenomena. In effect, physical conditions leading to extreme phenomena tend to persist over time. For instance, a high-water event can be caused by several successive days of intense rainfall, and heatwaves caused by

[2] Continuous-time random processes are not treated in this book.

high temperatures maintained over fairly long time periods. Extreme events therefore have a tendency to occur in groups.

The stationarity hypothesis provides us with a technical aid, which we will take advantage of in the book. Nevertheless, it relies on the supposition that the dependency structure between X_k and $X_{k-1}, X_{k-2}, \ldots$ is invariant after m iterations. Thus, the existence of trends and seasonal cycles will invalidate this underlying hypothesis, meaning that data selection and homogenization must be performed before any modeling can occur (Chap. 5).

The stochastic mechanism leading to the random process sometimes needs to be specified. This is true in particular when we are looking at whether the process converges to a stationary one as n increases. Suppose, for instance, that $X_1, \ldots, X_n, \ldots$ represents temperature measurements at very close time points, and that it is wished to select ranges of indicators corresponding to stabilized temperatures. To do this, we need to provide the probability distribution of X_k conditional on $X_{k-1}, X_{k-2}, \ldots, X_1$, and to use a *Markov chain* representation.

Definition 4.2 A random process $X_1, \ldots, X_n, \ldots$ is a Markov chain of order $r \in \mathbb{N}^*$ if for any $i \geq r$,

$$\mathbb{P}(X_i|X_{i-1}, \ldots, X_1) = \mathbb{P}(X_i|X_{i-1}, \ldots, X_{i-r}).$$

If furthermore $r = 1$ and this transition probability is independent of i, the process is said to be homogeneous.

Markov chains of order 1 are the easiest to work with, and provide an important generalization to the framework (an example is shown in Fig. 4.3). They also play an important role in inference and sampling situations. Indeed, processes $\psi_1, \ldots, \psi_n$ can be constructed as a way to explore a space Ψ, for instance, in a Bayesian framework (Chap. 11), and the exploration mechanism is often constructed using a Markov chain of order 1 which is known to converge to a stationary limit process,[3] whose properties (expectation, variance, etc.) can be estimated. We recommend the textbook [652] to readers who wish to study this theoretical probabilistic framework further.

4.1.6 Probabilistic and Statistical Modelings

The terms probabilistic modeling and statistical modeling are often mixed together, in particular in the engineering literature. They do, however, mean different things. A probabilistic model, defined by its probability density f_x, is intended to represent in this book the real physical phenomenon:

$$X \sim f_x,$$

[3] We also say *stationary distribution*.

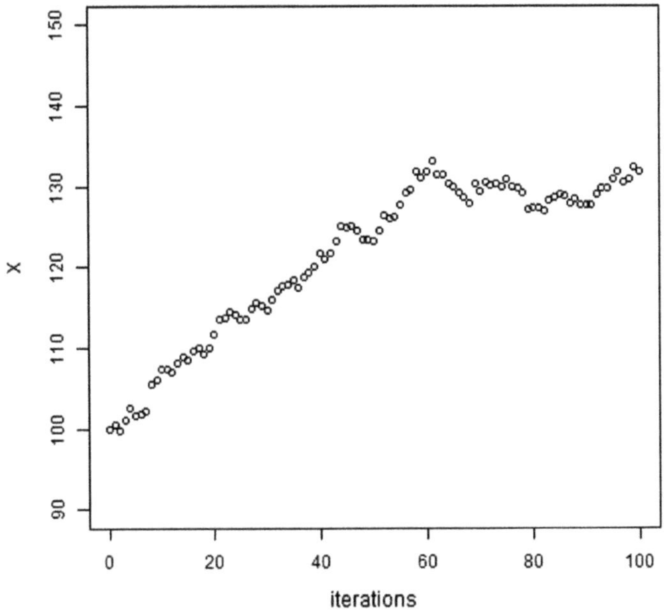

Fig. 4.3 Example of a non-stationary Markov chain of order 1

while the statistical model is meant to represent the fact that one or several observations of X, denoted by x^*, are linked to real examples x of X via a measuring device; for example,

$$x^* = x + \epsilon, \tag{4.4}$$

where ϵ is *noise* coming from an imprecise measurement, often supposed random with a Gaussian distribution. We denote by f_ϵ its density, which is generally known[4]. Knowledge of the relationship (4.4) allows us to characterize the probability distribution of the random variable X^* with samples x^* as a *convolution*, whose density f_{x^*} given by

$$f_{x^*}(u) = \int f_x(u+y) f_\epsilon(y)\, dy,$$

which can be used to determine the *statistical likelihood* of an observation x^* (see Sect. 4.2).

Nevertheless, it is essential for determining f_x, the main goal of the study, that the distribution characterizes the greater part of variation in the observed data (from X^*). Often, we will suppose that the influence of ϵ is negligible, which is the same as saying

[4] Notably due to specifications provided by the constructor of the measuring device in question, or repeated testing under controlled conditions.

$$f_{x^*}(u) \simeq f_x(u), \quad \forall u \in \Omega,$$

meaning that the probabilistic and statistical models are the same. This hypothesis is not always satisfied in practice, however, especially for historical data. Measurement noise can in fact be quite large, for several reasons:

- if measurements are not taken directly (for example, historical torrential rain records can be reconstructed using stratigraphic studies [299]);
- poor measurement precision (for example, a flood dating to the Middle Ages, written about qualitatively—it washed away a bridge, covered fields, etc., or quantitatively but with a lot of uncertainty—marks on a wall of a now-demolished house, etc.) [171, 601];
- measurements perturbed by an unknown bias linked to a poorly calibrated measuring device (or one damaged by the hazard itself, especially in extreme situations) [4].

Even recent and today's measurements can be disrupted by potentially high noise when they are derived from calculations—rather than direct measurements—subject to various sources of uncertainty (see also Sect. 5.1).

Example 4.7 (*Calibration curves in hydraulics*)

Measurements of river flow are usually calculated by mapping from the directly measured river height, using a height–flow curve called a *calibration* curve [11]. This curve is in general established from gauging assessments, performed in relatively smooth flow conditions under certain physical hypotheses [387], and suffers from increasing uncertainty [165, 218, 529] when environmental conditions change (e.g., in cases of high or low water levels). Broadly, we can suppose that flow measurements taken at the extremes have higher measurement error than those taken in normal conditions.

4.2 Bounding the Modeling Error

4.2.1 Convergence of Models

The reliability of statistical models depends on an approximation of reality whose error can be bounded under certain technical conditions. There are two types of approximation in this book:

1. approximation of the unknown behavior of a supposed random quantity X (for example, the maximum of a sample over a given time interval) by a theoretical one (for example, coming from extreme value statistics theory) that allows quantification and extrapolation;

2. approximation of a theoretical probabilistic model by an *estimated* statistical model, in the sense that the theoretical model involves a priori unknown parameters ψ which will be estimated using real observed data. As these observations $x_1, \ldots, x_n$ are considered drawn from a random variable X, the *estimated* parameters are also treated as drawn from a random variable $\hat{\psi}_n$, known as a *statistical estimator*.

The sequence $(\hat{\psi}_n)_n$ therefore represents an initial stochastic process which we would like to use to estimate ψ (fixed but unknown). The set of random variables $(X_n)_n$ produced by the estimated model represents a second stochastic process, which we would like to use to approximate the real behavior X (random variable with unknown distribution).

It is therefore essential to make sure that these two types of approximations do not prevent the *estimated statistical model*—the practical tool here—providing a meaningful diagnosis in terms of being able to reproduce data as seen in reality, nor interfere with its use in predictive studies. An absolutely necessary condition is to have *convergence* between reality and the theoretical model, then between the theoretical model and the estimated one. Convergence is expressed in the form of a difference between the models which must necessarily decrease as the quantity of information (number of observations n) increases, disappearing completely when $n \to \infty$.

In the probabilistic setting, this difference is random, so it is possible that it is zero except for a number k of given situations that forms a subset of the event space Ω with zero measure. Typically, this subset may be made up of a finite number of point values, or elements at the border of Ω. In effect, in the continuous case we know that, under independence,

$$\mathbb{P}(X \in \{x_1, \ldots, x_m\}) = \sum_{i=1}^{m} \mathbb{P}(X = x_i),$$

and that $\mathbb{P}(X = x_i) = 0$ for all x_i (as X is known). In this particular situation, we use the term: *almost surely* zero.

The notion of *almost sure convergence* follows almost automatically. It involves verifying that the probability that the limit of a stochastic process $\hat{\psi}_n$ (or X_n) is the target ψ (or X) equals 1. For X_n, this is written as

$$X_n \xrightarrow{p.s.} X \Leftrightarrow \mathbb{P}\left(\lim_{n\to\infty} X_n = X\right) = 1.$$

Another way to say the same thing is that the difference between the limit of the process and ψ (or X) is almost surely zero. This notion of convergence is the strongest and most useful in practice to show that a random process does indeed converge to a random variable, which could be simply a vector (or number). The term *strong consistency* is also used to refer to this idea.

Other weaker definitions of convergence—in the sense that they are implied by almost sure convergence but the reverse is not necessarily true—are also frequently used:

1. Convergence in *probability*:

$$X_n \xrightarrow{\mathbb{P}} X \Leftrightarrow \forall \epsilon > 0, \quad \lim_{n\to\infty} \mathbb{P}\left(|X_n - X| > \epsilon\right) = 0.$$

   This plays a major role in many proofs for convergence in distribution (see below), and also implies almost sure convergence of a subsequence of $(X_n)_n$; it allows X_n to deviate from X, but less and less significantly so as n increases.
2. Convergence in *distribution*, or *in law*, which is implied by convergence in probability, and corresponds to pointwise convergence[5] in the probabilist world:

$$X_n \xrightarrow{\mathcal{L}} X \Leftrightarrow \lim_{n\to\infty} F_n(x) = F(x),$$

   where (F_n, F) are the cumulative distribution functions of X_n and X, respectively, for any x where F is continuous. This notion of convergence does not characterize the values of the stochastic process, but uniquely its random behavior, i.e., whether that of X_n resembles a greater and greater degree that of X. Here, we say we have *weak consistency*.

Other notions of convergence (in L^p norm in particular) are also commonly used. Often, these can help to ensure useful *deterministic* convergence, of expectations (moments) for instance, as shown in the following two theorems.

Theorem 4.2 *Suppose that X_n converges in L^1 norm to X in $\Omega \in I\!\!R$, i.e.,*

$$\lim_{n\to\infty} \mathbb{E}\left[\|X_n - X\|^1\right] = 0.$$

Then $\lim_{n\to\infty} \mathbb{E}[X_n] = \mathbb{E}[X]$.

Theorem 4.3 *Suppose that $X_n \xrightarrow{\mathcal{L}} X$ with $(X_n, X) \in \Omega^2$, where $\Omega \subset I\!\!R$. Then, for any real continuous and bounded function g (and in particular, the identity function),*

$$\lim_{n\to\infty} \mathbb{E}[g(X_n)] = \mathbb{E}[g(X)].$$

A collection of technical results allows us to combine these different types of convergence and their mappings through continuous functions (*mapping theorems*), to study more complex models. For an in-depth look at this, as well as a generalization to multidimensional cases, we recommend the book [758].

---

[5] For specialists: in the sense of the *characteristic function* (Lévy's continuity theorem [207]).

4.2.2 Classical Statistical Estimation

Inference refers to the process of constructing one or several estimators of ψ,

$$\hat{\psi}_n = T(X_1, \ldots, X_n),$$

where T is a function of random variables associated with samples x_i obtained from the phenomenon under consideration. As mentioned earlier, $\hat{\psi}_n$ is itself therefore a random variable, whose *observed* value $T(x_1, \ldots, x_n)$ is called an *estimate*.

4.2.2.1 Important Properties of Statistical Estimators

There are an infinite number of possible estimators for a given parameter vector ψ, so it is important to be able to select one in particular. In classical statistics, the following principles are used to order desirable estimators:

1. *asymptotically*, there must be *consistency*, i.e.,

$$\hat{\psi}_n \overset{?}{\rightarrow} X,$$

where ? represents, at best, almost sure convergence;

2. the *quadratic error*

$$\text{LSE}(\hat{\psi}_n) = \mathbb{E}\left[(\hat{\psi}_n - \psi)^T(\hat{\psi}_n - \psi)\right], \tag{4.5}$$

also known as the *least square error* (LSE) or L^2 error, should be as small as possible. It can be written as the sum of the squares of the *bias*[6] (or *non-asymptotic* expected error)

$$\text{B}(\hat{\psi}_n) = \mathbb{E}\left[\hat{\psi}_n\right] - \psi,$$

and the determinant of the variance–covariance matrix of $\hat{\psi}_n$, which measures the non-asymptotic inaccuracy of $\hat{\psi}_n$. These two terms cannot be minimized simultaneously, and the minimization of (4.5) therefore necessarily proceeds from a *bias–variance equilibrium*.

We also remark that producing a weakly consistent estimator $\hat{\psi}_n$ for an unknown parameter ψ then allows us to do the same for any function $h(\psi)$ of that parameter, as long as it is differentiable. When the estimator converges asymptotically to a Gaussian distribution, the differentiation procedure is called the *delta method*.

Theorem 4.4 (Multivariate delta method [570]) *Let $\psi_1, \ldots, \psi_n$ be a stochastic process in $I\!R^d$ and let $g : I\!R^d \rightarrow R^q$ be a differentiable, non-null function in ψ. Denote by $J_g(\psi)$ the Jacobian matrix of g in ψ. Assume that $\sqrt{n}(\psi_n - \psi)$ converges in*

[6] An estimator $\hat{\psi}_n$ whose expected value is ψ is called *unbiased*.

distribution toward the multivariate normal distribution $\mathcal{N}_d(\mathbf{0}_d, \Sigma)$, with mean the null $d-$dimensional vector $\mathbf{0}_d$ and covariance matrix $\Sigma \in I\!R^{d\times d}$. Then

$$\sqrt{n}\left(g(\psi_n) - g(\psi)\right) \xrightarrow{\mathcal{L}} \mathcal{N}_q\left(0, J_g^T(\psi)\Sigma J_g(\psi)\right).$$

4.2.2.2 Least Squares Estimation

The class of *least squares estimators* (LSE), which tries to find a compromise between bias and variance, is thus naturally defined by rules of the type

$$\hat{\psi}_n = \arg\min_{\hat{\psi}} \widetilde{\text{LSE}}(\hat{\psi}), \tag{4.6}$$

where $\widetilde{\text{LSE}}$ is an empirical approximation of LSE, constructed as a function of $X_1, \ldots, X_n$. Estimates of this type often have good consistency properties, but may suffer from sensitivity to model choice and parameterization of ψ. Therefore, it is not always clear that the expectation and/or variance in criterion (4.6) exist(s).

4.2.2.3 Maximum Likelihood Estimation

The likelihood principle. A more general rule must necessarily be based on a more complete characterization of the information contained by $X_1, \ldots, X_n$ about the model's parameter ψ, and needs to be generic and always well-defined. One way to proceed is to work with the *statistical likelihood* ℓ, which is defined (for known X_i) as the joint density of the observations $X_i = x_i$ conditional on ψ. Thus, in the case where observations x_i are iid samples,

$$\ell(x_1, \ldots, x_n|\psi) = \prod_{i=1}^{n} f_X(x_i|\psi).$$

The likelihood can have a more complicated structure when samples are not independent or identically distributed, or have *missing* values replaced by thresholds when a measuring device's limit has been reached. See, for instance, [804] for details on building a likelihood of correlated extreme observations.

Example 4.8 (*Measuring wind speed*)

Some old anemometers cannot measure the wind speed above a certain value, and replace the observation that should have been made by a maximum measurable wind speed. This is referred to as a *right-censored* statistical observation. This type of partial observation is common in *survival analysis* [500].

The completeness of the information contained in the likelihood is a fundamental principle of classical statistical theory. In fact, the maximum likelihood estimator (MLE)[7]

$$\hat{\psi}_n = \arg\max \ell(X_1, \ldots, X_n|\psi) \tag{4.7}$$

defines the random variable with the *most likely value* of ψ given the observations $\{X_i = x_i\}_i$. Due to its quite general form, its significance, as well as its stellar consistency properties, it is the most common and indeed most natural statistical estimator we have.

Example 4.9 (*Interval-censored data*)

Very frequently, a historical unidimensional datum (see page xxx), or an incorrectly measured datum, can simply be described as a missing value x_i located between two known bounds $x_{i,\min} < x_{i,\max}$. The density term $f(x_i)$ must then be replaced in the statistical likelihood (4.7) by

$$\begin{aligned} & P\left(x_{i,\min} \leq X \leq x_{i,\max}|x_{i,\min}, x_{i,\max}\right) && (4.8)\\ &\qquad = P\left(X \leq x_{i,\max}|x_{i,\max}\right) - P\left(X \leq x_{i,\min}|x_{i,\min}\right), \\ &\qquad = F(x_{i,\max}) - F(x_{i,\min}). && (4.9) \end{aligned}$$

Moreover, if we make the assumption that the values $(x_{i,\min}, x_{i,\max})$ are themselves random data (e.g., noisy data), described as realizations of variables $(X_{i,\min}, X_{i,\max})$ with respective densities $f_{i,\min}, f_{i,\max}$), the corresponding likelihood term (4.9) becomes

$$\iint P\left(X_{i,\min} \leq X \leq X_{i,\max}|X_{i,\min} = y_1, X_{i,\max} = y_2\right) f_{i,\min}(y_1) f_{i,\max}(y_2)\, dy_1 dy_2.$$

When the datum is multivariate (Chap. 9), several situations can occur: one or several dimensions of X may be interval-censored, and extensive processing must be carried out to obtain useful statistical specifications (see [438] for survival analyses, and [676] for the specific case of multivariate extreme values).

Theorem 4.5 (Central limit theorem for the MLE) *Suppose that $X_1, \ldots, X_n$ are independent and identically distributed. Let q be the dimension of ψ. Then, under quite general regularity (Wald) conditions,*

$$\hat{\psi}_n \xrightarrow{\mathcal{L}} \mathcal{N}_q\left(\psi, I_\psi^{-1}\right), \tag{4.10}$$

[7] For practical reasons, we often replace ℓ by the log-likelihood, $\log \ell$, in definition (4.7).

where $\mathcal{N}_q$ is the q-dimensional multivariate normal distribution with variance–covariance matrix I_ψ^{-1}, where I_ψ is the Fisher information matrix, whose $(i, j) \in \{1, \ldots, q\}^2$ entry is given by (under the same regularity conditions)

$$I_\psi^{(i,j)} = -\mathbb{E}_X\left[\frac{\partial \log \ell(X_1, \ldots, X_n|\psi)}{\partial\psi_i\partial\psi_j}\right]. \tag{4.11}$$

More details about the notion of information and the Fisher information matrix are given in Sect. 4.2.2.4. One important property of the MLE is that it is *asymptotically efficient*, i.e., its asymptotic covariance, given by the inverse of the Fisher matrix, is *minimal* over all unbiased estimators of ψ.

4.2.2.4 Fisher Information

The notion of information was proposed in the 1920s by the English researcher Ronald A. Fisher. Fisher's approach is as follows: if we are interested in the characteristics of a large population, we can neither know nor deal with overly abundant information relating to each and every individual. The problem therefore becomes to be able to correctly describe the population by means of summaries that can be constructed from samples from the population to be studied. The more the summary that can be extracted from a sample correctly characterizes the reference population, the more informative it can be considered to be.

Based on this assumption, Fisher defined information as the mean square of the derivative of the log of the probability distribution being studied. Cramer's inequality then makes it possible to show that the value of this information is proportional to low variability—that is to say, a high degree of certainty—in conclusions that can be drawn from it. This idea, which is at the root of all statistical estimation and inference theory, is exactly the same one discovered 20 years later by Shannon, expressed this time in terms of probability rather than statistics. If X is a sample from a probability density $f(x|\psi)$, the Fisher information is defined by

$$I_\psi = \mathbb{E}\left[\left(\frac{\partial \log f(X|\psi)}{\partial\psi}\right)^2\right].$$

In the case where the probability distribution depends on several parameters, ψ becomes a vector. Then, the Fisher information is no longer defined as a scalar but as a covariance matrix, called the Fisher information matrix:

$$I_{\psi_i,\psi_j} = \mathbb{E}\left[\left(\frac{\partial \log f(X|\psi)}{\partial\psi_i}\right)\left(\frac{\partial \log f(X|\psi)}{\partial\psi_j}\right)\right].$$

4.2.2.5 Confidence Intervals

In practice, the asymptotic distribution of $\hat{\psi}_n$ is itself estimated by replacing the unknown term I_ψ by a consistent estimator $\hat{I}_n$ (in general, $\hat{I}_n = I_{\hat{\psi}_n}$), which allows us to define *confidence regions* $C_{\hat{\psi}_n,\alpha}$ associated with an estimator $\hat{\psi}_n$ such that as n tends to infinity,

$$\mathbb{P}\left(\hat{\psi}_n \in C_{\hat{\psi}_n,\alpha}\right) = \alpha. \tag{4.12}$$

In particular, when we are especially interested in a certain coordinate ψ_i, Theorem 4.5 helps us define the *(asymptotic) confidence interval of level* $1-\alpha$ associated with $\hat{\psi}_n$:

$$\mathbb{P}\left(\psi_i \in \left[\hat{\psi}_{n,i} - z_{\alpha/2}\sqrt{\sigma^2_{i,i}}\,,\ \hat{\psi}_{n,i} + z_{\alpha/2}\sqrt{\sigma^2_{i,i}}\right]\right) = 1-\alpha, \tag{4.13}$$

where z_α is the quantile of order α of the standardized normal distribution, and $\sigma^2_{i,i}$ the ith entry of the diagonal of the *estimate* of the inverse, $\hat{I}_n^{-1}$.

Equations (4.12) and (4.13) help us to evaluate the precision of the estimate of ψ using the sample $x_1,\ldots,x_n$. However, note that the probability measure $\mathbb{P}$ in equation (4.12) concerns $\hat{\psi}_n$ and not ψ (which is unknown but fixed). The confidence region is not therefore defined by the probability $1-\alpha$ that ψ stands inside, but as the region in which there is a priori a strong probability $1-\alpha$ of obtaining an *estimate* of ψ. By simulating a large number of samples similar to $x_1,\ldots,x_n$, the associated estimated confidence regions should contain the true value of ψ on average $100(1-\alpha)\%$ of the time. The confidence interval for the ith coordinate of ψ therefore aims to contain the true value ψ_i with a certain probability linked to the asymptotic distribution of the estimator $\hat{\psi}_n$, which presupposes that the statistical model is correct (and not in terms of an absolute probability that is independent of the model).

As is the case for the LSE, the MLE does not always have a closed-form expression (particularly for models used in extreme value theory), and often must be calculated using numerical methods. The LSE and MLE may not be unique for complex models, and the MLE may not even be defined (leading to an infinite likelihood). This is nevertheless rare in extreme value theory situations. Unlike the LSE, the MLE is always invariant under reparametrization, i.e., the MLE of $h(\psi)$ is $h(\hat{\psi}_n)$, provided h is bijective. This property is crucial for avoiding paradoxes and inconsistencies. Indeed, if we replace the observation x with a bijective mapping $y = d(x)$, the model parametrized by ψ is replaced by one parametrized by a bijective mapping $\psi' = h(\psi)$. However, the information delivered by x and y is the same, so any estimation rule $x \to \hat{\psi}(x)$ should be such that $y = d(x) \to \hat{\psi}'(y) = h(\hat{\psi}(x))$.

One final argument pleads in favor of the MLE: the maximized likelihood constitutes the fundamental ingredient of most *model selection* methods. Equipped with a penalty term linked to the number of degrees of freedom of a model [26, 694], the *estimate* of $\ell(x_1,\ldots,x_n|\hat{\psi}_n)$ provides a useful diagnostic tool for evaluating the pertinence of one model with respect to another, for a given data set. We suggest that readers wanting to learn more about this subject consult the textbook [620].

The vast majority of the tools and methods presented in this book come under the classical statistical framework, where the two estimators we have introduced are frequently used and compared. This choice is both a matter of theoretical properties and of a certain empirical robustness, the knowledge of which comes from practice. However, the classical statistical framework is an approximation to the larger, so-called *Bayesian* decision-making framework, which will be introduced in Chap. 11. In that framework, ψ is considered not fixed but random, which makes it possible to introduce, in addition to the statistical likelihood, *a priori* information, depending on the type of model and its use, and possibly expert knowledge about the phenomenon being examined.

4.3 Statistical Regression

We conclude this chapter by briefly recalling the main concepts of parametric and nonparametric *regression*, which will prove useful in some parts of the following chapters. The interested reader, confronted with the vast literature on this subject, may refer to Sen and Srivastava's recent book [696]. The objective of a stochastic regression model is to determine how the expectation of a random variable Y depends on the realizations x of a vector $X \in \chi$ of explanatory covariates, possibly also random, through the regression (or link) function g:

$$\mathbb{E}[Y|X = x] = g(x).$$

4.3.1 Parametric Approach

A parametric approach consists in placing on f a form a priori parameterized by a finite number of parameters to be estimated; linear regression presupposes as well as

$$g(x) = \beta_0 + \beta^T x. \tag{4.14}$$

The vector (β_0, β) can be estimated using different approaches. In this framework, maximum likelihood estimation is similar to a weighted least squares approach, provided the whole model can be written as

$$Y = \beta_0 + \beta^T X + \epsilon$$

where ϵ is a centered noise (with null expectancy), uncorrelated, and with a unique variance σ^2, and assuming that independent observations of Y and related values of X are available.

Depending on the dimension of X, Y, hypotheses on the correlation structure between covariates and/or between the dimensions of Y, and finally according to the number of available data, the parameter estimation of a parametric regression model (linear or not) constitutes a class of fundamental problems in statistics. The primary objective of the many estimation methods, which are generally based on a maximization of likelihood penalized by a smoothing term (e.g., Lasso regression and ridge), is to ensure a satisfactory comparison between interpolation power (low bias and possibly high variance) and smoothing (low variance but possibly high bias).

4.3.2 Nonparametric Approach

A nonparametric approach does not assume, e contrario, any analytical form as (4.14) or, necessarily, a limited number of parameters defining function g. Two main families of nonparametric regression methods are commonly used. Suppose we have a sample $\{x_i, y_i\}_i$ of realizations of(X, Y).

1. **Kernel regression** [543, 777] estimates g by a kernel estimator defined as

$$\hat{g}_h(x) = \sum_{i=1}^{n} \omega_i(x, h) y_i$$

with

$$\omega_i(x, h) = \frac{K\left(\frac{x_i - x}{h}\right)}{\sum\limits_{j=1}^{n} K\left(\frac{x_j - x}{h}\right)}$$

where $K(u) = K((x - i - x)/h)$ defines a kernel function—namely a probability density function with maximal value in 0 and depending only of the distance $|x_i - x|$ (therefore a symmetric function)—and h a smoothing parameter named *bandwidth*. Each weight $\omega_i(x, h)$ decreases as the distance $|x_i - x|$ increases, and increases when h becomes high. The selection of h results from both a bias–variance trade-off and an arbitration on the regularity of the curve $x \mapsto \hat{g}_h(x)$ (smooth versus irregular), which is very generally produced by approximating a criterion $C(h)$ defined as the *Mean Integrated Squared Error* (MISE)

$$C(h) = \mathbb{E}\left[\int_{\mathcal{X}} \left(\hat{g}(x, h) - g(x)\right)^2 \, dx\right].$$

This criterion being not directly usable (g being unknown), it is typically replaced by a cross-validation criterion between the y_i and each predictor $\hat{g}_{-i}(x_i, h)$ builds by using the whole sample from which the point (x_i, y_i) is removed.

2. **Local polynomial regression** is an alternative approach that reveals to be more robust to extreme values of X for estimating g. It postulates that the value $g(x)$ can be approximated by the value of the parametric function estimated in a neighborhood $\mathcal{V}(x)$ of x (*locally* estimated). It may be relevant, for instance, to approximate $g(x)$ with a polynomial of degree 1

$$\hat{g}(x) = \hat{\alpha}(x) + \hat{\beta}(x)^T x$$

where the estimators $\hat{\alpha}(x)$ and $\hat{\beta}(x)$ are determined by minimizing the weighted criterion

$$\sum_{x_i \in \mathcal{V}(x)} \omega_i' \left(y_i - \hat{\alpha}(x) + \hat{\beta}(x)^T x_i \right)^2$$

and where the weight ω_i' may be chosen constant (LOESS regression [168]) or be defined, again, through a kernel function (LOWESS regression [169]).

Chapter 5
Collecting and Analyzing Data

Marc Andreewsky and Nicolas Bousquet

Abstract This chapter focuses on the first steps of a statistical study of extreme values based on real hazard data. Illustrated by numerous examples, the qualitative and quantitative characteristics of these data are analyzed in order to come as close as possible to the theoretical conditions of application—in particular the notion of asymptotism—and to statistically characterize their nature (regular, truncated, censored...). This chapter is, therefore, an essential prerequisite for the concrete application of the theoretical tools presented in the following chapters.

5.1 Introduction

Extreme values are, by definition, rare and difficult to observe. Nevertheless, the more observations available, the clearer their statistical behavior becomes. The width of confidence intervals for statistical estimators is inversely proportional to the square root of the number of observations (Sect. 4.2.2.3). For this reason, data collection is a crucial and sensitive part of the study of extreme values. It is also very important to include the results of any study carried out in the past on the same problem, in order to compare them. The importance of this step should be particularly emphasized because, as mentioned earlier, validation is tricky, so any extra information that facilitates decision-making can be extremely useful.

In order to make it possible to apply theorems mentioned in Chap. 4, the most useful data has two forms:

- A sample of the phenomenon for a given time period (*Block Maxima* approach)—typically one data per month or year;
- A sample of events exceeding a threshold over a given time period (*Peaks Over Threshold* approach). This approach necessarily requires that we be able to directly

M. Andreewsky · N. Bousquet (✉)
EDF R&D, 6 quai Watier, 78401 Chatou, France
e-mail: nicolas.bousquet@edf.fr

N. Bousquet and P. Bernardara (eds.), *Extreme Value Theory with Applications to Natural Hazards*, https://doi.org/10.1007/978-3-030-74942-2_5

estimate the particular threshold above (or below) which we consider ourselves to be dealing with an extreme event; or alternatively, that we be able to estimate the duration of the phenomenon, i.e., a temporal threshold below (or above) which the extreme phenomenon can be considered ended.

Understanding the nature of these data, collecting them, sorting them according to their quality and representativeness, filling in the gaps, and ensuring a sufficient amount of information for conducting an extreme statistics study are all the issues addressed in this chapter.

5.2 Looking for Available Data

5.2.1 Building Databases

It is essential to construct the largest possible database of observations or estimates of the study variable, whether these be continuous, discrete, measured, historical, paleo-historical, simulated, etc. For each discipline, the sources and methods used to collect data may be very different. Here are some examples.

Example 5.1 (*Maritime surges (1)*)

Numerous sea (and ocean) level measurements on the French coast are provided by the Marine Hydrographic and Oceanographic Service* (SHOM) and can be downloaded from http://www.shom.fr. The investigation of historical extreme maritime tides has been the subject of the recent works [121, 350].

Example 5.2 (*Maritime surges (2)*)

Observations of sea level are generally available at hourly time-points [40]. Modeling of *instantaneous maxima* (i.e., peak sea level reached during high tides) cannot rely on this directly observed data, since the maximum level is unlikely to be produced at the exact time of measurement. This modeling can instead be developed using polynomial interpolation to reconstruct the signal between pairs of sea level observations located to either side of the supposed maxima, one hour apart. The diagram in Fig. 5.1 illustrates this calculation.

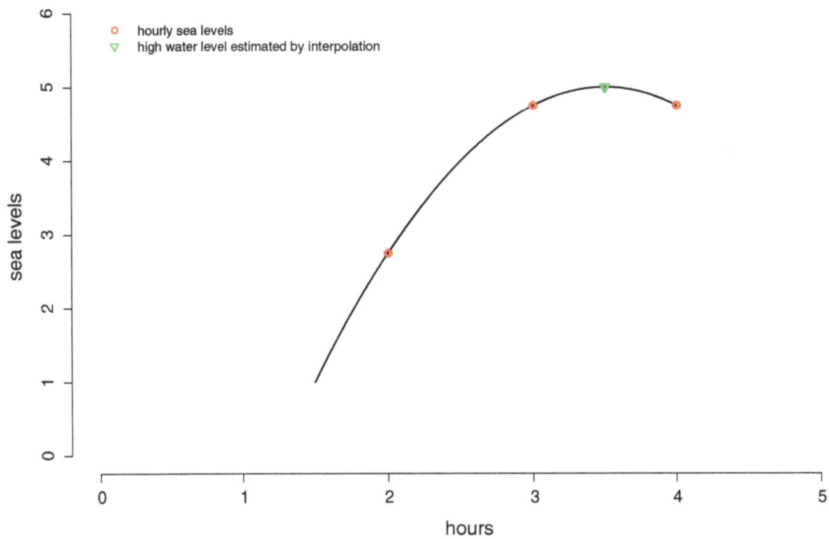

Fig. 5.1 Illustration of the interpolation allowing the calculation of the tide peak level

Example 5.3 (*Extreme tides*)

If the study variable is the extreme tide, it is necessary to subtract from the sea level calculated above, the *theoretical* extreme tide level [40], which can be *retro-predicted*; indeed, the theoretical tides are governed by the gravitational interaction between the earth and other large objects in the solar system, principally the sun and moon [636] (p. 249).

Example 5.4 (*Reconstructed hydrological data*)

The construction of a representative dataset (hydrological flow series) can be carried out with the help of measurements from a neighboring site [518]. A *flow naturalization* procedure can be used to account for the influence of hydraulic structures on flow (see Example 3.2).

Example 5.5 (*Sticky snow load data*)

Electrical cables can be subjected to a heavy load, potentially exceeding a safe-load criterion, due to accretion of sticky snow [214]. Load data, useful for this type of study, can be constructed using meteorological observations (see Example 3.3).

Missing data can be particularly detrimental to extreme value statistics studies. Certain measuring devices, usually old and somewhat fragile, can break down just when a particularly extreme event occurs. To give an example, tide gauges installed before 1990 frequently break down during storms [197]. Torrential rain can also be significantly underestimated (by more than 10%) depending on the type of rain gauge used [681], mainly due to evaporation. Observations of the same hazard at neighboring sites, or combined hazards, may indicate that an extreme data value was missed. This reaching of the limits of a measurement system characterizes what we call *censored* data (see Example 4.8).

Example 5.6 (*Meteorological data*)

It is common practice to replace missing values of a meteorological variable (temperature, wind speed, etc.), if limited in number, by the variable's mean (or by linear interpolation), while being cautious not to modify the effective time of observation. This, provided that the missing values do not correspond to the known occurrence of an extreme phenomenon (e.g., a missing wind speed measurement during a storm). In such cases, it is better to try to estimate the missing value from neighboring measurements [703].

Biases, and more generally measurement errors, are also important factors that can undermine the results of statistical studies. Ideally, each device should be calibrated before use, and each measurement accompanied by a description of the potential measurement error.

This is all more important since extreme value statistical studies are likely to have to resort to **historical data** (see the Preface and [72, 692]), to compensate for small sample sizes. Historical data is subject to measurement or reconstruction errors[1] which depend on the time period in question.[2]

Such data can generally be presented as:

[1] See [553, 603] to highlight, in hydrology, the systematic biases and random errors that appear when reconstructing historical data.

[2] For instance, satellite temperature surveys began in the late 1970s only, and sea level rise has only been monitored by satellite since 1992 [607].

- punctual and scattered values associated with the intensity of an extreme event [707];
- a sample measured by obsolete and possibly very imprecise processes [638];
- a threshold of perception over a historical period: between two major events observed in the past, it is assumed that the absence of data indicates that no events of at least the same intensity as the lower of the two "surrounding" events occurred (otherwise it would have been indicated, unless there was a known stop in the measurements, for example during a conflict).

The latter hypothesis allows historical data to be statistically treated as *censoring by threshold* data [588]. Alternatively, the first situation underlies the hypothesis that only a limited number of the most intense events of the past have been preserved in societal memory (*censoring by number* [707]). The statistical translation of these data has been particularly examined by [588, 638] in a context where only one hazard is considered, and then by [676] in cases where several extreme hazards are cumulated. A generic statistical specification of an interval-censoring situation is proposed in Example 4.9.

Since Le Roy Ladurie's pioneering work on climate history and phenology* [460], the study of a wide variety of historical sources (see Preface for details) has consistently revealed interesting qualitative and quantitative information for building historical datasets [172, 519]. Thus, the use of this type of data is very valuable, if not almost indispensable, for dealing with phenomena such as tsunamis [324] or high sea levels [72, 128]. We refer the reader interested in these methods of reconstruction to the works of specialized historians[3] [308, 311], who often deplore the rapid disappearance of the memory of disasters [307]. Surveys conducted after the 2010 Xynthia storm have until recently shown that historical lessons have often been overlooked [103, 319], despite a relatively close history.

Example 5.7 (*Reconstructed historical floods*)

The reconstruction of extreme historical floods must be the subject of meticulous historical work [581]. Payrastre and his co-authors [603] propose a list of desirable characteristics of these historical floods. A recent pedagogical example of such a reconstruction work, concerning the past floods of the Gardon (a river located in the French Languedoc-Roussillon-Midi-Pyrénées region), is carried out in [217].

Finally, potential problems due to **combinations of hazards** have to be detected. Do we have conjoint data? (e.g., swell and extreme tides), and if not, is it possible to

[3] The funding of scientific projects in the environmental sciences, including many historians, testify to the growing importance given to these data. See, for example, the OPHELIE project (PHenological Observations to Rebuild Europe's Climate), coordinated by the Pierre-Simon Laplace Institute: http://www.unicaen.fr/histclime/ophelie.php.

reconstruct certain conjunction scenarios or define conjoint data as a combination of "true measurements" and a probable interval in which a missing measurement value is likely to be?

5.2.2 *Analyzing the Quantity of Available Data*

Methods for checking that the length of an available data series is appropriate for a given study are not generic, and remain the subject of debate [174, 484, 645]. Each situation requires expert knowledge and input. Limit theorems underlying the use of extreme value statistical distributions are valid asymptotically. Demonstrating that we have attained this asymptotic validity "in practice" is no easy task. The available observed data can be used to guide this decision. In general, the length of the data series varies between 20 and 60 years. For example, on average, French extreme tide data from tide gauges is available from a 30-year period [780].

The consensus is that a sample of less than 20 maxima values is too small to estimate the value of a one-in-a-thousand event using GEV modeling. In general, it is good practice to systematically test whether the fit of a GEV distribution with a shape parameter of zero (Gumbel or exponential distribution) is not just as good a fit for the sample as the full GEV distribution. This comparison can be done using a likelihood ratio test, as described in Theorem 4.1. However, the test is only valid asymptotically, which contradicts the situation we find ourselves in. The estimation of the shape parameter ξ in the GEV distribution can be so uncertain that it seems more reasonable, in practice, to set it to zero.

While the POT approach can take advantage of more data than the MAXB approach, researchers working with this extreme value method still generally avoid samples that are too small. The fit to a GPD distribution is often not considered sufficiently reliable if the sample has fewer than 50 independent values.

Remark 5.1 The independence of annual maxima must be scrupulously checked by going back to the original data. An event may occur at the very end of a year and continue at the beginning of the next.

Some experts even advise not to estimate quantiles associated with return periods that exceed 100 or 200 years for a *local analysis* [727], i.e., for data recording periods generally below 50 years. This criterion is often too strict since a study of extreme values must necessarily proceed in a framework of extrapolation of observed data collected at very low probability levels (Sect. 3.1.1.5). Other experts believe that it is reasonable to estimate quantiles associated with return periods of up to five times the observation period [627]. Thus, some authors have proposed estimates of values associated with return periods of up to 1 000 years (sometimes even 10 000 years [803]). While some researchers point out that such extreme values may have little or no scientific [107, 173] or practical (e.g., [219] in a financial context) basis, others have accepted the use of these probability distributions as consensus tools to define particularly high dimensioning values [221].

While it has long seemed that there was no absolute rule in this area, recent work has made it possible to frame theoretically the validity of an extrapolation [29]. This validity remains in all cases conditional on the validity of the data production model, and for this reason, the "reasonableness" of an extrapolation must be constantly questioned. Clearly, it is not advisable to extrapolate a ten-year probability from ten years of data.

Remark 5.2 In the case of **joint hazards**, the amount of data and the information it contains must also be sufficient to allow estimation of the correlation structure between the hazards, and in particular, the correlation between extreme events. More details are provided in Chap. 9.

When we wish to estimate a return level, i.e., a quantile of the statistical distribution being used, it is possible to guard against high estimation uncertainty due to small sample sizes by calculating an upper bound to the quantile. For example, the ASN* guide [15] to the protection of standard nuclear installations from external flooding proposes the following approach:

After gathering available information thoroughly enough to ensure we have found all that can be found, it is necessary that the entire available statistical sample be used to determine the probability distribution of threshold-exceeding rare events. In addition to regularly recorded data from observation stations, any historical data from earlier times should be taken into account. Given the fact that these sources of data are limited, concerning generally only relatively short periods of time, it is essential to construct a confidence interval for the calculated mean value. In order to cover for uncertainty associated with such data samples, the extrapolated value used is the upper bound of a confidence interval associated with each flood risk reference scenario. In practice, using the 70% confidence interval for this task is generally considered "appropriate" (translated from [15]).

Remark 5.3 For a study conducted with the POT approach, the threshold is usually chosen so that the number of observations per year is around 2–10 [40]. Any value lower than this will have to be justified by an expert, or other sampling methods used instead, such as regionalization (Chap. 7), allowing, under certain conditions, more data to be collected for the desired analyses. See also Sect. 5.4.2 for formal threshold-selection approaches.

5.3 Checking for Data Quality

The quality of measurements must be examined in detail. In particular, noise in data must be characterized as it may influence statistical quantification (Sect. 4.1.6). Measurement errors need to be known in order to determine, among other things, the number of significant digits to keep in calculations. In some cases, the distribution of the measurement noise can be estimated by studying the phenomenon under normal conditions, either as indicated by the supplier, or via the work of experts (see for example [8] for the quantification of uncertainty and measurement errors for tide gauges, and [254] on the thickness of volcanic ash). It is possible, in such cases, to then appropriately take into account noise sources (see for example [62]).

In order to analyze the data quality, it is also necessary to verify its origin. Which organization did the measuring? Are they reliable? Were pre-treatment or quality analyses conducted? Verifying the quality of the measuring device's calibration is equally important: what method was used to calibrate it?

5.3.1 Relevance of Extreme Observations

Particular care must be taken when checking the larger values, because these will greatly influence the statistical fitting of the GEV or GPD distribution and the characterization of return levels. Such verification generally requires the input of several scientific disciplines.

Example 5.8 (*Extreme tides*)

It is standard practice to check that the ten (at least) largest extreme tides recordings can be linked with strong depressions (and hence severe storms) in the area where they were taken, using databases like those available at http://www.meteociel.fr.

In the event that no extreme event was detected in an area and on a date where a particularly large data value was recorded, it is permitted to remove the data point, specifically stating why it was removed. A major difficulty is that many methods for detecting such *outliers* are themselves based on the use of extreme values [132, 697], although solutions have recently been proposed to distinguish these aberrant values from truly "extreme" ones [300] and applied to the case of extreme rainfall. If, after analysis, there is doubt, it is better to simply keep the data. Data removal should not occur without particularly rigorous and transparent reasoning and justification.

Example 5.9 (*Extreme low tides*)

If we want to verify the validity of a sequence of values of extreme low tides (*undertides*), it may be useful to conduct an analysis of the associated meteorological conditions. In this example, recordings are from the Saint-Servan tide gauge (in the French county of Ille-et-Vilaine): the undertide observed on 22/11/1906 was large (96 cm) and among the widest undertides ever recorded at Saint-Servan. There is no information available that would allow us to remove

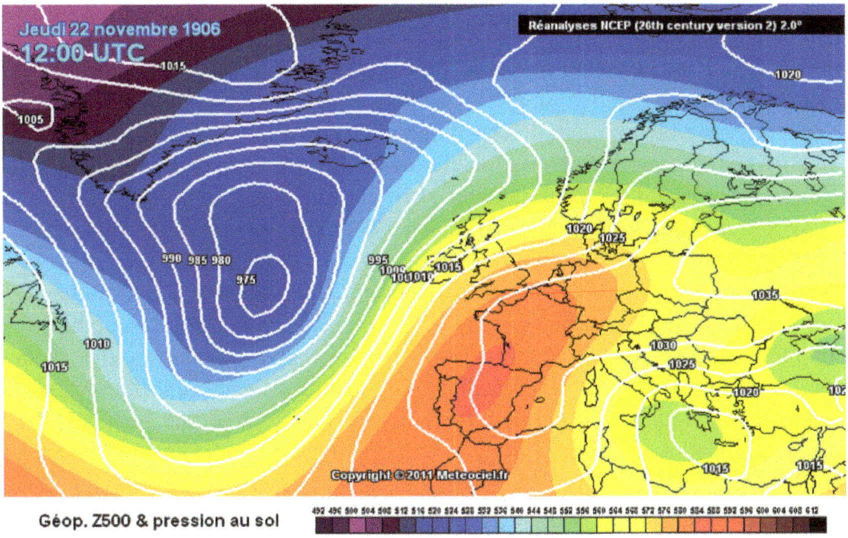

Fig. 5.2 Isobars of pressure at sea level (white lines) in hPa, with French annotations (*Source* http://www.meteociel.fr)

it. Indeed, the reconstruction of the atmospheric pressure and wind at the time,[4] certainly impacted by uncertainty in view of how long ago the measurements were taken, reveals the presence of a deep depression over the North Atlantic, with an anticyclone centered on France, with a high pressure of about 1030 hPa near Saint-Servan (the average pressure at sea level is about 1013 hPa) and wind coming from the south-west. Figure 5.2 shows the reconstructed pressure fields for 22/11/1906. This situation is a priori favorable in leading to a rather low sea level. Moreover, in the Saint-Servan low tide record, high excess (drops) levels (mostly greater than 30 cm) were observed between 09/11/1906 and 28/11/1906, i.e., in the temporal range of the 22/11/1906 event.
This situation corresponds rather well to a persistence of high pressure around 22/11/1906. These various points are in favor of maintaining the measurement of 22/11/1906 in the Saint-Servan time series. Nevertheless, the observed high pressure values are not really out of the ordinary; it is quite likely that the value of 96 cm, which is relatively large, was overestimated. As there is no evidence to prove this, the value for 22/11/1906 must be retained in the final sample used for the statistical study.

[4] The reconstruction can be found at http://www.meteociel.fr/modeles/archives/archives.php?mode=0&month=11&day=22&year=1906&map=0&hour=12

Other ways of checking the validity of data can be found in various disciplines

- *Hydrology.* It is advisable to check the coherence of flood volumes at one station with the measured volumes at nearby stations. It is also useful to compare sequences of daily *maximal* flows with pointwise *maximal* flows, and to study the quality and accuracy of calibration curves (see Example 4.7) to evaluate uncertainty in available flow measurements [14].
- *Maritime hydraulics.* In addition to what has been suggested in the previous examples, it is often useful to compare one site's tide level series with that from a neighboring tide gauge.[5] Furthermore, aberrant and isolated sea level values can be detected graphically when measurements are taken with a time step of the order of a few minutes (from 1–10 mins) [14]. Also, an analysis of meteorological conditions observed in the days preceding a suspicious measurement often makes it possible to determine whether a particularly extreme high or low tide could have been possible.

5.3.2 Detection of Changepoints and Outliers

Another characteristic of measuring systems must not be forgotten: they evolve with time. The quality of measurements when passing from one system to another must be examined closely. In this way, it becomes possible to map the most biased measurements onto the most reliable ones if, for example, we notice a constant bias between the two systems (see, for example, [465] for an illustration on old tide level measurements). More generally, statistical tests can be used to detect the presence of changepoints, unexplained trends, and aberrant values, due to poor-quality or erroneous measurements in the observed data time series. For example:

- *Alexandersson's test* was developed to detect changes in rainfall time series; it can be applied to sequences of *ratios* comparing observations from one measuring station to the average of several [32, 585];
- The *cumulative residuals* approach is a frequently-used method in hydrology for sequences of flows which consists of analyzing the cumulative difference between nearby stations so as to check the quality of a given sequence [518, 585, 640].

Possible trends can also be linked to the non-stationarity of phenomena. Tools for testing this can be found in Sect. 5.4.3. Finally, it is possible to find in [266] an analysis method to discriminate a measurement noise from a real extreme value of a climatic phenomenon (applied to temperatures).

[5] Information on sea level measurement quality control is provided by SHOM at http://www.sonel.org/_Controle-qualite-des-mesures.html.

5.3.3 Processing Roundness and Measurement Noise

Another important practical problem is raised by the **roundness** affecting the measurements. Since the statistical theory of extremes applies to continuous variables, roundness must be such that no data is repeated (because it is a zero probability event in the continuous frame). Another practical tip, indicated by [766] (for different statistical distributions) and taken from Sect. 4.3 of [484], is to choose an accuracy such that the existence of classical statistical estimators of quantities of interest, as a return level, remains assured. More precisely, it is a question of verifying that the numerical estimation procedure remains applicable. The validity of the calculations performed with sufficiently precise roundness is guaranteed by theoretical results obtained by [431, 484] (Sect. 4.4) then [267].

The presence of numerical roundness can be interpreted as the effect of a centered noise erasing a measurement accuracy [265], which can be assumed to be incorporated into a broader observational noise. In this sense, it is, therefore, essential to seek to "perturbate" the original data by a similar noise; the data thus produced remain extreme and constitute a legitimate sample that could have been observed. If the noise applied takes more precise numerical values than the roundness of the original data, the repetition mechanisms within the data vanish and it becomes possible to apply the theory of extreme values. This *jittering* approach, available in many software tools—usually to facilitate data representation [275]—has been successfully tested on many cases by the authors of this book.

In all rigor, the statistical quantification must be based on the convolution mechanism described in Sect. 4.1.6 and derived from an equation such as (5.1). An empirical way to implement this principle is to apply a bootstrap procedure by reproducing the study for a large number of simulated noises and then averaging the estimation results. The choice of the noise to be injected (generally uniform or Gaussian), if the observation error is not well known, depends on the stability of these results. However, some study variables remain poorly adapted to the application of such a process. Thus, EDF studies have shown that the jittering of temperature measurements generally posed little difficulty, in the sense that the traditionally calculated return levels remain stable, while wind speed measurements are often insufficiently precise to allow an addition of noise that does not significantly modify these return levels.

Note finally that the statistical quantification on the observed or jittered variable

$$X^* = X + \varepsilon$$

provides a *conservative* representation of real hazard X, in the sense that the estimated variability of this hazard is greater than that of the real hazard: under an independence assumption

$$\mathbb{V}[X^*] = \mathbb{V}[X] + \mathbb{V}[\varepsilon].$$

However, this does not automatically mean that the extreme return levels calculated assuming that X^* follows an extreme value distribution are always higher (and therefore, conservative) than the actual return levels on X^*. It is, therefore, essential to ensure the robustness of the jittered estimates when ε is simulated.

5.3.4 *The Case for Simulated Data*

Certain situations lead to the use of simulated data. For example, [780] uses the ANEMOC2* database [744] to obtain data on highly significant wave heights. The simulated data used needs to have been checked to make sure it also covers extreme events (it may only deal with data values close to the mean).

5.4 Statistical Characterization

Independence and stationarity (in frequency, intensity, and length) of observations are fundamental hypotheses underlying the use and fitting of extreme value distributions. We look closer at these in the following section using statistical tests (Sect. 4.1.3) such as those found in Excel, R, etc., in function of the chosen sampling scheme (MAXB or POT). For POT, a satisfactory (and necessary) choice of a threshold can be tested using a specific procedure.

5.4.1 *Testing Independence Between Data*

The hypothesis of independence of exceedance is crucial for the POT approach: the maxima have to be independent. Maxima found next to (or within) a major peak need to be removed. If we are working with annual maxima (MAXB approach), we can generally suppose that data are independent. However, in every case, we need to test this. The value of the autocorrelation coefficient, and testing the Poisson hypothesis (for the POT approach), allow us to accept (or indeed reject) the hypothesis.

5.4.1.1 Autocorrelation Indicators and Tests

The linear correlation coefficient (Pearson's) measures the quality of a linear relationship between two variables [207]. One can simply plot one variable in terms of the other (for instance, a variable at time t in terms of the same variable at time $t-1$) to get an idea of whether they are linearly related. A standard method for testing for independence uses—as hinted above—*autocorrelation*.

Table 5.1 Selected nonparametric tests of independence or correlation of univariate sample elements. Nonasymptotic distributions are provided between parentheses on purpose

Test	Spearman-Student	Modified Mann-Kendall	Durbin-Watson (DW)	C-statistic
Null hypoth.	Data independence	Data independence	Uncorrelated linear	Uncorrelated data
H_0	Independence	Independence	Model residuals	
Alternative hypoth.	Increasing or	Increasing or	Order 1	Order 1 correlated
H_1	Decreasing trend	Decreasing trend	Residuals	Correlated data
Minimal sample size *(in practice)*	$10 \geq n \geq 5$ (Spearman) $30 \geq n \geq 11$ (Student)	$10 \geq n \geq 5$ (Kendall)	$n \geq 8$	$n \geq 8$
Remarks			Underlying linear model hypothesis	More robust than DW
References	[339]	[376]	[236, 685]	[43, 208, 749, 793]

In statistics, the *autocorrelation* of a discrete time series is simply the correlation between the process and an offset version of itself. For real-valued time series, the empirical *autocorrelation function* (ACF) is defined by

$$\text{ACF}_n(h) = \frac{1}{ns_x^2} \sum_{i=1}^{n-h} (x_i - \bar{x})(x_{i+h} - \bar{x}),$$

where

$$s_x^2 = \frac{1}{n} \sum_{i=1}^{n} (x_i - \bar{x})^2,$$

and h is the time offset or *lag* at which the autocorrelation is calculated. For: variables, the autocorrelation function is given by

$$\text{ACF}_n(h) = \begin{cases} 1 \text{ if } h = 0, \\ 0 \text{ for } 0 < h < n. \end{cases}$$

Several nonparametric statistical tests, involving estimators of the autocorrelation function (or some related function), are selected on Table 5.1 for their relatively good behavior when applied to sample of small to moderate size. They can appropriately accompany the tests described hereafter, more specific to the extreme value context.

Example 5.10 (*Successive exceedance peaks*)

To test independence of n successive exceedance peaks, we can calculate the autocorrelation coefficients between the selected events, then the associated Student's variables. We then have to see whether these variables follow Student's distribution with $n-2$ degrees of freedom, with associated risk of error α. Other possible approaches are more domain-specific (see the following example).

Example 5.11 (*Extreme high tides*)

When looking at POT data for extreme high tides, it is acceptable to keep two high values close in time if and only if: *(a)* they are above the chosen threshold u_0 (above which we consider having seen an extreme event), and *(b)* if, between these two high values, there is another one equal to less than $u_0/2$. This criterion, called *redescent* [433, 434], allows us to suppose that the two close-together high values come from different storms and are thus independent.

An example of calculating the autocorrelation coefficient for a study of strong winds is given in Fig. 5.3. The confidence interval for the hypothesis $\text{ACF}(h) = 0$ can be found using the normal distribution with mean 0 and variance $(n-1)^{-1/2}$.

5.4.1.2 Testing the Poisson Hypothesis (POT Approach)

When modeling extreme values using thresholds (POT approach), it is necessary to test the *Poisson hypothesis*, i.e., that the distribution of the number of values per year above a threshold follows a Poisson distribution (Sect. 4.1.2.1). Testing this also means testing the absence of overly-strong dependency in the initial sample. The adequacy of the distribution of the number of observations per year retained in the POT sample with respect to the Poisson distribution can either be examined visually or using a classical statistical test, such as a χ^2 test (Sect. 4.1.3).

5.4.1.3 Detecting Dependency

If the previous tests do indicate dependency in the data, this can be dealt with in theory using stochastic processes (Sect. 4.1.5). As long as the dependency is not too strong, extreme value theory can still be used. However, it must not be forgotten that dependency automatically decreases the quality of asymptotic approximations.

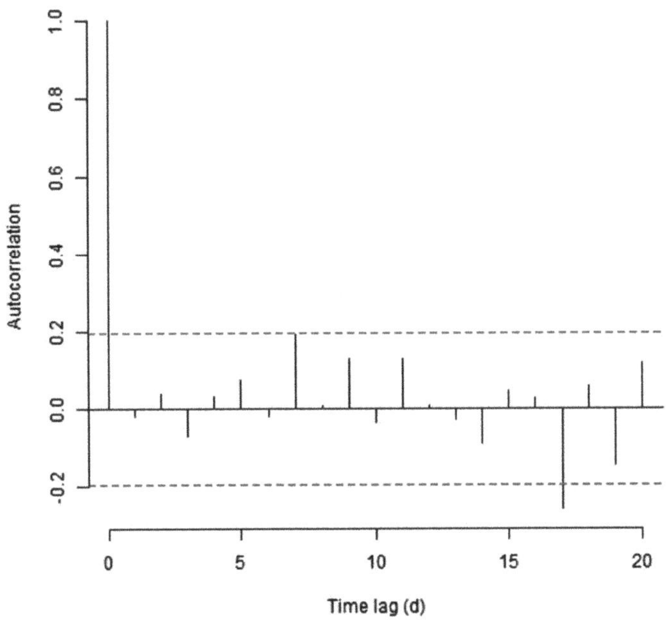

Fig. 5.3 Autocorrelation diagram for a wind strength time series. The dotted line is the confidence interval under the null hypothesis of autocorrelation 0. The time step is days; the small value calculated is thus compatible with the hypothesis

Following the criteria formally defined by [463], two events x_i and x_j, "above a certain threshold", for which the indices i and j are "sufficiently separated", can be considered independent even if they are found in a dependent dataset (which still needs to be stationary). The choice of a MAXB approach requires choosing the block size (periods), so that tests of independence, applied to extreme values, give results that do not refute independence. The choice of a POT approach requires **declustering*** techniques, which will be described in Chap. 7.

From a practical point of view, the concept of "sufficiently separated" data requires knowledge of the system's physics that could allow the definition of a distance above which data can be considered independent. This choice strongly depends on the phenomenon in question. For example, a time period of three days was used to define two independent extreme high tides by [57], whereas for certain water flow studies in large water basins, time periods of over ten days are often used.

In addition to calculating the autocorrelation coefficient, calculation of the *extremal index* provides an estimate of the distance at which two observations can be considered independent. This coefficient is used to overcome the fact that, in general, dependency in a time series can take different forms, and that it is impossible to develop a generalization of the behavior of extreme values without defining a precise practical framework. To define such a framework, it is necessary to introduce a measure of the temporal dependency between data points which are above high-valued thresholds (see Definition 6.9 for a formal characterization).

Definition 5.1 Extremal index. The extremal index (or coefficient) θ, such that $0 < \theta \leq 1$, is formally defined in Chap. 9 (Definition 6.9). The closer the value of θ is to 0, the more the extreme values have a tendency to cluster (or aggregate, as groups of associated values above a fixed threshold). The index can be interpreted and estimated in several ways:

1. as the inverse of the limit average size of cluster in the data series, the limit being defined by increasing the threshold; the original dataset of n dependent values brings as much information as an iid sample of size $n\theta$;
2. as the limiting probability that a point over the threshold will be followed by a sequence of values below it;
3. the proportion of null times between two threshold excesses.

We will focus on the first interpretation in this chapter. Therefore, the extremal index can be estimated by

$$\hat{\theta} = \frac{n_c}{n_u},$$

where n_u is the number of times a threshold u was exceeded, and n_c the number of clusters obtained with this threshold. It, therefore, suffices to calibrate (u, n_u, n_c) so as to obtain a sufficiently large equivalent number of iid data $n\hat{\theta}$.

5.4.2 *Peaks Occurrence Testing (POT Approach)*

The test for the occurrence of peaks works by comparing the empirical distribution of the occurrence of exceedance: to that of the theoretical Poisson (Sect. 4.1.2.1) and negative binomial distributions. Then, a χ^2 test is run to compare the observed frequency of exceeding the threshold n times in a reference period to the probability p of exceeding the same threshold under a Poisson $\mathscr{P}(n)$ or negative binomial $\mathscr{BN}(n, p)$ distribution, whose mass function is

$$P(N = k) = \frac{\Gamma(n+k)}{k!\Gamma(n)} p^n (1-p)^k.$$

Experience shows that in most cases, and if the threshold is well-chosen, the Poisson distribution works well for this.

This test, therefore, makes it possible to check whether the threshold has been chosen well and also provides a comparison with the representation of events proposed by the negative binomial distribution. If the Poisson fit is not satisfactory, it is advisable to start again and test with a new threshold. Another test, called the *uniform distribution* test, also makes it possible to judge the Poisson nature of the occurrence of maxima (see Sect. 5.4.3.3).

5.4.3 *Checking Stationarity*

Stationarity of the process generating extreme valued data (Sect. 4.1.5 and Definition 4.1) can be interpreted as meaning that the observed extreme values follow the same distribution and do not change their behavior and frequency over time.

5.4.3.1 Reasons for Non-stationarity

In general, potential non-stationarity can be detected either in the original full dataset or in the sample of extreme values. The absence of a predefined link between these two types of non-stationarity obliges us to check for both.

The presence of **cycles** (the typical example in meteorology being the seasons), **trends** (extreme values tend to increase or decrease over time), and **changepoints** (the extreme phenomenon begins after a certain date) are all signs of non-stationarity [641]. Appropriate statistical tests are described in Sects. 5.4.3.2 and 5.4.3.3.

The stationarity hypothesis can also be questioned if we go far back into the past and use **historical data**. The fact that the climate evolves over time makes this a legitimate concern [161, 517], but the apparent non-stationarity of a particular phenomenon cannot always be proven. For instance, no significant change in storm intensity in Europe has been observed in recent decades [272]. Another example from [641] has shown that most statistically significant trends detected in a large database of flows in France were due to measurement error or drift. Below is another example.

Example 5.12 (*Correcting excess high tide calculations*)

The open sea *high tide exceedance** or *residual* is the difference (if positive) between the observed level of high tide and the predicted one in open sea, the time at which these two occur being possibly different (see Fig. 5.4). For this reason, the meaning of *instantaneous* excesses, calculated as the difference *at the same time instant* between measured and predicted levels, adds confusion (it can potentially be much larger, as shown in Fig. 5.4). Undertaking an extreme value statistics study on high tide exceedances measured at a port requires both tide level measurements and retro-predictions of the theoretical tide level there [40].

To this end, Weiss [780] studied sea level data from Newlyn (Cornwall, England) and showed that there was a trend linked to the sea level. Here, eustatic change* is calculated according to the rules of the *Permanent Service for Mean Sea Level*[6] (PSMSL) using time averages from tide gauges over periods of one day, one month, and one year. The calculations showed that the mean sea level at Newlyn is rising at a rate of 1.8 mm/year (Fig. 5.5). This means that retro-predicted values need to be corrected for this eustatic change, so as not to bias the calculation of high tide exceedances. The tide level measurement at a date T, therefore, needs to be modified by a value of $(2010 - T) \times \alpha$, where $\alpha = 1.8$ mm/year. The corrected high tide exceedances are shown in Fig. 5.6. We see that the trend has disappeared (it would still be there if no eustatic change correction had been made), so now the time series can be used to provide extreme data (whether independent or not) for a statistical study.

[6] http://www.psmsl.org/data/supplying.

Other issues related to stationarity should also be taken into account. For instance, historical data may not correspond to samples of a random phenomenon similar to that currently observed, for reasons other than climate. For example, historical floods of a river whose bed has evolved over time may lose all or part of their significance; see [233] for an analysis applied to the Isère river (France). It is possible, for some historical data that seem to be outliers from the rest of the sample, that we don't necessarily find the reasons for the discrepancy, as illustrated in the following example.

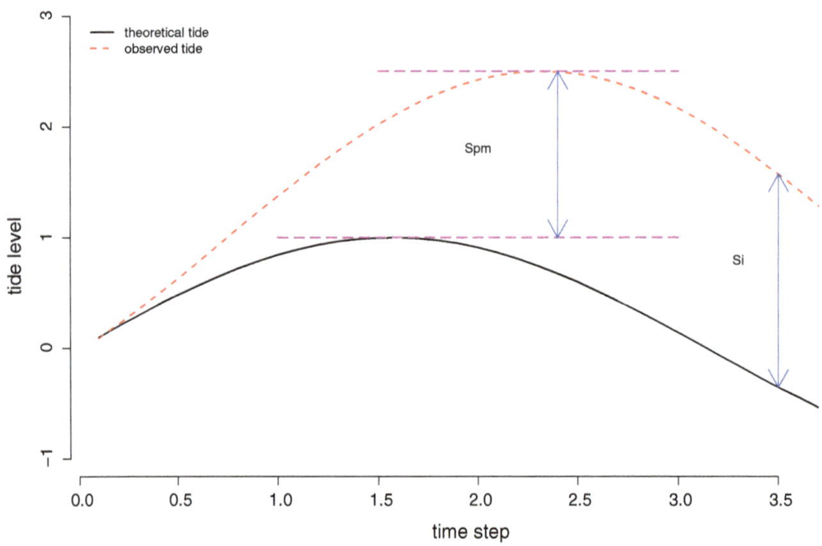

Fig. 5.4 Illustration of the calculation of high tide exceedances, adapted from [40]. S_{pm} indicates the high tide exceedance in open sea, while S_i indicates the instantaneous high tide exceedance

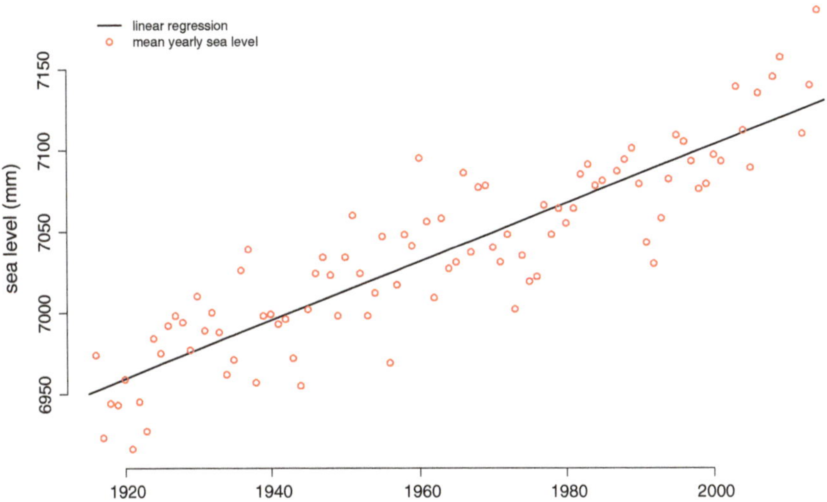

Fig. 5.5 The evolution of the annual mean sea level at Newlyn (adapted from [780])

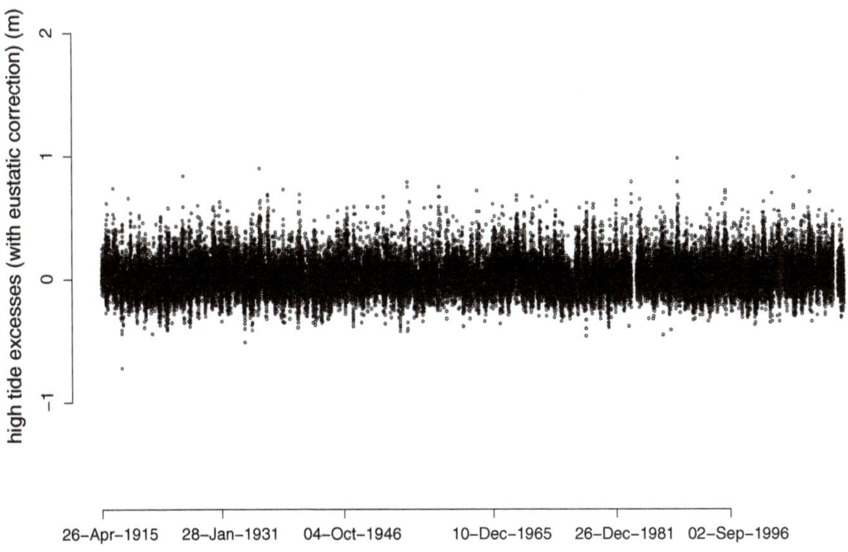

Fig. 5.6 High tide exceedances calculated with high accuracy at Newlyn using a corrected retro-projection of sea levels. This plot was constructed using tide level data from SHOM and an EDF software retro-predicting tide levels

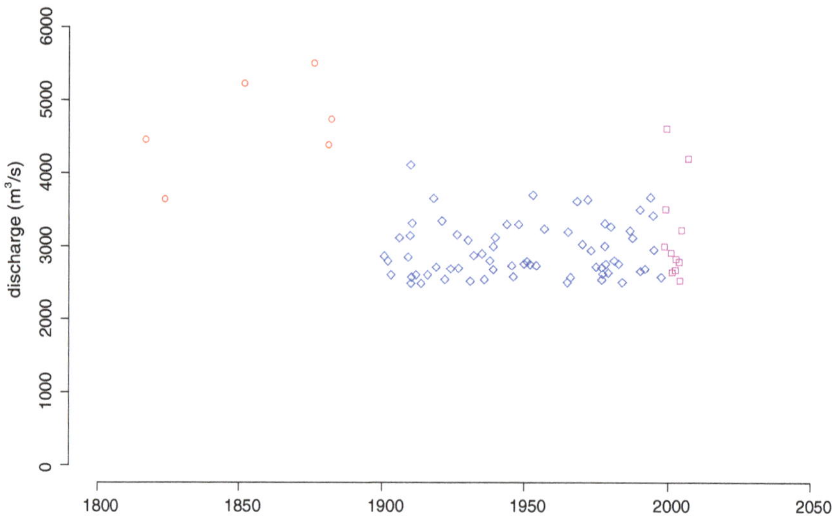

Fig. 5.7 The flow of the Rhine river at Kembs (*Source* EDF-DTG)

Example 5.13 (*Fluvial hazards*)

Changes affecting river basins and human usage of rivers and streams give rise to doubts as to the stationarity of flow distributions. Figure 5.7 illustrates the worries we should have about this: the sample showing the flow rate of the Rhine since 1999 seems stationary. But if we include historical floods of the nineteenth century, this appearance of stationarity disappears.

Is it because of a changing climate? Have the Swiss strongly influenced the behavior of the Rhine due to civil engineering works [771]? Is a century of data sufficient to consider building a statistical model? Has the measurement system changed? If now observations made between 1995 and 2010 are taken into account, it is legitimate to ask whether, if we wait a little longer, the appearance of stationarity might return. It is difficult to answer such questions!

5.4.3.2 Tests of Stationarity

The stationarity hypothesis excludes the possibility of any large-period cyclical variation and any systematic change in the phenomenon over time. Some important *nonparametric* tests for detecting trends or changepoints are presented below. Other tests, known as *parametric*, make it possible to compare models showing or not showing trends (e.g., likelihood ratio tests). These will be presented later in this book.

Testing means and variances (Fisher-Student). The data sample $\mathbf{x_n} = (x_1, \ldots, x_n)$ is divided into two sub-samples $\mathbf{x_{n_1}}$ et $\mathbf{x_{n_2}}$ (of separate measurement periods). The means μ and variances σ^2 associated with these two sub-samples are estimated by the classical empirical estimators $(\bar{x}_{n_1}, \bar{x}_{n_2})$ et (s_{n_1}, s_{n_2}). First, Hypothesis $H_{0(1)}$ of variance equality is tested (Fisher's test) by considering the statistic

$$T_1 = \frac{n_1(n_2-1)}{n_2(n_1-1)} \left(s_{n_1} \frac{\sigma_2}{s_{n_2}\sigma_1} \right)^2 \tag{5.1}$$

that follows a Fisher–Snedecor $\mathscr{F}_S(n_1-1, n_2-1)$ distribution under $H_{0(1)}$. If testing the empirical estimator of T_1 (defined by removing the σ_i in (5.1)) does not reject the equality of variances, then Student's procedure ($t-$test) testing the hypothesis $H_{0(2)}$ of equality of means is conducted. It considers the statistic

$$T_2 = \frac{\sqrt{n_1+n_2-2}}{\sqrt{1/n_1+1/n_2}} \frac{\left(\bar{x}_{n_1} - \bar{x_{n_2}}\right) - (\mu_1 - \mu_2)}{\sqrt{n_1 s_{n_1}^2 + n_2 s_{n_2}^2}} \tag{5.2}$$

following a Student $\mathscr{S}_t(n_1+n_2-2)$ distribution under $H_{0(2)}$. This test is conducted by comparing the observed statistic defined by removing the μ_i in (5.2). See [207] for more details.

Mann-Kendall test. The Mann-Kendall test [432, 499] is used to **detect a trend**. For each element x_i of a sequence of n terms, we calculate how many of the preceding elements are smaller than it, which we call C_i. The test then operates on the value of t, which is equal to the cumulative value

$$t = \sum_{j=1}^{i} C_j.$$

If there are at least 10 elements, then the value of t follows a normal distribution with mean

$$\mathbb{E}[t] = \frac{i(i-1)}{4}$$

and variance

$$\mathbb{V}[t] = \frac{i(i-1)(2i+5)}{72}.$$

The asymptotically Gaussian test statistic is given in terms of the difference $u(t)$ between t and $\mathbb{E}[t]$, i.e.

$$u(t) = \frac{t - \mathbb{E}[t]}{\sqrt{\mathbb{V}[t]}}.$$

With the help of a standardized normal distribution table, this difference is compared with the probability of randomly obtaining a value greater than $u(t)$.

Pettitt's test. This test [612] detects changepoints in time series. Let k be an integer between 1 and $n-1$. We calculate:

$$U(k) = \sum_{i=1}^{k} \sum_{j=k+1}^{n} \text{sign}(X_j - X_i).$$

The test statistic is then

$$Z = \max_{k=1,\dots,n-1} |U(k)|,$$

whose cumulative distribution function is known. We can, therefore, calculate the observed test statistic z and associate it with the level of significance required. A practical explanation of Pettitt's test is given by [641].

5.4.3.3 Test of Uniform Distribution (POT Approach)

This test examines whether there is a uniform distribution of occurrences of the random variable X. It is used only for POT-type samples [518] and provides back-up to the stationarity hypothesis for points exceeding the threshold. It also helps to test the Poisson nature of occurrences of exceedances.

With the help of this test, we can check whether the dates at which peaks occur over the observation period are uniformly distributed, i.e., that the cumulative number of peaks selected since the start of the observation period is linear with respect to time elapsed. The test, therefore, analyzes the occurrence dates of all [[sup-seuil]] peaks and compares the theoretical expectation or mean (*resp.* the standard deviation) of occurrence dates with the observed mean ((*resp.* standard deviation). A consideration of the distance between the two allows us to determine whether the distribution of peaks is uniform or not, for a given error risk (e.g., 5%). An illustration of this test on flood data is given in Fig. 5.8.

5.4.3.4 Detecting Non-stationarity

If, with the help of the previous tests, non-stationarity is found or suspected in study data, it is recommended to take it into account. Two approaches are possible; the exploration of each is advised in practice.

1. *Correcting for the non-stationarity in the original data*, for instance by extracting a trend and transforming these data in a stationary sequence. This is done, for example, for the correction of sea level eustatic change (Example 5.12), where a linear trend (eustatic change) is estimated on the series of annual sea level averages and then used to correct daily sea level data observed in the past. Note, however,

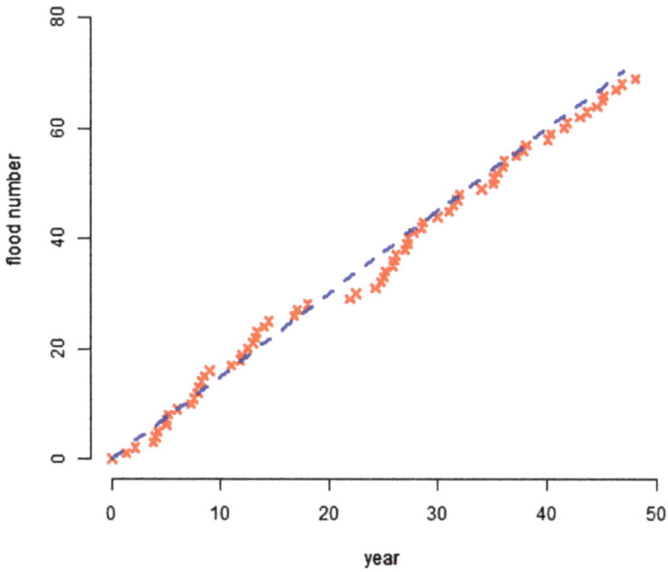

Fig. 5.8 Test of uniform distribution applied to flood data. The points represent the observations and the line indicates the uniformity of the flood distribution

that correcting for the "mean", non-stationarity cannot entirely correct the non-stationarity of the largest values in the sample. It may also be necessary to consider whether there is an evolution of the variance over time. The stationary extremes of the corrected series are used for the study, adding to them the previously extracted trend. This approach remains corrective, since it concerns all data without taking a specific interest in the deviations from the laws of extremes. Thus, it makes it difficult to extrapolate.

2. *Explicitly taking into account non-stationarity in the extreme values*, through the assessment of trends in the parameters of the extreme value distributions. This approach is detailed in Chap. 8.

Chapter 6
Univariate Extreme Value Theory: Practice and Limitations

Anne Dutfoy

Abstract This chapter presents the classic theory of univariate extreme values and its application to various natural hazards described in the previous chapters. The limitations of this approach are discussed, which pave the way for the second part of this book.

6.1 Introduction

6.1.1 Extreme Values for Natural Hazards

Extreme value studies conducted by engineers and researchers of large energy companies are generally aimed at quantifying the robustness of an electricity production structure against extreme conditions of climatic or meteorological origin. This already begins during the the design phase of the structure: the level of robustness required with regard to extreme natural hazards has a direct influence on the dimensioning of the structure. The natural hazards we consider are those which can damage the functioning of a structure, and also those which reduce its level of safety. For example, a structure located on a river needs to be protected against flood. Some structures also need to be robust against intense storms so as to protect power supply pylons.

In extreme value studies, extreme values of a natural hazard are defined as those which exceed some high threshold value. The definition of an extreme event thus requires the choice of this threshold. But which threshold should we use? Very often, the threshold is not set directly and is derived from the *return period* of the extreme event. Under certain conditions explained in the following sections, a return period can be interpreted as the average duration between two successive episodes of an extreme event. Thus, a centennial storm occurs, on average, every 100 years and a millennial flood occurs, on average, every 1000 years.

A. Dutfoy (✉)
EDF R&D, Boulevard Gaspard Monge, 91120 Palaiseau, France
e-mail: anne.duftoy@edf.fr

N. Bousquet and P. Bernardara (eds.), *Extreme Value Theory with Applications to Natural Hazards*, https://doi.org/10.1007/978-3-030-74942-2_6

The robustness of a structure can then be tested with respect to a particular return period of the extreme event. For instance, the height of a (flood protection) dike can be calculated from the centennial (or even millennial) river flow; the robustness of a structure to wind can be dimensioned with respect to the centennial wind speed, etc. To calculate such *return levels*, statisticians first look at the probabilistic behavior of the annual maxima of the quantity considered. This is called the *block maxima* (MAXB) approach, where each block generally corresponds to one year. In the flood example, they take into account the largest flow recorded every year over several decades, and in the storm example, they analyze the set of highest daily wind speeds to extract the maximum value for each year.

Extreme value theory allows us to determine the asymptotic distributions of such annual maxima. These are known as *generalized extreme value* (GEV) distributions and come in three flavors, corresponding to three possible *domains of attraction*, i.e., three types of limit distribution possible for the distribution of maxima of a sample: Gumbel, Fréchet, and Weibull. The return level can be easily deduced from the GEV distribution: it is simply the GEV's order $(1 - 1/T)$ quantile if the return period is T years.

Another approach is to consider *all* of the data above an extreme value threshold rather than reducing it to a single data point per period (block). This approach is called *peaks over threshold* (POT). Extreme value theory allows us, under certain conditions, to quantify the distribution of the magnitude the values of which exceed the threshold, which turns out to be a generalized Pareto distribution (GPD). Here the return level can again be easily deduced as a particular quantile of the distribution.

Inversely, it is also quite interesting to define an extreme event directly in terms of an extreme threshold value and then quantify the *annual frequency* of the event. Though the notion of a return period is well defined, that of the annual frequency is less so, and potentially corresponds to several definitions, depending on whether we consider the average annual number of days during which the extreme event occurs, the average annual number of episodes during which the event occurs, or the probability of having at least one episode per year.

6.1.2 Chapter Contents

This chapter first recalls two fundamental results in univariate extreme value theory without diving too deeply into the theory. A more in-depth treatment can be found in [65, 173]. First we present the main result on the asymptotic convergence of annual maxima, as well as a convergence result for the distribution of cluster maxima, stated in the unifying *colored point process* framework (Definition 6.8). As indicated in Sect. 3.1.1.2, the real-world extreme value study techniques we describe in what follows are based on these two theorems. We also recall practical difficulties in applying them, and show how in practice, faced with data that do not exactly satisfy the hypotheses of these fundamental theorems, practitioners try to move back to the stationary framework found in the theory.

The notions of a return period and its associated level, as well as their interpretations, are then examined in detail. We also explain the different ways of estimating these starting from GEV or GPD distributions, and examine various mathematical definitions and forms of the annual frequency of an event.

The final part of the chapter outlines the different steps involved in a study of extreme values, illustrated by two real examples. These steps help us, first of all, to preprocess the data in order to bring it back to the stationary framework required by the convergence theorems (where possible). Then, we give examples of the inference of extreme value models and validation steps for estimated models, before arriving at the calculation of actual return levels, as well as the associated annual frequencies.

6.2 Fundamental Results

This section recalls two fundamental convergence theorems for univariate extreme values. The first allows us to know the asymptotic distribution of annual maxima (MAXB framework) and the second is that of the exceedances of cluster maxima (POT framework). We will state these theorems in the unifying framework of colored point processes, which allows us to obtain results for both of the classical approaches: *block maxima* (block = year) and *threshold exceedance*.

6.2.1 Some Definitions

For clarity, in what follows, let us first recall several definitions.

Definition 6.1 (*GEV distribution*) The *generalized extreme value* distribution is defined in terms of its cumulative distribution function

$$F_{\mathcal{G}_{EV}}(x|\xi) = \left| \begin{array}{ll} \exp\left\{-(1+\xi x)^{-1/\xi}\right\} & \text{if } 1+\xi x > 0, \xi \neq 0 \\ \exp\{-\exp(-x)\} & \text{if } x \in I\!R, \xi = 0, \end{array} \right.$$

where $\xi \in I\!R$ is the shape parameter. More generally, we define

$$F_{\mathcal{G}_{EV}}(x|\xi,\sigma,\mu) = F_{\mathcal{G}_{EV}}\left(\frac{x-\mu}{\sigma}|\xi\right),$$

where $\mu \in I\!R$ and $\sigma \in I\!R^+_*$ are, respectively, the position and scale parameters.

The GEV distribution corresponds to three different distributions written using the same formulation. When $\xi > 0$, it is a Fréchet distribution defined for $x \geq \mu - \frac{\sigma}{\xi}$. When $\xi < 0$, it is a (reversed) Weibull distribution defined for $x \leq \mu - \frac{\sigma}{\xi}$. Lastly, it is a Gumbel distribution when $\xi = 0$, and is thus defined on $I\!R$. Figure 6.1 illustrates the behavior of the three GEV distributions; the evolution of the order $1-p$ quantile

Fig. 6.1 GEV distribution $F_{\mathscr{G}_{EV}}(.|\xi)$: evolution of the quantile of order $1-p$ as a function of $-\log\left[-\log(1-p)\right]$, where $\xi=-0.2$, 0 and 0.2

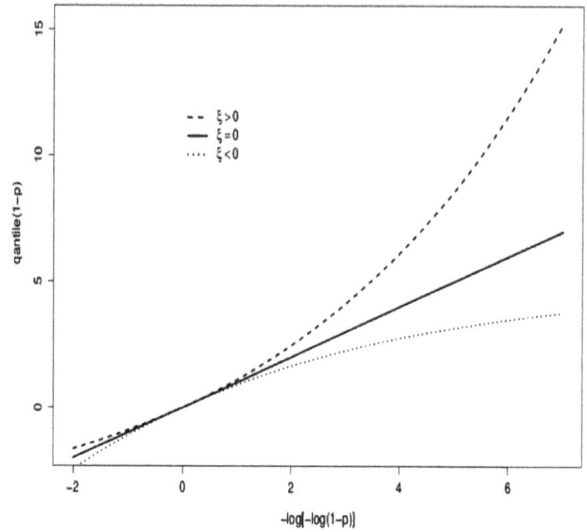

is plotted as a function of $-\log\left[-\log(1-p)\right]$. The straight line corresponds to the Gumbel distribution, the convex one to the Fréchet distribution, and the concave one to the Weibull distribution.

The GEV distribution has an important property: *max-stability*.[1] The interpretation of max-stability is the following: when we are interested in a sample of maxima of GEV variables, and also in the distribution of the maximum of such samples, the latter is still a GEV. This scale invariance can be written more formally in the unidimensional case as follows:

Definition 6.2 (*Max-stability for univariate distributions*) Let X be a real-valued random variable with cumulative distribution function F. Then, F is said to be *max-stable* if for any $n \geq 2$, there exist sequences $(a_n)_n \in I\!R$ and $(b_n)_n \in I\!R_*^+$ such that, for all $x \in I\!R$

$$F^n(a_n + b_n x) = F(x).$$

Definition 6.3 (*GPD distribution*) The *generalized Pareto distribution* is defined in terms of its cumulative distribution function

$$F_{\mathscr{G}_{PD}}(x|\xi,\sigma) = \begin{vmatrix} 1-\left(1+\xi\dfrac{x}{\sigma}\right)^{-1/\xi} & \text{if } 1+\xi\dfrac{x}{\sigma}>0, \xi\neq 0 \\ 1-\exp\left\{-\dfrac{x}{\sigma}\right\} & \text{if } x>0, \xi=0, \end{vmatrix}$$

where $\xi \in I\!R$ is the shape parameter and $\sigma \in I\!R^+$ the scale parameter.

[1] More precisely, this property exactly and uniquely characterizes the family of GEV distributions (Theorem 3.2 in [173]).

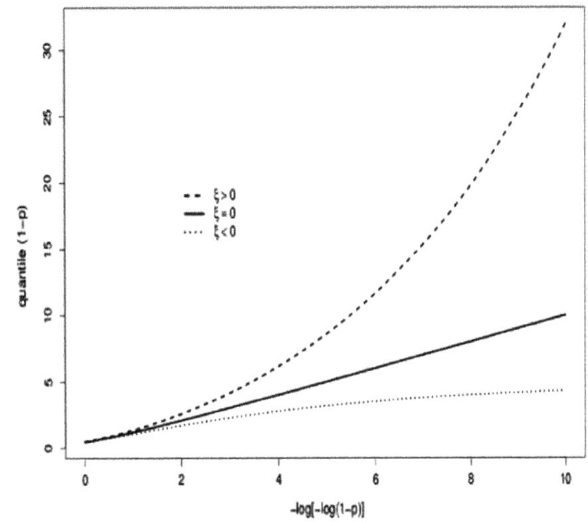

Fig. 6.2 GPD distribution $F_{\mathcal{G}PD}(.|\xi,\sigma)$: evolution of the quantile of order $1-p$ as a function of $-\log\left[-\log(1-p)\right]$, where $\xi=-0.2$, 0, and 0.2 and $\sigma=1$

More generally, we define

$$F_{\mathcal{G}_{PD}}(x|\xi,\sigma,u) = F_{\mathcal{G}_{PD}}(x-u|\xi,\sigma),$$

where $u \in I\!R$ is a threshold parameter.

Like for the GEV distribution, the GPD one brings together the Fréchet, Weibull, and Gumbel distributions into a single formula. Figure 6.2 illustrates this behavior: the evolution of the order $1-p$ quantile is plotted as a function of $-\log\left[-\log(1-p)\right]$; the straight line corresponds to the Gumbel, the convex one to the Fréchet, and the concave one to the Weibull.

The identification of independent clusters (or data *aggregates*) is a fundamental step in the analysis of extreme values.

Definition 6.4 (*Cluster*) A cluster is a group of consecutive and correlated values from the same process, sharing some feature. For example, a cluster of exceedances can be seen as an *upsurge* of exceedances of a given threshold.

In the case of river flow, a cluster might correspond to a river flooding: exceedances in the same cluster would all be due to the same flood. Similarly, in the case of a storm, a cluster might correspond to a strong wind gust during which the wind speed exceeds a given threshold over several consecutive (or almost consecutive) measurement times.

Recall also that the *stationarity* of a random process is a fundamental feature when dealing with extreme values.

Definition 6.5 (*Stationary processes*) A process is *stationary* if its probabilistic properties are the same over time; i.e., its probabilistic behavior—in other words, its probability distribution—over a fixed time window is the same, independent of where the window is placed in time. We say we have *second order stationarity* when the mean and variance (when they exist) do not depend on time. See Definition 4.1 for a formal characterization.

It is useful to remember that a stationary process can have *temporal correlation*, i.e., when the value taken by the process at a given time depends on the values it takes at earlier times. For example, the daily air temperature process is temporally correlated: the temperature on a given day is close to that of the day before, and likewise gives an idea of the probable temperature the following day without, however, determining it completely.

The *extremal index* θ, defined below, is a quite useful scalar variable as it makes it possible to quantify the correlation of a stationary process in its extreme values, and thus answer the following question: when a process exceeds a high threshold at a given time, will it be followed by others exceedances or remain an isolated event?

Definition 6.6 (*Extremal index* θ) Let $X_1, \ldots, X_n$ be a stationary process, and consider an iid sample $\tilde{X}_1, \ldots, \tilde{X}_n$ following the same distribution as the X_i. We note $X_n^* = \max(X_1, \ldots, X_n)$ and $\tilde{X}_n^* = \max(\tilde{X}_1, \ldots, \tilde{X}_n)$, and F and $\tilde{F}$ the cumulative distribution functions of X_n^* and $\tilde{X}_n^*$, respectively. Following [463], there exists a real number θ such that $0 < \theta \leq 1$, called the *extremal index*, for which

$$F = \tilde{F}^{\theta}.$$

This index quantifies the temporal correlation of a process in its extreme values. Beirlant [65, p 380] has shown that θ represents the limit probability that the exceedance of a threshold by a process be followed by another value above the threshold. Thus, when $\theta = 1$, values exceeding the threshold tend to occur in isolation as the threshold is raised higher and higher. When $\theta < 1$, values exceeding the threshold tend to occur in groups, the more so as θ gets smaller and smaller. By considering the distribution of times between independent clusters, [65] has shown that θ can also be interpreted as the mean cluster size. See also Definition 5.1 in a previous chapter for further interpretations and [133] for more technical details.

6.2.2 Fundamental Theorems

The principal theorems justifying the use of GEV and GPD distributions are due to Leadbetter. They apply to stationary processes whose temporal correlation in their extreme values is not too strong. We denote $D(u_n)$ [65, p 373] the technical condition which expresses this weak temporal correlation, where u_n is a high threshold value.

Definition 6.7 (*Condition* $D(u_n)$) Let $I_{j,k}(u_n)$ correspond to the event $I_{j,k}(u_n) = \bigcap_{j \leq i \leq k} \{X_i \leq u_n\}$. For any events A_1 from $I_{1,\ell}(u_n)$ and A_2 from $I_{\ell+s,n}(u_n)$ with $1 \leq \ell \leq n - s$, we have

$$|\mathbb{P}(A_1 \cap A_2) - \mathbb{P}(A_1)\mathbb{P}(A_2)| \le \alpha(n, s),$$

where $\alpha(n, s) \to 0$ as $n \to \infty$ for a sequence of positive integers s_n such that $s_n = o(n)$.

In other words, the events $\{\max_{k \in I_1} X_k \le u_n\}$ and $\{\max_{k \in I_2} X_k \le u_n\}$ become independent when n increases if the sets of indices $I_i \subset \{1, \dots, n\}$ are separated by a relatively small distance $s = o(n)$ with respect to n.

We now detail the theorem showing the existence of a limit distribution for M_n, the maximum of the stationary process X over blocks of size n

$$M_n = \max(X_1, \dots, X_n), \tag{6.1}$$

where X_i is the value of the process at time i.

Theorem 6.1 (Leadbetter 1974 [461]) *Let X be a stationary process and M_n defined by (6.1). Suppose that there exist two real-valued sequences $(a_n)_n$ and $(b_n)_n$ such that $a_n > 0$, and a nondegenerate cumulative distribution function F such that*

$$\lim_{n \to \infty} \mathbb{P}\left(\frac{M_n - b_n}{a_n} \le x\right) = F(x) \quad \textit{(convergence in distribution).}$$

If the condition $D(u_n)$ is satisfied for the threshold $u_n = a_n x + b_n$ at all x for which $F(x) > 0$, then F is a GEV cumulative distribution function $F_{\mathcal{G}_{EV}}(.|\xi)$.

By inserting the sequences a_n and b_n as position and scale parameters of G, it comes

$$\mathbb{P}(M_n \le x) \simeq F_{\mathcal{G}_{EV}}(x|\xi, \sigma, \mu).$$

Note that when the process has temporal correlation in its extreme values—as quantified by the extremal index θ—the parameters (ξ, σ, μ) of the limit distribution of the maxima of F are different to those—noted $(\tilde{\xi}, \tilde{\sigma}, \tilde{\mu})$—that we would have obtained for a process with the same marginal distribution but without temporal correlation. This difference is expressed in the following formulas:

- if $\xi \neq 0$, then $\left| \begin{array}{l} \xi = \tilde{\xi} \\ \mu = \tilde{\mu} - \tilde{\sigma} \frac{1-\theta^{\xi}}{\xi} \\ \sigma = \tilde{\sigma}\theta^{\xi}, \end{array} \right.$
- if $\xi = 0$, then $\left| \begin{array}{l} \mu = \tilde{\mu} + \sigma \log \theta \\ \sigma = \tilde{\sigma}. \end{array} \right.$

Other results can be stated in the framework of colored point processes, which record the moments at which events happen, as well as the value of a "color" at such moments.

Definition 6.8 (*Colored point process*) A point process is a stochastic process (cf. Sect. 4.1.5), where each draw is a set of points. It is said to be *colored* (or *marked*) if each point can be associated with some characteristic color (or mark).

For example, a point may represent a spatial location and the color the intensity of an event occurring at that location. A Poisson point process is the simplest example of a colored point process. It states that the number of events occurring in a spatial or temporal region follows a Poisson distribution and that the intensity of each event follows a particular distribution. This process will be encountered later when we will want to quantify events corresponding to exceedances of extreme-valued thresholds.

In order to understand how a behavior is distributed, it is valuable to look at the times at which its values exceed a high threshold and associate these with a color corresponding to the excess above the threshold. We can also extract the sequence of times corresponding to cluster maxima and associate these with cluster size.

The elegance of Leadbetter's work comes from how it unifies the MAXB and POT approaches. Indeed, the following theorem characterizes the number of clusters over a period of length n as well as the distribution of the exceedances of cluster maxima over u_n.

Theorem 6.2 (Leadbetter 1991 [462]) *Let X be a stationary process with extremal index θ and marginal cumulative distribution function F. Let u_n be a threshold for which the condition $D(u_n)$ (limiting temporal correlation in extreme values) is satisfied, such that*

$$n\,(1 - F(u_n)) \xrightarrow{n\to\infty} \tau < +\infty.$$

Then

(i) the number $N_n(u_n)$ of cluster maxima exceeding u_n, recorded over a period of length n, converges in distribution when $n \to \infty$ to a Poisson distribution with mean $\Lambda_c(u_n) = \theta\tau$;
(ii) when $n \to \infty$, the exceedances of cluster maxima over u_n converge in distribution to a generalized Pareto distribution $F_{\mathcal{G}_{PD}}(.|\xi, \sigma)$.

It is possible to link the intensity $\Lambda_c(u)$ of the threshold (u) exceedance process to the limit distribution of the maxima $F_{\mathcal{G}_{EV}}(.|\xi, \sigma, \mu)$ by noting the equivalence of the events

$$\{M_n \le u\} = \{N_n(u) = 0\}.$$

Then

$$F_{\mathcal{G}_{EV}}(u|\xi, \sigma, \mu) = \mathbb{P}\,(M_n \le u) \;=\; \mathbb{P}\,(N_n(u) = 0) \;=\; \exp\left(-\Lambda_c(u)\right), \quad (6.2)$$

which shows that $\Lambda_c(u) = -\log F_{\mathcal{G}_{EV}}(u|\xi, \sigma, \mu)$. Hence

$$\Lambda_c(u) = \begin{vmatrix} \left(1 + \xi \dfrac{u-\mu}{\sigma}\right)^{-1/\xi} & \text{if } \xi \neq 0, \\ \exp\left\{-\dfrac{u-\mu}{\sigma}\right\} & \text{if } \xi = 0. \end{vmatrix}$$

In other words, the exceedances of an extreme-valued threshold follow a Poisson process whose colors are Pareto distributed.

Theorem 6.2 proves to be very useful in the studies of extreme value, since the quantification of the distribution of cluster maxima exceeding a given threshold enables to model the tail distribution of the marginal distribution F of the process X. This is based on the following max-stability result.

Proposition 6.1 (Beirlant 2005 [65, p 387]) *The asymptotic distribution of the exceedances of cluster maxima over u_n is equal to the distribution of the exceedances over u_n of the process itself.*

Thus, the asymptotic distribution $F_{\mathcal{G}_{PD}}(.|\xi, \sigma)$ of the exceedances of cluster maxima over u_n is also that of $(X - u_n|X > u_n)$, which, setting $\alpha = \bar{F}(u) = 1 - F(u)$, allows to write

$$\bar{F}(x) \simeq \alpha \bar{F}(x - \bar{F}^{-1}(\alpha)) \quad \text{for } x > u. \tag{6.3}$$

Note that the interest in determining the distribution $F_{\mathcal{G}_{PD}}(.|\xi, \sigma)$ on cluster maxima rather than directly on process exceedances comes from the fact that these cluster maxima are independent by construction. For estimating the parameters of $F_{\mathcal{G}_{PD}}(.|\xi, \sigma)$, it is possible to use a maximum likelihood approach that requires this independence. Direct inference using the process's exceedances of a threshold certainly has the advantage of increasing the number of data points, but it also requires modifications to estimators and their confidence intervals to take into account the data temporal dependence. This approach was chosen in [127, 227].

Connections exist between the distribution $F_{\mathcal{G}_{PD}}(.|\xi, \sigma)$ of cluster maxima exceedances over a threshold u (and thus of the process exceedances above u too) and the limit distribution $F_{\mathcal{G}_{EV}}(.|\xi)$ of the process maxima. In particular, the shape parameters ξ are equal. However, the position and scale parameters of $F_{\xi,\sigma}$ and G are different. This is a consequence of the following proposition from Pickand's foundational article [615].

Proposition 6.2 (Pickands 1975 [615]) *There is equivalence between the convergence in distribution of the maximum towards a GEV distribution, noted $F_{\mathcal{G}_{EV}}(.|\xi)$, and that of an exceedance distribution towards a GPD, noted $F_{\mathcal{G}_{PD}}(.|\xi, \sigma)$. Furthermore, these two distributions have the same shape parameter ξ*

$$\lim_{n\to\infty} \mathbb{P}\left(\frac{M_n - b_n}{a_n} \leq x\right) = F_{\mathcal{G}_{EV}}(.|\xi)$$

if and only if

$$\lim_{u \to x_F} \sup_{y \in [0, x_F - u[} |\bar{F}_u(y) - F_{\mathcal{G}PD}(y|\xi, \sigma(u))| = 0,$$

where $x_F \in \bar{I\!R}$ is the upper bound of the support of F, and F_u the cumulative distribution function of exceedances X over u, i.e., of $(X - u|X > u)$.

An important property of GPD distributions, which is the counterpart of max-stability for GEV distributions, is shape-stability when the threshold u is modified. This property is sometimes called *heredity* and is the subject of the following proposition.

Proposition 6.3 (Stability of modified GPD parameters (heredity)) *Let X be a stationary process. If the exceedances of X over u_s follow a GPD distribution $F_{\mathcal{G}PD}(.|\xi, \sigma)$, then the exceedances of X over $u > u_s$ follow the distribution $F_{\mathcal{G}PD}(.|\xi, \sigma_u)$, where $\sigma_u = \sigma + \xi(u - u_s)$. The modified scale parameter $\sigma^*(u) = \sigma_u - \xi u$ is thus constant with respect to u.*

Incidentally, this result allows us to easily calculate an important function of interest: the expectation of the exceedances of X over the modified threshold. The plot of this function is usually called the *mean excess plot.*

Proposition 6.4 (Linearity of the conditional expectation of exceedances) *Let X be a stationary process. If the exceedances over u_s of X have the GPD distribution $F_{\xi,\sigma}$, then the expectation of the exceedances of $u > u_s$ by X is given by $\mathbb{E}[X - u|X > u] = \dfrac{\sigma_{u_s} + \xi u}{1 - \xi}$.*

In conclusion, the fundamental results recalled here guide all extreme value studies. Indeed, the study of cluster maxima will allow us to:

1. Evaluate the mean number of independent clusters exceeding the threshold u; this value will be essential when evaluating annual frequencies defined as the mean number of independent clusters exceeding u;
2. Model the tail of the marginal distribution of the process and thus estimate the probability of exceeding any threshold greater than u_n;
3. Estimate specific return period levels.

In the analysis of the extreme values of a process modeling environmental variables, the typical block considered is a year, often limited to a time period where the process is stationary.

6.2.3 *Practical Difficulties*

The theorems stated earlier, on which rest all of the results of extreme value studies, are asymptotic convergence results. They therefore suppose that the number of data points n tends to infinity, which is clearly not the case in practice. The finite nature of real data sets can lead to modeling errors and parametric estimation errors, which we detail below.

6.2.3.1 Choosing Data Representation

When practitioners want to model the distribution of block maxima using a GEV distribution, they are confronted with the following issues:

- Choosing large blocks (i.e., considering maxima over long time periods) minimizes the modeling error; the distribution of maxima indeed tends to a GEV distribution. On the other hand, this considerably reduces the number of data points available to estimate the model parameters, as only one point is retained per block. Thus, estimated GEV parameters threaten to lack of precision.
- Choosing small blocks increases the modeling error, as the block maxima distribution does not really converge to a GEV distribution. Nevertheless, more data is available for estimating the GEV parameters, which reduces estimation error.

When modeling the distribution of exceedances of a threshold by a GPD, practitioners face similar difficulties:

- Choosing a high threshold leads to a good fit of the exceedances distribution to a GPD, but drastically reduces the number of data available to estimate the GPD's parameters. Essentially, we have only the points above the threshold to work with.
- Choosing a lower threshold increases the modeling error for the exceedances distribution, but decreases parameter estimation errors.

In conclusion, we are faced with the following dilemma: *estimate poorly a good model*, or *estimate well a bad model*. Which should we choose? The notion of *bias-variance* tradeoff, introduced in Sect. 4.2.21 and applied to estimators of quantities of interest (return levels), could be a way to orient our choice, on the condition that we have a way to generate independent data from the chosen model (e.g., nonparametric *bootstrap* [244]) and sufficient data for validating the quality of return levels.

6.2.3.2 Minimal Quantity of Data Required

To the quantitative elements in Chap. 5, let us add the following conclusions drawn from a large number of case studies.

Depending on the environmental variable in question, the number of years of observations can vary from a dozen up to several decades, which makes very difficult—if not impossible—the statistical estimation of model parameters. For example, hourly river flow measurements often exist for periods of 80 years or more, while extreme tide ones generally exist over shorter periods, typically ten to twenty years.

Note that even in cases where measurements exist over long time periods, we have to be careful about their homogeneity, which may be impacted by a technological change in the measuring equipment or a significant change in the environment. In such cases, it is generally necessary to truncate the data series, unless information is available to statistically deal with such changes in *historical data* (cf. Sect. 5.2.2). In this context, the French national meteorological service—Météo-France—recommends not going further back than 1981 in historical wind speed studies due to changes in equipment, i.e., homogeneity cannot be guaranteed before this date.

Example 6.1 (*Hydroelectric impact on river data*)

The construction of hydroelectric power stations on the Garonne river significantly modified the river system, which can be seen in a sudden change in extreme flow values in the 1950s. This change point has until now restricted the analysis of historical data to records taken after 1950, even though flow measurements exist back to 1915.

In addition, temporal correlation found in physical phenomena amounts to further reducing the number of data that can actually be used for the statistical estimation of models. Indeed, it is easy to see that the greater the temporal correlation, the less information a dataset of size n contains. Thus, if we quantify correlation in terms of the index θ, a correlated dataset of size n is equivalent to an independent one of size $n\theta$ (cf. Definition 5.1). This is indeed what the asymptotic results show when the size n from an independent stationary process is replaced by $n\theta$ for a stationary process with temporal correlation. Such correlation thus amplifies the difficulties associated with having small quantities of data.

Hence, even if (as we have already emphasized) the POT approach stands on more data points available to the size used for adopting a MAXB approach, the problem of being short of data is not entirely resolved. In effect, in processes with temporal correlation, the data available to POT must be reduced to the set of cluster maxima which satisfy the independence hypothesis. This declustering* considerably decreases the amount of data that can be used to estimate the GPD distribution.

Generally speaking, the modeling of exceedances above a threshold with a GPD distribution is very sensitive to the choice of this threshold. Several techniques can help choose this threshold well—see Sect. 6.4.5.

Furthermore, estimation of the shape parameter ξ is quite tricky. Physical considerations on the boundedness of the magnitudes in question may help. For instance, river flow is bounded by the quantity of water found in its watershed. It is thus recommended to model the tails of flow distributions with $\xi < 0$. An error in the sign of ξ (or its nullity) would lead to a modeling error which could have significant consequences on the estimation of magnitudes calculated as levels of large return periods. Section 6.6 gives an idea of research efforts conducted about the estimation of this parameter. In practice, both the POT and MAXB approaches are used and the choice of one over the other mainly depends on the number of data available and temporal correlation considerations.

6.2.4 *Why the Choice of Generalized Distributions?*

Algebraically speaking, it might seem easier to simply use one of the the three possible distributions (Gumbel, Weibull, and Fréchet) rather than the GEV one. However, practice shows that in most cases, GEV estimators are more robust, in the sense that the estimated distribution provides more physically plausible results in terms of return levels and periods. The example of torrential rain that fell on Vargas (Venezuela), in 1999, illustrates this. In December 1999, two days of high rainfall (120 mm followed by 4104 mm the next day) provoked mudflows and the death of more than 30,000 people [203]. Coles and Pericchi [175] have studied, with GEV and Gumbel distributions, the distribution of annual rainfall maxima in this geographic region. Their GEV proposed a return period of 4280 years, while the Gumbel suggested 17.6 million years.

Remark 6.1 For a long time now, the Gumbel distribution, easier to estimate with small sample sizes (because it has one less parameter) has remained the most-used distribution for modeling the extremes of meteorological variables [204], notably rainfall [444]. Nevertheless, as shown by Jenkinson in 1955 [413], then Koutsoyiannis [445], this distribution comes with risks for structural engineers as it frequently underestimates the return levels associated with long periods. This behavior can be explained, in particular, by the extremely slow convergence of distributions of events occurring in Gumbel's domain of attraction, which necessitates the use of large-scale blocks (periods) [445]. A further argument supporting the use of GEV distributions in place of Gumbel and Fréchet ones is the non-negligible risk of not being able to differentiate between the two when sample sizes are small [53]. Other practical and more general reasons for working with GEVs are provided in [184].

6.3 What Does *Extreme* Mean for a Natural Hazard?

As we stated in the introduction, in the conception phase, structures are dimensioned with respect to the extreme values of some hazardous event. Similarly, the robustness of an existing structure is very often defined in terms of its ability to resist a certain level of extremality, defined a priori, of a given hazard. This level is described in terms of a return level or annual occurrence frequency.

Before continuing, we recall the definitions of return level and frequency of extreme values of a hazard, along with interpretations to help better understand these concepts.

6.3.1 Return Levels

In this section, consider a process X that is stationary over a period Δ of each year, and suppose that the extreme values of X which interest us occur during Δ. In practice, Δ corresponds to the season in which extreme values of X occur, e.g., the summer for heatwaves (high air temperature), fall and winter for high river levels (large flows), etc.

6.3.1.1 The GEV Case

The annual maximum of a process occurs during the period Δ. We note its value M_n, where n is the number of measures taken during Δ. Return period levels are then defined as particular quantiles of the asymptotic distribution of the maxima M_n.

Definition 6.9 (*Return period levels*) If F, the marginal distribution of the stationary process X, is in the domain of attraction of the GEV distribution of the cumulative distribution function $F_{\mathscr{G}_{EV}}(.|\xi, \sigma, \mu)$, then the level x_T of the "T years" return period of X is defined by

$$F_{\mathscr{G}_{EV}}(x_T|\xi, \sigma, \mu) = 1 - 1/T. \tag{6.4}$$

For large n we have:

$$\mathbb{P}(M_n > x_T) \simeq \frac{1}{T}.$$

The return level x_T is therefore the order-$(1 - 1/T)$ quantile of the asymptotic distribution of the annual maxima M_n

$$x_T \simeq \left| \begin{array}{ll} \mu - \dfrac{\sigma}{\xi}\left\{1 - \left[-\log(1 - 1/T)\right]^{-\xi}\right\} & \text{if } \xi \neq 0 \\ \mu - \sigma \log(\log(1 - 1/T)) & \text{if } \xi = 0. \end{array} \right. \tag{6.5}$$

Therefore, the use of asymptotically Gaussian estimators $(\xi, \sigma, and \mu)$ with asymptotic variance V (such as those produced by the methods described in Sect. 6.4.2, where V is the matrix V_1 or V_2), allows us—with the help of the delta method (cf. Theorem 4.4)—to construct the also asymptotically Gaussian estimators $\hat{x}_T$

$$\sqrt{m}\left(\hat{x}_T - x_T\right) \xrightarrow{\mathscr{L}} \mathscr{N}(0, KVK^T) \tag{6.6}$$

when $m \to +\infty$, where K is defined by

$$K^T = \left[\frac{\partial x_T}{\partial \sigma}, \frac{\partial x_T}{\partial \xi}, \frac{\partial x_T}{\partial \mu} \right],$$

where

$$\begin{aligned}
K_{1,1} &= \frac{1}{\xi} \left[(-\log(1 - 1/T))^{-\xi} - 1 \right], \\
K_{1,2} &= -\frac{\sigma}{\xi} \left[K_{1,1} + (-\log(1 - 1/T))^{-\xi} \log(-\log(1 - 1/T)) \right], \\
K_{1,3} &= 1.
\end{aligned}$$

Now let us give a simpler interpretation of the level corresponding to a return period of T years. It is based on the convergence of the cluster maxima which exceed the level u to a Poisson process with intensity $\Lambda_c(u)$. Indeed, convergence of cluster arrivals to a Poisson process implies that the time between the arrival of two cluster maxima is a random variable $Y(u)$, which follows an exponential distribution with parameter $\Lambda_c(u)$. Thus, the mean duration between the arrival of two independent clusters (or exceedances) is $\mathbb{E}[Y(u)] = 1/\Lambda_c(u)$. The asymptotic approximation (6.2)

$$\Lambda_c(u) \simeq n\theta \bar{F}(u) \simeq -\log F_{\mathscr{G}_{EV}}(.|\xi, \sigma, \mu),$$

applied at the level x_T, shows that

$$\mathbb{E}[Y(x_T)] = -\frac{1}{\log(1 - 1/T)} \simeq T \text{ for } T \gg 1,$$

which leads to the following—very common—interpretation. Recall that Proposition 6.5 on return levels is a characterization and not a definition (which is provided in (6.4)).

Proposition 6.5 (Interpretation of a return level of T years) *A return level of T years for an extreme-valued hazard can be interpreted as follows:*

- *On average, there are T years between extreme episodes of a process with a return period of T years.*
- *An episode exceeding the level with a return period of T years occurs on average once every T years.*

6.3.1.2 The GPD Case

However, practitioners do not always use (6.4) for estimating a return level x_T since it depends on an estimation of the asymptotic distribution of the annual maxima. In order to make best use of their data, they prefer to study the cluster maxima process, which provides an estimate of the tail of the marginal distribution F.

This approach is straightforward since it is easy to link a level x_T with a particular quantile of F using the relationship $F_{\mathscr{G}_{EV}}(.|\xi, \sigma, \mu) \simeq F^{n\theta}$ for large n, which comes from Definition 6.6, where θ is the process's extremal index. Applied to x_T, we obtain

$$F(x_T) = (1 - 1/T)^{1/n\theta},$$

which turns the return level x_T into the quantile of order $\alpha = (1 - 1/T)^{1/n\theta}$ of F.

Then, if the excess of maxima above u converges towards a Pareto distribution, from (6.3), it comes

$$x_T \simeq \left| \begin{array}{ll} u + \dfrac{\sigma}{\xi}\left\{\left(\dfrac{1-\alpha}{\zeta_u}\right)^{-\xi} - 1\right\} & \text{if } \xi \neq 0 \\ u - \sigma \log\left(\dfrac{1-\alpha}{\zeta_u}\right) & \text{if } \xi = 0, \end{array} \right.$$

where $\zeta_u = \mathbb{P}(X_t > u)$.

When T is large, we can make the approximation $1 - \alpha \simeq 1/n\theta T$, which leads to the well-known equations [173]

$$x_T \simeq \left| \begin{array}{ll} u + \dfrac{\sigma}{\xi}\left\{(n\theta T\zeta_u)^{\xi} - 1\right\} & \text{if } \xi \neq 0, \\ u + \sigma \log(n\theta T\zeta_u) & \text{if } \xi = 0. \end{array} \right. \tag{6.7}$$

The probability ζ_u can be estimated by the fraction n_u/n, where n_u is the number of exceedances of u by the process (though the variance of this estimator increases with temporal correlation in the process). Similarly, the parameter θ, whose inverse corresponds to an estimate of the mean cluster length, can be estimated by the fraction n_c/n_u, where n_c is the number of clusters exceeding u.

As in the GEV case, the use of asymptotically Gaussian estimators, with asymptotic variance V (such as those resulting from the methods in Sect. 6.4.2, where V is the matrix V_3 or V_4), for the parameters (σ, ξ) enables us, using the delta method (cf. Theorem 4.4), to construct estimators $\hat{x}_T$ which are themselves Gaussian

$$\sqrt{N_u}\left(\hat{x}_T - x_T\right) \rightarrow \mathscr{N}(0, LVL^T) \quad \text{in distribution when } m \rightarrow +\infty, \tag{6.8}$$

where the matrix L is defined by

$$L^T = \left[\frac{\partial x_T}{\partial \sigma}, \frac{\partial x_T}{\partial \xi}\right],$$

with

$$L_{1,1} = \frac{1}{\xi}(p^{-\xi} - 1),$$
$$L_{1,2} = -\frac{\sigma}{\xi^2}(p^{-\xi} - 1) - \frac{\sigma}{\xi} p^{-\xi} \log p,$$

where we have set $p = \left(\frac{1-\alpha}{\zeta_u}\right)$ and $\alpha = \left(1 - \frac{1}{T}\right)^{1/n\theta}$ for clarity.

6.3.2 *Annual Frequency*

Let us now return to the stationarity hypotheses on the process X over the period Δ of size n, and consider the extreme-valued hazard $\mathscr{E} = \{X > x\}$, where x is a high threshold value. By the stationary of X, the probability $p = \mathbb{P}(\mathscr{E})$ is independent of the time t for any t in Δ.

Remark 6.2 We often see the use of the term *daily probability* for p, which is surprising in the sense that a probability is dimensionless. Nevertheless, continuous processes X are often projected onto a daily grid. For example, flow measurements are mean daily ones, wind speeds are the daily maxima of measurements taken every 10 min, etc. Hence, the time t corresponds to a day. The term *daily probability* therefore simply reminds us that this probability corresponds to events defined on one day long periods.

The stationarity hypothesis on X does not preclude the process from having temporal correlation. However, when it does, it turns out that the annual frequency of the event $\mathscr{E}$ can be defined in different ways. In this book, we propose three possible definitions, which correspond to whether we consider all of the occurrences of $\mathscr{E}$ over the year, or only the independent ones in which $\mathscr{E}$ occurs.

6.3.2.1 Several Possible Definitions

Suppose that observations are daily and we want to estimate the annual frequency of the event $\mathscr{E}$. The explanations below generalize well to other units of time.

Definition 6.10 (*Annual frequency—1*) The annual frequency of the event $\mathscr{E}$ is the mean number of days per year where $\mathscr{E}$ occurs.

This point of view is based on a mean magnitude. Thus, from one year to the next, the effective number of days where $\mathscr{E}$ occurs can vary greatly around the mean in question. In particular, if the temporal correlation in X is strong, the number of days in a year where $\mathscr{E}$ occurs can be much larger or much smaller than the estimated mean.

In general, the stronger the correlation, the less the mean value is representative of the number of days that $\mathscr{E}$ effectively occurs.

Definition 6.11 (*Annual frequency—2*) The annual frequency of the event $\mathscr{E}$ is the mean number of episodes per year where $\mathscr{E}$ occurs.

This definition introduces the notion of *episode*, similar to that of *cluster*. An *episode* groups together the sets of times at which $\mathscr{E}$ occurs which are correlated. When X exceeds the threshold x at time t, the correlation between time instants means that the process will likely exceed x during a certain number of subsequent instants. Each such wave of correlated exceedances is counted as "1".

Here, the annual frequency thus corresponds to the mean number of waves of exceedances per year, irrespective of their lengths.

Definition 6.12 (*Annual frequency—3*) The annual frequency of the event $\mathscr{E}$ is the probability that $\mathscr{E}$ occurs at least once per year.

In other words, it is the probability that there is at least one episode of exceedances in a given year, or equivalently, the probability that the annual maximum exceeds the threshold x. This definition again uses the notion of *episode*. On the other hand, the annual frequency estimated here is simply the probability of having an episode of exceedances at least once in a given year, no matter the mean annual number of episodes. This definition is especially relevant if the occurrence of the episode in question is critical for the industrial installation that is affected by it. By analogy with faulty materials, the installation will be beyond *repair* to its initial state after such an episode occurs. Thus, all that interests us is the occurrence or not of the episode.

Remark 6.3 Note that Definitions 6.10 and 6.11 are mean numbers of days or episodes. These frequencies are thus positive numbers that may be greater than 1. In the case of extreme values, the event $\mathscr{E}$ has a very small probability of occurring, which will lead to small annual frequencies, likely less than 1 for Definitions 6.10 and 6.11—though not necessarily. On the other hand, Definition 6.12 defines the annual frequency as a probability. By definition, it is, therefore, a positive number no greater than 1.

6.3.2.2 Calculating the Annual Frequency

Depending on the definition chosen, the calculation is different. Expressions for these annual frequencies can be easily obtained with the help of results on Poisson processes for exceeding thresholds. They are given in the following proposition.

Proposition 6.6 (Annual frequencies) *Denoting* $\mathbb{F}_a^i(\mathscr{E})$ *the annual frequency of the event* $\mathscr{E}$ *for Definitions 6.10, 6.11, and 6.12, respectively, we have that:*

$$\begin{cases} \mathbb{F}_a^{6.13}(\mathscr{E}) = np, \\ \mathbb{F}_a^{6.14}(\mathscr{E}) = np\theta, \\ \mathbb{F}_a^{6.15}(\mathscr{E}) = 1 - \exp(-np\theta). \end{cases} \tag{6.9}$$

Remark 6.4 The expressions $\mathbb{F}_a^{6.13}(\mathscr{E})$ and $\mathbb{F}_a^{6.14}(\mathscr{E})$ are the same when $\theta = 1$, i.e., when the process has no temporal correlation in its extreme values. Similarly, for extremely rare events, $np\theta \ll 1$, and thus $\mathbb{F}_a^{6.15}(\mathscr{E}) \simeq \mathbb{F}_a^{6.14}(\mathscr{E})$.

If the tail of the distribution has been modeled by studying the distribution of cluster maxima, we then have access to approximation (6.3), which lets us write, for the level x of the event $\mathscr{E}$ being considered (supposed higher than the threshold u of the GPD $F_{\mathscr{G}_{PD}}(.|\xi, \sigma, u)$)

$$p = \bar{F}(x) \simeq \alpha \left(1 - F_{\mathscr{G}_{PD}}\left[x - \bar{F}^{-1}(\alpha)|\xi, \sigma, u\right]\right). \tag{6.10}$$

Remark 6.5 (*Links with the reliability of materials*) The reliability of materials is a domain in which there is often no temporal correlation ($\theta = 1$) and/or extremely low failure rates. In such cases, the three definitions lead to the same result. In fact, we often use one of these definitions to define the annual failure rate of materials.

6.4 Model Inference

Numerous methods can be found in the literature for estimating GEV model parameters under the MAXB approach, and GPD parameters under the POT approach. Here we present the *maximum likelihood*, *weighted moments*, and *profile likelihood* methods. These are easy to implement and lead to estimators with known asymptotic distributions, which means we can construct confidence intervals for each estimated parameter (see Sect. 4.2.2.3 for more details).

Estimation of the shape parameter ξ is extremely important, particularly in terms of getting its sign (or nullity) correct. Note that when its confidence interval contains 0, we generally retain the model $\xi = 0$. It is essential to support this decision with physical arguments from domain experts. In this chapter, in addition to the previous estimators, we will also describe the most famous one, Hill's estimator [389].

6.4.1 Parameter Estimation for GEV Distributions

Let us for clarity denote Y the maximum of a statistical sample $X_1, \ldots, X_n$, a set of independent copies of the same random variable X. We then note $Y_1, \ldots, Y_m$ the statistical sample of such maxima. The aim of this section is to estimate the parameters (σ, ξ, μ) of the GEV distribution of the maxima.

6.4.1.1 Maximum Likelihood

The *log-likelihood* function of a sample of maxima can be written—with the help of the density of the GEV distribution—when $\xi \neq 0$

$$\log L(\sigma, \xi, \mu) = -m \log \sigma - \left(\frac{1}{\xi} + 1\right) \sum_{i=1}^{m} \log\left(1 + \xi \frac{Y_i - \mu}{\sigma}\right) - \sum_{i=1}^{m} \left(1 + \xi \frac{Y_i - \mu}{\sigma}\right)^{-\frac{1}{\xi}}, \tag{6.11}$$

if $1 + \xi \dfrac{Y_i - \mu}{\sigma} > 0$ for each $i = 1, \ldots, m$. When $\xi = 0$, the log-likelihood is given by

$$\log L(\sigma, 0, \mu) = -m \log \sigma - \sum_{i=1}^{m} \exp\left(-\frac{Y_i - \mu}{\sigma}\right) - \sum_{i=1}^{m} \left(\frac{Y_i - \mu}{\sigma}\right). \tag{6.12}$$

For $\xi \neq 0$, the support of the GEV distribution depends on the estimated parameters; the distribution is defined for all real-valued x for which

$$1 + \xi \frac{x - \mu}{\sigma} > 0.$$

Thus, the usual conditions of regularity of the model underlying the asymptotic convergence properties of maximum likelihood estimators are no longer satisfied. However, in the case where $\xi > -0.5$, we recover the convergence, asymptotic efficiency, and asymptotic normality properties of the estimators

$$\sqrt{m}\left((\hat{\sigma}, \hat{\xi}, \hat{\mu}) - (\sigma, \xi, \mu)\right) \xrightarrow{\mathscr{L}} \mathscr{N}(0, V_1) \tag{6.13}$$

when $m \to +\infty$ and $\xi > -0.5$. V_1 is the inverse of the Fisher information matrix defined by

$$I(\theta) = -\mathbb{E}\left(\frac{\partial^2 \log g(Y)}{\partial\theta \partial\theta^T}\right), \tag{6.14}$$

where $\theta = (\sigma, \xi, \mu)^T$. Thus, the components of $I(\theta)$ are:

$$\begin{aligned}
I_{1,1}(\theta) &= \frac{1}{\sigma^2 \xi^2}(1 - \Gamma(2 + \xi) + p) \\
I_{1,2}(\theta) &= -\frac{1}{\sigma \xi^2}\left(1 - \gamma_* - q + \frac{1 - \Gamma(2 + \xi)}{\xi} + \frac{p}{\xi}\right) \\
I_{1,3}(\theta) &= -\frac{1}{\sigma^2 \xi}(p - \Gamma(2 + \xi)) \\
I_{2,2}(\theta) &= \frac{1}{\xi^2}\left[\frac{\pi^2}{6} + \left(1 - \gamma_* + \frac{1}{\xi}\right)^2 - \frac{2q}{\xi} + \frac{p}{\xi^2}\right] \\
I_{2,3}(\theta) &= -\frac{1}{\sigma \xi}\left(q - \frac{p}{\xi}\right) \\
I_{3,3}(\theta) &= \frac{p}{\sigma^2},
\end{aligned}$$

where γ_* is Euler's constant, $p = (1+\xi^2)\Gamma(1+2\xi)$, $q = \Gamma(2+\xi)\left(\psi(1+\xi) + \frac{1+\xi}{\xi}\right)$, and $\psi(x) = d\log\Gamma(x)/dx$.

6.4.1.2 Weighted Moments

This method, introduced in [357], considers the following weighted moments, defined for a random variable Y with cumulative distribution function F:

$$M_{p,r,s} = \mathbb{E}Y^p\,(F(Y))^r\,(1-F(Y))^s,$$

for real values $p, r,$ and s. The special case of GEV distribution parameters uses moments of type $M_{1,r,0}$ with $r = 0, 1, 2$, written as

$$M_{1,r,0} = \frac{1}{r+1}\left\{\mu - \frac{\sigma}{\xi}\left[1-(r+1)^{\xi}\Gamma(1-\xi)\right]\right\}.$$

The underlying principle of the method is to express each parameter (σ, ξ, μ) as a function of these moments by inverting equations (6.15)–(6.17)

$$M_{1,0,0} = \mu - \frac{\sigma}{\xi}\left(1-\Gamma(1-\xi)\right) \tag{6.15}$$

$$2M_{1,1,0} - M_{1,0,0} = \frac{\sigma}{\xi}\Gamma(1-\xi)\left(2^{\xi}-1\right) \tag{6.16}$$

$$\frac{3M_{1,2,0} - M_{1,0,0}}{2M_{1,1,0} - M_{1,0,0}} = \frac{3^{\xi}-1}{2^{\xi}-1}. \tag{6.17}$$

Then, each weighted moment $M_{1,r,0}$ is estimated by the unbiased *Landwehr* estimator

$$\hat{M}_{1,r,0} = \frac{1}{m}\sum_{j=1}^{m}\left(\Pi_{\ell=1}^{r}\frac{(j-\ell)}{(m-\ell)}\right)Y_{j,m},$$

where $Y_{j,m}$ is the j-th value of the sample $Y_1, \ldots, Y_m$ ordered from smallest to largest. Equation (6.17), when solved numerically, gives an estimate $\hat{\xi}$ of ξ. Lastly, (6.15) and (6.16) allow us to estimate the two other parameters as follows:

$$\hat{\sigma} = \frac{\hat{\xi}\left(2\hat{M}_{1,1,0} - \hat{M}_{1,0,0}\right)}{\Gamma(1-\hat{\xi})\left(2^{\hat{\xi}} + -1\right)}$$

$$\hat{\mu} = \hat{M}_{1,0,0} + \frac{\hat{\sigma}}{\hat{\xi}}\left(1-\Gamma(1-\hat{\xi})\right).$$

The advantage of this method is that it can make use of well-known properties of moment estimators to construct estimators of (σ, ξ, μ) with the same convergence properties. In effect, let us now show that $M = (\hat{M}_{1,r,0}, \hat{M}_{1,r,0}, \hat{M}_{1,r,0})^T$ is asymptotically efficient and Gaussian provided that $\xi < 0.5$, i.e.,

$$\sqrt{m}\left(\hat{M} - M\right) \xrightarrow{\mathscr{L}} \mathscr{N}(0, V) \tag{6.18}$$

when $m \to +\infty$ and $\xi < 0.5$, where the entries of V are written

$$\begin{aligned}
v_{r,r} &= \left[\frac{\sigma}{\xi}(r+1)^{\xi}\right]^2 \left(\Gamma(1-2\xi)K(r/(r+1)) - \Gamma^2(1-\xi)\right)\\
v_{r,r+1} &= \frac{1}{2}\left(\frac{\sigma}{\xi}\right)^2 \left\{(r+2)^{2\xi}\Gamma(1-2\xi)K(r/(r+2)) + (r+1)^{\xi}\right.\\
&\quad \left.\left[(r+1)^{\xi} - 2(r+2)^{\xi}\right]\Gamma^2(1-\xi)\right\}\\
v_{r,r+s} &= \frac{1}{2}\left(\frac{\sigma}{\xi}\right)^2 \left\{(r+s+1)^{2\xi}\Gamma(1-2\xi)K(r/(r+s+1))\right.\\
&\quad -(r+s)^{\xi}\Gamma(1-2\xi)K((r+1)/(r+s))\\
&\quad \left.+2(r+1)^{\xi}\left[(r+s)^{\xi} - (r+s+1)^{\xi}\right]\Gamma^2(1-\xi)\right\}, \quad s \geq 2,
\end{aligned}$$

with $K(x) = \mathscr{HG}(-\xi, -2\xi; 1-\xi; -x)$, where $\mathscr{HG}$ is the hypergeometric function.

Denoting $\theta = (\sigma, \xi, \mu)^T$ and $\hat{\theta}$ its estimator, Eqs. (6.15)–(6.17) can be written $\theta = k(M)$. With G the 3×3 matrix with entries $g_{i,j} = \partial k / \partial M_{1,j,0}$, we obtain the convergence properties of $\hat{\theta}$ from those of $\hat{M}$, i.e.

$$\sqrt{m}\left(\hat{\theta} - \theta\right) \xrightarrow{\mathscr{L}} \mathscr{N}(0, V_2) \tag{6.19}$$

when $m \to +\infty$ and $\xi < 0.5$, where $V_2 = GVG^T$.

6.4.1.3 Confidence Intervals

Using the Gaussian convergence property, we can construct asymptotic confidence intervals of level $100(1-\alpha)\%$ for each parameter of the GEV distribution, centered on the estimates

$$\hat{\sigma} \pm \Phi(1-\alpha/2)\frac{\hat{\upsilon}_{1,1}}{m}$$
$$\hat{\xi} \pm \Phi(1-\alpha/2)\frac{\hat{\upsilon}_{2,2}}{m}$$
$$\hat{\mu} \pm \Phi(1-\alpha/2)\frac{\hat{\upsilon}_{3,3}}{m},$$

where $\hat{\upsilon}_{i,i}$ is the i-th diagonal entry of the matrix V_1 or V_2—depending on the estimation method used—when (σ, ξ, μ) is replaced by its estimate. Recall that Φ is the cumulative distribution function of the standard Gaussian distribution.

6.4.1.4 Profile Likelihood Method

We can usually obtain faster convergence of estimators using the profile likelihood method, introduced in [58]. The profile likelihood of ξ is written as

$$L_p(\xi) = \max_{\sigma,\mu|\xi} L(\sigma, \xi, \mu).$$

The ratio defined by

$$\Lambda = \frac{L_p(\xi_0)}{L_p(\hat{\xi})}$$

allows testing of the hypothesis $H_0 : \{\xi = \xi_0\}$ against the alternative $H_1 : \{\xi \neq \xi_0\}$. Thus, for $m \to \infty$ and under H_0

$$-2\log\Lambda \xrightarrow{\mathscr{L}} \chi_1^2,$$

where χ_1^2 is the χ^2 distribution with one degree of freedom. H_0 is rejected at level α if $-2\log\Lambda > \chi_1^2(1-\alpha)$. The level $100(1-\alpha)\%$ confidence interval for ξ is thus given by

$$IC_\xi = \left\{\xi : -2\log\frac{L_p(\xi)}{L_p(\hat{\xi})} \leq \chi_1^2(1-\alpha)\right\},$$

which can also be written as

$$IC_\xi = \left\{\xi : \log L_p(\xi) \geq \log L_p(\hat{\xi}) - \frac{\chi_1^2(1-\alpha)}{2}\right\}.$$

When the 95% confidence interval contains 0, if $\alpha = 5\%$ it is not possible to reject $H_0 : \xi = 0$. In this case, we select the Gumbel distribution.

6.4.2 Parameter Estimation for GPD Distributions

Given a sample $X_1, \ldots, X_n$, we first extract the sample corresponding to exceedances of the threshold u, which we call $Y_1, \ldots, Y_{N_u}$, where $Y_i = X_i - u$ if $X_i > u$, with N_u the number of times u is attained.

6.4.2.1 Maximum Likelihood

The log-likelihood of the sample $Y_1, \ldots, Y_{N_u}$ can be written—with the help of the density of the GPD distribution—by reparameterizing the problem as a function of (τ, σ), where $\tau = \xi/\sigma$

$$\log L(\tau, \xi) = -N_u \log \xi + N_u \log \tau - \left(\frac{1}{\xi} + 1\right) \sum_{i=1}^{N_u} \log(1 + \tau Y_i) \qquad (6.20)$$

if $1 + \xi \dfrac{Y_i}{\sigma} > 0$ for each $i = 1, \ldots, N_u$. Thus, the maximum likelihood estimator $(\hat{\tau}, \hat{\xi})$ is given by

$$\frac{1}{\hat{\tau}} - \left(\frac{1}{\hat{\xi}} + 1\right) \frac{1}{N_u} \sum_{i=1}^{N_u} \frac{Y_i}{1 + \hat{\tau} Y_i} = 0,$$

where

$$\hat{\xi} = \frac{1}{N_u} \sum_{i=1}^{N_u} \log\left(1 + \hat{\tau} Y_i\right).$$

The estimator $\hat{\sigma}$ can be deduced from $(\hat{\tau}, \hat{\xi})$

$$\hat{\sigma} = \frac{\hat{\xi}}{\hat{\tau}}.$$

The estimator $(\hat{\tau}, \hat{\xi})$ is asymptotically Gaussian if $\xi > -1/2$, which if true implies that $(\hat{\sigma}, \hat{\xi})$ is too

$$\sqrt{N_u}\left((\hat{\sigma}, \hat{\xi}) - (\sigma, \xi)\right) \xrightarrow{\mathcal{L}} \mathcal{N}(0, V_3)$$

when $N_u \to +\infty$ and $\xi > -0.5$, where

$$V_3 = (1 + \xi) \begin{bmatrix} 2\sigma^2 & -\sigma \\ -\sigma & 1 + \xi \end{bmatrix}.$$

6.4.2.2 Weighted Moments

In the GPD case, the weighted moments method involves moments of type $M_{1,0,s}$ for $s = 0, 1$, given by

$$M_{1,0,s} = \frac{\sigma}{(s+1)(s+1-\xi)} \quad \text{for} \quad \xi < 1,$$

and replacing $M_{1,0,s}$ by its empirical estimator

$$\hat{M}_{1,r,0} = \frac{1}{N_u} \sum_{j=1}^{N_u} \left(\Pi_{\ell=1}^{r} \frac{(N_u - j - \ell + 1)}{(N_u - \ell)} \right) Y_{j,N_u},$$

where Y_{j,N_u} is the j-th value of the sample $Y_1, \ldots, Y_{N_u}$ put in order of increasing values. Inverting the system of equations defined by $M_{1,0,0}$ and $M_{1,0,1}$ leads to the following estimators:

$$\hat{\xi} = 2 - \frac{\hat{M}_{1,0,0}}{\hat{M}_{1,0,0} - 2\hat{M}_{1,0,1}}$$
$$\hat{\sigma} = \frac{2\hat{M}_{1,0,0}\hat{M}_{1,0,1}}{\hat{M}_{1,0,0} - 2\hat{M}_{1,0,1}}.$$

The estimator $(\hat{\sigma}, \hat{\xi})$ is asymptotically Gaussian if $\xi < 1/2$:

$$\sqrt{N_u}\left((\hat{\sigma}, \hat{\xi}) - (\sigma, \xi)\right) \xrightarrow{\mathscr{L}} \mathscr{N}(0, V_4) \tag{6.21}$$

when $N_u \to +\infty$ and $\xi < 0.5$, where

$$V_4 = C \begin{bmatrix} \sigma^2(7 - 18\xi + 11\xi^2 - 2\xi^3 & -\sigma(2-\xi)(2 - 6\xi + 7\xi^2 - 2\xi^3) \\ -\sigma(2-\xi)(2 - 6\xi + 7\xi^2 - 2\xi^3) & (1-\xi)(2-\xi)^2(1 - \xi + 2\xi^2) \end{bmatrix},$$

with $C = \dfrac{1}{(1-2\xi)(3-2\xi)}$.

6.4.2.3 Confidence Intervals

Due to their asymptotic convergence to Gaussian distributions, we can construct asymptotic confidence intervals of level $100(1-\alpha)\%$ for each of the GPD's parameters, centered at the estimates

$$\hat{\sigma} \pm \Phi(1-\alpha/2)\frac{\hat{v}_{1,1}}{m}$$
$$\hat{\xi} \pm \Phi(1-\alpha/2)\frac{\hat{v}_{2,2}}{m},$$

where $\hat{v}_{i,i}$ is the i-th diagonal element of the matrix V_3 or V_4, depending on which estimation method is used when replacing (σ, ξ) by its estimate.

6.4.2.4 Profile Likelihood Method

As for the estimation of GEV parameters, we can generally obtain faster convergence of estimators using the profile likelihood method, detailed in the previous section. For GPD models, the profile likelihood of ξ is given by

$$L_p(\xi) = \max_{\sigma|\xi} L(\sigma, \xi).$$

The level $100(1-\alpha)\%$ asymptotic confidence interval for ξ is thus defined by

$$IC_\xi = \left\{\xi : \log L_p(\xi) \geq \log L_p(\hat{\xi}) - \frac{\chi_1^2(1-\alpha)}{2}\right\}.$$

The special case where $\xi = 0$ is treated in [502].

6.4.3 *Details on Hill's Estimator for ξ*

Let us now take a closer look at the estimation of ξ. Hill's estimator [389] is the most well-known estimator for this task. It is defined by

$$\hat{\xi}_n = \frac{1}{k}\sum_{j=1}^{k} \log X_{n-j+1,n} - \log X_{n-k,n},$$

with $X_{1,n}, \ldots, X_{n,n}$ the order statistics associated with the sample $(X_1, \ldots, X_n)$, i.e., $X_{1,n}$ is the smallest value and $X_{n,n}$ the largest of the full sample. Hill's estimator is used often for reasons such as those listed below.

- If we denote $E_{j,t}$ the relative excess with respect to t, i.e., $E_{j,t} = \dfrac{X_j}{t}$ with $X_j > t$, it can be shown that

$$\mathbb{P}\left(E_{j,t} > x | E_{j,t} > 1\right) \xrightarrow{t\to\infty} x^{-1/\xi} \quad \text{when } x > 1.$$

Using the statistical likelihood based on this limit distribution, we can show that Hill's estimator is none other than the maximum likelihood estimator in the case where $t = X_{n-k,n}$ (i.e., the k-th largest threshold).

- Furthermore, Hill's estimator can be interpreted visually, which is of particular importance to practitioners. Indeed, if we look at the plot with the points

$$\left(\log \frac{n+1}{j}, \log X_{n-j+1,n}\right),$$

which is called the *Pareto quantile plot*, then—when the distribution is Pareto—this plot will be approximately linear in its extreme points, with slope ξ. Hill's estimator is then nothing more than a naive estimator of this slope, and thus of ξ.

Asymptotic properties of Hill's estimator were established, for consistency, by [210, 507], and for asymptotic convergence to a Gaussian by [67, 188].

The main disadvantage of this estimator is that it only works for positive ξ, and is therefore restrained—theoretically speaking—to the Fréchet domain of attraction. Various generalizations have therefore been proposed. Among these, let us highlight, in particular, the moments estimator from [211] and the UH estimator from [69]. Readers can also find in the thesis [746] goodness-of-fit test statistics for the Pareto distribution, in particular for the parameter ξ.

Recent work on the properties of Hill's estimator has shown that under certain regularity conditions, it can be written as a function of the largest order statistics of a sample from the exponential distribution [104]. The choice of the number of order statistics to use in the calculation can be determined as a function of the precision of the estimator of ξ [742]. This precision is *non-asymptotic* and can be calculated using *concentration inequalities* [508].

6.4.4 Estimating the Extremal Index θ

When data has temporal correlation, it is necessary to estimate the extremal index θ measuring the strength of this correlation.

Several estimators have been proposed for the extremal index. Ferro and Segers [273] have shown that T_θ, representing the duration between successive exceedances, converges in distribution—when the threshold corresponds to a larger and larger quantile—to a distribution with mass $1-\theta$ at $t = 0$ and an exponential distribution in its continuous part $t > 0$. Its coefficient of variation ν, defined as the ratio of its standard deviation to its expectation, then satisfies

$$1 + \nu^2 = \frac{\mathbb{E}[T_\theta^2]}{(\mathbb{E}[T_\theta])^2} = \frac{2}{\theta}.$$

If N is the number of exceedances of a threshold u by the data series and $T_1, \ldots, T_{N-1}$ the inter-event times, the proposed estimator is given by

$$\bar{\theta}_n(u) = \begin{cases} \min(1, \hat{\theta}_n(u)) \text{ if } \max\{T_i : 1 \leq i \leq N-1\} \leq 2 \\ \min(1, \hat{\theta}_n^*(u)) \text{ if } \max\{T_i : 1 \leq i \leq N-1\} > 2, \end{cases} \tag{6.22}$$

with

$$\hat{\theta}_n(u) = \frac{2\left(\sum_{i=1}^{N-1} T_i\right)^2}{(N-1)\sum_{i=1}^{N-1} T_i^2}$$

and

$$\hat{\theta}_n^*(u) = \frac{2\left[\sum_{i=1}^{N-1}(T_i-1)\right]^2}{(N-1)\sum_{i=1}^{N-1}(T_i-1)(T_i-2)}.$$

Unlike many other estimators, this one has the important property of guaranteeing an estimation of θ between 0 and 1.

6.4.5 *Model Validation*

Validation of an inferred model requires to use numerous informative plots. It is always useful—if not essential—to look carefully at the following ones.

- ***QQ-plot***: the empirical quantiles from the data are compared with the theoretical quantiles of the proposed distribution. A good fit corresponds to a straight line across the whole range of quantiles.
- ***Probability-plot***: this plot compares the empirical ranks of the data with the theoretical ones from the proposed model. A good model fit again corresponds to a straight line.
- ***Return levels*** graph: on this graph is plotted how changes the link between return periods and associated return levels, estimated using the proposed model or the data.
- ***Density*** graph: this graph compares the data histogram with the density distribution of the estimated (*inferred*) model.

An illustration of each of these plots is given in Sect. 6.6.2 for an analysis of extreme wind speeds.

6.5 Study Steps

In this section, we describe the main steps of a typical univariate extreme values. The aim is not to list in full the set of statistical tools available at each step, but rather to detail a few common methods. This description is illustrated with two examples

- a time series of daily flow measurements of the Loire river from 01/08/1958 to 31/12/2011, and
- a time series of daily wind speed measurements from 01/08/1981 to 30/04/2011.

6.5.1 *Obtaining a Stationary Process*

Using modeling theorems requires a stationary process. However, especially when dealing with processes related to environmental variables, this hypothesis is unlikely to be satisfied. For example, a river flow is clearly seasonal, which leads to a non-stationary process over the course of a year. In general, non-stationarity in a process may be due to a change point (e.g., the date at which measurements began, or a change in the environment leading to a significant impact on the process), a trend (e.g., air temperature increases due to a warming planet), or seasonality.

The presence of a trend is specifically dealt with in Chap. 8 of this book.

As for seasonality, it suffices to restrain the study of the process to a period of the year in which it can be considered stationary. This period must necessarily contain the extreme values. Hence, for instance, in the study of rivers flooding, we retain only the fall and winter flow measurements.

A precise determination of the stationary period can be made using physical and statistical criteria, e.g., the calculation of monthly empirical means and variances; the stationary period will be that in which these values are the most similar, thus ensuring second order stationarity. Statistical tests can also be used to help validate stationarity quantitatively (see Chap. 3 for more details). Particular examples are the uniform distribution test for threshold exceedance dates, and the Mann-Kendall test for detecting trends. Other tests are specifically designed to compare the likelihood of a model with constant parameters to one with time-dependent parameters. These tests are based on the distribution of the *deviance* D, defined by

$$D = 2(\ell_1 - \ell_0),$$

where ℓ_1 is the value maximizing the log-likelihood of model M_1, and ℓ_0 that of model M_0. The deviance can then be compared with a $\chi_2(k)$ distribution, where k is the difference in the number of parameters between the two models (cf. Theorem 8.1). Example 4.2 in Chap. 2 details the use of this test for deciding on the best model. For a recent look at constructing tests adapted to detecting models with variable parameters, see [264].

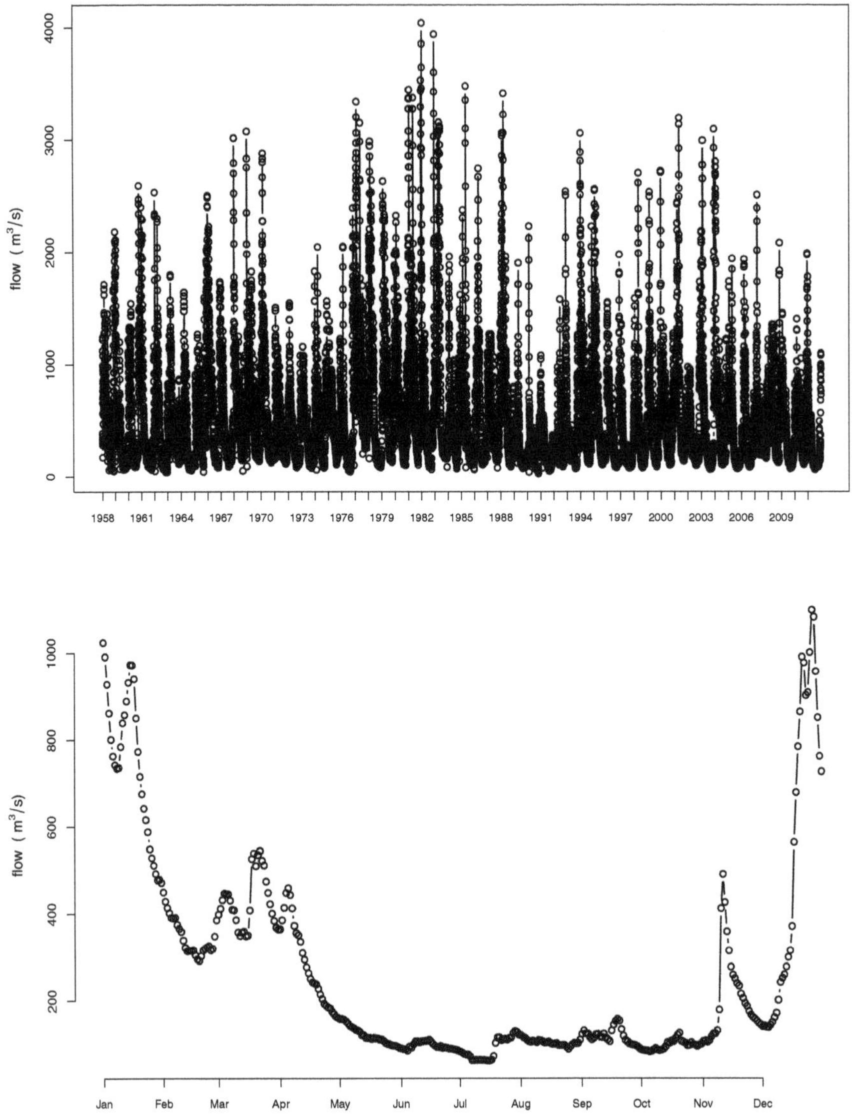

Fig. 6.3 Daily river flow measurements from 01/08/1920 to 31/12/2011 (top) and a zoom on year 1981

Let us illustrate this step to determine the stationary period using the time series of river flows. Figure 6.3 shows these, along with a zoom in to the year 1981 to better show the annual river flow cycle.

Figure 6.4 provides a visual representation of the monthly river flow values. The boxes contain 50% of the values, spread out around the median—the solid line inside each box. Extreme values are shown as circles. A second order stationary process

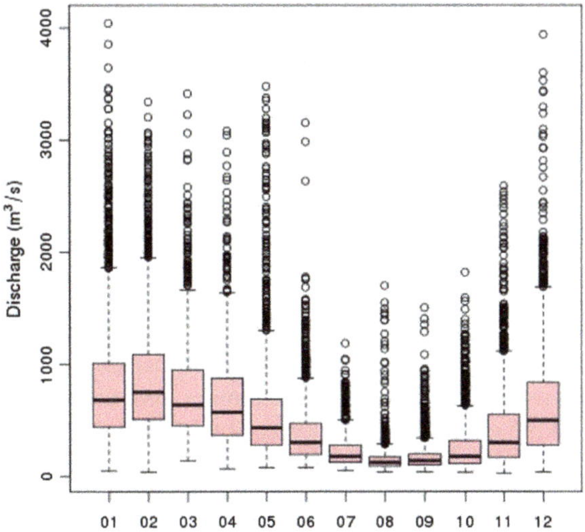

Fig. 6.4 Daily river flow measurements, boxplotted per month; the process can be considered stationary from December to May

Table 6.1 Monthly distribution of the five highest flows between 1958 and 2011

Month	J	F	M	A	M	J	J	A	S	O	N	D
Number of flows	75	55	28	14	28	2	0	0	0	2	21	45

corresponds to constant median and variance—whose value is reflected by the height of the boxes. This graph suggests that the river flow can reasonably be considered a stationary process when restricted to the months December to May.

It is also useful to study how extreme values of interest are spread out over the year by calculating, for instance, the monthly distribution of the ten most extreme values from each year of measurements. This can help better see which months extreme natural events occur in, which complements expert knowledge on the subject (Table 6.1).

6.5.2 *Extracting Annual* Maxima *and Cluster* Maxima

Once the process has been restricted to a period where it can *reasonably be considered stationary*, the annual maxima subprocess—whose values are asymptotically distributed according to one of the three GEV distributions—can be extracted.

In the POT approach, the analysis takes place on the cluster maxima. It is, therefore, necessary to extract independent clusters from the initial data time series, i.e., groups of close-together values exceeding some high threshold. In this way, we obtain the cluster maxima subprocess. This step is called *declustering* and must be performed carefully because the rest of the study strongly depends on it. An overly-low choice of threshold will lead to a modeling error, while an overly-high one means poor parameter estimation (cf. Sect. 6.2.3.1).

6.5.2.1 Declustering Procedure

For a chosen threshold u_s, a practitioner has to identify independent clusters. Several methods exist to do so [88, 686, 737], generally based on bias-variance tradeoffs in estimated functions of interest [646]. However, the most common is based on the so-called *redescent* criterion [433, 434], whereby two exceedances are considered independent when separated by at least r values below the threshold. The value of r may be dictated by physical considerations. For example, for certain high-water events, experts suggest a redescent value of $r = 10$ days. See also Example 5.11.

Choosing the threshold. In practice, validation of the declustering step (choice of threshold u_s and formation of independent clusters) is performed by checking the probabilistic properties stemming from the Poissonian nature of cluster maxima and the max-stability property of GPD distributions. Thus, the sequence of cluster maxima must be such that:

- they are independent,
- their number per year follows a Poisson distribution, and
- the waiting time between consecutive ones follows an exponential distribution.

Validating declusterized data. A great number of quantitative and graphical statistical tools are available to do this—see Chap. 5 for more details. In particular, the *Kolmogorov–Smirnov* test, which quantifies the data likelihood under the null hypothesis H_0 of a continuous distribution $F_0(x)$ is often used to test for an exponential distribution on the waiting times between successive cluster maxima (see Example 4.1), with the help of the empirical distribution (Fig. 6.5). The equivalent test for discrete distributions is the χ^2 test, which is used to test the pertinence of a Poisson distribution for the number of cluster per year.

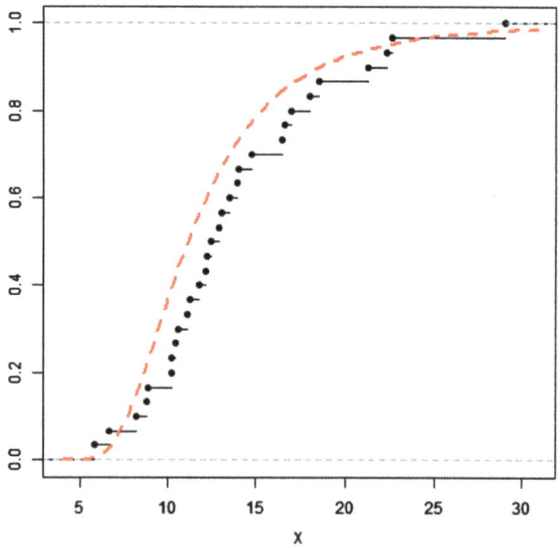

Fig. 6.5 Empirical cumulative distribution function of a dataset simulated under a GEV distribution

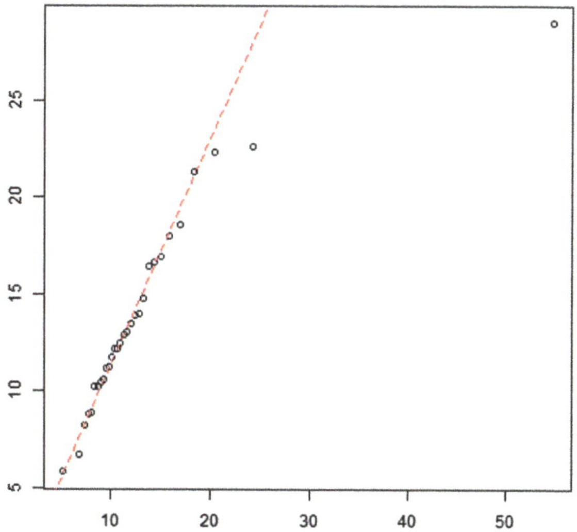

Fig. 6.6 QQ-plot of a dataset simulated under a GEV distribution. X−axis are empirical quantiles while Y−axis are theoretical quantiles

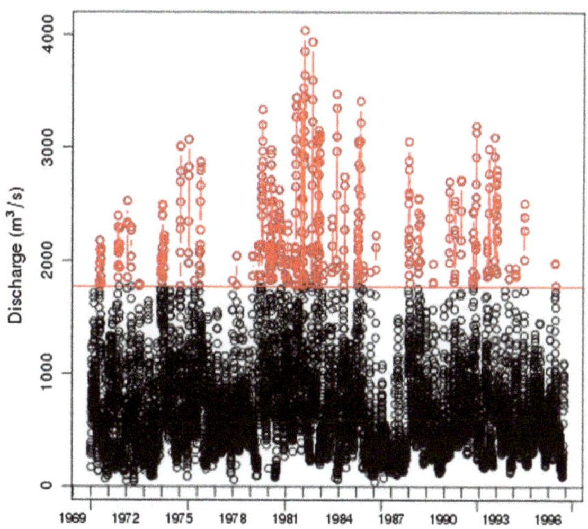

Fig. 6.7 Modeling extreme flows. Selecting cluster maxima over the threshold $u_s = 1773\,\mathrm{m}^3\mathrm{s}^{-1}$

Graphical goodness-of-fit tests for a given model are easy to interpret. The most well-known one is the *QQ-plot* (i.e., "quantile-quantile") which plots the empirical quantiles from the data and compares them with the theoretical ones of the proposed distribution. A good fit between the data and the distribution would mean an essentially straight line across the whole range of quantiles (Fig. 6.6).

The autocorrelation function (cf. Sect. 5.4.1.1) plots the temporal empirical correlation of the data between times t and $t + k$ for different lags k. In the example considered on Fig. 6.8, the confidence interval delimited by the two horizontal lines gives a region inside which the correlation coefficient is not statistically different to 0. Independent data should lead to correlations found in this interval. This graphical test is generally used to verify independence in cluster maxima.

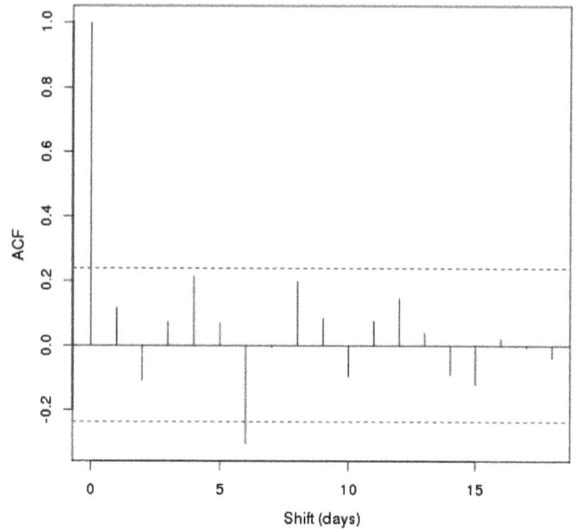

Fig. 6.8 Modeling extreme flows. Independence of cluster maxima: autocorrelation function

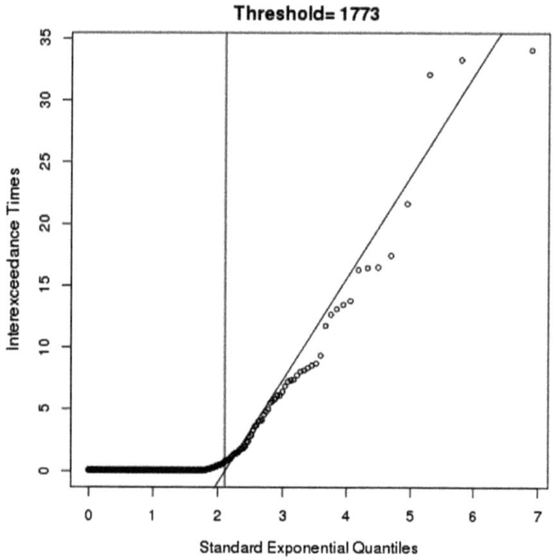

Fig. 6.9 Modeling extreme flows. Waiting times between successive cluster maxima (QQ-plot for the exponential distribution)

6.5.2.2 Practical Example

Flow data. We now illustrate the types of results that can be obtained using the flow time series shown in Fig. 6.3. Figure 6.7 plots the maxima of the clusters retained. Results on Fig. 6.8 thus allow us to test for cluster maxima independence using the autocorrelation function; the estimated correlations between data at times t and $t+k$ for several lags k are all indistinguishable from 0 except one ($k=6$). It is, therefore,

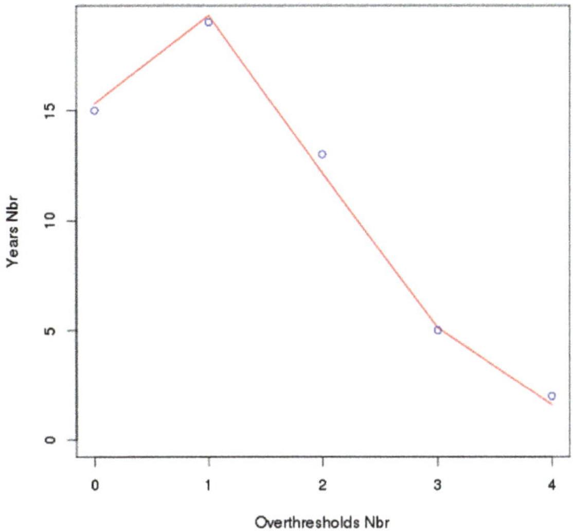

Fig. 6.10 Modeling extreme flows. Annual number of cluster maxima (empirical distribution and estimated Poisson distribution)

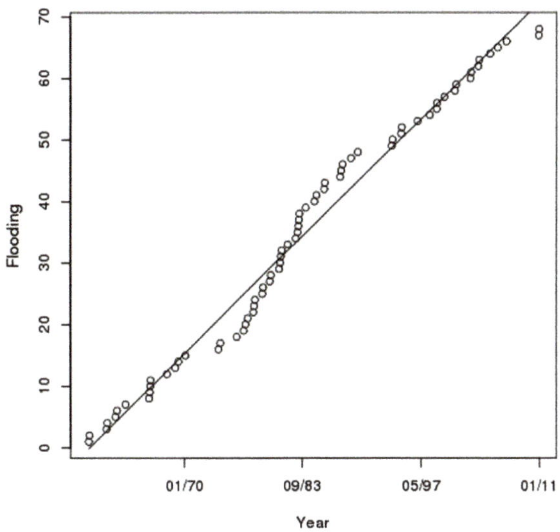

Fig. 6.11 Modeling extreme flows. Uniform distribution test on the cluster maxima

reasonable to suppose that the hypothesis of independent cluster maxima is plausible. The QQ-plot, which allows us to test whether the waiting times between successive cluster maxima follow an exponential distribution, does not invalidate the above choice (Fig. 6.9): the plot clearly shows a good fit between the empirical quantiles of the data and those from the proposed exponential distribution.

As well as this, the Poisson distribution for the annual number of cluster can also be validated: Fig. 6.10 plots the observed values as well as those from the estimated distribution. The test for a uniform distribution of the dates at which cluster maxima occur (Fig. 6.11) also appears to show that this hypothesis is acceptable.

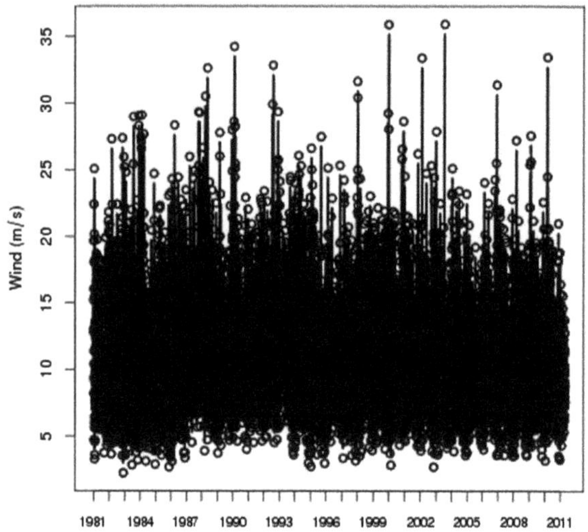

Fig. 6.12 Daily wind speed measurements from 01/08/1981 to 30/04/2011

Wind speed data. We illustrate now the stability properties of GPD distribution parameters on the wind speed data. This time series is shown in Fig. 6.12 and covers the period 1981–2011. The data has been analyzed in the same way the flow data was. We concentrate on other aspects of the methodology here; notably, we focus on the choosing the threshold above which exceedances will be taken into account.

First, we illustrate the stability property of GPD distribution parameters given in Proposition 6.3. Figure 6.13 plots the scale $\sigma^*(u)$ and shape ξ parameters as a function of u from 29 m.s^{-1} to 31 m.s^{-1}. The stability of the curves implies that $u_s = 29$ m.s^{-1} is an acceptable choice. Next, Fig. 6.14 plots the conditional expectation of exceeding u for $u \geq 29$ m.s^{-1}. A certain amount of linearity can be seen starting from 29 m.s^{-1}, which allows us to validate the choice of u_s, according to Proposition 6.4. Figure 6.15 plots the cluster maxima retained above $u_s = 29$ m.s^{-1}.

6.6 Fitting a GPD Model to Wind Speed Data

Remember that any estimate must be accompanied by a confidence interval allowing us to quantify the potential statistical error due to having a limited number of data (see Sect. 4.13). However, be careful: such intervals are based on the asymptotic distributions of estimators, which supposes that the number of data used in the estimation is relatively large.

The study of the wind speed time series is summarized in Table 6.2, giving the estimated parameters for the GPD distribution of cluster maxima exceedances of the threshold $u_s = 29$ m.s^{-1}, along with the 90% confidence intervals, noted $[a_{90}, b_{90}]$.

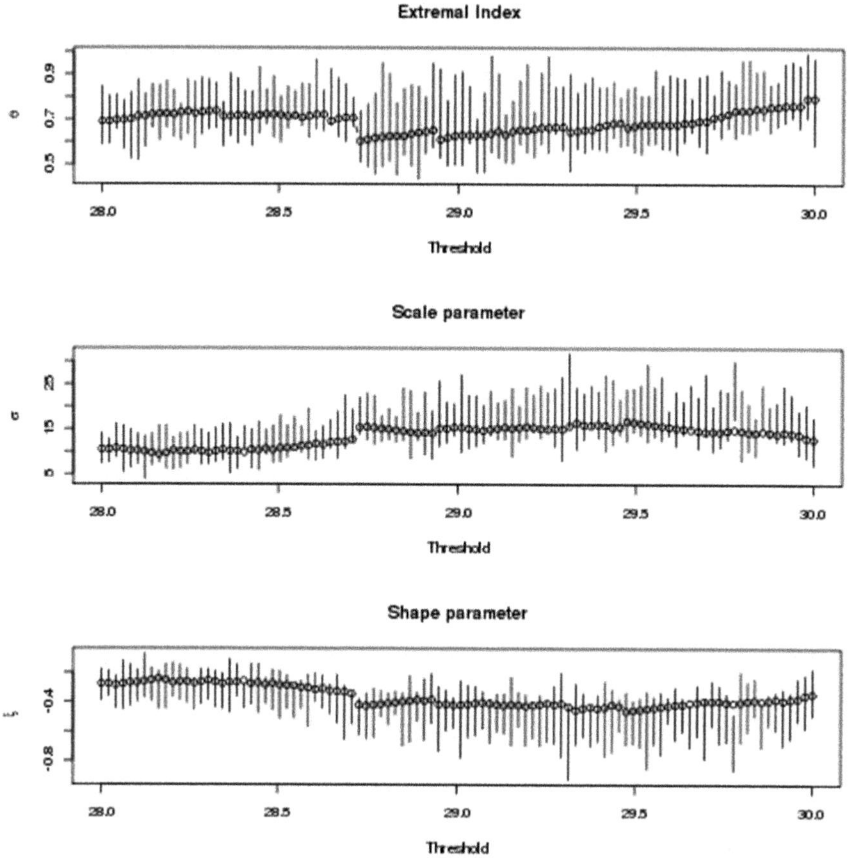

Fig. 6.13 Modeling extreme wind speeds. Stability of the extremal index and parameters of the generalized Pareto distribution

6.6.1 *Estimating the Extremal Index θ*

We estimate θ with the help of the estimator proposed by [273]—see Eq. 6.22. The relative stability in the estimation of $\hat{\theta}_n^*(u)$ seen in the first plot in Fig. 6.13 for thresholds around 29 m.s^{-1}, validates the value $\theta \simeq 0.63$ for the wind speed data. Cluster size is therefore estimated at $1/\theta \simeq 1.6$ days, i.e., less than two days.

6.6.2 *Validating the Inferred Model*

To validate the choice of model, we look at the *probability-plots* and *return level* and *density* graphs, whose definitions were given earlier.

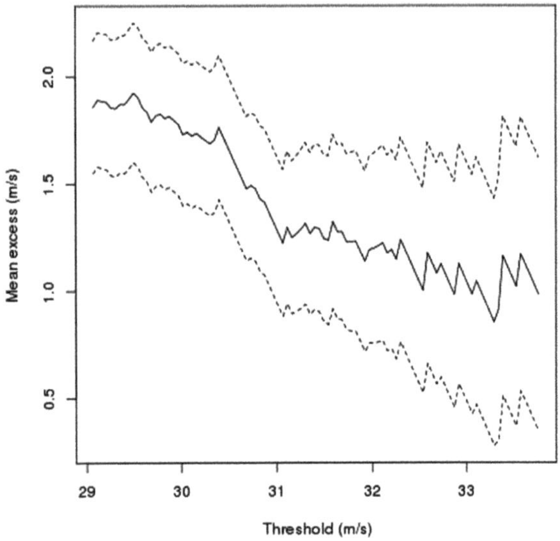

Fig. 6.14 Modeling extreme wind speeds. Evolution of the conditional expectation of exceeding the threshold u by independent cluster maxima above the value $v_1 = 29\,\text{m.s}^{-1}$

Table 6.2 Fitting a generalized Pareto distribution to independent cluster maxima exceeding the threshold $u_s = 29\,\text{m.s}^{-1}$ for the wind speed data

Parameter	Estimation	St. dev.	a_{90}	b_{90}
ξ	−0.315	0.0879	−0.488	−0.143
σ	2.546	0.334	1.892	3.201

For illustration's sake, the graphs for the wind speed analysis are given on Fig. 6.16. The quite linear form of the first two, and the good fit to the data seen in the others, means that the model selection and inference performed is credible.

6.6.3 Estimating Return Periods and Annual Frequencies

The distribution of cluster maxima allows us to determine the levels of T-year return periods using (6.7). These estimates comes with confidence intervals denoted IC, based on the asymptotic convergence result of the maximum likelihood estimator recalled in Eq. 6.8.

For the wind speed time series, we find the following centennial level, along with its 95% confidence interval: v_{100}=36.6 m.s^{-1}, and $\text{IC}_{95\%} = [34.6\,\text{m.s}^{-1};\, 38.5\,\text{m.s}^{-1}]$.

The annual frequencies as defined earlier and calculated according to formulas (6.9) for the event of exceeding the centennial level are the following:

- the frequency defined by the mean yearly number of days exceeding v_{100}: $\mathbb{F}_a^{6.13}(\mathscr{E}) = 1.59 \times 10^{-2}$,

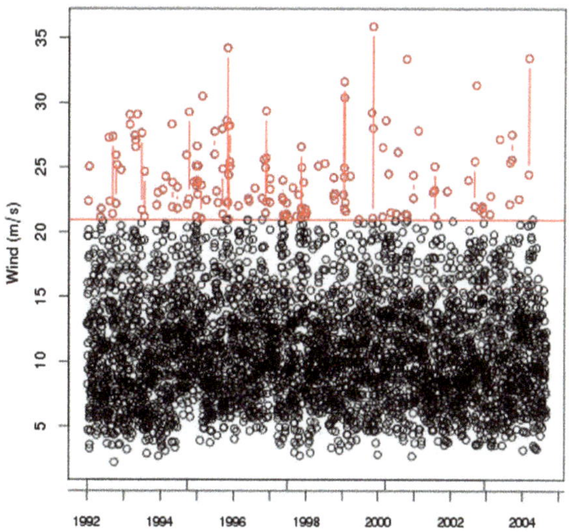

Fig. 6.15 Modeling extreme wind speeds. Selecting cluster maxima above the threshold $u_s = 29\,\text{m.s}^{-1}$

- the frequency defined by the mean yearly number of episodes in which v_{100} is exceeded: $\mathbb{F}_a^{6.14}(\mathscr{E}) = 10^{-2}$, and
- the frequency defined by the probability of having at least one episode in which v_{100} is exceeded in a given year: $\mathbb{F}_a^{6.15}(\mathscr{E}) = 9.95 \times 10^{-3}$.

6.7 Conclusions and Perspectives

This chapter has presented the main results in classical univariate extreme value theory, illustrated with recent studies undertaken by engineers and researchers at EDF. We have emphasized the "engineering" side of things and the data preprocessing required to access the theoretical framework for stationary processes. Such analyses can also be found in tutorials made freely available by engineers and researchers (see for example [239]).

The theory is, however, much broader that what we have shown, and contains numerous other results. In particular, the study of the colored Poisson process, where the color is the cluster size, makes it possible to estimate the distribution of cluster lengths. This model gives an idea of the severity of the natural hazard, quantified by the duration of above-threshold episodes. Moreover, the study of specific stationary processes such as Markov chains, where the current instant depends on only a finite number of past instances, is also of interest. Indeed, this type of modeling is frequently used for variables related to the environment (e.g., autoregressive processes such as ARMA for studying how the temperature evolves).

Even though it is a well-studied setting, univariate extreme value theory is still a subject of active research, notably in the following areas—the importance of which

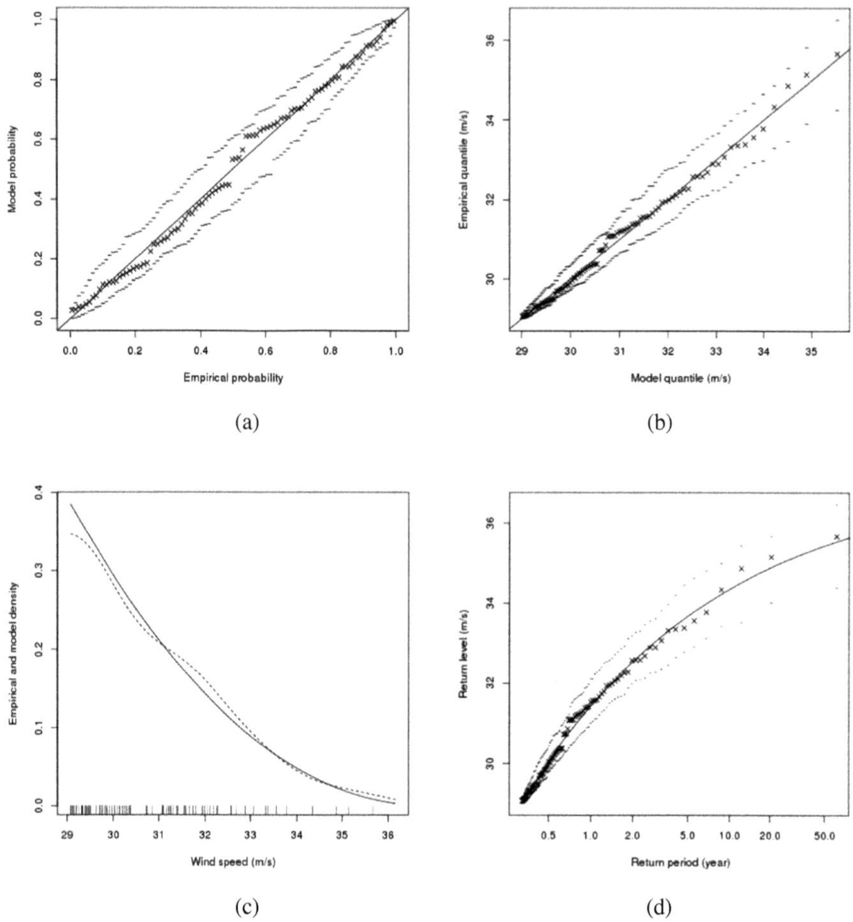

Fig. 6.16 Modeling extreme wind speed data. Graphical validation of the modeling of cluster maxima exceedances: **a** probability-plot, **b** QQ-plot, **c** empirical and model density graph, **d** return levels graph

is clear from the current chapter: estimating the shape parameter ξ, estimating the extremal index θ, declustering methods [288, 369]), etc. If interested, readers can find in Chap. 12 more information on this subject. The study of non-stationary processes is also an active domain, especially the extension of the point processes to this setting [402, 403].

The shortage of data is a major problem which strongly impacts the precision of model inference. To overcome this, a variety of methods can be tried. One possibility is the regional approach presented in Chap. 7. This method permits us to combine data coming from different locations using a data homogenization approach. This strongly increases the quantity of data available and improves the convergence of estimators and models.

An alternative method is to take potentially very old historical data into account. Such data holds precious information on extreme events which are not part of a systematic time series, and can be used in model inference. However, the best way to include it is not always obvious. A Bayesian approach (Chap. 11) is a possible solution to this.

Yet another strategy is to artificially increase the number of data by simulating it using a coded model of the physical processes involved. The challenge is, therefore, to be able to model the physics well and produce simulated data that looks like it comes from the true process, including its extreme values, both in terms of actual values and frequency. The second part of Chap. 10 provides some details on this.

To conclude, we now arrive at the question asked by all extreme values analysts: "Up to what point is it reasonable to extrapolate an extreme value distribution"? In other words, is it reasonable to calculate—starting from a GPD or GEV model estimated with the help of n data points—the level of *any* return period?

The answer to this question is not immediately obvious. Clearly, it depends on the level of precision underlying the reasoning used, i.e., the precision in estimated return levels, generally in terms of the length of confidence intervals (or *credibility intervals* if this notion is correctly interpreted—see Chap. 11).

It also depends on the quantity of data, its quality, the rate of convergence (e.g., of the marginal distribution—a priori unknown—of the process) to asymptotic distributions, etc. A number of research projects including [29], supervised by EDF engineers and researchers, are underway on this subject. Their aim is to improve the credibility of extrapolations of extreme value distributions, and in the long run reinforce certain demonstrations of robustness of industrial facilities.

Part II
Elements of Extensive Statistical Analysis

Chapter 7
Regional Extreme Value Analysis

Jérôme Weiss and Marc Andreewsky

Abstract This chapter focuses on the occurrence of extreme natural hazards in non-instrumented locations, whose neighborhood contains sites for which measured data are available. The spatial regionalization approach developed here, which is based on the detection of regions that are homogeneous in terms of risk, allows to propose a localized quantification of these events.

7.1 Introduction

Recall that extreme value theory provides a theoretical and rational framework to estimate the intensity of events like significant wave heights with a return period of 100 years (or 1 000 or even 10 000 years) at a site of interest. In environmental engineering, these return levels are usually estimated using local statistical analyses, whose principle is to apply extreme value theory to only observations from the site of interest. A direct application of this theory can sometimes pose difficulties, because the length of time over which observations have been made at a given site is generally low (e.g., around 10–30 years in the case of sea-based data). In such situations, uncertainty in extrapolations associated with long return periods is often far too high to assess risk in a meaningful quantitative way, i.e., to provide real decision support. The aim of this chapter is therefore to examine the potential of regional analysis to obtain less uncertain estimates of extreme levels.

Indeed, regional approaches [196] have been introduced with the aim of taking advantage of information available in several time series (or at several sites) near to the site of interest, so as to help forecast return levels of extreme hazards at a local level. This type of approach is of particular interest today since increasingly large spatial databases have become available, providing project managers with large amounts of useful information. This chapter examines the possibility of taking advantage of a *regional* enlargement of the extreme hazard zone of interest to: (a) group together

---

J. Weiss · M. Andreewsky (✉)
EDF R&D, Boulevard Gaspard Monge, 91120 Palaiseau, France
e-mail: marc.andreevsky@edf.fr

N. Bousquet and P. Bernardara (eds.), *Extreme Value Theory with Applications to Natural Hazards*, https://doi.org/10.1007/978-3-030-74942-2_7

more extreme events and thus obtain larger sample sizes; and: (b) build a predictive field at specific sites.

7.1.1 General Principle

Regional approaches are therefore based on the rather natural assumption that at similar sites—labeled "homogeneous"—characterized by the same natural phenomena, the occurrence probabilities of extreme events are similar. In such cases, all collected information can be used to estimate so-called "regional" probability distributions, leading to more certainty in estimations. Logically, more robust estimates of return levels could be derived.

More technically, the statistical concept of regionalization is based on the following hypothesis: homogeneity of a region, for the type of event being considered, results in equality between the distributions of the frequency of maximum amplitude events (over a fixed period, typically a year) at each measuring station (site), up to some scale factor. This factor embodies, or rather summarizes, the precise features of each site. In the case of river flow data, the scale factor is defined as the *index flood*. In the motivating case of marine data considered in this chapter (wave height, storm surge), it is called the *sea index*.

7.1.2 Historical Setting

The term *index flood* actually comes from the application domain where regional studies were originally conducted: hydrology. The regional analysis methods presented here have a more general scope, though they essentially originate in the approach introduced by Dalrymple [196] to estimate the occurrence frequency of floods using an *index flood model*. Stedinger et al. [720] subsequently showed that the index flood model, using information at the regional scale, effectively reduces uncertainty in return level estimates relative to local analyses.

Since the work of Dalrymple [196], regional analysis has become popular in hydrology and meteorology [397]. For example, the use of this type of model in the estimation of extreme flows in watercourses can be found in the works of Cunnane [189], De Michele et al. [205], Javelle et al. [410], Madsen et al. [494], Merz et al. [512] and Stedinger [719]. The modeling of extreme rainfall has been studied by Borga et al. [101], Hosking et al. [397] and Schaefer [687], and extreme winds by Escalante-Sandoval [256], Goel et al. [344] and Sotillo et al. [714]. Note also that UK regulations for estimating extreme rainfall recommend the use of such regional models [3].

As for maritime applications, Van Gelder et al. [761] have used regional methods to estimate extreme sea levels at 13 sites located on the Dutch North Sea coast. Extreme high sea levels were modeled at 9 sites in the North Sea by [760] and at 11 sites on the eastern coast of the Sea of Japan by [342, 343]. Extreme storm surges out to sea have been modeled by Bernardara et al. [86] at 18 sites on the French Atlantic coast and the English Channel, and by Bardet et al. [57] at 21 sites in the same area. In addition, Hosking [395] considered tsunami *run-ups*[1] at 114 sites in the Pacific Ocean. A common characteristic of these studies is a homogeneity of maritime hazards over rather broad areas. Note also that the regulatory framework for estimating storm surges in France recommends statistical studies at local or regional levels [15].

Lastly, recent research results [780], conducted at EDF and applied to wave, storm surge, rainfall, and wind data, are able to take into account the spatial dimension of phenomena. In particular, spatio-temporal *declustering** algorithms specific to regional analyses have been proposed, including techniques that deal with the correlation between extreme values from neighboring sites. Also introduced was the concept of *regional return period* of an event, in contrast to the *local return period* of an extreme value.

7.1.3 Principle of Regional Analysis

The regional analyses proposed in this chapter are based on a POT approach to extreme values, and thus on modeling with a GPD distribution. Indeed, this generally allows better use of information available on extreme events [433]. If, for example, several particularly intense events occurred in the same year, then the annual maxima method cannot really take advantage of this information. Conversely, in a year with no significant events, including the annual maximum when estimating extreme events may skew the results. In both cases, the POT approach avoids these pitfalls (although, as mentioned in the previous chapters, it is still affected by correlation problems in the data). By comparing the performance of these two methods on extreme value estimates, [492] has shown the superiority of the POT approach for regional analyses.

7.1.3.1 Modeling

The following modeling is directly derived from the previously introduced index flood model. Consider an area defined by N sites of interest, each with a measuring station. For $i \in \{1, \ldots, N\}$, the random variable X_i denotes an extreme hazard characterizing site i, with cumulative distribution function F_{X_i}. The hypothesis of regional homogeneity supposes that the probability distribution of extreme values is the same at each site, up to normalization with a local index μ_i at each site i. This is written:

[1] The run-up is the maximal amplitude of the tsunami its contact with the coast, i.e., the height of the wave above the mean high tide level.

$$\frac{X_i}{\mu_i} \sim Y \sim F_r, \tag{7.1}$$

where Y is the so-called *regionalized* random variable for the hazard, with cumulative distribution function F_r. Thus, $\forall i \in \{1, \ldots, N\}$, $\forall y \in I\!R$,

$$F_{X_i}(\mu_i y) = F_r(y). \tag{7.2}$$

In terms of return levels, Formula (7.1) is equivalent to

$$x_i^{(T)} = \mu_i y^{(T)}, \tag{7.3}$$

where

- $x_i^{(T)}$ denotes the return level with period T at site i,
- $y^{(T)}$ is the *regional return level* with period T.

The definition of the local index μ_i depends on whether we choose to model the actually values X_i recorded over a threshold u_i (or *exceedances*), written

$$X_i = S_i | S_i \geq u_i,$$

or the *threshold deviations* instead:

$$Z_i = S_i - u_i | S_i \geq u_i.$$

Though it is equivalent to work with the exceedances X_i or the threshold deviations Z_i in local analyses, this is no longer the case for regional ones. Indeed, the choice of one or the other has an impact on both the structure of the local index and the regional distribution.

The exceedance-based approach. In this framework, extreme values at site i are those which exceed a threshold u_i. Extreme value theory recommends modeling these using a generalized Pareto distribution:

$$\forall i \in \{1, \ldots, N\}, \quad X_i \sim \mathcal{G}_{PD}(u_i, \sigma_i, \xi_i).$$

The regional homogeneity hypothesis (7.2) gives

$$\begin{aligned} \forall x \geq \frac{u_i}{\mu_i}, \quad \mathbb{P}\left(\frac{X_i}{\mu_i} \leq x\right) &= \mathbb{P}\left(S_i \leq \mu_i x | S_i \geq u_i\right), \\ &= F_r(x). \end{aligned}$$

Following [669], and noting that $\forall i$, $F(u_i/\mu_i) = 0$ and $F_r(u_i/\mu_i + \varepsilon) > 0 \; \forall \varepsilon > 0$, we have that the ratio u_i/μ_i is the lower bound of the support of the distribution F_r. In particular, as this bound should not depend on i, the local index μ_i is constrained to be proportional to the threshold u_i. Without loss of generality, we can therefore choose the local index μ_i to be equal to the threshold:

$$\mu_i = u_i. \tag{7.4}$$

Thus, under the regional homogeneity hypothesis,

$$Y \sim \mathcal{G}_{PD}(1, \gamma, \xi),$$

where the scale and shape parameters (γ, ξ) satisfy

$$\begin{aligned} \gamma &= \frac{\sigma_i}{u_i}, \\ \xi_i &= \xi. \end{aligned}$$

Renormalizing by the local index, this implies that

$$X_i \sim \mathcal{G}_{PD}(u_i, \gamma u_i, \xi). \tag{7.5}$$

In particular, the shape parameter of the generalized Pareto distribution is constant for all sites of the homogeneous region, and the threshold u_i alone contains all local details for site i. Ways to determine this threshold will be discussed in Sect. 7.2.

The threshold-deviation-based approach. For the variable Z_i, the local index can take any form. In effect, the regional homogeneity hypothesis becomes

$$\begin{aligned} \forall x \geq 0, \quad \mathbb{P}\left(\frac{Z_i}{\mu_i} \leq x\right) &= \mathbb{P}\left(S_i \leq u_i + \mu_i x | S_i \geq u_i\right), \\ &= F_r(x). \end{aligned}$$

There are now no constraints on the value of μ_i since the support $I\!R^+$ of the variable Z_i/μ_i does not depend on i now. Thus, giving values to the local index is a (subjective) choice to be made. For instance, numerous studies define an empirical local index as the mean or median of the observed values of Z_i. The mean seems usually a reasonable choice that may avoid potential regional heterogeneity and the effects of the dependence between the sites [780, 781].

If we make the classical choice that $\mu_i = \mathbb{E}[Z_i] = \gamma_i/(1-\xi_i)$, the regional homogeneity hypothesis becomes, in terms of distributions,

$$Y \sim \mathcal{G}_{PD}(0, 1-\xi, \xi).$$

The only regional parameter is the shape ξ, shared by all sites. This fairly rigid regional distribution (even more so in the Gumbel/exponential setting with $\xi = 0$) is a real problem. Indeed, a good level of flexibility of the regional distribution is wished for representing correctly the most extreme values, and incorporating the effect of external covariates.

Therefore, it is generally better to adopt the threshold-exceeding-based approach rather than the second approach; doing so provides two regional parameters rather than one and avoids a subjective assessment.

7.1.4 Main Steps of Regional Analyses

Regional analysis, deeply studied in [780], involves the following steps.

1. *Sampling extreme values X_i at various sites.* In local analyses, temporal criteria are typically used to transform observed extreme values into independent events. However, in a regional approach, sampling extreme values require not only taking into account the temporal flow of events generating extreme values, but also their spatial structure. Section 7.2 focuses on this question and provides a method guaranteeing spatio-temporal independence of extreme values occurring in the study area.
2. *Grouping together sites into homogeneous regions.* How best to divide up the study area into homogeneous regions is still an open question. A method for finding a good partition, proposed in [780], involves identifying typical storm footprints generating extreme measurements and is detailed in Sect. 7.3.
3. *Grouping together observed values x_i normalized by the local index μ_i* in the regional sample (this is called *pooling*) using Eq. (7.1).
4. *Calculating the effective regional duration.* After grouping together the extreme values from different sites in the same region as a new sample, we obtain a new observed time series over an "effective" regional time period. The value of this effective duration is a function of the level of spatial dependency found in the region. Section 7.3.3 deals with taking into account spatial dependency in order to deduce various quantities, such as the effective duration or return period of a storm.
5. *Estimating the regional distribution $\mathcal{G}_{PD}(1, \gamma, \xi)$* using the regional sample and standard parameter estimation techniques.
6. *Calculating the regional return level y_T* after extrapolating the previously estimated regional distribution.
7. *Calculating return levels $x_i^{(T)}$ at each site* after normalizing (7.3) by the local index.
8. *Comparing performance with local analyses.* Section 7.5 shows a comparison of local and regional analyses in terms of uncertainty in return level estimates, as well as in terms of capacity to model the occurrence of supposed outliers*.

7.2 Sampling Extreme Values

In this section, we present a sampling procedure for extreme values which can be used for performing regional analyses. In the classical time series framework, the declustering principle is based on retaining a unique maximum from a dependent group of values exceeding a threshold. The idea is to extract from a series of exceedances (POT approach) a subset of supposedly independent terms, so as to respect the theoretical underpinning of extreme value statistics (see Definition 6.4 and Theorem 6.2). In the spatio-temporal setting for regional analyses, this sampling procedure must be modified.

The sampling method we present can be seen as a regional extension of the traditional POT approach (at any given site). However, in the POT context, Bernardara et al. [87] recommend using a "double threshold" approach when processing environmental data, which is used in the remaining of this chapter. The idea is to separate the uncertainty on the physical mechanisms of a phenomenon from a purely statistical analysis of the extreme values:

(i) identify independent events, starting from an initial physical ("natural process") threshold, where the variable being studied is largely out of its typical range;
(ii) with these events, determine a second "statistical" threshold to use in estimating extreme values.

One of the notable advantages of this "double threshold" approach is the possibility of considering high statistical thresholds without altering the physically based part of extreme events, the first threshold having already identified it.

Though this is a general approach, it is easily understandable with the help of an example. We have therefore illustrated it throughout the text using a case study on marine storms (significant wave heights).

7.2.1 Spatio-Temporal Declustering

7.2.1.1 Constructing Physically Plausible Samples

In order to build physically plausible samples within the regional framework of a "double threshold" POT approach, we first define *storm* objects. The aim is to trace the spatio-temporal dynamics of environmental extreme values generated by large-scale meteorological phenomena, solely from available recorded data.

Here, *storms* are directly characterized in terms of a variable of interest (for example, significant wave heights or wind speed). In particular, we define them as meteorological phenomena generating extreme events (maritime in this case) in at least one site in the study area. The procedure used to detect them amounts to defining them as coherent spatio-temporal overshoots of a high-valued quantile. A similar approach for detecting storms from wind speed measurements has been proposed by [469, 561].

Let us note q_p the order p quantile (where p is close to 1) of the original observed time series at a given site. We suppose that a storm has hit the site if q_p is exceeded. The values q_p represent the physical thresholds defined in the "double threshold" POT approach. Storms propagate in space and time, and detecting them depends on a spatio-temporal *declustering*. The main idea is to suppose that neighboring extreme values in space and time were generated by the same storm. More precisely, two extreme values are spatio-temporal neighbors if the following conditions are both satisfied:

- the events happened less than Δ hours apart,
- the events were at sites within the η-nearest neighbors of each other.

Thus, three parameters are required to detect a storm: p (defining the impact), and (Δ, η) related to spatio-temporal propagation. A more detailed description of these can be found in [782].

- The choice of Δ, related to the propagation time of extreme values between neighboring sites, notably allows us to detect successive storms impacting the same region (as for example the storms Martin and Lothar, which occurred between 26 and 28 December 1999, or the series of storms which hit Brittany between December 2013 and March 2014),
- The choice of η, which depends in part on the distance between sites, allows us to distinguish between different storms leading to extreme values at different places.

The parameters (p, Δ, η) need to be chosen so as to best represent the dynamics of typical storms. For example, if p is too large and Δ (or η) too small, it is likely that the same extreme natural event will be incorrectly divided into two or more events. Conversely, if p is too small and Δ (or η) too large, two distinct events may be incorrectly combined as a single storm.

Furthermore, the values chosen for (p, Δ, η) may be quite different, depending on the actual variable being studied. Factors influencing the choice of values include propagation dynamics to other sites, spatio-temporal resolution of available data, and the presence of missing data. It is therefore recommended to perform a sensitivity study when choosing (p, Δ, η) to ensure, for example, that the dynamics of the most extreme events are correctly reproduced. Due to how they have been constructed, identified storms are considered independent of each other, both from spatial and temporal points of view, making the choice of parameters even more important. This step corresponds to identifying physically plausible storms.

7.2.1.2 Constructing Statistical Samples

The second step of the "double threshold" approach consists in building statistical samples which will be used only to examine statistical aspects of the extreme values. This involves looking at the strongest recorded storms; at each site, a new threshold u greater than the physical threshold q_p is then determined.

The value of u can be chosen so that λ storms per year occur on average at each site. From an annual point of view, at a site i, the statistical threshold u_i thus corresponds to on average λ storms. The value of λ needs to be determined in terms of the application in mind and is often the result of a bias-variance trade-off for best-estimating extreme quantiles.

Storms are then re-defined from the statistical point of view. If a site i was impacted by a storm, the storm is retained if and only if u_i was exceeded. This step corresponds to the identification of "statistical" storms.

7.2.2 *Application: Marine Storms*

7.2.2.1 Available Data

Throughout this chapter, various features of regional analysis methods are illustrated on an application involving extremely high waves. These measurements come from the ANEMOC-2 marine database described in [744] and constructed using retrospective simulation over the Atlantic ocean between 1979 and 2010. Significant wave heights were extracted for 1847 sites (Fig. 7.1), with a spatial resolution of approximately 120 km in the North Atlantic, 20 km on the European coast, with a more precise 10 km on the French coast. For each of the 1847 sites, the hourly time series of significantly high waves were extracted for the period 1979–2010.

7.2.2.2 Extracting Physical Storms

Here, marine storms are therefore described as natural events generating significant wave heights in at least one of the 1847 sites in Fig. 7.1. After various tests, the values

$$p = 0.995, \ \Delta = 2\,h, \ \eta = 10$$

appear to provide a satisfactory representation of the natural processes behind storms, where (Δ, η) are connected respectively to the propagation of extreme waves between neighboring sites and the grid's density in Fig. 7.1.

This set of parameters leads to the extraction of 5939 marine storms. The thresholds $q_{0.995}$ for detecting storms (99.5% quantiles of the hourly significant wave heights) are shown in Fig. 7.2. The lowest quantiles (located near the English Channel and North Sea coasts) are between 2 and 6 m, while the largest overall are around 14 m.

To illustrate the procedure for extracting marine storms with the chosen parameters, the propagation of the storm Daria from 24–26 January 1990 is shown in Fig. 7.3. Figure 7.5 shows several of the most well-known marine storms that occurred during the study period, in addition to Daria. In particular, the storms Lothar and Martin,

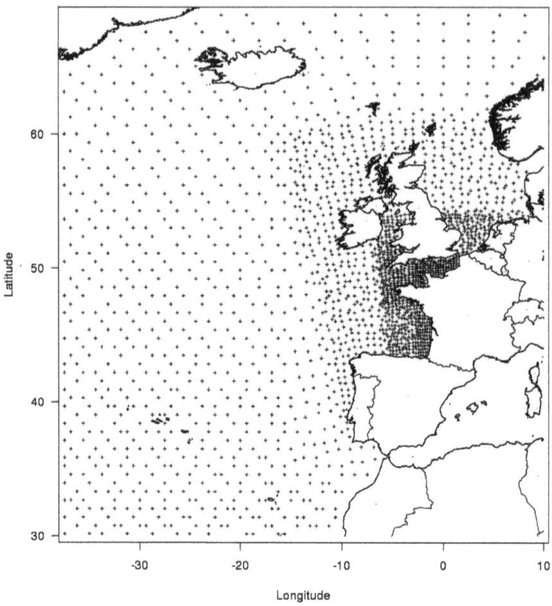

Fig. 7.1 Locations of the 1847 sites extracted from ANEMOC-2. Figure taken from [782] with editorial permission

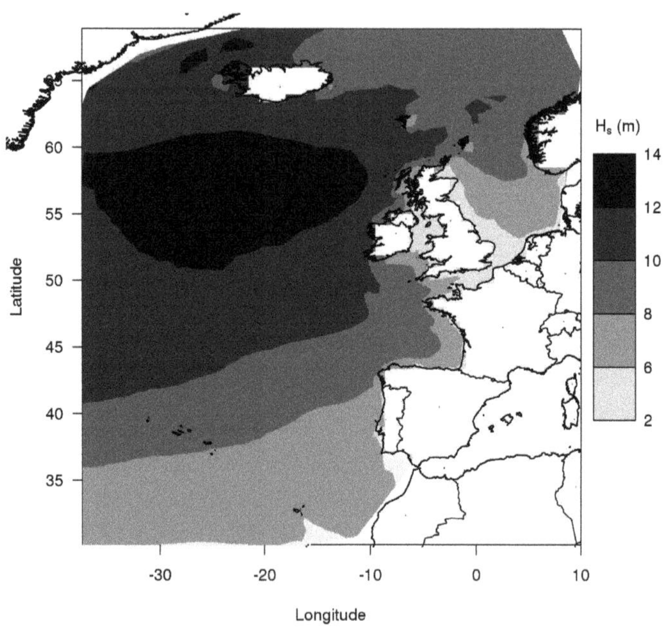

Fig. 7.2 Detection thresholds for marine storms: 0.995 quantiles for wave height time series (m). Figure taken from [782] with editorial permission

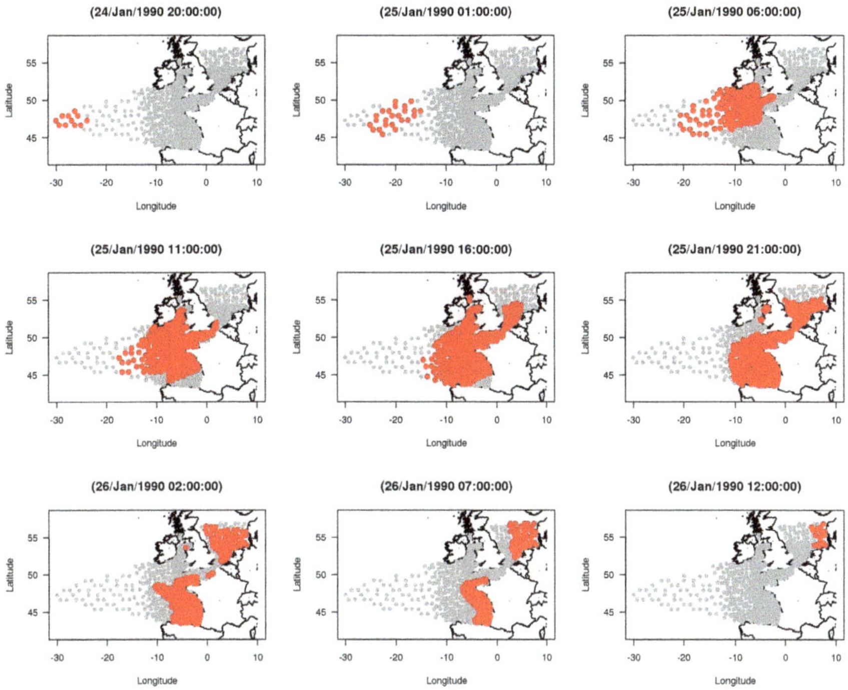

Fig. 7.3 Propagation of the storm Daria (24–26 January 1990) at 5 h intervals. The red points indicate affected sites (exceeding the 0.995 quantiles of the wave height time series) and the gray points the full footprint of the storm. Figure taken from [782] with editorial permission

which occurred in the same region and less than 36 h apart, are correctly separated into two distinct events (Fig. 7.4).

In addition, the exploratory study of 5939 marine storms extracted over the period 1979–2010 shows that:

- on average, 192 storms occurred per year in the study region,
- on average, each storm affected 38 of the 1847 sites,
- on average, storms lasted 12.5 h at any given site,
- storms were much more frequent in winter than in summer: 89% occurred from October to March, with 20% in the month of January alone.

7.2.2.3 Extracting Statistical Storms

Here as well, after various tests, the value $\lambda = 1$ was selected to extract statistical storms from among the physically plausible ones. This determined the statistical thresholds u_i, all of which are greater than the physical ones $q_{0.995}$, as well as the local indices given in Eq. (7.4). These thresholds are shown in Fig. 7.4, with values

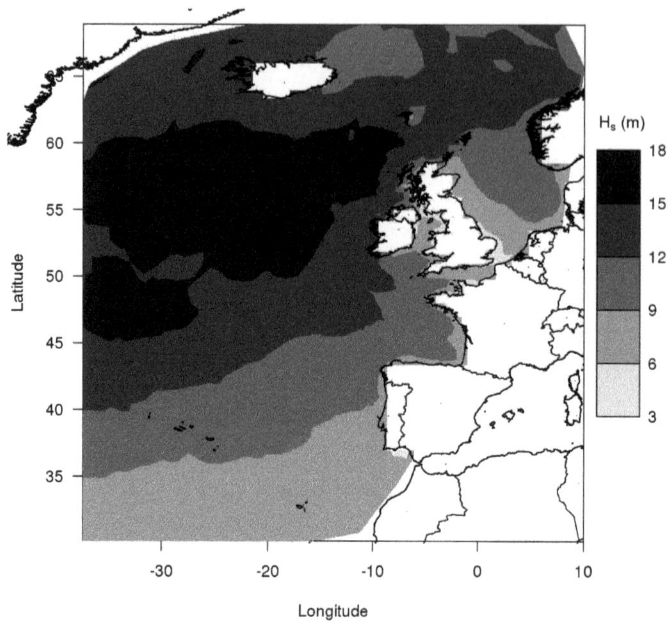

Fig. 7.4 Marine storms: statistical thresholds for wave heights (exceeded on average $\lambda = 1$ times per year) in meters. Figure taken from [783] with editorial permission

ranging from 3.26 to 16.84 m. From the 5939 physical storms, 1340 statistical storms were identified.

7.3 Selecting Homogeneous Regions

Equation (7.1) supposes that sites can be grouped into homogeneous regions, in each of which hazards at each site behave statistically similarly. This grouping of homogeneous regions must answer above all to a physical reality that validates this fundamental hypothesis of homogeneity. For instance, which zones share conditions (*covariates*) that can explain the extreme phenomenon? Does this physical hypothesis correspond to statistical homogeneity?

7.3.1 Identifying Physical Homogeneity

The identification of potentially homogeneous regions may be based on a physical criterion involving the storms extracted in the previous section. In particular, regions can be defined as the typical footprints of storms generating extreme events, which

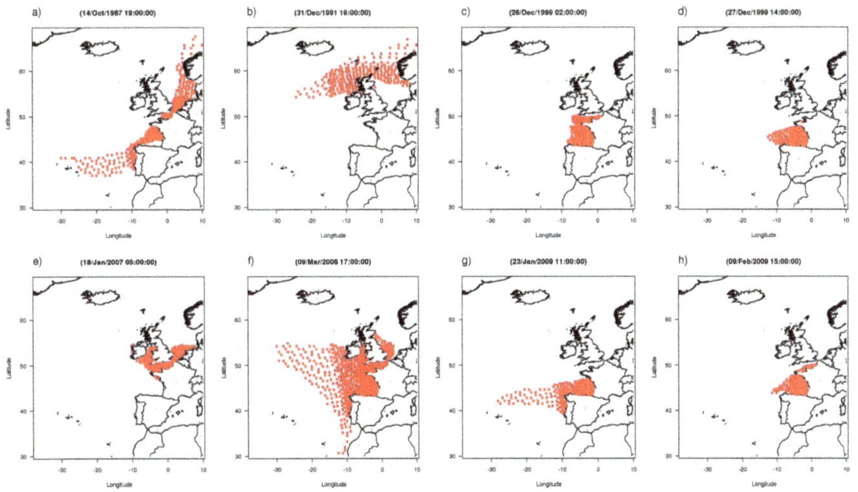

Fig. 7.5 Examples of marine storm footprints. **a** 15–16 October 1987, **b** New Year's Day Storm (31 December 1991—1 January 1992), **c** Lothar (26 December 1999), **d** Martin (27 December 1999), **e** Kyrill (17–23 January 2007), **f** Johanna (10 mars 2008), **g** Klaus (23–24 January 2009) and **h** Quentin (8–13 February 2009). Points represent affected sites

has the advantage of giving a physical interpretation to the results: sites in a given region will tend to be impacted by the same storm. It is therefore expected that extreme values observed in a region, coming from the same natural events, will behave similarly from a statistical point of view, and thus satisfy the homogeneity hypothesis. We now summarize the method, which is described in more detail in [782].

7.3.1.1 Detecting Typical Storm Footprints

The aim is to partition the set of sites so that each subset corresponds to a typical storm impact region. The use of classification procedures based on a storm propagation criterion is one way to achieve this.

For any two sites i and j in the study region, let p_{ij} be the probability that extreme values are produced during a storm at both sites, given that either i or j is affected. These probabilities can be empirically estimated for all pairs of sites (i, j), given a sample of previously-extracted storms. The storm propagation criterion is then defined as

$$d_{ij} = 1 - p_{ij}. \tag{7.6}$$

If $d_{ij} = 0$, then any storm impacting one of the sites also impacts the other. Conversely, if $d_{ij} = 1$, storms hitting one site never hit the other.

The next step involves grouping together sites into R disjoint regions, where the distance between sites is defined directly in terms of the storm propagation criterion (7.6). The resulting partition can then be seen as a set of storm footprints. In particular, ascending hierarchical clustering using Ward's method (or one of its extensions [731]), based on maximizing inter-class variance, can be used. The resulting hierarchy of regions can be shown as a *dendrogram* [122] (see Fig. 7.7) in which heights are defined in terms of inter-class distances which depend on the distance criterion (7.6) chosen.

Ascending hierarchical clustering.

1. Each site is initially considered a region to itself.
2. The "closest" regions to each other in terms of criterion (7.6) are combined to make a new region.
3. Stop when only one region remains (the whole study region).

Given the range of storm footprint configurations obtained from the dendrogram as R changes, the goal is to identify the "best" one. The optimal number of regions may be determined using some measure of clustering quality. A good partition should contain well-separated regions, with sites similar to each other found in each region. In particular, a *significant jump* in height as R changes indicates that two dissimilar sites have been joined together. Visual detection of this kind of jump can help choose the final number of regions to keep, as well as identify the most representative storm footprints. Sites grouped together as a region will tend to be impacted by the same storms (though not always), and a storm impacting a certain region will tend to stay within that region.

7.3.1.2 Application: Marine Storms

From the 5939 marine storms extracted from the period 1979–2010, the storm propagation criterion (7.6) was calculated for each pair of sites out of the 1847. Hierarchical clustering using Ward's method provides us with several possible storm footprint configurations as a function of the number of regions asked for (Fig. 7.6). It is interesting to note that a form of geographic contiguity between regrouped sites is obtained without having to fiddle with the algorithm in any way.

The dendrogram is shown in Fig. 7.7a. Its sequence of heights (Fig. 7.7b) suggests we cut the tree at $R = 5$, giving a subdivision of the study area into five smaller regions, corresponding to the most common storm footprints. These supposedly homogeneous regions are shown in Fig. 7.8: the South Atlantic (Region 1, 399 sites), North Atlantic (Region 2, 497 sites), North Sea (Region 3, 241 sites), English Channel and its approaches (Region 4, 392 sites), and the Bay of Biscay (Region 5, 336 sites).

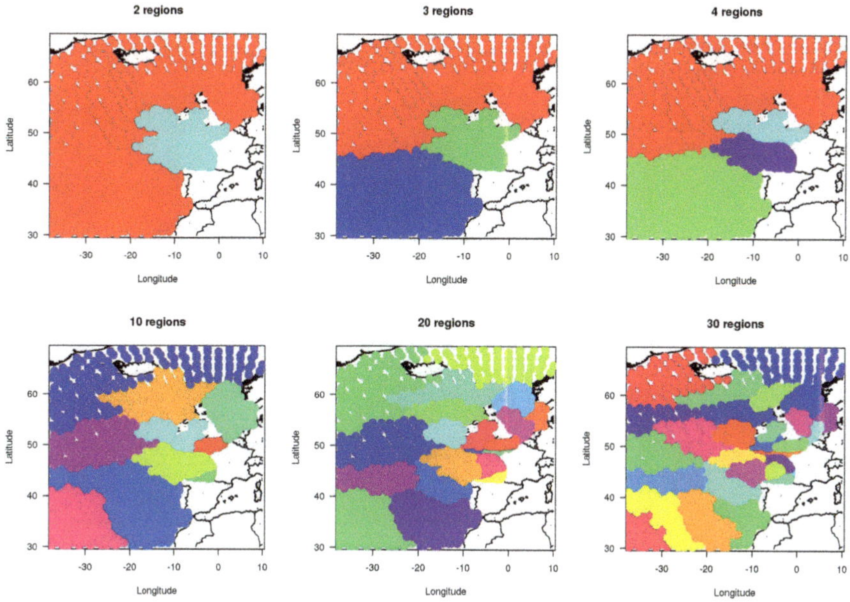

Fig. 7.6 Marine storms: Storm footprint configurations obtained for $R = 2, 3, 4, 10, 20$, and 30 regions. Figure taken from [782] with editorial permission

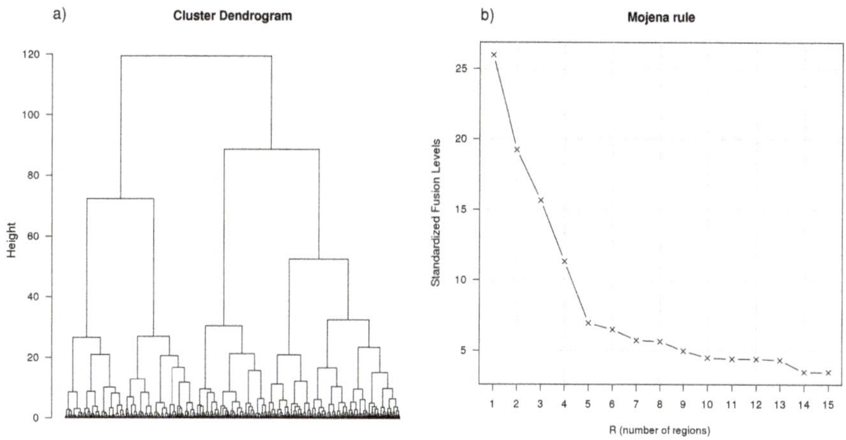

Fig. 7.7 **a** Dendrogram. **b** Evolution of dendrogram heights as a function of the number of regions. Figure taken from [782] with editorial permission

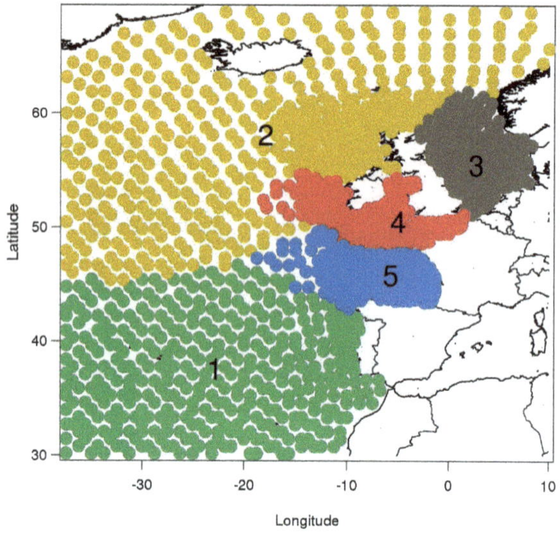

Fig. 7.8 Marine storms: partition of the study area into 5 regions corresponding to representative storm footprints. Figure taken from [782] with editorial permission

7.3.2 *Testing for Statistical Homogeneity*

The previous section showed how to construct regions that are homogeneous from a physical point of view; it only remains to now test their statistical homogeneity, i.e., whether the distribution of extreme values normalized by the local index is the same across the whole region.

7.3.2.1 Methods

One of the most often-used methods for testing a region's statistical homogeneity was proposed by Hosking et al. [396]. It is based on the principle that the ratios of $L-$moments characterizing the probability distributions should in theory be identical at each site in a homogeneous region.

Definition 7.1 (*$L-$moment*) Let X be a random variable. Its $L-$moment of order $r \in I\!N^*$ is defined by

$$\lambda_r = r^{-1} \sum_{k=0}^{r-1} (-1)^k \binom{r-1}{k} \mathbb{E}\left[X_{(r-k):r}\right],$$

where $X_{k:r}$ is the kth smallest order statistic of an independent r-sample of values of X.

According to [396], problematic sites within a region can be first detected using a *discordance* criterion D, which measures whether a given site is significantly different to the others in the same region in terms of $L-$moments. At site i, the criterion is defined by

$$D_i = \frac{1}{3} (\omega_i - \bar{\omega})^T S^{-1} (\omega_i - \bar{\omega}),$$

with $\bar{\omega} = N^{-1} \sum_{i=1}^{N} \omega_i$ and $S = \sum_{i=1}^{N} (\omega_i - \bar{\omega})^T (\omega_i - \bar{\omega})$, where $\omega_i = (\tau_i, \tau_{i,3}, \tau_{i,4})$ is the vector of the first three empirical $L-$moment ratios observed at site i:

$$\tau = \frac{\lambda_2}{\lambda_1},$$
$$\tau_3 = \frac{\lambda_3}{\lambda_2},$$
$$\tau_4 = \frac{\lambda_4}{\lambda_2}.$$

In particular, a site i is considered discordant if $D_i \geq 3$ for regions with more than 15 sites. If $4 < N < 15$, a table of critical values for D_i is provided in [396].

Following this, a region's *degree of statistical homogeneity* can be evaluated using a heterogeneity criterion H, which indicates whether the dispersion of $L-$moments observed across sites is comparable to that which would be expected in a statistically homogeneous region. This criterion is defined by

$$H = \frac{V - \mu_V}{\sigma_V},$$

where (V, μ_V, σ_V) corresponds to the empirical dispersion, mean, and standard deviation of the coefficients of variation (called $L-$CV) t_i at site i. As the regional $L-$CV is defined by

$$t = \frac{\sum_{i=1}^{N} n_i t_i}{\sum_{i=1}^{N} n_i},$$

where n_i is the sample size at site i, the empirical dispersion is given by

$$V = \sqrt{\frac{\sum_{i=1}^{N} n_i (t_i - t)^2}{\sum_{i=1}^{N} n_i}}.$$

The test itself (called the *Hosking–Wallis* test) is based on the fact that under the homogeneity hypothesis, the random variable H follows a 4-parameter Kappa distribution, whose cumulative distribution function is given by

$$F(x) = \left[1 - h\left\{1 - k\frac{x - \tilde{\mu}}{\alpha}\right\}^{1/k}\right]^{1/h},$$

where $\tilde{\mu}$ is a location parameter, α a scale parameter, and (k, h) shape parameters. This distribution, which contains the GEV ($h = 0$) and GPD ($h = 1$) ones as special cases, is sufficiently flexible to approximate most types of variables seen in environmental science [782]. According to the test, a region is considered homogeneous if $H < 1$, possibly heterogeneous if $1 \leq H \leq 2$, and definitely heterogeneous otherwise.

Thus, for each of the previously identified representative storm footprints, the following procedure is used, leading to regions that are homogeneous physically and statistically.

Homogeneous clustering procedure.

(a) Calculate the heterogeneity criterion H. If $H < 2$, go to **(d)**, otherwise go to **(b)**.

(b) Calculate the discordance criterion at each site.

- If no site is discordant, go to **(c)**;
- otherwise, remove discordant sites and again calculate the heterogeneity criterion H_0.
  - If $H_0 < 2$, go to **(d)**;
  - otherwise, go to **(c)**.

(c) Divide the region into two new sub-regions according to the hierarchical clustering dendrogram calculated previously. For each sub-region, go to **(a)**.

(d) The region is homogeneous both statistically and physically.

7.3.2.2 Application to Marine Storms

We first calculate the heterogeneity criterion H for each of the 5 regions. According to Table 7.1, Regions 1, 2 and 3 can be considered statistically homogeneous. However, statistical heterogeneity is present in Regions 4 and 5, so these need to be re-defined before proceeding with a regional analysis.

Looking closer at Region 4, we find that 15 of the 392 sites are discordant. Once these have been removed, the region becomes statistically homogeneous ($H = 0.68$).

Table 7.1 Marine storms: the heterogeneity criterion H for each of the five regions

Region 1	Region 2	Region 3	Region 4	Region 5
1.60	−0.07	−2.36	4.22	7.36

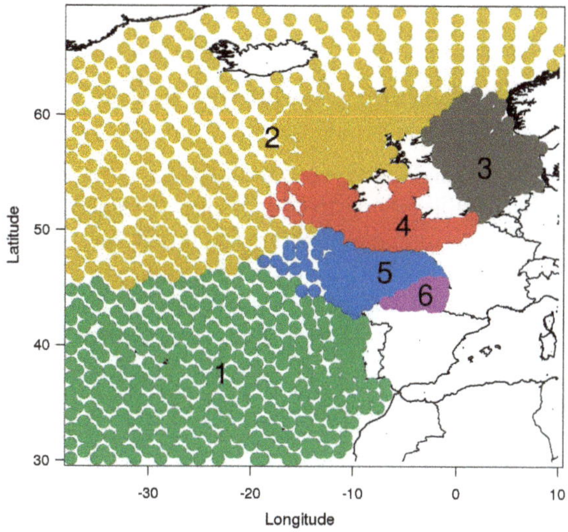

Fig. 7.9 Marine storms: dividing the study area into six homogeneous regions (both in the physical and statistical sense). Figure taken from [782] with editorial permission

The removed sites—which do not form a homogeneous region themselves—are found close to the coast, where local effects can significantly influence extreme wave heights. Thus, the English Channel region is restricted to the 377 non-discordant sites.

As for Region 5, removing the discordant sites does not improve its statistical homogeneity. We, therefore, divide it into two storm footprints with the help of the hierarchical clustering shown in the dendrogram: northern Bay of Biscay ($H = -0.45$, 234 sites) and southern Bay of Biscay ($H = -4.69$, 102 sites). These two regions are now statistically homogeneous. In particular, strong statistical asymmetry can be seen in the northern compared to the southern region.

Thus, we end up with six homogeneous (both in the physical and statistical sense) regions, with which we can now estimate extreme value quantiles using regional analysis. The regions are shown in Fig. 7.9.

7.3.3 Accounting for Spatial Dependency

Spatial dependency can be modeled by the choice of a correlation structure within a statistical model. This choice can be guided by the use of non-parametric tools like *correlograms*, which are autocorrelation diagrams; see Sect. 5.4.1.1. Here, we build on the work in [783], where a model for characterizing spatial dependency between storms was proposed. Based on the tendency of sites to act similarly during a storm, dependency is characterized using a function involving both the propagation of storms and their regional strength. This model helps estimate the effective duration of a regional sample (and thus predict the merit of extrapolation) and compare local and regional return periods of a given storm.

7.3.4 Modeling Dependency Between Sites

The homogeneous regions constructed earlier, corresponding to typical storm footprints, exhibit by construction strong dependency between their sites. Therefore, it is necessary to find a way to model this at the regional level in a way relevant to the *pooling* strategy.

Weiss et al. [783] have proposed a model that captures spatial dependency during storms by linking the probability distribution of the regional maximum with the regional distribution F_r. Let M_s be the regional maximum occurring during storm s:

$$M_s = \bigvee_{i=1,\ldots,N} Y_s^i. \tag{7.7}$$

For $x \geq 1$, the distribution M_s is given by

$$\mathbb{P}(M_s > x) = (1 - F_r(x))\frac{\lambda}{\lambda_r}\varphi(x), \tag{7.8}$$

where λ and λ_r represent respectively the mean number of storms per year at each site and in the region, and φ the regional dependency function:

$$\varphi(x) = 1 + \sum_{i=1}^{N-1} P\left(\bigvee_{j=i+1,\ldots,N} Y_s^j | Y_s^i > x\right), \quad \text{for } x \geq 1. \tag{7.9}$$

This model expresses the fact that a region's sites tend to act similarly during a storm that hits it. In particular, dependency between sites is encoded by the function φ, which simultaneously characterizes a storm's propagation and regional strength.

The values of φ are theoretically between 1 (total dependency) and N (total independence). The lower the value of φ, the greater the regional dependency, indicating that most sites are affected by each storm and tend to behave similarly in terms of generated extreme values. In order to compare φ across regions, the effect of a region's size N can be removed with the help of the dimensionless function

$$\phi = \frac{N - \varphi}{N - 1}$$

with a value between 0 (total independence) and 1 (total dependency). Weiss et al. [783] show that under the hypothesis that the distribution of the regional maximum M_s is identical to that of the regional distribution F_r, the function φ simplifies to a constant:

$$\varphi(x) = \frac{\lambda_r}{\lambda}. \tag{7.10}$$

Therefore, under this hypothesis, each storm hitting the region generates extreme values with the same (normalized) distribution as the maximum value observed during the event. Though in practice this hypothesis can usually be accepted, it should really be checked using statistical tests. In particular, it implies that the way in which sites behave does not depend on the storm's intensity and that there is no dominant storm trajectory across the region: any observed storm can potentially affect any site within it. This is coherent with the way we form homogeneous regions by identifying typical storm footprints.

This regional dependency model can then help us to perform certain tasks, like estimating the effective regional duration and establishing a link between local and regional return periods of a storm. Moreover, the regional sample used to estimate the regional distribution is constructed according to this model hypotheses.

7.4 Local and Regional Diagnosis

7.4.1 Constructing Regional Samples

The *pooling* method consists in grouping together the normalized observed values at sites across the region into the same sample, before estimating the parameters of the regional distribution from it. However, the totality of these observed values should not be directly grouped into the same sample, due to a certain level of redundant information caused by spatial dependency.

Storms—containing spatial footprints of extreme values from the same event—can nevertheless help us to filter out this dependency. To obtain independent data, only the normalized regional maximum of each storm can be retained. This approach echoes the simplification hypothesis in the regional dependency model (7.10). Bernardara et al. [86] used a similar filtering method, keeping only the maximum from the set of extreme values observed over a 72 h period in a given region. Thus, if n_r storms are observed in a region, the regional sample will be made up of n_r normalized regional maxima.

7.4.2 Effective Regional Duration

As well as constructing the regional sample using pooling, we need to know its effective duration, i.e., the duration of observed values after having grouped together extreme values from several sites. This effective regional duration D essentially refers to the fact that on average, λ storms occur per year over a period of D years, where λ is the mean number of storms per year at each site of the region. This quantity can then be used to help calculate empirical regional return periods, as well as quantify the gains in using a regional rather than local analysis.

Calculation of this effective duration is closely linked to the question of spatial dependency from Sect. 7.3.4: the stronger the dependency, the shorter the effective duration.

(i) If all sites are completely independent of each other, then a given storm will only affect one of the region's sites, no matter how strong it is. In this case, each observed value brings new information, and D can be given as the sum of local observed durations d_i across all sites i:

$$D = N\bar{d} = \sum_{i=1}^{N} d_i.$$

(ii) At the other extreme, if all sites are entirely dependent on each other, then a given storm will affect all of them. In this case, D corresponds to the local observed duration at one site only, as the information coming from the others is purely redundant. Thus, we have that

$$D = \bar{d} = \frac{1}{N}\sum_{i=1}^{N} d_i.$$

Intuitively, the likely value of D is found somewhere between these two extremes. Weiss et al. [783] have therefore proposed the more general form:

$$D = \varphi\bar{d}, \tag{7.11}$$

where φ represents the degree of regional dependency as defined in Sect. 7.3.4. This way of writing D covers the total dependency ($\varphi = 1$) and total independence ($\varphi = N$) cases too. Under the hypothesis of Equation (7.10), D can therefore be rewritten as

$$D = \frac{\lambda_r}{\lambda}\bar{d}. \tag{7.12}$$

A natural estimate of the mean number of storms λ_r per year in the region is given by $n_r\bar{d}$, n_r being the number of recorded storms there. Thus, the effective duration

D can be estimated by

$$\hat{D} = \frac{n_r}{\lambda}.$$

7.4.3 Local and Regional Return Periods

Regional analysis provides a framework for understanding risk at different spatial scales. In particular, the spatial dependency model described earlier can help to distinguish between a storm's regional and local return periods. For example, a storm with a 100-year return period in a given region will likely not lead to extreme values with a 100-year return period at each of the region's sites. We now take a closer look at the link between these two return periods.

Let s be a storm and x its maximal (normalized) intensity in a region. The regional return period of s, denoted T_r, is defined as the mean time between two storms affecting *at least one site* in the region with a normalized intensity greater than x:

$$T_r = \frac{1}{\lambda_r \mathbb{P}(M_s > x)}. \tag{7.13}$$

The local return period of s, denoted T, is defined as the mean length of time between two storms impacting *any given site* in the region with an intensity greater than x:

$$T = \frac{1}{\lambda\,(1 - F_r(x))}. \tag{7.14}$$

From Eq. (7.8), T_r and T are connected in the following way:

$$T_r = \frac{T}{\varphi(x)}. \tag{7.15}$$

Furthermore, under the simplification hypothesis for the regional dependency model, we have that

$$T_r = \frac{\lambda}{\lambda_r} T. \tag{7.16}$$

7.4.4 Application to Marine Storms

For each of the six homogeneous regions, a regional sample is obtained by pooling the normalized regional maxima of each storm. Anderson–Darling tests [723] are run to validate the simplification hypothesis for the regional dependency model, and

Table 7.2 Marine storms: indices showing spatial dependency in each region. N is the number of sites in each region, λ_r the mean number of storms per year per region, ϕ the dimensionless regional dependency function, D the estimated effective regional duration (in years, with 95% confidence intervals), and 100_r the regional return period (in years) of storms generating an extreme event with a 100-year return period in at least one of the sites of the region

	Region 1	Region 2	Region 3	Region 4	Region 5	Region 6
N	399	479	241	377	234	102
λ_r	10.3	25.2	9.0	8.8	4.1	2.7
ϕ	0.98	0.95	0.97	0.98	0.99	0.98
D	320 [284maxima356]	782 [698maxima 866]	279 [234maxima 324]	274 [232maxima 316]	128 [106maxima 150]	84 [68maxima 100]
100_r	9.7	4.0	11.1	11.3	24.2	36.9

show that the distribution of the maxima could be assimilated to that of the standard regional distribution.

Table 7.2 provides values summarizing the spatial dependency within each region. Storms appear to be most frequent in Region 2 and least frequent in Region 6, with respective occurrences of 25.2 and 2.7 storms per year on average. These quite different levels of storm activity were strongly influenced by the sizes of these regions, which were respectively the largest and smallest. The dimensionless dependency value can instead be used to compare the regions, irrespective of the influence of size. Thus, spatial dependency is largest in Region 5, meaning that the sites of this region have a strong tendency to behave similarly during storms. Indeed, on average, 57 of the 234 sites (24%) were impacted by each storm that hit the region. In contrast, the lowest level of spatial dependency was found in Region 2. There, on average, 19 of the 479 sites (4%) were impacted by each storm that hit the region.

As for the effective regional duration, grouping together, for example, the observed values at the 234 sites of Region 5, led to a regional sample whose effective duration was estimated at 128 years, with a 95% confidence interval of [106 yr, 150 yr]. It is of interest to note that taking into account spatial dependency considerably reduced the effective duration that would have been found under the independent sites hypothesis (here, $D = 254 \times 31 = 7\,254$ years).

Furthermore, the quantity 100_r corresponds to the regional return period (in years) of a storm generating an extreme value with a local return period of 100 years in at least one of a region's sites (Eq. (7.16) with $T = 100$). For example, $100_r = 4$ in region 2, which means that on average, every four years, a storm in this region generates an extreme value with a return period of 100 years in at least one site. Also, while Region 1 is much larger than Region 3, their regional return periods are similar (around 10 years). Under the hypothesis of independence between sites, these quantities would have been much larger proportionally in Region 3 than in Region 1. However, this is not the case here due to a higher degree of spatial dependency in region 1. Taking this into account, therefore, leads to a more realistic evaluation of regional risks.

Table 7.3 Marine storms: empirical return periods of the strongest storms in each region at the local and regional scales

Region	Storm	Empirical regional return period (yr)	Empirical regional return period (yr)
1	15–18 Feb. 1986	31	321
2	11–13 Feb. 1979	31	783
3	11–13 Dec. 1990	31	280
4	8–10 Feb. 1988	31	275
5	15–17 Dec. 1989	31	129
6	23–24 Jan. 2009	31	85

For each region, we have estimated the return periods of the most powerful storms at the local and regional scales. The results are shown in Table 7.3. For example, in Region 3, as the normalized strength of the storm of 11–13 December 1990 has never been beaten over the 31 year period, the regional return period is naturally estimated at 31 years. Applying Eq. (7.16), the local return period is estimated at 280 years.

7.5 Comparing the Performance of Methods

Let us look at what we gain from using the regional approach instead of the local one. This comparison is run both in terms of uncertainty affecting estimated return levels, and in how outliers are taken into account, i.e., sample values that are significantly different from the others. Essentially, this means looking at the ability of the two approaches to detect the strongest storms.

7.5.1 Dealing with Uncertainty in Estimated Return Levels

Several major authors, such as [720], have shown that regional analyses lead to more reliable estimation of extreme quantiles than local ones. Indeed, while estimations based on the latter are generally unbiased, they may suffer from high variance [397]. Therefore, our comparison between the two approaches will be based on the variance of return level estimates, rather than their corresponding pointwise ones.

The method used here consists in comparing the confidence intervals of the following estimates:

- Local return levels estimated using regional analyses: in each homogeneous region, we fit a GPD distribution using the regional sample. Normalizing by local indices then brings us back to the local scale.

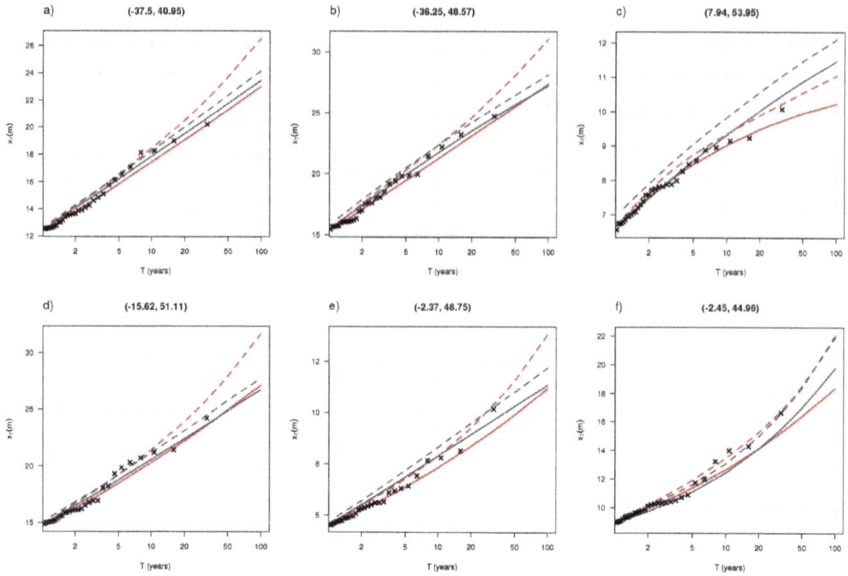

Fig. 7.10 Marine storms: comparison of estimated return levels using regional (gray) and local (red) analyses, with confidence interval upper bounds set at 95% (dotted line) for the 6 specified geographic locations. Crosses represent observed extreme values

- Local return levels estimated at each site using local analysis: at each site, we fit a GPD distribution using local threshold exceedances.

Confidence intervals in both cases are calculated using parametric bootstrap [244]. We then apply this method comparing the two types of analysis to our marine storm case study.

Figure 7.10 shows the return levels for significant wave heights estimated by regional and local analysis at 6 sites, one from each of the homogeneous regions. We see that the confidence intervals for local analysis are generally wider than those for regional analysis.

Figure 7.11 shows the relative evolution (in %) of the width of the 95% confidence intervals for centennial wave heights (100 year return levels), performing first a local, then regional analysis. The darker the color, the better the improvement due to regional analysis. Thus, lighter colored regions are those where regional analysis was less precise than local analysis. These relative values go from +196% (near the Orkney Islands in the North Sea) down to −91% (near the western tip of Norway in the North Sea), with a median value of −56%. Lighter colored regions are in the minority, indicating that regional analysis generally leads to a decrease in uncertainty.

This decrease in uncertainty due to regional analysis is illustrated by the concept of effective regional duration D (see Sect. 7.4.2). Indeed, as local analysis is based on observation durations $d_i < D$, uncertainty in the extrapolation of extreme quantiles is reduced with regional analysis.

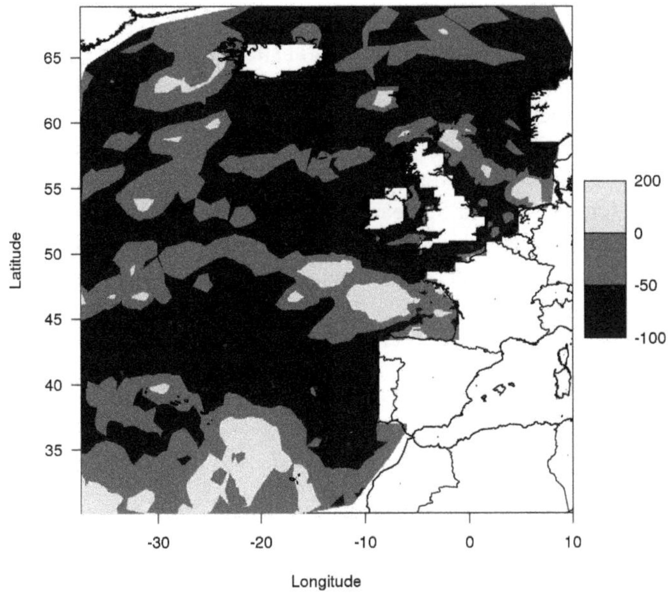

Fig. 7.11 Marine storms: relative evolution (in %) of the width of 95% confidence intervals of estimated centennial wave heights, first by local, then regional analysis

7.5.2 *Taking Outliers into Account*

By its nature, regional analysis helps to reduce the apparent rarity of storms seen at a local scale, and to better model their likelihood of occurrence.

Take for example the marine storm context. Suppose that we have recorded at a given site in Region 1 (South Atlantic) a millennial wave height (1 000 year return level). The probability of having observed this height at this site during the period of recorded measurements (31 years) is $1 - 0.99931^{31} = 3\%$. Now consider this storm at the regional level. As the effective duration of observations in Region 1 was estimated at 320 years (see Table 7.2), the probability of having observed a millennial height over a 320-year sample is now $1 - 0.999320^{320} = 27\%$.

Furthermore, the storm of 15–18 February 1986 did generate, in this region, the highest normalized wave height. Therefore, while a local analysis would only show that such a storm was observed once in 31 years (duration of observations made at each of the region's sites), taking into account regional information leads to a re-evaluation of this empirical frequency as once in 320 years (effective observation duration of Region 1), and this for the whole region. Indeed, the underlying principle of the regional approach used here is that there is no preferential storm path within a region; each observed storm could have occurred anywhere else in the region, consistent with the formation of homogeneous regions based on the identification of representative storm paths.

The storm Klaus of 23–24 January 2009 generated, in Region 6 (southern Bay of Biscay), high wave height outliers at several sites, showing the extreme nature of the storm. At a given site in the region, the wave height associated with Klaus has been estimated to reoccur on average every 85 years (empirically, i.e., based on the effective regional duration) or 157 years (theoretically, based on an estimate of the regional GPD distribution). However, a similar strength storm is estimated to reoccur every 31 years (empirically) or 58 years (theoretically) in this region.

These simple examples show that an extreme event at a given site (which could even be categorized as an outlier) may no longer be considered extreme at the regional scale. Looking at storms from a regional point of view can therefore potentially reduce their "exceptional" nature when observed locally.

7.6 Conclusions

The regional analysis approach developed in this chapter is based on distinguishing between:

(i) regional effects (using a marine storm case study, where storm strengths and spatial footprints were observed at the synoptic scale);
(ii) purely local effects (via a local index increasing or reducing the regional impact of storms).

This supposed separation of effects can help to understand processes generating observed extreme values, while simultaneously reducing complexity. Several possible paths can be considered to improve this regional approach—some of which are described below.

7.6.1 Formation of Homogeneous Regions

In this chapter, homogeneous regions are obtained using representative storm footprints. However, the use of covariates involving natural processes (e.g., wind strength and direction, atmospheric pressure, bathymetry, etc.) combined with multivariate analysis methods (such as principal components analysis (PCA) [467]) could either help improve the current method or provide an alternative for constructing homogeneous regions, where sites are grouped together based on their similarity with respect to the covariates.

The formation of homogeneous regions could also benefit from an approach based on weather patterns, i.e., distinct and typical synoptic situations. For example, Garavaglia et al. [303] have identified weather patterns in France so as to relate rainfall events with the processes generating them. Each observed extreme value is then linked with a corresponding weather pattern, which can notably help homogenize samples. A similar approach could be developed to construct homogeneous regions in a regional analysis study. For instance,

(i) each site could be associated with the weather pattern that most frequently leads to extreme values at this site;
(ii) sites could then be clustered based on these types of dominant weather pattern.

It would also be possible to define homogeneous regions with the help of correlograms specifically dedicated to extreme values. These are called *extremograms* [201]: pairs of sites are considered to be part of the same region if their observed extreme values are at least partially correlated.

7.6.2 Residual Heterogeneity

It is possible that the Hosking–Wallis test of statistical homogeneity [397] fails to detect a regional heterogeneity. This is unfortunate because a source of heterogeneity—even minor—can have a significant effect on the final results. One way to deal with this is to include the "traces" of heterogeneity that the test was unable to detect. The bootstrap technique for calculating uncertainty could, for example, integrate these suspicions of heterogeneity: each simulated bootstrap region could be slightly heterogeneous, and the whole uncertainty would then include uncertainty due to undetected heterogeneity. This remains difficult to do in practice since heterogeneity can come in many forms, and there are an infinite number of ways to artificially simulate heterogeneous regions [780].

7.6.3 Historical Regional Analysis

Uncertainties associated with estimating extreme-valued events have pushed the development of methods looking to increase the sample size, like regional analysis does. Taking advantage of information available on historical severe events is thus an alternative approach to constructing regional samples. The main hypothesis for including historical data in the statistical model is that there exists some threshold above which all exceedances were recorded during the historical period of interest. In a local analysis of extreme rainfall, Payrastre et al. [602] provided a method to include historical information in the likelihood function. Also, Gaume et al. [322] and Nguyen et al. [557] developed a regional Bayesian model that can take into account severe rainfall occurring at sites where measurements were not systematically recorded (see Chap. 11).

However, collecting historical data is usually a complicated task and the resulting dataset is often imprecise. Subsequent statistical analyses must therefore take this into account so as not to introduce bias into estimates. Although tricky, the extension of regional analysis to take into account historical information is a promising avenue for further reducing uncertainty in extreme value estimates.

7.6.4 Non-stationary Regional Analysis

In this chapter, the question of non-stationarity (temporal trends due to climate change and/or the effect of physical covariates influencing regional distributions) has not been studied. With regards to the extreme values statistics for storms presented in this chapter, non-stationarity is not really a problem, as many studies indicate that the overall storm situation in Europe has not shown any significant trends (see the MICORE[2] project [272] and the Preface of this book for further details).

Nevertheless, in the case of proven or hypothesized non-stationarity, Hanel et al. [378] and Roth et al. [669] have, for example, introduced a non-stationary regional model allowing temporal trends to be incorporated into the regional distribution's parameters. Techniques for dealing with extreme values of non-stationary time series are presented in the following chapter.

7.6.5 Multivariate Regional Analysis

This chapter has only looked at the univariate case. However, extreme environmental conditions are often characterized by different combinations of several interdependent variables (e.g., wind and extreme wave heights), so simultaneously taking into account all of these variables could certainly help to better characterize the risk of extreme events. In this context, Chebana et al. [158] have extended the regional index flood model to the multivariate case using copulas. Methods and tools for dealing with multivariate extreme values are presented in Chap. 9.

[2] http://www.micore.eu.

Chapter 8
Extreme Values of Non-stationary Time Series

Sylvie Parey and Thi-Thu-Huong Hoang

Abstract This chapter considers the problem of quantifying extreme natural hazards in non-stationary situations, namely outside the traditional framework. For certain hazards, this framework makes it possible to take into account the influence of climate change. Two main approaches are considered: the first deals with the incorporation and selection of trends in the parameters of the laws of extremes, while the second, nonparametric, considers trend variations over the first two moments of the extreme distribution.

One of the underlying hypotheses of the classical theory of extreme values is that variables are identically distributed or *stationary*, i.e., the distribution is the same for any subsample of the original sample. This hypothesis is rarely satisfied by natural phenomena, if only because the latter are often seasonal (see Chap. 5). In general however, by restricting ourselves to the season in which events of interest occur, it is usually possible to obtain a stationary setting. However, in cases where a phenomenon evolves over time, it is necessary to use other techniques to take non-stationarity into account. These are the subject of the present chapter, and are introduced in the univariate setting in Chap. 6.

Remember nevertheless that taking into account detected non-stationarity is not necessarily easy in practice. Indeed, this is an ongoing subject of discussion, especially in hydrology. Montanari and Koutsoyiannis [521] and Serinaldi and Kilsby [698] consider that it has not been proven that doing so brings real improvement. Their main arguments are the following:

- Estimating a non-stationary model implies estimating more parameters, which necessarily increases incertitude. It seems all the more necessary to compare the confidence intervals and not simply the point estimates obtained.
- The evolution in variables whose extreme value behavior we are trying to estimate is generally poorly understood (projected values of climate parameters are still not terribly reliable for certain variables, e.g., rainfall [779]) or even unpre-

S. Parey (✉) · T.-T.-H. Hoang
EDF R & D, Boulevard Gaspard Monge, 91120 Palaiseau, France
e-mail: sylvie.parey@edf.fr

N. Bousquet and P. Bernardara (eds.), *Extreme Value Theory with Applications to Natural Hazards*, https://doi.org/10.1007/978-3-030-74942-2_8

dictable. These cannot be inferred from observations alone, and the extrapolation of a deterministic trend over long time periods is not a particular good idea unless the trend represents a stationary exogenous process—in which case it would need to be determined.

These authors consider that stationarity should remain the default hypothesis, even in cases of proven non-stationarity, and that in any case, it is always necessary to compare confidence intervals obtained with or without the stationarity hypothesis.

This chapter first presents two of the main methods for modeling non-stationarity, either by adding trends to the parameter values (which thus become functions) of extreme value distributions, or nonparametrically using a *mean-variance trend* approach. Note that any non-stationary modeling choice requires the notion of *return level* to be redefined, because as it stands, it a priori supposes a certain regularity in phenomena. Besides, other risk indicators could be more adapted. A section is dedicated to this question in this chapter. Lastly, we study several meteorological hazards in order to illustrate the two approaches.

8.1 Trends in Extreme Value Distribution Parameters

The first approach—proposed among others by Coles [173]—consists in supposing that the parameters of extreme value distributions are not constant over time and depend on some covariate. The most natural covariate here is the time t, but others may be considered, such as for instance the concentration of greenhouse gases in the atmosphere, or an index related to variation in the climate system, e.g., the North Atlantic Oscillation (NAO). This approach, supported by numerous examples and asymptotic results [173], can be formalized under the MAXB approach through the modeling hypothesis:

$$X_t \sim \mathcal{G}_{EV}\left(\mu(t), \sigma(t), \xi(t)\right), \tag{8.1}$$

where $(X_t)_t$ is a sequence of independent but not identically distributed maxima. The estimation, selection, and verification of the well-foundedness of the simple parametric function for each parameter, e.g., linear as in:

$$\mu(t) = \mu_0 + \mu_1 \cdot t,$$

is the main task of the modeler. This needs to be a compromise between increasing estimation complexity (due to an increasing number of parameters) and a realistic representation of the non-stationary behavior of the sequence of observed extreme values. The first step in this work is to limit the number of dimensions by asking whether it is necessary to make all three parameters non-stationary, or whether non-stationarity is sufficiently well modeled when only concentrating on one (or two) of the parameters. A variety of techniques can be used to perform an exploratory analysis of the temporal evolution of the parameters and indicate the form of the evolution if it exists.

The choice of such forms means we can then conduct classical parametric estimation (e.g., maximum likelihood) and run statistical tests that work in the functional context in order to obtain one (or several) pertinent models.

Example 8.1

In the GEV case (Eq. (8.1)), once pertinent forms have been chosen for $\psi(t) = (\mu(t), \sigma(t), \xi(t))$, so that for each observation x at time t_i the following condition holds:

$$\xi(t_i)\left(\frac{x_{t_i} - \mu(t_i)}{\sigma(t_i)}\right) + 1 \geq 0,$$

then under the independence hypothesis on the data, the statistical likelihood of the problem:

$$L(x|\psi(t_{1:n})) = \prod_{i=1}^{n} f\left(x_{t_i}|\mu(t_i), \sigma(t_i), \xi(t_i)\right) \tag{8.2}$$

is well defined and can be maximized using numerical [173] or Bayesian (see Chap. 11) approaches.

8.1.1 Detecting and Modeling Trends

8.1.1.1 Semiparametric or Nonparametric Trend Estimation

Suppose we have n extreme values denoted $x_t = (x_i)_{i=1,\ldots,n}$, renormalized so that $t_i = i/n$. Detecting trends in the classical parameters of the distribution of the associated random variable X_t is a task that can be conducted using various approaches. Detection of a trend supposes that the question of what form it might take has been asked. For the most general setting, it is preferable not to choose the form a priori and let the data reveal the trend to us (under a few regularity hypotheses, the most usual being the existence of the second derivative). Semiparametric and nonparametric approaches can be used to do this. In the following, we describe two of the main ones. Previously, adapting the approach of preliminary independence tests provided in Table 5.1, we provide on Table 8.1 a summary of preliminary stationarity tests, that are selected by their good power for moderate sizes. They can appropriately accompany the tests described hereafter.

Table 8.1 Selected nonparametric tests of stationarity of univariate time series for samples of moderate size

Test	**Kwiatkowski-Phillips-Schmidt-Shin** (KKPS)	**Augmented Dickey-Fuller** (ADF)	**Philipps-Perron** (PP)
Null hypoth. H_0	stationarity (with null mean or constant bias)	non-stationarity (with or without trend)	non-stationarity (with or without trend)
Alternative hypoth. H_1	non-stationarity	stationarity	stationarity
Minimal sample size *(in practice)*	$n \geq 30$	$n \geq 30$	$n \geq 20$
References	[190]	[297]	[614]

1. The first is a semiparametric approach called *local models* (LM) [202, 371, 635] based on a weighted polynomial representation of the given extreme value distribution. This approach starts from the hypothesis that the data (x, t) form a point process which is *marked* (or *colored*, cf. Definition 6.11) by the **temporal index**, whose associated counting process (with n values) is itself random—possibly Poissonian [371]. The approach consists in selecting iteratively a set of observations containing information on the distribution at time t via a *kernel* $\omega(.)$ which is bounded, typically on $[-1, 1]$ [202] (cf. Sect. 4.3.2). At each time t, the corresponding extreme value is then associated with a vector of weights $(\omega_i(t))_{i=1,\ldots,n}$ defined by

$$\omega_i(t) = \omega\left\{(i/n - t)/h\right\},$$

where h is a smoothing (*bandwidth*) parameter. A weight $\omega_i(t)$ is 0 when i is outside the interval $[n(t-h), n(t+h)]$. The combination of this weights vector and a set of polynomials with statistical likelihood $\exp(\ell(x_i|\psi))$ at x_i then allows a global statistical log-likelihood to be defined at time t [263]:

$$\ell(t|\psi) = \sum_{i=1}^{n} \omega_i(t)\ell(x_i|\psi). \tag{8.3}$$

Maximizing (8.3) with respect to ψ then allows a *local estimator* $\hat{\psi}_t$ to be constructed. It is straightforward to modify this approach to associate ψ and the parameters of a GEV distribution (for example), then interpolate the functions obtained for $\hat{\mu}_t, \hat{\sigma}_t$ and $\hat{\xi}_t$ in terms of t, to finally deduce whether a trend is present or not, and the type of trend if present. However, the choice of bandwidth

h, polynomial degree, and kernel[1] will all influence the results. The fact that in general, the behavior of extreme values changes slowly over time, often encourages the choice of low-dimensional polynomials—often with degree 1.

2. The second method is a nonparametric *penalized maximum likelihood* (PML) approach [155, 156, 600, 668], which supposes, very generally, that in each dimension j ($\in \{1, 2, 3\}$ for classical GEV and GPD models), the jth component ψ_j of the parameter vector for an extreme value distribution can be functionally estimated using a *generalized additive model*

$$\psi_j(t) = g_j\left(\mathbf{y}^T \eta_j + h_j(t)\right),$$

where g_j is a link function, η_j a parameter vector, $\mathbf{y}$ a covariate vector and h_j a nonparametric smoothing parameter whose support is limited to a compact set $\mathscr{A}$ containing all of the observed values of t (typically, h_j is a *spline* [360]). The estimation of each function $t \to \psi_j(t)$ can then be obtained by maximizing the penalized log-likelihood [156] (defined here by pursuing Example (8.2)):

$$\log L(x|\psi(t_{1:n})) - \sum_{i=1}^{n}\left[\gamma_i \int_{\mathscr{A}} (h_i''(t))^2 \, dt\right]. \tag{8.4}$$

The introduction of the penalty term in the estimation rule is a standard technique that helps to avoid overfitting by selecting smooth functions h_i [355, 380]. More precisely, the integral terms provide a measure of the roughness of the second-order derivative terms, while the parameters γ_i are selected in a way that limits the impact of the estimation of the h_i on the regularity of the estimated functions. For more details, as well as a recent application in the POT context, see [157].

In both settings, confidence intervals can be obtained using the delta method (Theorem 4.4) or bootstrap approaches (see Sect. 8.1.2.3), which remain valid in the non-stationary framework under convergence conditions on the statistical estimators. Panagoulia et al. [584] recommend using bootstrap methods when sample sizes are small.

The thesis [391] compared the two approaches described above and judged that the nonparametric PML one was generally more amenable than LM when it came to the study of non-stationary extreme values of natural hazards. In effect, the former is less computationally intensive and leads to better automatic selection when there is local linearity [635], despite improvements to the LM approach proposed by [371].

For this reason, a nonparametric estimation algorithm for the PML approach—the results of which are shown in Figs. 8.1 and 8.2—has been proposed for both the MAXB and POT frameworks, and its convergence to the maximum of the penalized likelihood studied in detail in Chap. 3 of [391]. This algorithm is based on the principle that the optimal estimation of the smoothing parameters is obtained using

[1] To a lesser extent: In practice, the Epanechnikov or Gaussian kernel are often used.

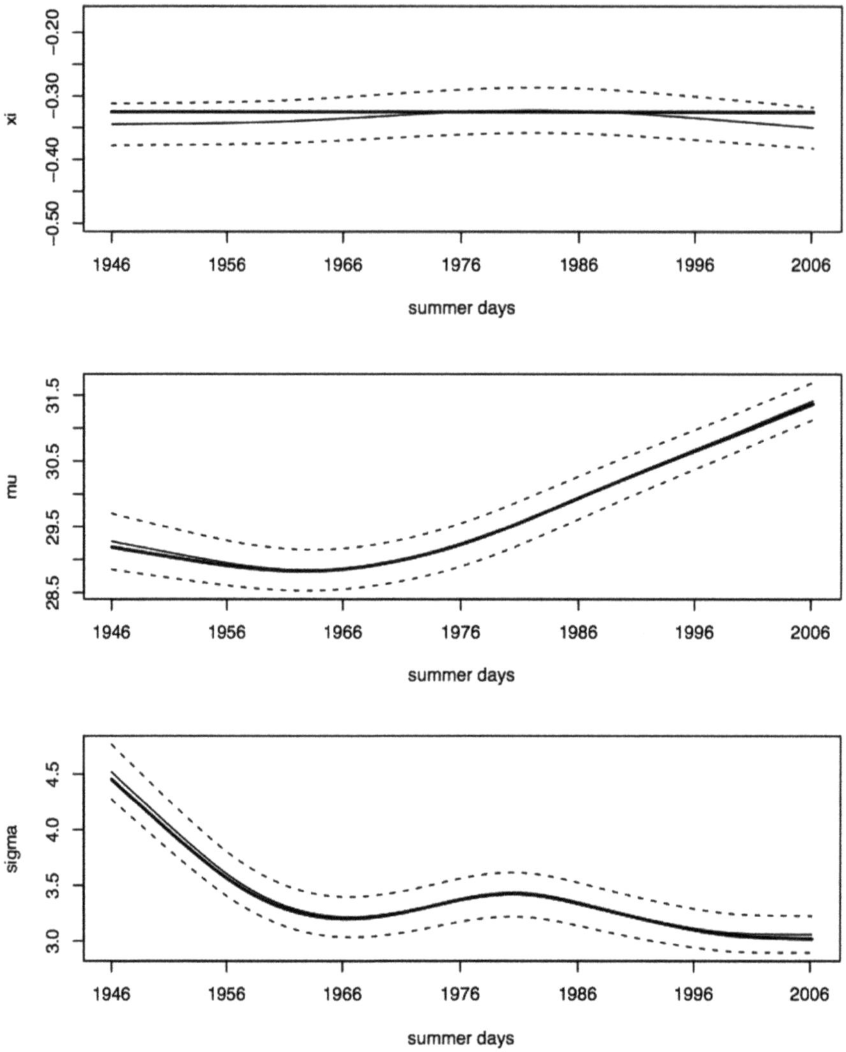

Fig. 8.1 Nonparametric evolution of the parameters of the GEV distribution fitted to maximum daily temperature data at La Rochelle (France). The dotted lines show the 90% confidence interval; the thick curves are the estimates obtained when supposing that the three parameters are variable, the thin ones when supposing the shape parameter ξ is constant

cross-validation [579], and in the spline case, these calculations requiring high computational cost can be approximated using an iterative approach proposed by [360]. This *iterative cross-validation* method is based on rewriting (8.4) in the form of a least squares criterion involving the Fisher information matrix $(I_{\psi_j\psi_k})_{j,k}$ of the extreme value model (cf. Eq. 4.2.2.4) and some *pseudo-data*.

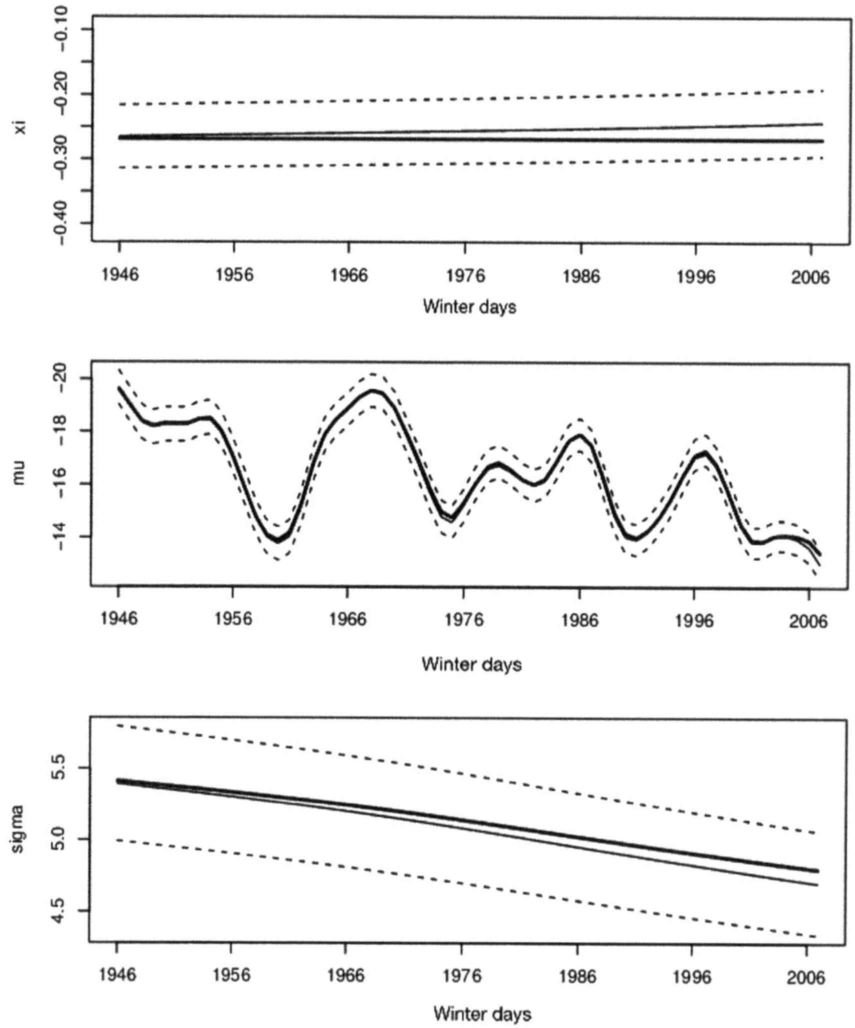

Fig. 8.2 Nonparametric evolution of the parameters of the GEV distribution fitted to minimum daily temperature data at Uman (Ukraine). The dotted lines show the 90% confidence interval; the thick curves are the estimates obtained when supposing that the three parameters are variable, the thin ones when supposing the shape parameter ξ is constant

8.1.1.2 The Case for Shape Parameter ξ

In the dedicated literature, it is commonly accepted that the shape parameter ξ for extreme value models (GEV and GPD) can be considered constant over time, i.e.,

$$\xi(t) = \xi \quad \forall t.$$

Indeed, Zhang et al. [800] exclude the possibility of variation in this parameter by asserting that there is no reason for the shape of the tail of climate variable distributions to change over typical periods of observation, which rarely exceed 100 years. This assertion uses for support numerous estimation results obtained by Coles [173] as well as from simulations run in [202]. In [596], estimations were made over different periods of a long temperature time series from Paris-Montsouris (since 1873), and indeed they showed that the shape parameter did not change significantly. For temperature time series from La Rochelle (France) and Uman (Ukrainia) corresponding to different climates, the recent thesis [391] estimated the nonparametric evolution of the parameters of the GEV distribution fitted to the annual maxima. Both the case, where the three parameters were considered variable and that where the shape parameter was constant, were examined. This work showed that the constant value of the shape parameter remained in the confidence interval of the varying estimate of the same parameter, while the evolution of the two others remained similar whether or not the shape parameter was considered variable (Figs. 8.1 and 8.2). For this reason, in the following it is supposed that the shape parameter does not change significantly over time. Readers interested in the technical details and a more in-depth historical treatment are recommended to consult [391].

8.1.2 Estimating and Selecting a Global Functional Model

The most frequently obtained functional forms for the parameters $\mu(t)$, $\sigma(t)$, and sometimes $\xi(t)$, are polynomial (quadratic at most, to obtain a monotonic trend that can be extrapolated) or piecewise linear [596]. It is, therefore, feasible to define nested models and test the pertinence of a more complicated model with respect to a less complicated one using a likelihood ratio test (LRT)—see Theorem 4.1. Let us recall, in the present case, the following theorem which uses the concept of *deviance* [713].

Theorem 8.1 *If a model $\mathcal{M}_0$ with parameters $\psi^{(2)}$ is a submodel of a model $\mathcal{M}_1$ with parameters $\psi_0 = (\psi^{(1)}, \psi^{(2)})$, where $\psi^{(1)} = 0$ (of dimension k), then the* deviance *statistic*

$$D = 2\{\ell(\psi_0) - \ell(\psi^{(2)})\},$$

where $\ell(\psi_0)$ and $\ell(\psi^{(2)}$ are the maximized log-likelihoods for each model, tends asymptotically to a χ^2 distribution with k degrees of freedom under the hypothesis that the model $\mathcal{M}_0$ is valid.

The more complex model with parameters ψ_0 is then rejected for a given level $1-\alpha$ if $D \leq c_\alpha$, where c_α is the order $1-\alpha$ quantile of the χ^2 distribution with k degrees of freedom. For example, if we want to test for the existence of a linear trend in the location parameter of the GEV distribution:

$$\mu = \mu_0 + \mu_1 t,$$

then $\psi^{(2)} = (\mu, \sigma, \xi)$ and $\psi_0 = (\mu_0, \mu_1, \sigma, \xi)$, and the linear trend is considered significant at 95% if D is greater than $c_\alpha = 3.841$ ($\alpha = 0.05$, $k = 1$).

It is often necessary to perform several tests to obtain the optimal model, since all possible comparisons need to be made.

8.1.2.1 The MAXB Framework

If we wish to identify the optimal model for the block maxima distribution, without any prior assumption on which parameters potentially include trends (except for the shape parameter ξ, supposed constant), conducting the following tests is required:

$$\begin{cases} (\mu_0, \mu_1, \sigma) \text{ against } (\mu, \sigma), \\ (\mu, \sigma_0, \sigma_1) \text{ against } (\mu, \sigma). \end{cases}$$

Then, if (μ_0, μ_1, σ) is retained, test $(\mu_0, \mu_1, \sigma_0, \sigma_1)$ against (μ_0, μ_1, σ). Else, if $(\mu, \sigma_0, \sigma_1)$ is retained, then test $(\mu_0, \mu_1, \sigma_0, \sigma_1)$ against $(\mu, \sigma_0, \sigma_1)$.

Note that an alternative goodness-of-fit χ^2 test well adapted to the specific case of the functional Weibull distribution, which can be easily derived in the Gumbel and Fréchet cases, is provided in [609].

8.1.2.2 The POT Framework

In the POT framework, retaining the hypothesis that the shape parameter ξ is constant, two approaches can be considered.

1. In the first situation, we can fix a threshold and look for a trend in the intensity λ of the Poisson process (i.e., in the frequency of exceeding the threshold) and/or the scale parameter σ of the Pareto distribution. The required tests are then:
    - H_0: (λ constant, σ variable) against H_1: (λ constant, σ constant); if H_0 is retained, we then test H_2: (λ variable, σ variable) against H_0,
    - H_3: (λ variable, σ constant) against H_4: (λ constant, σ constant); if H_3 is retained, we then test H_4 against H_5: (λ variable, σ variable).

2. In the second situation, we vary the threshold so that the frequency at which it is exceeded remains constant, then look to see whether it is possible to identify an additional trend in the scale parameter σ of the Pareto distribution. The variable threshold retained is generally obtained by quantile regression [794] for a very high quantile threshold for the distribution of the variable being studied (e.g., 95% or 98%). See Sect. req.quantile.page for more details.

In addition to the use of the deviance via the LRT theorem, two specific statistical criteria based on *parsimony*, otherwise known as a **model complexity versus fit tradeoff,** are often used in model selection: AIC (*Akaike Information Criterion* [25]) and BIC (*Bayesian Information Criterion* [693]). If $\hat{\ell}$ is the maximized log-likelihood for a model with p parameters, then for a sample of size n,

$$\begin{aligned} \text{AIC} &= -2\hat{\ell} + 2p, \\ \text{BIC} &= -2\hat{\ell} + p \log n. \end{aligned}$$

If the sample size n is small (typically $n < 11p + 9$), a modified AIC can be used [399]:

$$\text{AIC} = -2\hat{\ell} + 2p + \frac{2p(p+1)}{n-p-1}.$$

These criteria are typically used to choose the optimal number of parameters by selecting the model that minimizes them. Note however that they have different meanings. As it performs a bias-variance tradeoff [97], AIC is designed to be good at selecting the "least bad" model in a list that does not contain the true model. In contrast BIC, which is known for selecting (possible underfitted) models with a lower dimension than AIC [466], is consistent in the sense that it (asymptotically) selects the true model in a list, given that the model is in the list and unique [130]. See [131] for more details.

AIC would thus seem to be more theoretically sound, and the preferential criterion of choice. However, simulations by Panagoulia et al. [584] show that for samples of size 20–100, both criteria are equally capable of detecting non-stationarity, and BIC selects good models more often, due to its general ability to retain the most parsimonious ones.

8.1.2.3 Bootstrap and Confidence Intervals

Three types of methods are typically proposed and tested in the literature for estimating confidence intervals for extreme levels using bootstrap. Bootstrapping means generating a large number of equivalent samples starting from the original dataset in order to construct a distribution for the quantity whose confidence interval we wish to estimate.

All of the approaches we consider here involve a *full sample* bootstrap, i.e., all generated samples have the same size n as the original sample. The specialized literature (e.g., [630]) suggests that it is necessary to use subsampling (drawing samples of size $m < n$ from the original sample of size n) to obtain good nonparametric estimates of extreme quantiles, but Panagoulia et al. [584] consider that this restriction does not concern the specific case of extreme quantiles since they are estimated as functions of parameters of fitted distributions.

Of the three approaches, two are nonparametric. Let us denote $(x_{t_1}, \ldots, x_{t_n})$ the sample of non-stationary extreme values.

1. *Random-t resampling*. Here samples are formed from random draws with replacement from the initial sample.
2. *Fixed-t resampling*. This consists not in resampling from the initial values, but from the stationary residuals obtained from them. In [138], this method is given for the block maxima approach, in the steps described below (after supposing that $\xi(t) = \xi$ as suggested in Sect. 8.1.1.2):

Let $\mu(t_i)$, $\sigma(t_i)$, and ξ be the parameters of a non-stationary GEV model fitted to the initial sample, $i \in \{1, \ldots, n\}$.

(i) Calculate the independent so-called *Cox-Snell* residuals, exponentially distributed

$$\varepsilon(t_i) = \left[1 + \frac{\hat{\xi}}{\hat{\sigma}(t_i)}\left(x_{t_i} - \hat{\mu}(t_i)\right)\right]^{-1/\hat{\xi}}, \tag{8.5}$$

(ii) Resample the residuals $\varepsilon(t_i)$ as $\varepsilon_b(t_i)$ using random sampling with replacement,

(iii) Obtain thus a new sample: for $i = 1, \ldots, n$,

$$\tilde{x}_{t_i} = \hat{\mu}(t_i) + \frac{\hat{\sigma}(t_i)}{\hat{\xi}}\left(\varepsilon_b^{-\hat{\xi}}(t_i) - 1\right), \tag{8.6}$$

(iv) Fit a non-stationary GEV model to the new sample and calculate quantities of interest (parameters, return levels).

(v) Repeat steps **(ii)–(iv)** a large number of times to obtain a distribution for each quantity of interest.

3. *Parametric bootstrap.* Here samples are not generated by resampling from the initial sample, but rather by simulating a large number of samples of the same size using the fitted non-stationary GEV distribution. For each time t_i, a new value $\tilde{x}_{t_i}$ is generated from a $\mathcal{G}_{EV}(\hat{\mu}(t_i), \hat{\sigma}(t_i), \hat{\xi}(t_i))$ distribution.

An intermediate method between *random-t* and *fixed-t* turns out to be useful in practice by limiting the increase in uncertainty related to the handling of estimators in Eqs. (8.5) and (8.6). Instead of resampling from the original sample, it is possible to produce draws with a stationary distribution,

$$\varepsilon(t_i) = \frac{x_{t_i} - \hat{\mu}(t_i)}{\hat{\sigma}(t_i)},$$

then sample values $\varepsilon_b(t_i)$ (with replacement) within the sequence of $(\varepsilon(t_i))_i$, then directly construct a new sample:

$$\tilde{x}_{t_i} = \hat{\mu}(t_i) + \hat{\sigma}(t_i)\varepsilon_b(t_i).$$

With a bootstrap distribution in hand, several techniques for calculating confidence intervals for the *debiased* estimator $\hat{\psi} - (\mathbb{E}[\hat{\psi}] - \psi)$ of the parameter (or function of interest) ψ are available.

(a) **Normal approximation.** Based on a classical asymptotic approximation by a normal distribution (cf. Chap. 4) and an estimation of the bias, the $100(1-\alpha)\%$ confidence interval can be estimated by

$$\left[\hat{\psi} - \left(\hat{\psi} - \bar{\psi}_b\right) + z_{\alpha/2}\tau(\hat{\psi}_b)\, ,\; \hat{\psi} - \left(\hat{\psi} - \bar{\psi}_b\right) + z_{1-\alpha/2}\tau(\hat{\psi}_b)\right],$$

where $\hat{\psi}$ is the value estimated using the initial sample, $\bar{\psi}_b$ and $\tau(\hat{\psi}_b)$ respectively the mean and standard deviation of the bootstrap estimates, and $z_{\alpha/2}$ the order $\alpha/2$ quantile of the standard normal distribution.

(b) **Nonparametric approach.** The boundaries of the $100(1-\alpha)\%$ confidence interval are estimated by the order $\alpha/2$ and $1-\alpha/2$ empirical quantiles of the bootstrap distribution. We denote $\hat{\psi}_b$ the variable that has this distribution. As in the previous method, this approach can be corrected by supposing that the unknown bias $\hat{\psi} - \psi$ can be estimated by the difference $\bar{\psi}_b - \hat{\psi}$. The original bootstrap distribution is thus modified by replacing $\hat{\psi}_b$ with $\hat{\psi}_b + \hat{\psi} - \bar{\psi}_b$.

(c) **Bias-corrected and accelerated (BCa) method.** This method, proposed by Efron [245], allows confidence intervals obtained using the nonparametric approach (empirical quantiles) to be corrected so as to obtain the correct skewness* and kurtosis* of the distribution. The boundaries of the $100(1-\alpha)\%$ confidence interval are estimated by the q_{α_1} and q_{α_2} quantiles of the bootstrap distribution, with

$$\alpha_1 = \Phi\left(\hat{z}_0 + \frac{\hat{z}_0 + z_{\alpha/2}}{1 - \hat{\alpha}\left(\hat{z}_0 + z_{\alpha/2}\right)}\right),$$
$$\alpha_2 = \Phi\left(\hat{z}_0 + \frac{\hat{z}_0 + z_{1-\alpha/2}}{1 - \hat{\alpha}\left(\hat{z}_0 + z_{1-\alpha/2}\right)}\right),$$

where Φ is the cumulative distribution function of the standard normal distribution, and

$$\hat{z}_0 = \Phi^{-1}\left(\frac{\mathbf{1}_{\{\hat{\psi}_b < \hat{\psi}\}}}{M}\right),$$

where M is the number of bootstrap samples and $\hat{\alpha}$ a *bootstrap acceleration constant* defined as $1/6$ of the skewness of set of n estimates of $\hat{\psi}$ produced by *jackknife* or *leave-one-out*, i.e., $1/6$ of the skewness of the distribution of the estimators $(\hat{\psi}_{(i)})_i$ for $i = 1, \ldots, n$, where the ith one of them is obtained by removing the ith observation from the initial sample:

$$\hat{\alpha} = \frac{\sum_{i=1}^{n}\left(\hat{\psi}_{(i)} - \bar{\psi}_{(.)}\right)^3}{6\left[\sum_{i=1}^{n}\left(\hat{\psi}_{(i)} - \bar{\psi}_{(.)}\right)^2\right]^{3/2}},$$

where $\bar{\psi}_{(.)} = n^{-1}\sum_{i=1}^{n}\hat{\psi}_{(i)}$.

With the help of a simulation study, Panagoulia et al. [584] showed that, like in the stationary case [448], the different methods perform similarly, but the parametric bootstrap with the *BCa* method produces better confidence intervals. The *fixed-t* method has a tendency to produce wider confidence intervals. In terms of computational time, the methods are globally equivalent, though the *BCa* method by construction is slightly more costly than the other approaches.

8.1.3 Limitations

One of the limits to these approaches is already clearly visible in their descriptions. Indeed, the necessity to extrapolate requires the choice of a monotonic trend—often linear—which is not necessarily the one that best describes the true evolution of the phenomenon. Aside from this, as the future will certainly not look exactly like the past, such trends are themselves likely to change. This is why it is not advisable to extrapolate too far into the future based on trends detected in the past. One way to overcome this type of limitation, favored, for example, in [229], is to consider as covariates one or several meteorological variables whose evolution over time can be determined using climate simulations, instead of time. In doing so however, the problem becomes strongly dependent on uncertainties associated with the climate modeling and removal of bias from these models.

Generally speaking, maximum likelihood estimation turns out to be quite sensitive to large values found toward the end of a sample, whether observed or simulated. Trends identified can, therefore, be overly dependent on such features and become less meaningful. If return levels are estimated by applying these approaches, then it is strongly recommended not to target an overly distant future, and secondly to update the estimates regularly based on newly available observations—typically every 5 to 10 years. Confidence intervals provide only a limited indication of the actual incertitude in extrapolating since they are estimated with respect to the model selected (using the delta method or parametric bootstrap). Further work is still needed on the study of model uncertainty to obtain practical rules for determining periods over which extrapolation is meaningful.

8.2 Mean-Variance Trend Approach

The *mean-variance trend* approach originated from a detailed study of temperature evolution undertaken in the founding work [391], which focused notably on moments of the distribution and their links with the extreme values. The aim was to find an alternative approach to the parametric one, suffering from limitations identified above.

By nonparametrically estimating the evolution over time of the daily temperature (daily mean, minimum, or maximum) in France, Europe, or the Northern Hemisphere as a whole, it has been shown that trends in the mean and variance are not independent [593]. For instance, in France, when the mean increases in winter, the variance decreases; the opposite is true in summer: an increase in the mean is associated with an increase in the variance. So as to evaluate the impact of such changes on trends in extreme values, a test statistic was constructed to help decide whether the extreme values of the standardized variable (i.e., after having removed nonparametric trends in the mean and variance) could be considered stationary.

More generally, let us denote $(X_t)_t$ a sequence of non-stationary random variables such that $\mathbb{E}[X_t] = m_t$ and $\mathbb{V}[X_t] = s_t^2$. Extending Formula (4.1), the associated standardized variable is

$$Y(t) = \frac{X(t) - m(t)}{s(t)},$$

which can be approximated by

$$\hat{Y}(t) = \frac{X(t) - \hat{m}(t)}{\hat{s}(t)},$$

where $(\hat{m}_t, \hat{s}_t^2)$ are estimators of $(\hat{m}_t, \hat{s}_t^2)$.

8.2.1 *Methodological Principles of Testing and Modeling*

8.2.1.1 Estimation of the Standardized Variable

To test the stationarity hypothesis for the extreme values of the variable Y_t, it is first necessary to remove the effects, estimated by $(\hat{m}_t, \hat{s}_t^2)$, corresponding to trends in the mean and variance. The decision to nonparametrically estimate (m_t, s_t^2) is pushed by wanting to make as few assumptions as possible on the trends, and to find the best compromise between smoothing and fitting to observed trends. This nonparametric smoothing can be performed using a variety of approaches (splines, penalized least squares, wavelets, local regression, etc., see Sect. 8.1.1.1). After several tests showed that there was not a great difference in results, the LOESS method for local regression (Sect. 4.3.2) was retained for looking at temperatures in [593].

The subsequent question is which smoothing parameter to use. Usually the optimal one is found using cross-validation. However, due to high autocorrelation in the daily temperature time series, and the fact that the variance is not constant over time, this approach needed to be modified. Indeed, Hoang [391] has proposed the *modified partitioned cross-validation* method for this setting.

Constructing approximately standardized data [391].

(i) Select the dates of the season to be considered across the whole period of data collection.
(ii) Estimate $\hat{m}(t)$ using LOESS and an optimal smoothing parameter found using modified partitioned cross-validation.
(iii) Estimate $\hat{s}^2(t)$ using LOESS starting from $(X(t) - \hat{m}(t))^2$, with the same smoothing parameter.

8.2.1.2 Testing Stationarity

We wish to test the stationarity of the extreme values of the standardized sequence $Y(t)$, but there is no unique alternative to the stationarity hypothesis to test. In effect, if the extreme values of $Y(t)$ are non-stationary, this can take different forms, e.g., a trend (monotone or other), a cycle, or even both at the same time. For this reason, the hypothesis has to be tested using simulation. Consider a distance between time functions $u(t)$ and $v(t)$, e.g., the L_2 norm:

$$\mathcal{D}(u, v) = \left(\int_{t \in [0,T]} (u(t) - v(t))^2 dt \right)^{1/2},$$

where $[0, T]$ is the period examined in the study. Thus, if a function u is estimated by v, $\mathcal{D}(u, v)$ is a measure of the quality of v as an estimator of u.

The parameters of the distribution fitted to the extreme values of $Y(t)$ can then be estimated in two ways: either as nonparametric functions of time as described in Sect. 8.1.1, or as constants. Suppose that we estimate the extreme values of $Y(t)$ using the maxima in blocks method, fitting a GEV distribution. Then the parameters μ and σ of this distribution (the shape parameter ξ is still considered constant) can be estimated either as functions $\tilde{\mu}(t)$ and $\tilde{\sigma}(t)$, or constants $\hat{\mu}$ and $\hat{\sigma}$. Whether the stationary extreme values hypothesis is true or not, the nonparametric estimates of $\tilde{\mu}(t)$ and $\tilde{\sigma}(t)$ converge to the true values μ and σ when the sample size tends to infinity, at a rate which depends on the degree of smoothing.

The situation is different when estimating with constants $\hat{\mu}$ and $\hat{\sigma}$. If the stationarity hypothesis is true, then they converge to the true values at the rate $\sqrt{n}$, where n is the sample size. In this case, $\mathcal{D}(\hat{\mu}, \tilde{\mu})$ is very close to $\mathcal{D}(\mu, \tilde{\mu})$ for very large sample sizes. On the other hand, if the hypothesis is false, $\hat{\mu}$ converges to a constant which is different to the true function $\mu(t)$, and the distance $\mathcal{D}(\hat{\mu}, \tilde{\mu})$ does not converge to 0, remaining above some positive constant A.

This result has also been shown for $\mathcal{D}(\hat{\sigma}, \tilde{\sigma})$ by simulation [391]. The idea is to generate the distributions of $\mathcal{D}(\hat{\mu}, \tilde{\mu})$ and $\mathcal{D}(\hat{\sigma}, \tilde{\sigma})$ when the hypothesis is true. For this, we simulate a large number of samples of the same size as the sample of maxima of $Y(t)$ using the distribution $\mathcal{G}_{EV}(\mu_y, \sigma_y, \xi)$ fitted to the maxima of $Y(t)$. For each simulated sample, we fit a GEV distribution with constant parameters $(\hat{\mu}, \hat{\sigma}_b)$ and another with nonparametrically time-evolving parameters $(\tilde{\mu}_b, \tilde{\sigma}_b)$, so as to eventually obtain distributions for $\mathcal{D}(\hat{\mu}_b, \tilde{\mu}_b)$ and $\mathcal{D}(\hat{\sigma}_b, \tilde{\sigma}_b)$. These distributions correspond to the errors made by estimating constant parameters using time-evolving ones.

Then, if the distances estimated by fitting to the extreme values of $Y(t)$ are within these distributions, it means that the distribution of extreme values of $Y(t)$ cannot be distinguished from a stationary one, and we accept the stationarity hypothesis for the extreme values of $Y(t)$, or more precisely, we cannot reject it. The same procedure can be applied in the POT setting for the threshold exceedance frequency and the scale parameter of the Pareto distribution.

Applying this test to numerous temperature time series has shown that in most cases, the hypothesis is not rejected [594].

8.3 Redefining Return Levels and Other Risk Criteria

According to the definition in Sect. 3.1.1.3, a level x_p associated with a return period T (with $p = 1/T$) corresponds to a level of the natural hazard observed on average every T years, or equivalently, the average probability per year of exceeding that level is p. However, this definition supposes that the frequency of extreme events remains the same, independent of the period under consideration, i.e., the stationary case. How then can we define a return level of T years in the presence of a trend?

Several alternative definitions appropriate to the non-stationary setting have been proposed in the literature. However, it is not clear that these are satisfactory for decision-makers. Thus other criteria have been proposed to aid more practical decision making.

8.3.1 *Return Levels in a Non-stationary Setting*

A first definition, proposed by [596], consists in getting as close as possible to the definition from the stationary setting. In a general setting, suppose that Δ is a duration corresponding to the granularity required in the stationary context for a phenomenon measured by X_t. Typically Δ is 365 d in the common framework when annual, centennial, millennial, etc., levels are wished indicators.

Definition 8.1 (Return level in a non-stationary context) The return level in the non-stationary context, defined starting from time t_0, is the level x_p such that the expectation of exceeding x_p in the next T periods of length Δ is 1:

$$\mathbb{E}\left[X_t > x_p | t_0 \leq t \leq \Delta \cdot T\right] = 1. \tag{8.7}$$

In the two classical frameworks for extreme values (MAXB and POT), detailed below, the estimation of the level x_p must be performed numerically, which is also the case for estimating a confidence interval for the estimator $\hat{x}_p$ produced. As before, it is possible to use the delta method or a bootstrap approach to perform these calculations.

8.3.1.1 MAXB Framework

In MAXB framework, where each maxima distribution is asymptotically GEV, the return level is calculated by integrating the distribution over the upcoming period using the extrapolation of trends identified in the observed past, such as the level x_p satisfying the following properties.

- *GEV parameters as functions.* When $\xi \neq 0$, the level x_p is such that

$$\frac{1}{n_b}\sum_{t=t_0}^{t_0+\Delta T}\left\{1-\exp\left(-\left[1+\frac{\xi}{\sigma(t)}\left(x_p-\mu(t)\right)^{-1/\xi}\right]\right)\right\}=1, \tag{8.8}$$

  where n_b is the number of values by block, and t_0 the start date of the extrapolated period. When $\xi = 0$, Eq. (8.8) becomes, by continuity,

$$\frac{1}{n_b}\sum_{t=t_0}^{t_0+\Delta T}\left\{1-\exp\left(\exp\left[-\frac{x_p-\mu(t)}{\sigma(t)}\right]\right)\right\}=1.$$

- *Mean-variance trends*. Equation (8.8) can be modified to

$$\frac{1}{n_b}\sum_{t=t_0}^{t_0+\Delta T}\left\{1-\exp\left(-\left[1+\frac{\xi}{\sigma_y}\left(\frac{x_p-m(t)}{s(t)}-\mu_y\right)^{-1/\xi}\right]\right)\right\}=1.$$

However, the limitation related to extrapolating trends from the past is still present (Sect. 8.1.3), even though the fact that these trends were identified using the whole set of values and not only the highest ones makes them more robust. To overcome this constraint, it is possible—using the mean-variance trend approach—to estimate a return level for a specific future period using $x_p = s_f y_p + m_f$, where y_p is the stationary return period of the standardized variable and m_f and s_f the mean and standard deviation of the variable over that specific period. In the temperature case, these values can be obtained using climate simulations [593].

8.3.1.2 POT Framework

In this setting, where each distribution of exceedances is a GPD, the return level x_p is calculated as follows:

- *GPD parameters as functions*. When $\xi \neq 0$,

$$\sum_{t=t_0}^{t_0+\Delta T}\left[1+\frac{\xi}{\sigma(t)}\left(x_p-u\right)\right]^{-1/\xi} I(t)=1 \tag{8.9}$$

in the case of a fixed threshold, or

$$\sum_{t=t_0}^{t_0+\Delta T}\left[1+\frac{\xi}{\sigma(t)}\left(x_p-u(t)\right)\right]^{-1/\xi} I=1 \tag{8.10}$$

in the case of a variable one. When $\xi = 0$, Eqs. (8.9) and (8.10) become, respectively,

$$\sum_{t=t_0}^{t_0+\Delta T}\left[1+\exp\left(-\frac{x_p-u(t)}{\sigma(t)}\right)\right] I(t)=1$$

and

$$\sum_{t=t_0}^{t_0+\Delta T}\left[1+\exp\left(-\frac{x_p-u(t)}{\sigma(t)}\right)\right] I=1.$$

- *Mean-variance trends*. Eq. (8.9) can be modified to

$$\sum_{t=t_0}^{t_0+\Delta T} \left[1 + \frac{\xi}{\sigma_y}\left(\frac{x_p - m(t)}{s(t)} - u\right)\right]^{-1/\xi} I_y = 1.$$

8.3.1.3 Stationary Return Level for a Given Year

Alternatively, we can suppose that each year constitutes a stationary period and estimates a different return level for each using the formula given in the stationary case for each of the approaches. This is how the *return.level* function from the *R* package `extRemes` is coded; if a trend is identified and taken into account, a return level is estimated for each year in the observed time series. This function will not extrapolate trends beyond the observed period. However, we can still do this anyway and estimate—by extrapolating detected trends—parameter values for a chosen future year, and deduce the corresponding return level of interest.

8.3.2 Other Risk Criteria in Non-stationary Settings

Various authors of recent publications have returned to these notions of return levels and periods in the non-stationary setting, with an insistance that the protection of installations cannot occur when different values are calculated for each year [792]. They, therefore, point to the need to define a notion of risk over the whole lifetime (or period of operation) of installations requiring protection. This period is denoted $[T_1, T_2]$.

In this setting, several definitions have been proposed, besides Definition 8.1, and occasionally compared [593, 596]. We summarize them in the list below. Suppose we are in the functional setting of Sect. 8.1, where ψ_t designates a vector of parameters of a non-stationary extreme valued distribution of the maxima in blocks type, and recall the notation $F(x|\psi_t)$ for the cumulative distribution function of the extreme values of the phenomenon under investigation. The probability of exceeding a level x_q is then written:

$$p_t = 1 - F(x_q|\psi_t).$$

Suppose finally that the data are independent.

1. ***Expected Waiting Time*** (EWT) [575, 682]: the idea is to estimate the mean waiting time between successive exceedances of an extreme level. If a value x_{p_0} is exceeded in a given year t_{p_0}, the (discretized) time until the next exceedance follows a geometric distribution with mass function (density)

$$f(t) = P(T = t_{p_0}) = (1 - p_0)^{t_{p_0}-1} p_0,$$

   and the mean waiting time is then

$$\mathrm{EWT} = \mathbb{E}[T] = \sum_{t=1}^{\infty} t f(t) = \sum_{t=1}^{\infty} t p_t \prod_{t'=1}^{t_{p_0}-1} (1 - p_{t'}).$$

Estimation of the EWT requires however an estimate of the evolution of the exceedance probability up to infinity, which limits the application of this definition in practice. Cooley [657] and Salas and Obeysekera [682] proposed stopping the sum at a value $t_{\max}$ corresponding to the time at which the probability of exceeding x_T is 1 for time series with increasing trends or 0 for those with decreasing trends.

2. ***Reliability*** (RE) [682]: this concept is based on the probability of not exceeding the value x_q over the operating period $[T_1, T_2]$, denoted $\mathrm{RE}^{ns}_{[T_1,T_2]}$, where ns signifies *non-stationary*. This value can be estimated by

$$\mathrm{RE}^{ns}_{[T_1,T_2]} = \prod_{t=T_1}^{T_2} F(x_q | \psi_t).$$

3. ***Design Life Level*** (DLL) [665]: the principle is the same as above, but the value is used to estimate a dimensioning value for a future project, rather than to verify the robustness of an existing installation. The value is thus estimated as the level whose probability of non-exceedance during $[T_1, T_2]$ is equal to $1 - 1/T$, where T is a period of interest, equated to a classical return period in the stationary case. Thus,

$$x^{DLL}_{[T_1,T_2]} = F^{-1}_{[T_1,T_2]}\,(1 - 1/T)\,, \quad \text{with } F^{-1}_{[T_1,T_2]}(x) = \prod_{t=T_1}^{T_2} F(x | \psi_t).$$

4. ***Equivalent Reliability*** (ERE) [479]: inspired again by concepts used in the reliability of structures, [479] has proposed estimating—over the operating period of an installation and in the non-stationary context—the level whose probability of non-exceedance is the same as that for the stationary context. In the latter, the probability of non-exceedance of a given level is the same over each unit of time (typically a year), and as extreme value distributions are fitted on independent events, its value is

$$p^{\mathrm{ERE}}_{[T_1,T_2]} = (1 - 1/T)^{T_2 - T_1 + 1}\,.$$

The level $x^{ERE}_{[T_1,T_2]}$ being looked for thus solves the following equation:

$$\prod_{t=T_1}^{T_2} F\left(x^{ERE}_{[T_1,T_2]} | \psi_t\right) = p^{\mathrm{ERE}}_{[T_1,T_2]}.$$

5. ***Average Design Life Level*** (ADLL) [479]: these authors also define the level $x^{ADLL}_{[T_1,T_2]}$ exceeded on average $1/T$ times over the period $[T_1, T_2]$, estimated by the solution of the equation:

$$\frac{1}{T_2 - T_1 + 1} \sum_{t=T_1}^{T_2} F\left(x^{ADLL}_{[T_1,T_2]}(T)|\psi_t\right) = 1 - 1/T.$$

This definition is equivalent to (8.7) if the operating period is equal to T. With this notation, the return level x_p proposed in Sect. 8.3.1 can be estimated by solving

$$\sum_{t=T_1}^{T_2} \left\{1 - F\left(x_p|\psi_t\right)\right\} = 1.$$

8.4 Case Studies

8.4.1 *Extreme Temperatures*

To illustrate the different approaches, we consider the time series of maximum daily temperatures observed in Rennes (Brittany, France) over the period 1951–2013, made available by the *European Climate Assessment and Dataset* project (http://eca.knmi.nl). The aim of this project was to construct a database of observations across Europe for studying climate change (notably in terms of homogeneity), which has been regularly updated ever since. We use this database here to estimate a 30-year return level for high temperatures for the period 2014–2043, taking into account observed trends.

8.4.1.1 Analysis of the Time Series

Before applying extreme value theory, it is necessary to get an idea of the range of temperatures seen. The minimum was −7.5 °C on 12 January 1987, and the maximum was 39.5 °C on 5 August 2003 (remember that these are daily maxima; the winter of 1987 was particularly cold!). The highest extreme values occur of course in summer, as it can be seen in Table 8.2, so only the temperatures recorded in the months of June, July, and August will be taken into account in our estimations.

Can we then accept that measurements made at each summer (from June to August) are equidistributed? Figure 8.3 summarizes the result that linear trends in the evolution of the means and maxima can be seen over time. We can test the significance of these trends using a nonparametric Mann-Kendall test (to avoid making any hypotheses on the type of distribution) with a 90% confidence level (cf. Sect. 5.4.3.2). In both cases, the increasing trend is significant. There is thus a warming trend: the

Table 8.2 For each month, the number of annual maxima (first row) over the period of observation, two largest values from each year (second row), and three largest values from each year (third row). The numbers between brackets are the expected number in the case of an equal distribution across all months

	January	February	March	April	May	June
Max (5.25)	0	0	0	0	0	8
2 largest (10.5)	0	0	0	0	2	22
3 largest (15.75)	0	0	0	0	4	33
	July	August	September	October	November	December
Max (5.25)	29	24	2	0	0	0
2 largest (10.5)	54	43	5	0	0	0
3 largest (15.75)	78	64	10	0	0	0

temperatures reached at the end of the period are not similar to those which were observed at the beginning of this period.

8.4.1.2 Parametric Trend Approach

The first approach consists in fitting an extreme value distribution to the highest values while supposing that this distribution's parameters may be time-dependent (excluding the shape parameter ξ). If we apply the maxima in blocks approach, taking into account each summer's maximum, we test using the *likelihood ratio*[2] successively the following cases:

- a linear trend for the position parameter, with a constant scale parameter versus a choice of constant parameters: (μ_0, μ_1, σ_0) against (μ_0, σ_0);
- a constant position parameter and a linear trend in the scale parameter against a choice of constant parameters: $(\mu_0, \sigma_0, \sigma_1)$ against (μ_0, σ_0);
- if the linear trend is accepted for the position parameter, this is tested against a linear trend in each of the parameters: $(\mu_0, \mu_1, \sigma_0, \sigma_1)$ against (μ_0, μ_1, σ_0);
- if a linear trend is accepted for the scale parameter, it is tested against a linear trend in each of the parameters: $(\mu_0, \mu_1, \sigma_0, \sigma_1)$ against $(\mu_0, \sigma_0, \sigma_1)$.

Table 8.3 summarizes the results obtained for the negative log-likelihood of each model and the AIC and BIC criteria. All of the methods select a model with a linear position parameter and a constant scale parameter (μ_0, μ_1, σ_0).

If we adopt the POT approach for exceeding thresholds, it is necessary to test the hypothesis: "constant threshold and a search for a trend in the intensity λ of the Pois-

[2] For example with the *lr.test* function in the *R* package `extRemes`.

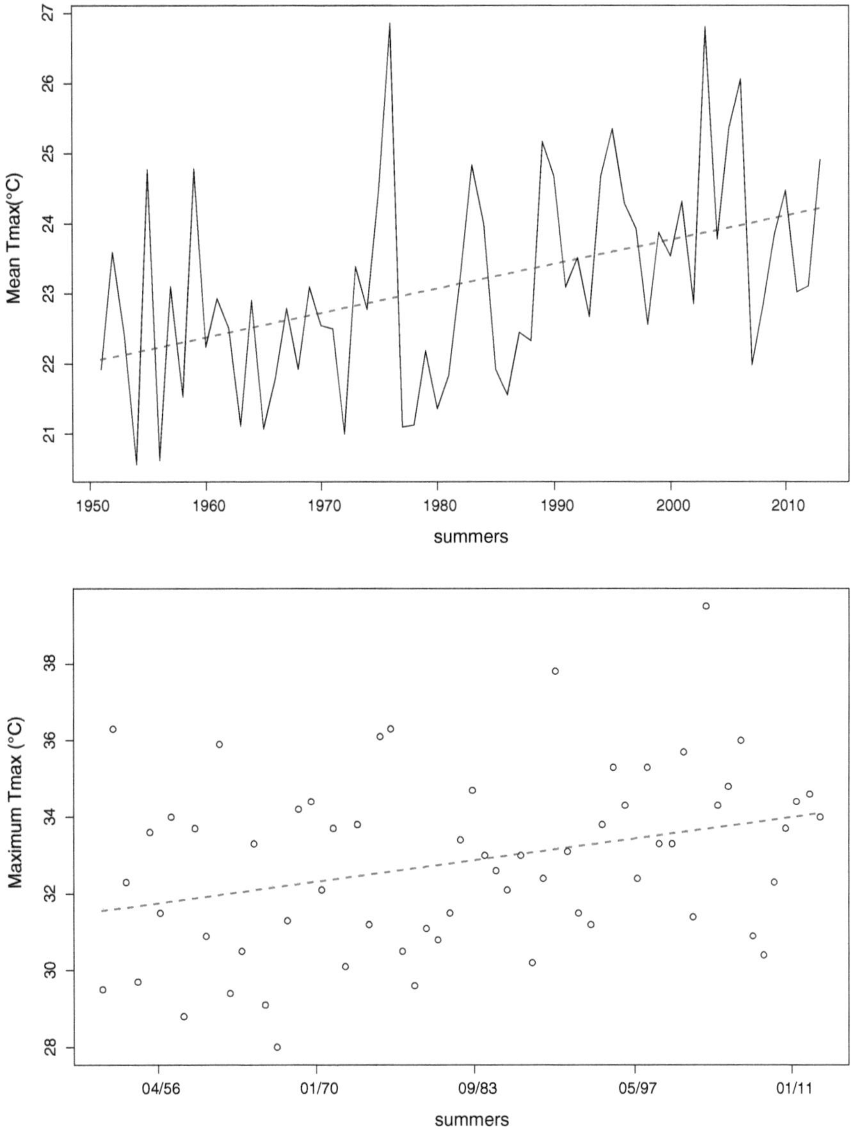

Fig. 8.3 Evolution of the means (top) and maxima (bottom) of maximum daily temperatures in June, July, and August from 1951–2013. The word "significant" in the top-left of each plot means that the trends are significant at the 90% level according to Mann-Kendall tests

Table 8.3 Comparison of different trend models for the parameters of the GEV distribution fitted to the maxima of each summer: negative log-likelihood (*nllh*), deviance D compared with the χ^2 value for 1 degree of freedom (=3.841) and a type I error rate of 0.05, AIC and BIC criteria.

	(μ_0, σ_0)	(μ_0, μ_1, σ_0)	$(\mu_0, \sigma_0, \sigma_1)$	$(\mu_0, \mu_1, \sigma_0, \sigma_1)$
nllh	141.43	136.76	141.31	136.47
$D/\chi_1^2(5\%)$		9.34	0.24	0.58
AIC	288.86	281.52	290.62	282.94
BIC	295.29	290.10	299.20	293.66

son process and the scale parameter of the Pareto distribution" (assuming again that the scale parameter is constant).[3] With a threshold of 31.5 °C (corresponding to the 97% quantile of the summer temperature distribution) considered as suitable according to the constant shape parameter and linearity in the mean of the exceedances criteria, the various models can be tested in the same way:

- constant threshold, linear intensity: $(u_0, \lambda_0, \lambda_1)$ against (u_0, λ_0);
- constant threshold, linear scale parameter: $(u_0, \sigma_0, \sigma_1)$ against (u_0, σ_0);
- linear threshold, linear scale parameter: $(u_0, u_1, \sigma_0, \sigma_1)$ against $(u_0, u_1\sigma_0)$; the linear threshold (obtained using quantile regression) being retained if the intensity λ of the Poisson process is constant (found by testing $(u_0, u_1, \lambda_0, \lambda_1)$ against (u_0, u_1, λ_0)).

The test results, summarized in Table 8.4, show that the optimal models either have a constant threshold, linear Poisson process intensity, and constant Pareto scale parameter, or a linear threshold and a constant scale parameter.

As we saw in Sect. 8.3.1, there are several ways to estimate the return level in the presence of identified trends, and several ways to calculate confidence intervals.

First, let us consider the GEV setting with the model (μ_0, μ_1, σ_0) and the definition of the targeted return level as the level reached at least once in expectancy in the next 30 years. This leads to a return level of 39.4 °C. The 95% confidence interval can be estimated by *random-t* bootstrap (using the stationary residuals) or parametric bootstrap. The results are quite similar:

- nonparametric bootstrap: [37.6 °C 41.5 °C];
- parametric bootstrap: [37.5 °C 41.1 °C].

Table 8.5 shows various estimates of the targeted return level depending on the approach taken (MAXB or POT) and the definition of the return level used. The 95% confidence intervals have been estimated using nonparametric bootstrap. The competing approaches do not lead to the same values. As shown by [593], evolution in the position parameter $\mu(t)$ of the GEV distribution has more impact on the return level than evolution in the intensity $\lambda(t)$ of the Poisson process in the POT approach. In the latter, it is the threshold that plays the equivalent role to the position parameter,

[3] This hypothesis is not available in the *R* package `extRemes` and was coded specifically for this example.

Table 8.4 Comparison of various trend models for parameters of the Poisson process and Pareto distribution fitted to exceedances of a threshold corresponding to the 97% quantile each summer: negative log-likelihood (*nllh*), deviance D compared with the χ^2 value for 1 degree of freedom (=3.841) and a type I error rate of 0.05, AIC and BIC criteria, in the constant and linear threshold cases.

Constant threshold	(u_0, I_0)	(u_0, I_0, I_1)	(u_0, σ_0)	$(u_0, \sigma_0, \sigma_1)$
nllh	480.14	472.58	158.06	157.49
$D/\chi_1^2(5\%)$		15.13		1.13
AIC	962.29	949.15	320.12	320.99
BIC	968.94	962.46	325.20	328.62
Linear threshold	(u_0, u_1, I_0)	(u_0, u_1, I_0, I_1)	(u_0, u_1, σ_0)	$(u_0, u_1, \sigma_0, \sigma_1)$
nllh	488.37	487.63	153.05	152.69
$D/\chi_1^2(5\%)$		1.478		0.72
AIC	978.74	979.27	310.10	311.38
BIC	985.39	992.57	315.23	319.08

Table 8.5 30-year return levels and 95% confidence intervals obtained by nonparametric bootstrap for the two extreme value methods (block maxima MAXB and threshold exceedance POT) for different non-stationary models and different definitions of future return levels

MAXB approach		**POT approach**		
(μ_0, μ_1, σ_0)		$(u_0, I_0, I_1, \sigma_0)$		$(u_0, u_1, I_0, \sigma_0)$
Once in the next 30 years	Stationary in 30 years	Once in the next 30 years	Stationary in 30 years	Stationary in 30 years
39.4°C	**40.0°C**	**38.4°C**	**39.2**	**40.4°C**
[37.6°C 41.5°C]	[38.0°C 42.2°C]	[36.5°C 38.8°C]	[36.4°C 40.8°C]	[38.2°C 42.5°C]

and taking into account a variable threshold leads to results that are more coherent with an evolution in the position parameter of the GEV distribution.

8.4.1.3 Mean-Variance Approach

To apply the alternative approach presented in Sect. 8.2, it is necessary to construct the standardized variable by estimating the evolution of the mean and variance of summer temperatures nonparametrically. Using local LOESS regression, Fig. 8.4

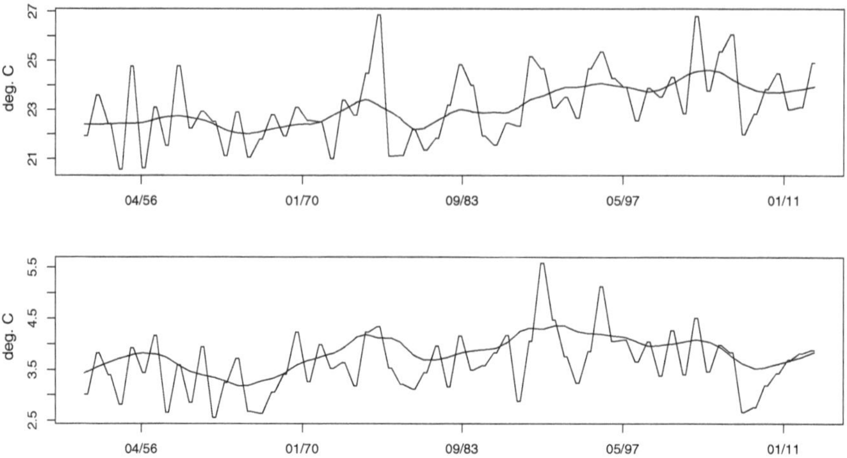

Fig. 8.4 Evolution of the means (top) and standard deviations (bottom) of the maximum daily temperature recorded in Rennes from 1951–2013. The smoothing obtained using LOESS is superimposed on the curve obtained by attributing to each day the mean temperature or the standard deviation of the temperature of the summer it came from

displays these trends, where the smoothing parameter has been chosen by optimizing a modified partitioned cross-validation criterion proposed in [391].

The test for stationarity in the extreme values of the standardized variable suggests that no significant evolution can be detected in the parameters of the fitted extreme value distributions for either the maxima in blocks or threshold exceedance approaches.

The return levels estimated by extrapolating the linear trends seen according to one or the other of the previously proposed definitions (level exceeded in expectation at least once in the next 30 years, or return level estimated under stationarity in 30 years) are summarized in Table 8.6. The confidence intervals are estimated using:

- nonparametric bootstrap for the level exceeded in expectation once every 30 years;
- the delta method by adding the contributions of the mean and standard deviation to the variance-covariance matrix and the Jacobian matrix in the case of the stationary return level in 30 years.

The results obtained are very similar regardless of the approach and definition chosen. The sensitivity seen in the previous section, due to the estimation of trends in the parameters of extreme value distributions, is no longer visible. It should be noted that this method can be used to estimate future means and variances for more distant horizons—such as the end of the 21st century—from climate simulations run in international scientific projects supporting the *Intergovernmental Panel on Climate Change* (IPCC).

Table 8.6 30-year return levels and 95% confidence intervals for the two proposed definitions, estimated using stationarity of the extreme values of the standardized variable and evolution in the mean and standard deviation of summer temperatures in terms of the linear trend identified in the time series.

MAXB approach		**POT approach**	
Once in the next 30 years	Stationary in 30 years	Once in the next 30 years	Stationary in 30 years
40.0 °C	**40.2** °C	**40.2** °C	**40.4** °C
[38.5 °C 43.0 °C]	[38.6 °C 41.8 °C]	[38.2 °C 42.1 °C]	[39.3 °C 41.4 °C]

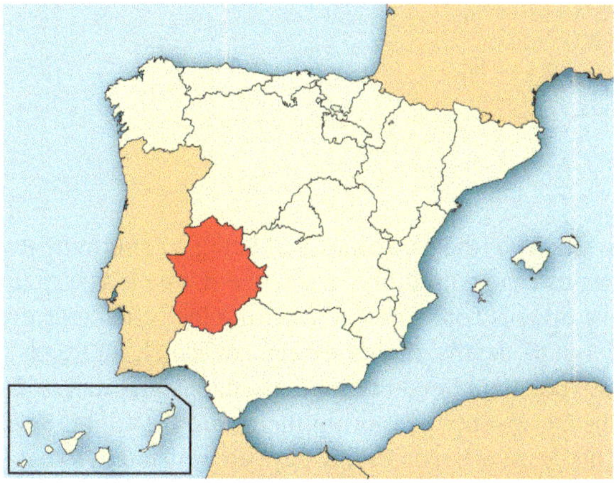

Fig. 8.5 The Extremadura region, Spain

8.4.2 *Extreme Rainfall*

These approaches can be applied to other variables that potentially have trends. Rainfall in France does not as yet exhibit significant trends, unlike Spain. Let us consider a time series of measurements from the Jaraíz de la Vera station in the Cáceres province, in the North East of the Extremadura region (Fig. 8.5), provided by the Spanish State Meteorological Agency (AEMET). This data involves daily cumulative rainfall from 1961–2010, with no missing values. Its quality has been verified by Javier Acero, researcher at the University of Extremadura, and the following results come from a study run in collaboration with EDF R&D [22].

8.4.2.1 Analysis of the Time Series

The largest observed cumulative daily rainfall was 176 mm on 2 February 1972. The distribution of the maxima (annual maximum along with two or three of the largest values) shows that October to February are the rainiest months (Table 8.7). This part of the year was, therefore, chosen to study the extreme values.

Table 8.7 Number of yearly maxima (first row), number of two largest values each year (second row), and number of three largest values each year (third row) for each month. The numbers in brackets are the expected numbers if the results were equidistributed over the set of months

	January	February	March	April	May	June
Max (4.17)	9	5	4	2	2	2
2 largest (8.33)	18	13	7	5	4	3
3 largest (12.5)	23	19	8	6	7	3
	July	**August**	**September**	**October**	**November**	**December**
Max (4.17)	0	0	3	7	11	5
2 largest (8.33)	0	0	3	17	15	15
3 largest (12.5)	0	0	7	28	23	26

Rain has the particularity of being an intermittent variable. Two (stochastic) processes can be used to characterize it: one is binary (it rains or does not rain), while the other is continuous (measuring the actual quantity of rainfall). See Sect. 10.2.2 for more details. In this setting, the search for trends is more tricky. For example, if we look for a trend in the seasonal mean rainfall using the daily data, i.e., including the 0s (absence of rain) in the mean, the slight downward trend is not significant. On the other hand, if rainy days are separated from days with no rainfall, significant trends can be seen in the mean, variance, and number of days with rain. Incidentally, these are opposing trends in that the mean and variance decrease, while the number of days with rain increases (Fig. 8.6). It, therefore, rains less but more often (hence the absence of a significant trend when we take into account all of the days).

For studying the extreme values, we will thus prefer to use the threshold exceedance (POT) approach as it can better take into account such intermittent data. In effect, threshold exceedances obviously only occur on days with rain, but it is easy to pass from an estimated return level using only days with rain to the same return level estimated using all days of the season. The difference is in the frequency of threshold exceedance, which either involves only days with rain:

$$\lambda = \frac{n_u}{n_r},$$

where n_u is the number of independent exceedances of a threshold u and n_r the number of days it rained, or instead involves all days in the season:

$$\lambda = \frac{n_u}{n}.$$

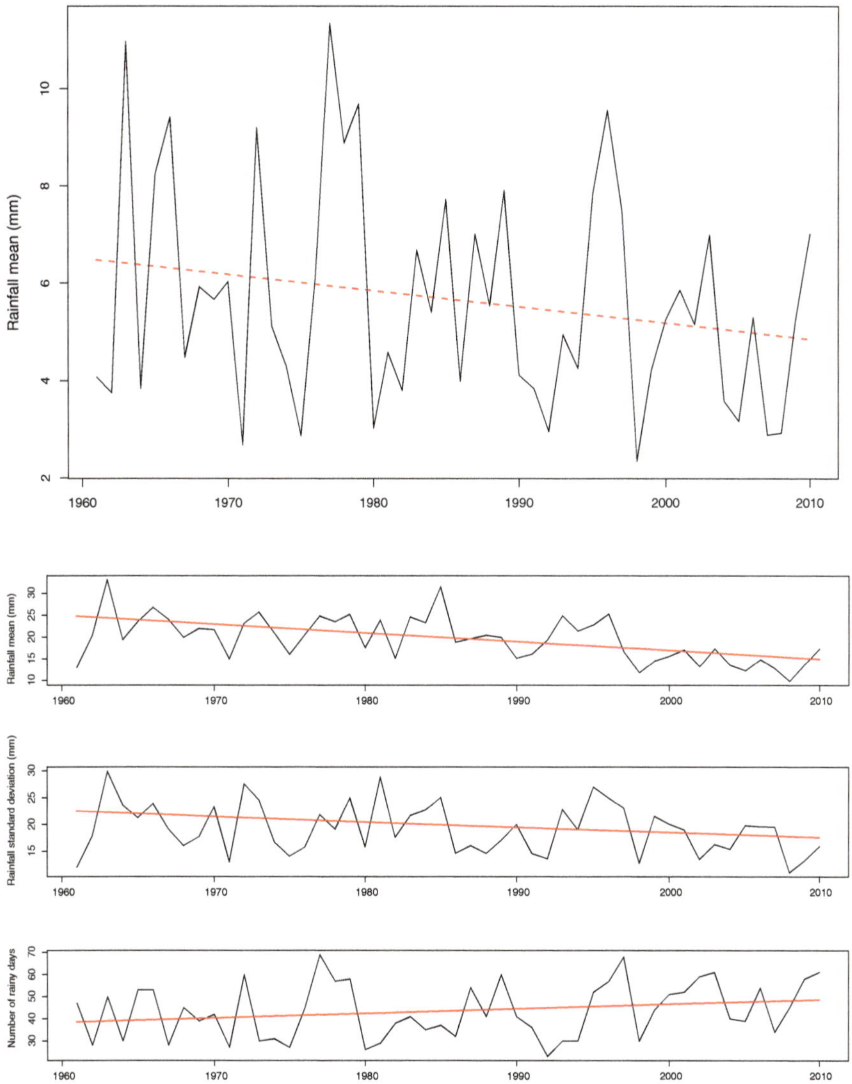

Fig. 8.6 Evolution of the seasonal means calculated using all data (top), and seasonal means, variances, and numbers of days with rain when considering only days where it rained (bottom)

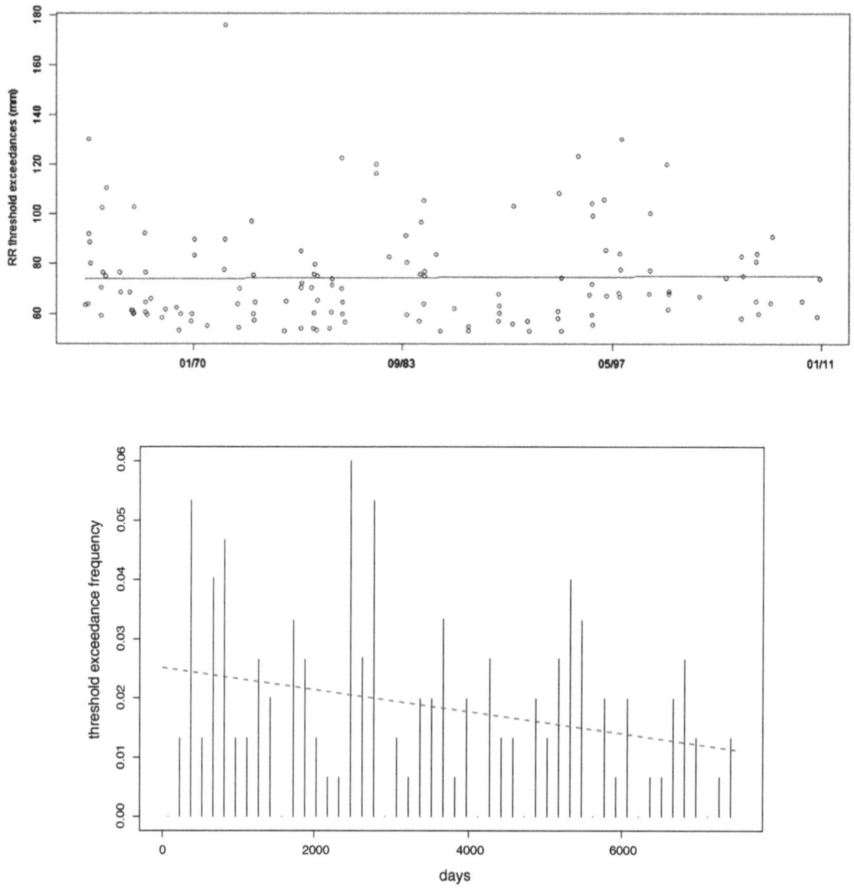

Fig. 8.7 Observed evolution in the amount of rainfall over the threshold (top) and the frequency of threshold exceedance (bottom) between October and February

Independence is dealt with as before for the temperature data; we consider that two events are independent if separated by at least one day with rainfall below the threshold. This choice has been justified by several tests. As the number of exceedances is the same in both cases, we can pass from one approach to the other by multiplying by the fraction of days it rained: n_r/n. Then, if we examine the evolution in exceedances of a high threshold—here set at the 98% quantile of the distribution of the values from the season in question (= 52.3 mm), no trend is detected in the quantity of rain above the threshold; however, the frequency of exceeding the threshold decreases (Fig. 8.7). It is therefore valid to implement methods that can deal with trends when examining the extreme values of this time series. Here, we will try to estimate a 20-year return level in 2020 (10 years after the end of the time series).

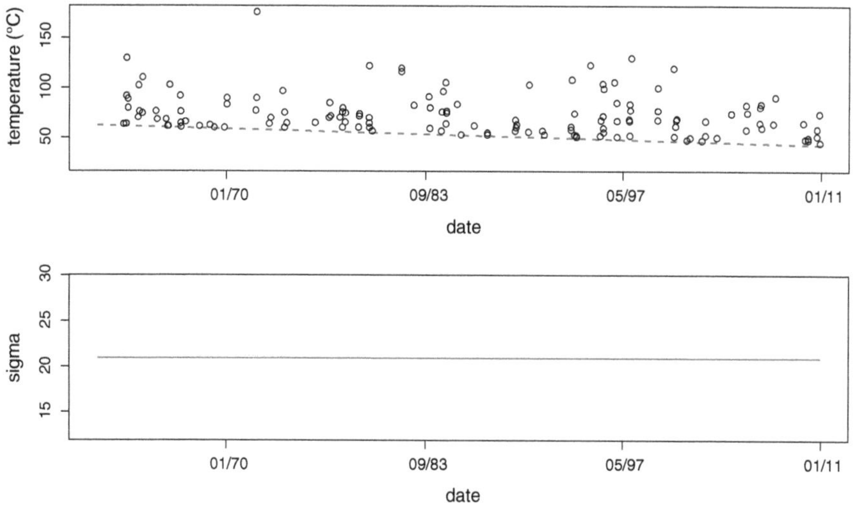

Fig. 8.8 Variable threshold (top) when keeping independent exceedances, and the optimal evolution identified for the scale parameter of the exponential distribution (bottom)

8.4.2.2 Trends in Extreme Value Distribution Parameters

A trend showing a decrease in the frequency of threshold exceedance was identified in the previous section. As for temperature, identifying this trend can be done using likelihood ratio tests. However, as it is a decreasing trend, there is a risk when extrapolating to obtain frequencies of 0 or even less than 0. For this reason, it is judged preferable to have a variable threshold (thus rendering the exceedance frequency stationary) and look for a possible trend in the scale parameter of the GPD distribution. As an aside, an LRT test can also be used to check whether the GPD distribution can be simplified to a standard exponential one (i.e., a shape parameter of $\xi = 0$). This test indicates that in this case, the shape parameter (-0.09) cannot be significantly (at 5%) distinguished from 0 (unlike the temperature example, where this test actually concludes that the shape parameter is non-zero—details not shown). The test results indicate that with a variable threshold estimated by quantile regression (98%) and working with all of the days, the scale parameter of the exponential distribution can be considered constant. Figure 8.8 illustrates this.

For this situation, a 20-year return level in 2020 was estimated using the second definition (stationary level in 2020) at 123.5 mm, for a 95% confidence interval [95.5 ; 151.7 mm].

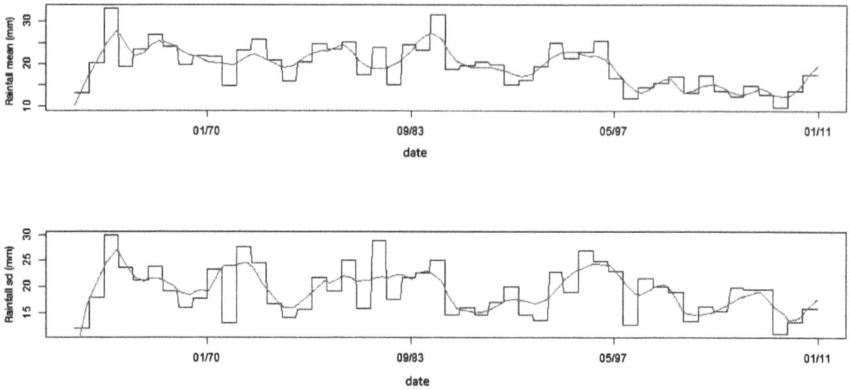

Fig. 8.9 Empirical evolution in the seasonal means and standard deviations of number of days with rain (dashed line) and the estimated nonparametric trend (solid line)

8.4.2.3 Mean-Variance Approach

So as not to mix up the two processes described in Sect. 8.4.2.1 when estimating the mean and variance, these two will be estimated using only the non-zero rainfall values, in the presence of a supplementary variable giving the number of days of rainfall per season. The mean number of days of rainfall per season is only 43.5—far from the 151 d in total. Note in passing that the probabilistic approximation made if we were to use a maxima in blocks approach would be misleading since the maximum supposedly taken over 151 d is in reality only over 43.5 d. This further justifies our preference for a threshold exceedance approach in this specific case.

The threshold is then chosen as the 95% quantile of the set of values from days with rain and is equal to 60.2 mm. The shape parameter of the GPD distribution ($= -0.03$) is not significantly different from 0, so the simple exponential distribution will be retained. Finally, 101 independent values are found over the threshold.

Nonparametric trends in the mean and variance are identified in the same way as for the temperature data and plotted in Fig. 8.9. These allow us to calculate the standardized variable Y_r (where r is a reminder that we are only using days on which it rained), which we test for stationarity in its extreme values. More precisely, the intensity of the Poisson process and the scale parameter of the exponential distribution are tested for being constants, following the procedure described in Sect. 8.2.1.2. The p-values obtained are 0.48 for the Poisson process intensity and 0.88 for the scale parameter of the exponential distribution; these results support the stationary hypothesis.

The future return level can then be estimated using the *hereditary property* of GPD distributions (Proposition 6.3). Indeed, if v is the threshold defined for the standardized variable Y_r, then $Y_r > v$ is equivalent to $X_r > w$, with

$$w = s_r \cdot v + m_r,$$

where m_r and s_r are respectively the mean and variance at the targeted date. Then, if Y_r follows a GPD above the threshold v with parameters ξ_v and σ_v and an intensity λ_v for the Poisson process, X_r also follows a GPD above the threshold w with parameters:

$$\begin{aligned}\xi_w &= \xi_v = \xi, \quad (= 0 \text{ in our case})\\ \sigma_w &= \sigma_v \cdot s_r,\\ \lambda_{rw} &= \lambda_v,\end{aligned}$$

where λ_{rw} is the exceedance frequency calculated using only the days on which it rained.

To estimate the return level in 2020, we must estimate the mean and variance at that future time, as well as the number of days with rain, to then be able to get back to a return level over all days in the season. These estimates are made here by extrapolating the linear trends identified earlier in the mean and variance in the amount of rainfall (on days with rain), as well as in the number of days with rain (alternatively, they could have been obtained using climate simulations). Here we find $m_f = 12.9$, $s_f = 16.7$, and $nb_f = 50.6$.

The return level is then estimated using Formula (6.7):

$$x_N = w + \frac{\sigma_w}{\xi}\left[\left(N n_y \lambda_w\right)^{\xi} - 1\right],$$

where N is the required return period (20 years here), n_y the number of values per year (per season here, i.e., 151), and λ_w the exceedance frequency of the threshold w when considering all of the days. The latter is calculated using:

$$\lambda_w = \lambda_{Rw}\frac{nb_f}{n_y}.$$

We implement a nonparametric bootstrap method consisting of (in this specific case) repeating the following steps a great number of times:

1. Generate a sequence of the number of days with rain each season by adding a random perturbation around the linear trend identified.
2. Associate with each day of rain a quantity estimated using a bootstrap sample by blocks (of 10 d to control temporal correlation) of Y_r and non-parametric trends in the mean and variance.
3. Deduce a 20-year return level for 2020 using the same method used for the original sample.

This procedure allows us to obtain a distribution for 20-year return levels in 2020 with which we can deduce a confidence interval for the return level estimator. The estimate of this return level is 107 mm, and the 95% confidence interval is [81.3 maxima 117.7 mm]. This interval is much smaller than the one obtained earlier (with trends in the distributions' parameters). The impact of the decrease in the mean and—especially—the variance, overpowers the increase in the number of days with rain, and more strongly influences the estimated future return level than the decreasing value of the threshold.

These examples show that the estimation of return levels in the non-stationary setting cannot take advantage of solid theoretical underpinnings, and that different ways to treat the same problem can least to fairly different estimates, even for relatively non-extreme levels (here, 20 or 30 years). It, therefore, is illusory to hope to estimate, in this context, extremely rare events (e.g., those associated with exceedance probabilities of 10^{-4}); this reality has been recognized by certain industrial actors like WENRA (the *Western European Nuclear Regulators Association*).

Chapter 9
Multivariate Extreme Value Theory: Practice and Limits

Anne Dutfoy

Abstract This chapter presents the most recent developments in extreme value theory in cases of joint, or multivariate, hazards. This type of phenomenon is increasingly being studied, as it can be more destructive than a single hazard. Introducing, in particular, the notion of copula, this chapter takes up the notations and main concepts introduced in Chap. 6.

9.1 Introduction

9.1.1 Why Multivariate Extreme Values?

Univariate extreme value analyses concern the probabilistic modeling of a phenomenon in its extreme values. However, especially in the natural hazards setting, natural processes are often correlated, and their extreme values arrive not in isolation but in clusters. For example, flood season is often also storm season, while heatwaves occur in summer at the same time that rivers are at their lowest (baseflow) levels.

Detecting the potential for groups of extreme values to occur is a fundamental problem in industrial dimensioning and civil engineering. A tragic illustration of this was the nuclear accident at Fukushima in March 2011, a consequence of the combined effects of an earthquake (which destroyed the nuclear power plant's electricity cables as well as several paths of communication) and a tsunami generated by the same earthquake (which flooded the emergency power sources). Even though the structures in question were conceived to resist each of these events independently, it might not be able to withstand the simultaneity of those events.

Thus, looking at each natural hazard in isolation is not always sufficient, and it may be necessary to jointly analyze the extreme values of several phenomena, where the choice of relevant combinations (i.e., those which together could potentially impact a structure's robustness) must be made by experts in the domain.

A. Dutfoy (✉)
EDF R&D, Boulevard Gaspard Monge, 91120 Palaiseau, France
e-mail: anne.dutfoy@edf.fr

N. Bousquet and P. Bernardara (eds.), *Extreme Value Theory with Applications to Natural Hazards*, https://doi.org/10.1007/978-3-030-74942-2_9

Consider, for example, the probability that extreme values from two natural hazards occur at the same time. When these processes are independent, i.e., when the occurrence of one has no influence on the occurrence of the other, the probability of both producing extreme values is simply the product of the probabilities. Formally, if A and B are the extreme events under consideration, then $\mathbb{P}(A \cap B) = \mathbb{P}(A)\,\mathbb{P}(B)$. In such cases, multivariate analysis reduces to several univariate analyses in which the goal is to calculate the marginal probabilities $\mathbb{P}(A)$ and $\mathbb{P}(B)$.

In contrast, when natural hazard processes are correlated, the combined probability cannot be determined from the two probabilities alone, and $\mathbb{P}(A \cap B) \neq \mathbb{P}(A)\,\mathbb{P}(B)$. In particular, if the dependence is such that the occurrence of extreme event A leads to or favors that of B, then the combined probability is larger than the product of the marginal ones: $\mathbb{P}(A \cap B) > \mathbb{P}(A)\,\mathbb{P}(B)$. Thus, supposing that these extreme events were independent would lead to a non-conservative estimate of the probability of the events occurring together, namely, this probability would be underestimated.

This multivariate analysis which we allude to is constructed on a base of results pulled from the rich theory of multivariate extreme values. With such calculations of the combined probability of extreme events occurring for different levels of extremeness in hand, it is possible to construct mappings of these probabilities. With the help of appropriate visualizations, these tools can help get an overall idea of the occurrence of groups of extreme events for different levels of risk.

Joint analysis also allows us to model the joint probabilistic behavior of extreme-valued hazards. The multivariate probability distribution thus obtained can then be used in multiple ways. Notably, we can use it to model conditional extreme event scenarios and determine the distribution of a phenomenon given that other phenomena have exceeded some fixed extreme level. For example, what is the distribution of a river's flow given that a violent storm is taking place? Or given that a river is in flood, what is the distribution of flows in a neighboring river in the same watershed?

9.1.2 Difficulties in the Multivariate Setting

Numerous concepts in the univariate setting lose their meaning when transferred to the multivariate one. It begins with the question of what being a *multivariate extreme value* even means. Is it when all of its components are extreme values, i.e., when all of the hazards have extreme values simultaneously? Or is it when at least one component is extreme, i.e., when at least one process has an extreme value, but not necessarily all of them?

In what follows, we will see that models for multivariate extreme value theory describe the probabilistic behavior of phenomena which are all simultaneously extreme valued. On the other hand, model inference takes place jointly using all of the data. According to the number of its extreme-valued components, to each datum is associated a different weight.

This question naturally leads to another: how should the *maximum of a set of multivariate data points* be defined? Is it the data point containing the most extreme value (no matter the component) in all of the data? Or rather, is it some mathematical construct, such as, for instance, the vector containing the *maximum* from each component?

It turns out that the theory of multivariate extreme values deals with the component-wise vector of maxima $\mathbf{M}_n$ over periods of length n. Interpreting and working with such vectors is a tricky business. First, these vectors $\mathbf{M}_n$ are not actual observed data points. Second, the event $\{\mathbf{M}_n > \mathbf{x}\}$, where comparisons are component-wise and $\mathbf{x}$ is a vector of extreme-valued thresholds, means that all of the extreme-valued events occurred over the period in question, but not necessarily at the same time! Thus, when jointly analyzing a flood and a storm, the limit distribution of $\mathbf{M}_n$ allows us to calculate the probability that both extreme events occurred during fall, without making clear whether the events happened the same day or a month apart, two situations which do not at all correspond to the same overall severity. As a consequence, modeling the tail of the multivariate distribution will be the focus of study, rather than modeling $\mathbf{M}_n$.

Following the same thread, the notion of *return period* of a multivariate event must also be clarified. Is it the mean waiting time between arrivals in a domain defined in terms of extreme-valued thresholds? Or rather, defined by the quantile of the multivariate distribution associated with the event in question?

De facto, the question of how to define what the *level of a given return period* arises, as the relationship between levels and return periods is no longer bijective: a given return period may correspond to a hypersurface on which—using an additional criterion—we must select a particular point, which will be the dimensioning combination of the protection structure or civil work.

As brought up above, the simultaneous study of several extreme-valued hazards brings the additional notion of dependence to the table. It will be necessary to study, on top of the marginal behavior of each component, the dependence structure between the extreme values of different phenomena. Several types of dependence are possible. Between total independence and total dependence lies *asymptotic independence*, an intermediate situation in which real dependence diminishes for more and more extreme values. The variables can be either *positively associated* or *negatively associated* with each other.

Finally, multivariate analyses are based on multivariate observed values, i.e., simultaneous measurements of several phenomena, obtained over a time period that is as long as possible. In practice, however, data for different phenomena are obtained independently of each other by groups such as Météo-France, but also EDF—which itself records the values of certain phenomena. These data sources are then brought together to form a multivariate dataset, which is only possible if measurements were made at the same frequency (daily, hourly, etc.) over the same period. In practice, joint analyses are restricted to the period in which data is available from all sources, which can be a real problem if the period in question turns out to be short.

9.1.3 Chapter Contents

In this chapter, we start by explaining with precision the most important notion in multivariate analyses: dependence between processes with extreme values. Then, we recall, without going into detail, the main fundamental theorems of multivariate extreme value theory. In [65, 173], readers can find an in-depth exposition of this theory.

The first structural theorem concerns the limit distribution of multivariate maxima and defines multivariate extreme value distributions (GEVM), whose marginal distributions are univariate GEV distributions linked by a *extreme value copula*. The second theorem provides an approximation of the tail distribution of the multivariate distribution of the phenomena. This approximation can either be deduced from the limit distribution of the multivariate maxima, or come from some parametric model. We will describe Ledford and Tawn's model, which allows all types of asymptotic dependence to be modeled.

A section is then dedicated to multivariate inference of both the joint distribution of the phenomena and the probability of combinations of extreme-valued events occurring. Notably, we make clear different possible strategies for inferring multivariate distributions, e.g., should inference occur simultaneously or successively on the marginal distributions and the dependence structure?

Lastly, we will give several examples showing how to understand and exploit results obtained by multivariate extreme value analyses, including modeling conditional scenarios, iso-quantile lines, iso-frequency contours, and bounding the probability of multiple extreme values occurring simultaneously.

As much as possible, we will illustrate the theoretical results using French real data:

- **Case 1**: conjunction of flooding of the Loire and Vienne rivers at their confluence. These two rivers are found in the same watershed. The data is daily maximum flows reported jointly from 1958 to 2000 (source: HYDRO databank*).
- **Case 2**: conjunction of a windstorm and extreme cold weather in the central region of France. The data are daily maximum wind speeds and minimum air temperatures, jointly recorded by Météo-France from 1971 to 2008.
- **Case 3**: conjunction of baseflow of the Garonne river and extreme cold weather. The data are daily measures of the maximum flow and minimum temperature, jointly recorded by Météo-France from 1967 to 2012.
- **Case 4**: conjunction of flooding of the Rhone River and windstorms. The data are daily maximal flows and maximal wind speeds jointly recorded from 1920 to 2011.

Remark 9.10 In addition, for French-speaking readers interested in technical details of the theoretical results shown and used in this chapter, we suggest reading [154] and [285], which deal, respectively, with multivariate modeling (in general, for studying risk) and multivariate extreme values.

9.2 Nature of Dependence

9.2.1 Asymptotic Independence

Dependence between several random variables can be modeled by a function from $\mathbb{R}^d \longrightarrow [0, 1]$ called a *copula*. This is an essential ingredient of any multivariate distribution, as made clear by Sklar (1959) in the following theorem.

Theorem 9.9 (Sklar [704]) *Let F be the multivariate cumulative distribution function of a random vector of dimension d: $\mathbf{X} = (X_1, \ldots, X_d)$. Then there exists a function $C : \mathbb{R}^d \to [0, 1]$ called a* copula *defined for all $\mathbf{x} \in \mathbb{R}^d$:*

$$F(x_1, \ldots, x_d) = C(F_1(x_1), \ldots, F_d(x_d)).$$

If the $F_1, \ldots, F_d$ are continuous, the copula C is unique; otherwise, it is uniquely determined on the image product space $I_m(F_1) \times \cdots \times I_m(F_d)$.

Any property of a random vector relating only to dependence corresponds to a property of the copula function alone. For example, for independent random variables X_i, the copula is the product copula $C(u_1, \ldots, u_d) = \Pi_{i=1}^d u_i$. This corresponds to saying that the cumulative distribution function F of the vector $\mathbf{X}$ is the product of the marginal cumulative distribution functions F_i: $F(x_1, \ldots, x_d) = \Pi_{i=1}^d F_i(x_i)$.

We now introduce the notion of asymptotic dependence of a random d-dimensional vector. This definition requires only the copula C_F of the distribution of the random vector.

Definition 9.18 (*Copula with asymptotic independence ($d \geq 2$)*) A multivariate distribution F with copula C_F is *asymptotically independent* if C_F satisfies, for all $\mathbf{u} \in [0, 1]^d$,

$$\lim_{t\to\infty} C_F^t(u_1^{1/t}, \ldots, u_d^{1/t}) = u_1 \ldots u_d.$$

Asymptotic independence of a random vector is equivalent to pairwise asymptotic independence of its components [84]. This result allows us to restrict analyses to asymptotic independence in the bivariate case, for which we now give a definition easier to understand.

Definition 9.19 (*Asymptotic independence ($d = 2$)*) The scalar variables X_1 and X_2 with marginal distributions F_1 and F_2 are said to be *asymptotically independent* if and only if:

$$\lim_{u \to 1^-} \mathbb{P}\left(X_1 > F_1^{-1}(u) \mid X_2 > F_2^{-1}(u)\right) = 0. \tag{9.1}$$

In other words, the probability that the variable X_1 exceeds its order u quantile (the term $F_1^{-1}(u)$) given that the variable X_2 has already exceeded its order u quantile (the term $F_2^{-1}(u)$) tends to 0 as we consider higher and higher quantiles u. This relationship implies that the probability of the simultaneous occurrence of two processes' extreme values is negligible with respect to the marginal probability of one of them occurring.

There are several ways to check whether (9.1) holds.

- The first way is to see whether the variables are completely independent, because, in this case,

$$\mathbb{P}\left(X_1 > F_1^{-1}(u), X_2 > F_2^{-1}(u)\right) = (1-u)^2,$$

and thus

$$\mathbb{P}\left(X_1 > F_1^{-1}(u) \mid X_2 > F_2^{-1}(u)\right) = 1 - u.$$

Hence, complete independence is a special type of asymptotic independence.
- The second way is more subtle: when the variables are dependent but with an *evanescent* dependence as u tends to 1, i.e., when[1]

$$\mathbb{P}\left(X_1 > F_1^{-1}(u), X_2 > F_2^{-1}(u)\right) = O\,(1-u)^{\frac{1}{\eta}},$$

where $0 < \eta < 1$. Then,

$$\mathbb{P}\left(X_1 > F_1^{-1}(u) \mid X_2 > F_2^{-1}(u)\right) = (1-u)^{\frac{1}{\eta}-1}.$$

We further distinguish between three sub-cases:

1. $0.5 < \eta < 1$: here we say there is *positive association*. The combination of extreme events $\left\{X_1 > F_1^{-1}(u)\right\}$ and $\left\{X_2 > F_2^{-1}(u)\right\}$ happens more often than when the variables are totally independent. Since $1 < 1/\eta < 2$, we have

$$\mathbb{P}\left(X_1 > F_1^{-1}(u), X_2 > F_2^{-1}(u)\right) > \mathbb{P}\left(X_1 > F_1^{-1}(u)\right)\mathbb{P}\left(X_2 > F_2^{-1}(u)\right).$$

2. $\eta = 0.5$ corresponds to exact asymptotic independence as described earlier, which means we can write

$$\mathbb{P}\left(X_1 > F_1^{-1}(u), X_2 > F_2^{-1}(u)\right) = \mathbb{P}\left(X_1 > F_1^{-1}(u)\right)\mathbb{P}\left(X_2 > F_2^{-1}(u)\right).$$

[1] Recall that $O(.)$ means *is of the order of*.

3. $0 < \eta < 0.5$: here we say there is *negative association*. The combination of extreme events $\{X_1 > F_1^{-1}(u)\}$ and $\{X_2 > F_2^{-1}(u)\}$ happens less often than when the variables are totally independent. Since $1/\eta > 2$, we have

$$\mathbb{P}\left(X_1 > F_1^{-1}(u), X_2 > F_2^{-1}(u)\right) < \mathbb{P}\left(X_1 > F_1^{-1}(u)\right)\mathbb{P}\left(X_2 > F_2^{-1}(u)\right).$$

These considerations suggest that making the hypothesis that the phenomena are independent leads to overestimation of the probability of simultaneous extreme events in the case of negatively associated asymptotic independence, or underestimation in the case of positively associated asymptotic independence.

It is the latter that engineers working on risks to structures fear, since the independence hypothesis puts in danger their claims to the robustness of their constructions. On the other hand, in the former case, the independence hypothesis—which strongly simplifies analyses—is not detrimental to their claims of robustness since, in reality, the extreme events occur simultaneously with a lower probability than the one obtained under the independence hypothesis.

9.2.2 Data Visualization

Visualization of multivariate data in various spaces is a qualitative technique to help understand the behavior of physical processes in the extreme values setting. The type of dependence between phenomena may be detectable in this way.

Types of visualization (used for each of our examples) include those in the physical (original sample) space, in the rank space, in the Fréchet space, and in the pseudo-polar space. Our examples will help illustrate the different types of asymptotic dependence:

- **Case 1**: asymptotic dependence of the two rivers when in flood;
- **Case 2**: negatively associated asymptotic independence of storms and extreme cold;
- **Case 3**: positively associated asymptotic independence of river baseflow and extreme cold;
- **Case 4**: exact independence between extreme flood and storm events.

9.2.2.1 Plotting Points in the Physical Space

The physical space, or *original sample space*, corresponds to that of the raw data in its given units. This is both the most natural yet hardest space in which to interpret the data. In effect, the distribution of the points in this space depends on the data's units as well as its marginal cumulative distribution functions. It is thus very difficult to estimate how extreme valued the points actually are on this graph. The rank space is preferable to this, bringing with it more directly interpretable information.

Figures 9.1, 9.2, 9.3 , 9.4 illustrate, in their first columns, the data obtained for each case study.

9.2.2.2 Plotting Points in the Rank Space

Recall that the rank of a observed data point corresponds to its position in the list of the points after ordering them.

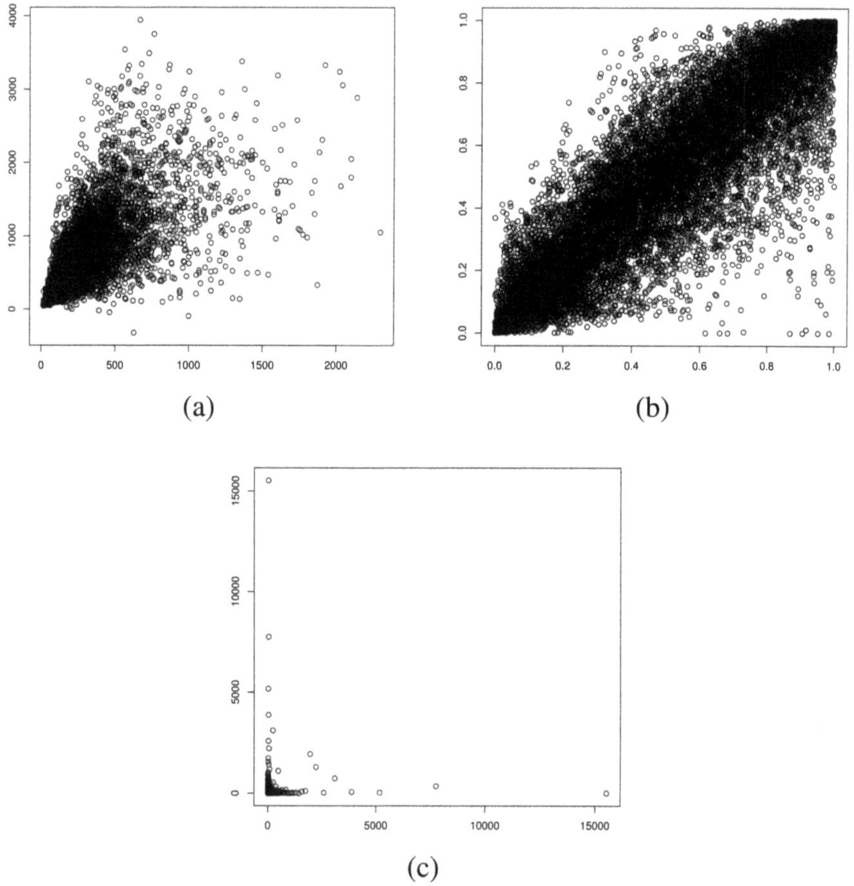

Fig. 9.1 Case 1: asymptotic dependence for flow measurements. Plotting the data (**a**) in the physical space, (**b**) in the rank space, (**c**) in the Fréchet space. The x-axis corresponds to the flow of the Loire River, the y-axis to that of the Vienne River. Values in the physical space are in m^3/s

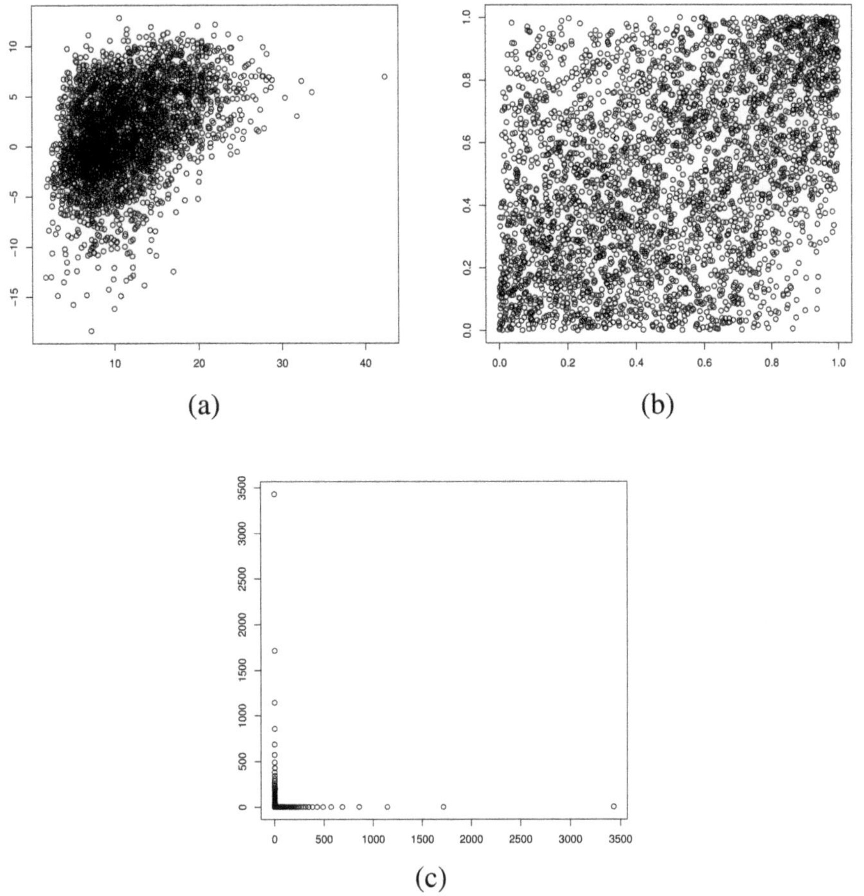

Fig. 9.2 Case 2: negatively associated asymptotic independence for high wind speed and low-temperature measurements. Plotting the data (**a**) in the physical space, (**b**) in the rank space, (**c**) in the Fréchet space. The x-axis corresponds to wind speeds (m/s in (a)), and the y-axis to temperatures (°C in (**a**))

Definition 9.20 (*Rank space*) Let X be a random d-dimensional vector. Let F_i be the marginal cumulative distribution function of the distribution associated with the i-th component of X. The rank space of a multidimensional sample $x_1, \ldots, x_n$ of draws of X is the space whose points are vectors of (normalized) *rank statistics* $s_1, \ldots, s_n$ defined for each dimension i and for $1 \leq k \leq n$ by

$$s_{i,k} = r_{i,k}/n,$$

where

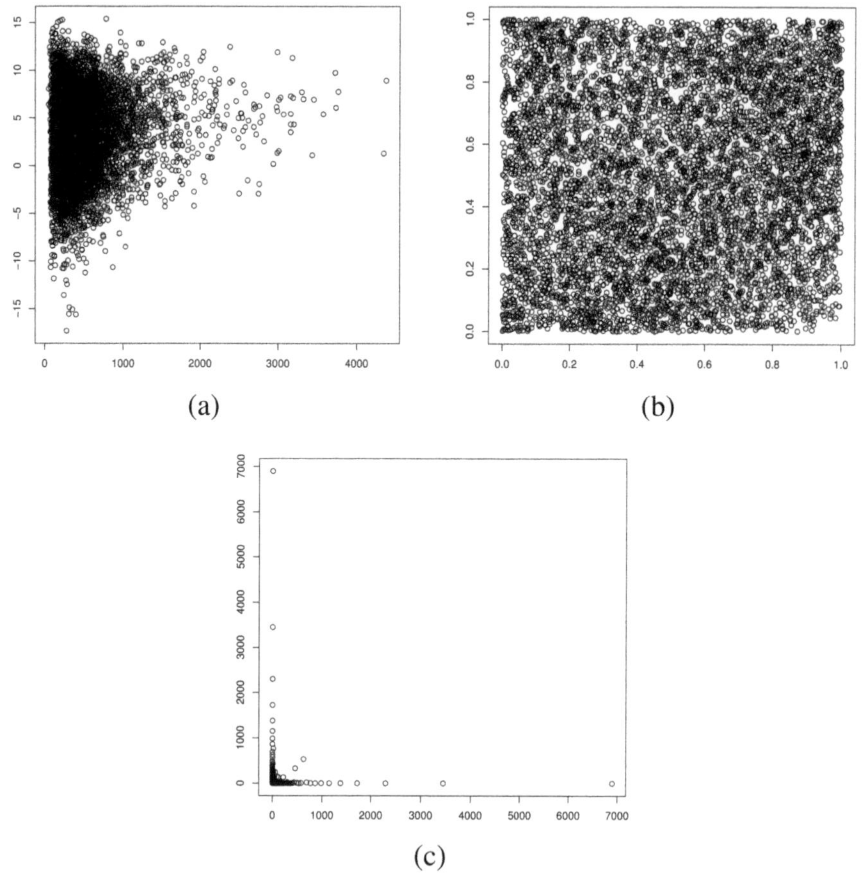

Fig. 9.3 Case 3: positively associated asymptotic independence for low-flow and low-temperature measurements. Plotting the data (**a**) in the physical space, (**b**) in the rank space, (**c**) in the Fréchet space. The x-axis corresponds to flows (m^3/s in (**a**)), and the y-axis to the temperature (°C in (**a**))

$$r_{i,k} = 1 + \sum_{j=1}^{n} \mathbf{1}_{(0,\infty)}(x_{i,k} - x_{j,k})$$

defines the *rank* (position) of the i-th component of the k-th vector in its marginal sample.

Each vector of normalized ranks $s_1, \ldots, s_n$ is, therefore, found in $[0, 1]^d$, and the raw data is transported to this space by the component-wise mapping:

$$F_1^{-1}(X_1), \ldots, F_d^{-1}(X_d),$$

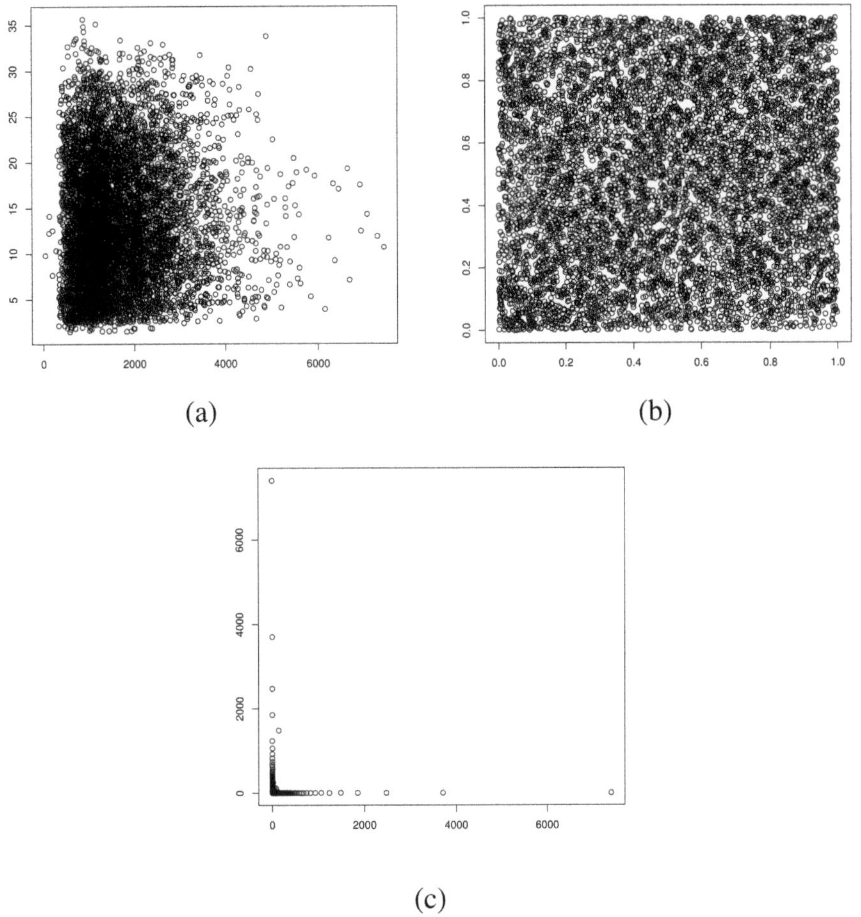

(a) (b) (c)

Fig. 9.4 Case 4: total independence for high-flow and high wind speed measurements. Plotting the data (**a**) in the physical space, (**b**) in the rank space, (**c**) in the Fréchet space. The x-axis corresponds to flows (m^3/s in (**a**)), and the y-axis to wind speeds (m/s in (**a**))

as each function $F^{-1}(.)$ is increasing. This mapping does not alter the dependence structure in the data.

The positioning of data in this new space is much easier to interpret than in the physical coordinate space since the mapped data $s_1, \ldots, s_n$ all follow uniform distributions on [0, 1], given that the effect of the marginal distributions F_i has been removed. This rank space makes it easy to spot the dependence that exists between the extreme values of different variables.

Figures 9.1, 9.2, 9.3 , 9.4 show, in their second columns, the results for each example. The zones of interest appear to be top-right for Cases 1 and 4, bottom right for Case 2, and bottom left for Case 3.

These graphs show that asymptotic dependence corresponds to a point cloud with clearly visible trends in it—which persist for high quantiles. Negatively associated asymptotic independence corresponds to a point cloud that thins out for the most extreme quantiles. Positively associated asymptotic independence corresponds to a point cloud which becomes slightly denser when moving toward the more extreme quantiles. Finally, total independence corresponds to a point cloud spread out uniformly in the extreme quantiles zone.

9.2.2.3 Plotting Points in the Fréchet Space

Here, the data undergoes a transform so that each component follows a standard Fréchet distribution with a cumulative distribution function F_F defined by $F_F(y) = e^{-1/y},\ y > 0$. Therefore, this transform is given by, in the bidimensional case $X = (X_1, X_2)$,

$$(Z_1, Z_2) = (F_F^{-1} \circ F_1(X_1), \Psi^{-1} \circ F_2(X_2)), \tag{9.2}$$

where F_1 and F_2 are estimated using the empirical cumulative distribution functions. Here again, this transform operates component-wise with an increasing function that again does not alter dependence structure in the data.

So that the extreme values studies correspond to high quantiles of the Fréchet distribution, the data are flipped by artificially multiplying them by -1 when the extreme values of interest correspond to lows rather than highs. This is what happens for the temperature data in Cases 2 and 3, and the flow data in Case 3.

Figures 9.1, 9.2, 9.3 , 9.4 show, in their third component, the resulting plots. Note that only the plot showing dependence in the conjunction between the two flood events has points outside of the axes shown.

9.2.2.4 Plotting Points in the Pseudo-polar Coordinate Space

Here we perform the Fréchet change of variable as in (9.2), followed by the transform:

$$(R, W) = \left(Z_1 + Z_2, \frac{Z_1}{Z_1 + Z_2}\right). \tag{9.3}$$

The random variable R is the sum of two positive variables. Bivariate data points corresponding to large values of R are, therefore, those for which at least one of them has a high rank. As for W, it takes values in $[0, 1]$. The event $\{W = 1/2\}$ brings together all bivariate data points in which both of the components have the same rank, whatever it may be.

To study combined occurrences of extreme values, we are thus interested in data points for which R is large and W is close to $1/2$. More precisely, we select the

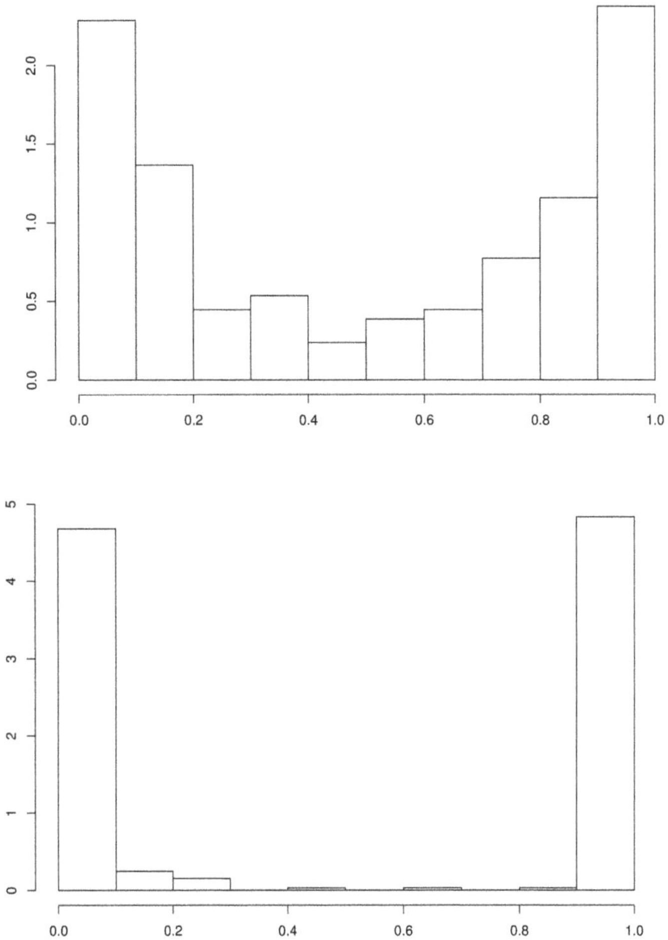

Fig. 9.5 Histograms of data in pseudo-polar coordinates. Case 1 (top): asymptotic dependence between flow measurements. Case 2 (bottom): negatively associated asymptotic independence between strong wind and cold temperature measurements

bivariate data points corresponding to the largest values of R. We then plot the histogram of values of W for these selected points in order to analyze its central part.

Figure 9.5 shows the histogram of W obtained in the cases of asymptotic dependence (Case 1) and negatively associated asymptotic independence (Case 2). For Case 1, the histogram has a non-negligible set of values around $w = 1/2$, thus showing that the simultaneous occurrence of extreme values is not negligible. For Case 2, the histogram has large peaks around 0 and 1, and very few values around $w = 1/2$, showing that here extreme values essentially occur in isolation.

9.2.3 *Measuring and Testing Dependence*

Complementing the qualitative analysis of dependence, numerous quantitative tools are available to help better characterize the type and strength of dependence, including the scalar coefficients χ and $\bar{\chi}$, which we present here. Interested readers can also find an innovative measure of asymptotic independence in [738].

9.2.3.1 Coefficients and Functions of Asymptotic Dependence

The asymptotic dependence coefficient χ, with $0 \leq \chi \leq 1$, is defined by

$$\chi = \lim_{u \to 1^-} \mathbb{P}\left(X_1 > F_1^{-1}(u) \mid X_2 > F_2^{-1}(u)\right).$$

Therefore, by definition, asymptotic independence is equivalent to $\chi = 0$.

This χ can also be obtained in the limit as $u \to 1^-$ of the value of the function $u \to \chi(u)$ defined for each $u \in [0, 1]$ as

$$\begin{aligned} \chi(u) &= 2 - \frac{\log \mathbb{P}\left(F_{X_1}(X_1) < u, F_2(X_2) < u\right)}{\log \mathbb{P}\left(F_{X_1}(X_1) < u\right)} \\ &= 2 - \frac{\log C_F(u, u)}{\log u}. \end{aligned} \tag{9.4}$$

In effect, it can be shown that

$$\chi = \lim_{u \to 1^-} \chi(u). \tag{9.5}$$

We also introduce the $\bar{\chi}$ coefficient, with $-1 \leq \bar{\chi} \leq 1$, defined by

$$\bar{\chi} = \lim_{u \to 1} \bar{\chi}(u),$$

where the function $u \longrightarrow \bar{\chi}(u)$ is defined for each $u \in [0, 1]$ by

$$\begin{aligned} \bar{\chi}(u) &= \frac{2 \log \mathbb{P}\left(F_1(X_1) > u\right)}{\log \mathbb{P}\left(F_1(X_1) > u, F_2(X_2) > u\right)} - 1 \\ &= \frac{2 \log(1-u)}{\log \bar{C}_F(u, u)} - 1. \end{aligned}$$

The form the dependence takes can be analyzed as follows:

(i) For asymptotically dependent variables, $\bar{\chi} = 1$ and $0 < \chi \leq 1$. In this case, χ measures the strength of the dependence.

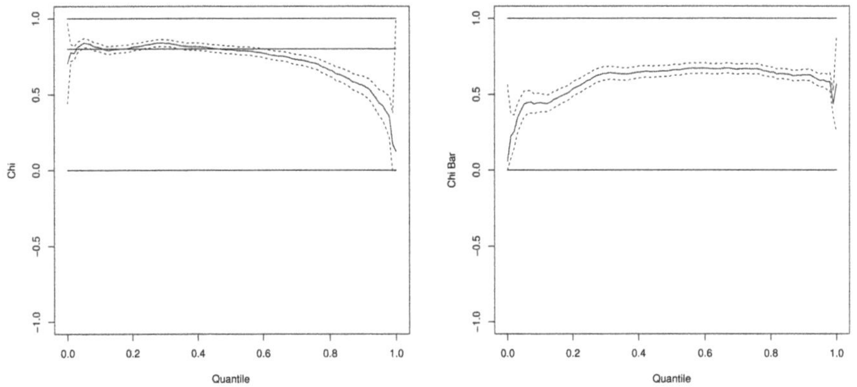

(a) (b)

Fig. 9.6 Case 1: asymptotic dependence of flow measurements; (**a**) $u \to \chi(u)$, and (**b**): $u \to \bar{\chi}(u)$

(ii) For asymptotically independent variables, $\chi = 0$ and $-1 \leq \bar{\chi} < 1$; the greater the value of $\bar{\chi}$, the stronger the dependence. A negative association corresponds to $\bar{\chi} < 0$, a positive one to $\bar{\chi} > 0$.

Figures 9.6, 9.7, 9.8, and 9.9 illustrate the behavior of these coefficients for each example, showing plots of $u \to \chi(u)$ and $u \to \bar{\chi}(u)$. The limiting values seen as u tends to 1 leading to the following conclusions:

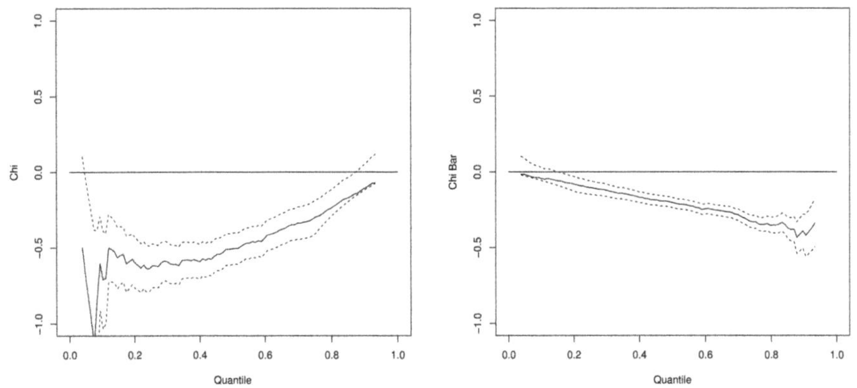

(a) (b)

Fig. 9.7 Case 2: negatively associated asymptotic independence of strong wind and low-temperature measurements; (**a**) $u \to \chi(u)$, and (**b**): $u \to \bar{\chi}(u)$

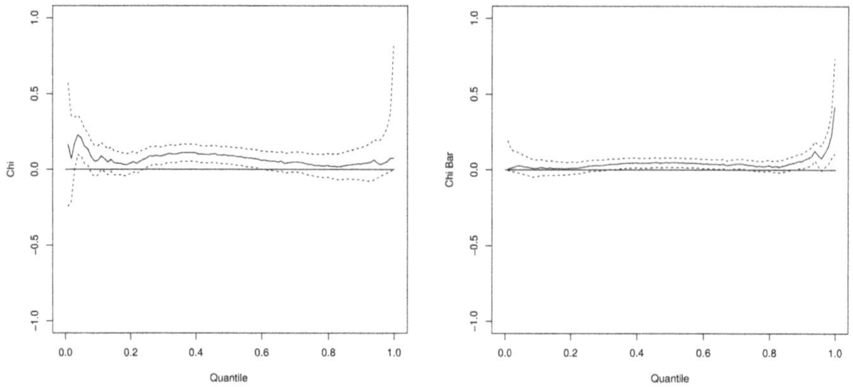

(a) (b)

Fig. 9.8 Case 3: positively associated asymptotic independence of low-flow and low-temperature measurements; (**a**): $u \to \chi(u)$, and (**b**): $u \to \bar{\chi}(u)$

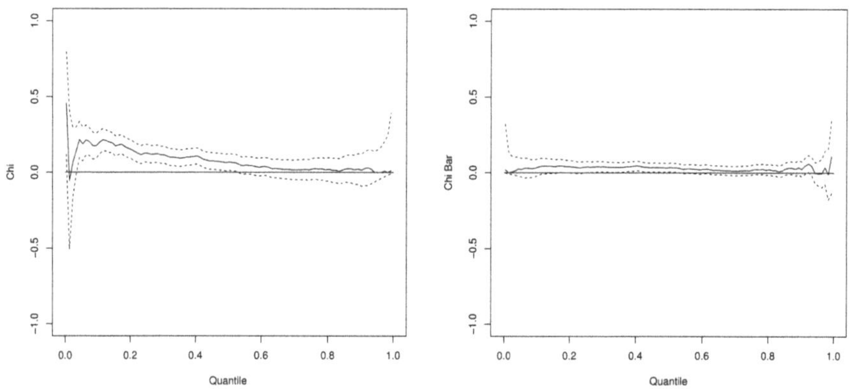

(a) (b)

Fig. 9.9 Case 4: total independence of high-flow and strong wind measurements; (**a**): $u \to \chi(u)$, and (**b**): $u \to \bar{\chi}(u)$

- Case 2 corresponds to negatively associated asymptotic independence: $(\chi, \bar{\chi}) \simeq (1, -0.4)$.
- Case 3 corresponds to positively associated asymptotic independence: $(\chi, \bar{\chi}) = (0, 0.1)$. The level of dependence is thus very low.
- Case 4 corresponds to total independence: $(\chi, \bar{\chi}) = (0, 0)$.

The graph for Case 1 is not conclusive and decisions will have to be made using other statistical tools.

9.2.3.2 Detecting Asymptotic Independence

Detecting cases of asymptotic independence is crucial as this prohibits the subsequent use of certain modeling methods, as we will see in the following. This is the case, for instance, when the modeling is based on the limit distribution of multivariate maxima.

Numerous tests for detecting asymptotic independence can be found in the literature. Readers can turn to [401, 750, 751, 801] for technical details and comparisons. Recently, [143] has comprehensively reviewed the state of the art when it comes to these tests, which include—among others—score tests, the gamma test, and the madogram test. We now describe the idea behind one of the most powerful such tests, proposed by Ledford and Tawn [473].

Ledford and Tawn [473] have introduced a new scalar coefficient η known as the *tail dependence coefficient*, with $0 < \eta \leq 1$, described in more detail in Sect. 9.3.4. This scalar value allows us to test for asymptotic independence in data. With a weak hypothesis on the tail of the multivariate distribution F, Ledford and Tawn show that asymptotic dependence implies $\eta = 1$. They also suggest testing for asymptotic dependence by testing the simple hypothesis $H_0 = \{\eta = 1\}$ against the composite alternative $H_1 = \{0 < \eta < 1\}$. Theoretically, H_0 is not equivalent to asymptotic dependence, but in practice we can equate the two [65]. Several estimators of η can be found in the literature, some of which are described in Sect. 9.3.4.

Figure 9.10 illustrates the results obtained on our examples. For Case 1, the decision made is to conserve the hypothesis H_0 since although the estimation of η is more or less constant and below 1, the 95% confidence interval always contains 1. In contrast, for Cases 2 and 3, the decision is made to reject H_0. Although again η is more or less constant and below 1, the 95% confidence intervals for various thresholds do not contain the value 1.

9.3 Fundamental Results

9.3.1 Partial and Marginal Ordering

The vectorial order relation used in the analysis of multivariate extreme values is *marginal ordering*, the definition of which is as follows.

Definition 9.21 (*Marginal ordering*) Let $(\mathbf{x}, \mathbf{y}) \in \mathbb{R}^d$. We say that $\mathbf{x}$ is less than or equal to $\mathbf{y}$, and write $\mathbf{x} \leq \mathbf{y}$, if and only if:

$$\mathbf{x} \leq \mathbf{y} \Longleftrightarrow x_i \leq y_i, \quad \forall i = 1, \ldots, d.$$

This order relation is a partial one as it cannot order all pairs of vectors; for each $\mathbf{x}$, we can find some $\mathbf{y}$ such that neither $\mathbf{x} \leq \mathbf{y}$ nor $\mathbf{y} \leq \mathbf{x}$ is true. As for the maximum, it is defined component-wise:

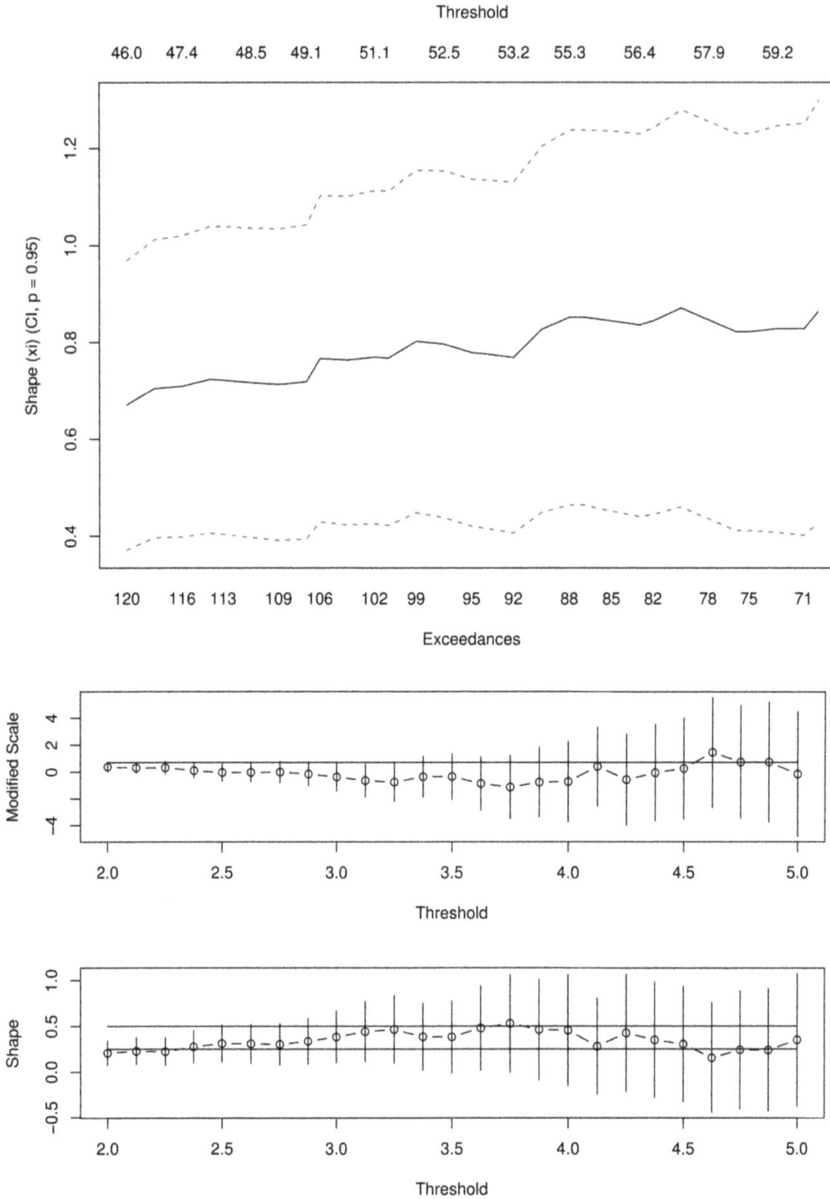

Fig. 9.10 Estimation of the shape parameter η of the GPD of T for several thresholds. Case 1 (top): H_0 is not rejected. Cases 2 and 3 (center and bottom): H_0 is rejected

Definition 9.22 (*Maximum of two vectors*) Let $\mathbf{x}, \mathbf{y} \in I\!R^d \times I\!R^d$. The *maximum* of $\mathbf{x}$ and $\mathbf{y}$, written $\mathbf{x} \vee \mathbf{y}$, is defined as

$$\mathbf{x} \vee \mathbf{y} = (x_1 \vee y_1, \ldots, x_d \vee y_d).$$

Note that $\mathbf{x} \vee \mathbf{y}$ may be equal to neither $\mathbf{x}$ nor $\mathbf{y}$. Thus, in the context of n data points $(\mathbf{x}_1, \ldots, \mathbf{x}_n)$, the maximum $\mathbf{M}_n = \mathbf{x}_1 \vee \cdots \vee \mathbf{x}_n$ may not be equal to any of them.

9.3.2 Limit Distributions of Multivariate Maxima

Theorem 9.10 extends Leadbetter's univariate result from 1974 [461]. It proves that a limit distribution for multivariate maxima exists for multivariate stationary processes under a condition $D(\mathbf{u}_n)$ bounding the temporal dependence in the process, where $\mathbf{u}_n$ is a high threshold. Details of this highly technical condition are not given here, but can be found in [65, p 420], as well as in Definition 6.10 in the present book for the univariate case only.

Theorem 9.10 *Let* $\mathbf{X}$ *be a* d*-dimensional stationary process with marginal distribution* F. *We say that* F *is in the domain of attraction of a limit distribution of multivariate maxima* G *(GEV), written* $F \in \mathscr{D}(G)$, *if there exist sequences* $(\mathbf{a}_n)_n > 0$ *and* $(\mathbf{b}_n)_n$ *in* $I\!R^d$, *and a nondegenerate distribution* G, *such that*

$$\mathbb{P}\left(\mathbf{a}_n^{-1}(\mathbf{M}_n - \mathbf{b}_n) \le \mathbf{x}\right) \longrightarrow G(\mathbf{x}), \quad n \to +\infty$$

and if condition $D(\mathbf{u}_n)$ *is satisfied for* $\mathbf{u}_n = \mathbf{a}_n\mathbf{x} + \mathbf{b}_n$ *for all* $\mathbf{x}$ *such that* $G(\mathbf{x}) > 0$.

Considering this for each marginal distribution, $F \in \mathscr{D}(G)$ implies that $F_i \in \mathscr{D}(G_i)$, which means that each marginal distribution G_i is a limit distribution of univariate maxima (GEV) of the type Fréchet, Weibull, or Gumbel.

The limit distribution of the maxima of G has the following characteristics:

- G is a *max-stable* distribution, meaning that it is in its own domain of attraction (cf. Definition 6.5);
- G is a *max-infinitely divisible* distribution, meaning that for each $k \in \mathbb{N}$, $G^{1/k}$ remains a multivariate extreme value distribution.

The dependence structure of the limit distribution of the maxima is known thanks to the help of the following result.

Definition 9.23 (*Extreme value copulas*) If C_F is the copula of F, then $F \in \mathscr{D}(G)$ implies that for each $\mathbf{u} \in [0, 1]^d$:

$$C_G(\mathbf{u}) = \lim_{t \to \infty} C_F^t(u_1^{1/t}, \ldots, u_d^{1/t}),$$

where C_G is the copula of G.

We can show that for each $\mathbf{u}$ for which each component u_i is close to 1, the *max-stability* of G implies that

$$C_F(\mathbf{u}) \sim C_G(\mathbf{u}). \tag{9.6}$$

Thus, the tail of the copula of F is that of the limit distribution of the maxima G when $F \in \mathscr{D}(G)$.

9.3.2.1 Tail Dependence Function

The properties of the distribution G also imply that it can be defined in the following way: for each $\mathbf{x} \in \mathbb{R}^d$,

$$G(\mathbf{x}) = \exp\left[-\ell(-\log G_1(x_1), \ldots, -\log G_d(x_d))\right],$$

where $\ell : \mathbb{R}^d \longrightarrow \mathbb{R}$ is the *tail dependence function* satisfying:

$$\begin{aligned} \forall s > 0, \ell(s.) &= s\ell(.), \\ \forall i = 1 \ldots d, \ell(\mathbf{e}_i) &= 1, \\ \text{and } \forall\, \mathbf{v} > \mathbf{0}, v_1 \vee \ldots \vee v_d \leq \ell(\mathbf{v}) &\leq v_1 + \ldots + v_d. \end{aligned}$$

The function ℓ is defined for each $\mathbf{v} \in \mathbb{R}^d$ by

$$\ell(\mathbf{v}) = \lim_{t \to +\infty} t\mathbb{P}\left(F_1(X_1) > 1 - \frac{v_1}{t} \text{ or } \ldots \text{ or } F_d(X_d) > 1 - \frac{v_d}{t}\right).$$

The tail dependence function only provides information about the dependence structure of G, and is linked to the copula C_G as follows:

$$C_G(\mathbf{u}) = \exp\left[-\ell(-\log u_1, \ldots, -\log u_d)\right].$$

Remark 9.11 Unlike the univariate case where the class of limit distributions of maxima is a GEV distribution parameterized by three scalars (μ, σ, ξ), the class of limit distributions of multivariate maxima has no finite parameterization since it is indexed by the tail dependence function ℓ.

We now give the expression for ℓ in the limit cases of dependence.

(i) The case where $\ell(\mathbf{v}) = v_1 \vee \ldots \vee v_d$ corresponds to $C_G = u_1 \wedge \cdots \wedge u_d$ (where $\wedge = \min$). This characterizes total asymptotic dependence: components are connected via a deterministic formula in their extreme values.
(ii) The case where $\ell(\mathbf{v}) = v_1 + \ldots + v_d$ corresponds to $C_G(\mathbf{u}) = \prod_i^d u_i$, which characterizes total independence.

(iii) For intermediate cases, [65] provides a detailed list of models for the tail dependence function ℓ. For a well-chosen set of parameters, these models allow all types of asymptotic dependence to be modeled.

9.3.2.2 Pickands' Function

Much attention has been paid to the two-dimensional case. In this setting we prefer, in place of ℓ, *Pickands' dependence function*, denoted A, which is defined on $[0, 1]$, takes values in $[1/2, 1]$, and is linked to ℓ by the formula

$$A(t) = \ell(1-t, t) \quad \forall t \in [0, 1],$$

or equivalently,

$$\ell(v_1, v_2) = (v_1 + v_2) A\left(\frac{v_2}{v_1 + v_2}\right) \quad \forall (v_1, v_2) > \mathbf{0}.$$

Pickands' function is convex and satisfies $\max\{t, (1-t)\} \le A(t) \le 1$ for $t \in [0, 1]$ [616]. The case $A(t) = 1$ corresponds to total asymptotic independence, and $A(t) = (1-t) \vee t$, to total asymptotic dependence. However, the bounding and convexity constraints on A imply that

$$A(t) = 1 \Longleftrightarrow A(1/2) = 1$$

and

$$A(t) = (1-t) \ \vee t \Longleftrightarrow A(1/2) = 1/2.$$

Thus, the real number $A(1/2)$ alone can characterize the type of dependence seen. Finally, G and C_G can be written as function of A for each $\mathbf{x} = (x_1, x_2) \in I\!R^2$ and $\mathbf{u} = (u_1, u_2) \in [0, 1]^2$:

$$G(x_1, x_2) = \exp\left\{\log\left[G_1(x_1)G_2(x_2)\right] A\left(\frac{\log\left(G_2(x_2)\right)}{\log\left(G_1(x_1)G_2(x_2)\right)}\right)\right\}$$

and

$$C_G(u_1, u_2) = \exp\left\{\log(u_1 u_2) A\left(\frac{\log(u_2)}{\log(u_1 u_2)}\right)\right\}. \tag{9.7}$$

Specialists in this domain have proposed several parametric (e.g., the logistic model in (9.16)) and nonparametric models for $A(t)$, and numerous authors have provided statistical estimators of this function. See [125] and [363] for the most recent results and bibliographies on the subject.

9.3.3 Limitations of GEV Modeling

In this section, we show that the limit distributions of multivariate maxima cannot model all types of dependence in extreme values, notably certain types of asymptotic independence.

Remark 9.12 Formula (9.7) implies that any copula C_G with a bivariate maxima distribution satisfies

$$C_G(u, u) = u^{2A(1/2)}. \tag{9.8}$$

Formulas (9.4) and (9.6) mean that we have $\chi = 2[1 - A(1/2)]$. Thus, the only way to model asymptotic independence is to set $A(1/2) = 1$, which corresponds to modeling total independence and thus forcing the limit distribution of the maxima to be of the form: $G(u_1, u_2) = G_1(u_1)G_2(u_2)$.

De facto, the only multivariate GEV distribution that models asymptotic independence is the total independence one. *Thus, GEV distributions cannot be used to model positively or negatively associated asymptotic independence.*

The use of the limit distribution of multivariate maxima to estimate the probability of simultaneous exceedances of high thresholds can have fatal consequences if the marginal processes are asymptotically independent. In effect, the independent copula underestimates the weight of simultaneous extreme values in the positive association case, and overestimates it in the negative one.

In the case of asymptotic independence, it is, therefore, necessary to model the tail of the distribution F with the help of models other than those coming from the limit distributions of multivariate maxima. For instance, the Ledford and Tawn model [473] is an alternative which can model asymptotic dependence just as well as asymptotic independence.

9.3.4 Tails of Multivariate Distributions

9.3.4.1 Non-temporally Correlated Process

If $\mathbf{X}$ is a process with no temporal correlation, then

$$\mathbb{P}\left(\mathbf{a}_n^{-1}(\mathbf{M}_n - \mathbf{b}_n) \leq \mathbf{x}\right) = F^n(\mathbf{a}_n\mathbf{x} + \mathbf{b}_n),$$

and under the hypotheses for the existence of the nondegenerate limit distribution of normalized maxima G, Theorem 9.10 implies that, for large n,

$$F^n(\mathbf{x}) \simeq G(\mathbf{x}), \quad \forall \mathbf{x} \in \mathbb{R}^d,$$

where the constants $(\mathbf{a}_n, \mathbf{b}_n)$ have been folded into the parameters of G due to its max-stability.

With the latter formula, we can approximate the tail of the distribution F, valid for x_i such that $F_i(x_i) \simeq 1$:

$$F(\mathbf{x}) \sim \exp\{-\ell(-\log F_1(x_1), \ldots, -\log F_d(x_d))\}, \tag{9.9}$$

where the F_i are the marginal distributions of the process.

In contrast, the relationship between the multivariate distributions F and G is difficult to obtain when the process exhibits temporal correlation.

9.3.4.2 Time-Correlated Processes

In the case where each marginal process X_k has temporal correlation characterized by an extremal index θ_k (see Definition 6.9), we have that

$$G_k = F_k^{n\theta_k}.$$

As the indices θ_k are not necessarily all equal, the link between G and F is not self-evident. Nandagopalan [545] has proposed extending the definition of the extremal index θ given in the univariate context to the multivariate one by defining the *multivariate extremal index*

$$\theta(\mathbf{v}) = \frac{\log G\left\{G_1^{-1}\left(e^{-\theta_1 v_1}\right), \ldots, G_d^{-1}\left(e^{-\theta_d v_d}\right)\right\}}{\log \tilde{G}\left\{\tilde{G}_1^{-1}\left(e^{-v_1}\right), \ldots, \tilde{G}_d^{-1}\left(e^{-v_d}\right)\right\}}, \quad \forall \mathbf{v} \in (\mathbf{0}, \infty).$$

It was necessary to wait until 2005 for Beirlant [65] and Martins and Ferreira [506] to propose theoretically motivated estimators of the function $\theta(\mathbf{v})$, despite a brave earlier attempt by [709]. A nonparametric solution to the inherently difficult inference of the multivariate extremal index function [247] was nevertheless recently provided by [653].

Note, however, that the tail model obtained from a GEV model of the limit distribution of maxima can only model total independence or total dependence (cf. Sect. 9.3.3). Thus, for a process with temporal correlation, just as in the asymptotic independence case, it is necessary to model the tail of F differently to in (9.9).

9.3.4.3 Ledford and Tawn Model

The Ledford and Tawn model is an interesting alternative. For clarity, we describe it here in two dimensions even though it is defined for any number.

Ledford and Tawn transform the data to obtain *Fréchet marginals* as in (9.2), and propose the following tail model on the transformed data (Z_1, Z_2):

$$\mathbb{P}(Z_1 > z_1, Z_2 > z_2) = \tilde{\mathscr{L}}(z_1, z_2)(z_1 z_2)^{-1/(2\eta)}, \quad \forall z_i > 0, \tag{9.10}$$

where η is the *tail dependence coefficient* with $0 < \eta \leq 1$, and $\tilde{\mathscr{L}}$ a *slowly varying* bivariate function. The latter hypothesis—whose mathematical definition is not shown here—is quite general and unconstraining. It implies that the probability of simultaneous extreme threshold exceedances decreases in a regular way and faster than the corresponding marginal probabilities.

Ledford and Tawn demonstrated the use of their model in several examples, and [382] has shown its quite broad applicability on an even greater number. However, [688] has provided several counter-examples, though these seem to be somewhat pathological cases [65].

Model (9.10) allows us to give the extreme value dependence coefficients χ and $\bar{\chi}$ as functions of $\tilde{\mathscr{L}}$ and η. From the expression for the survival copula $\bar{C}$ of (X_1, X_2),

$$\begin{aligned}\bar{C}(u,u) &= \mathbb{P}(Z_1 > -1/\log u, Z_2 > -1/\log u), \\ &= \tilde{\mathscr{L}}(-1/\log u, -1/\log u)(-\log u)^{1/\eta},\end{aligned}$$

we can deduce that

$$\bar{\chi} = \lim_{u \to 1} \bar{\chi}(u) = 2\eta - 1 \quad \text{and} \quad \chi = \lim_{z \to \infty} \tilde{\mathscr{L}}(z,z) z^{1-1/\eta}. \tag{9.11}$$

Therefore, Ledford and Tawn can model:

- asymptotic dependence when $\eta = 1$ if $\lim_{z\to\infty} \tilde{\mathscr{L}}(z,z) = c$, with $0 < c \leq 1$, and in this case, $\chi = c$ measures the strength of this dependence;
- positively associated asymptotic independence when $1/2 < \eta < 1$, or with $\eta = 1$ and $\lim_{z\to\infty} \tilde{\mathscr{L}}(z,z) = 0$;
- *quasi-independence* when $\eta = 1/2$, or even total independence if furthermore $\tilde{\mathscr{L}}(z,z) = 1$;
- negatively associated asymptotic independence when $0 < \eta < 1/2$.

Example 9.25 (*Farlie-Gumbel-Morgenstern copula*)

Suppose that F has a Farlie-Gumbel-Morgenstern copula C_F:

$$C_F(u_1, u_2) = u_1 u_2 \left(1 + \alpha(1 - u_1)(1 - u2)\right)$$

with parameters $-1 \leq \alpha \leq 1$. This copula is the independent one for $\alpha = 0$. It can be shown that

$$\chi(u) = 2 - \frac{\log\left(u^2\left\{1 + \alpha(1-u)^2\right\}\right)}{\log u}, \quad 0 < u < 1,$$

which implies that $\chi(u) \to \chi = 0$ as $u \to 1$, as long as $\alpha \neq 0$. Thus, all distributions F whose copulas belong to this family are asymptotically independent.

For $z \to \infty$, the formula

$$\mathbb{P}(Z_1 > z, Z_2 > z) = \frac{\alpha+1}{z^2} - \frac{3\alpha+1}{z^3} + \frac{55\alpha+7}{12z^4} + o(z^{-4})$$

allows us to obtain η and $\mathscr{L}$ as functions of α:

- for $\alpha > -1$, we have $\eta = 1/2$ and $\tilde{\mathscr{L}}(z, z) = (\alpha+1) - (3\alpha+1)z^{-1} + o(z^{-1})$ for $z \to \infty$; this is quasi-independence,
- for $\alpha = -1$, we have $\eta = 1/3$ and $\tilde{\mathscr{L}}(z, z) = 2 - 4z^{-1} + o(z^{-1})$ for $z \to \infty$: this corresponds to the negatively associated case.

Example 9.26 (*Gaussian bivariate copula*)

If F has a Gaussian bivariate copula C_F with correlation coefficient $\rho < 1$, then for $z \to \infty$

$$\mathbb{P}(Z_1 > z, Z_2 > z) \sim c_\rho (\log z)^{-\rho/(1+\rho)} z^{-2/(1+\rho)},$$

where $c_\rho = (1+\rho)^{3/2}(1-\rho)^{-1/2}(4\pi)^{\rho/(1+\rho)}$. For any value of ρ, the distribution is asymptotically independent: $\chi = 0$, with $\bar{\chi} = \rho$ and $\eta = (1+\rho)/2$. The positively associated case is modeled by $\rho > 0$, the negatively associated one by $\rho < 0$, and the total independence one by $\rho = 0$.

Remark 9.13 These copulas are not those of limit distributions of multivariate maxima since $C_F(u, u) \neq u^{2A(1/2)}$.

9.4 Study Steps

9.4.1 Overview

A study of multivariate extreme values begins, like in the univariate case, by looking for evidence of a trend and/or seasonal effects in the data. After estimating these and removing their influence from the data, what remains corresponds to the standard stationary process setting, which is necessary for the validity of all of the results given earlier in the chapter.

In a general sense, the non-stationarity of a process may be due to the presences of a break point (e.g., the date at which the process began to be measured, or some change to the environment that significantly impacted the process), the presence of a trend (e.g., increase in air temperature due to climate change), or a seasonal effect.

The study of non-stationary multivariate processes is extremely complex, and a lot of research has been undertaken in this direction. In the case of a simple seasonal effect, practitioners can simply constrain their analyses to a period of the year during which each marginal phenomenon is stationary. This strategy presupposes that the dependence structure between phenomena is itself stationary over the relevant period. This hypothesis seems reasonable and is satisfied if, for example, dependence between the given processes is constant over time. In this case, the techniques to use are those of the univariate analysis detailed in Chap. 6 of this book.

Once a common stationary period has been found, it is essential, as we have seen before, to determine the type of dependence existing between the extreme values of different processes. Indeed, on this depends whether to use a model obtained from the limit distribution of the maxima, or an alternative one modeling the tail of the multivariate distribution—like the quite general Ledford and Tawn model.

This step begins with plotting the data in the various spaces described in Sect. 9.2.2. This qualitative analysis can be supplemented with the quantitative one given in Sect. 9.2.3, using statistical tests to accept or reject the asymptotic independence hypothesis. Calculation of the coefficients χ and $\bar{\chi}$ can also help in this decision-making.

Depending on the type of dependence and the goals of the study, a practitioner can then estimate an approximation of the tail of the distribution F or the probability of simultaneous extreme events directly.

Modeling the limit distribution of the maxima can also be used to approximate the tail of the distribution with Eq. (9.9) in the total independence or extreme value dependence cases, and when the phenomena in question have no temporal correlation. This is, for example, the case in the wind speed example, where the daily maximums can be considered independent of each other. This is not, however, the case for many other phenomena, such as air temperature and river flows. In the latter cases, one can model the tail distribution of F using an alternative model that can take into account the range of types of extreme value dependence.

Finally, there exist several estimators of the probability of simultaneous extreme values that can be applied directly to the data with passing by an estimation of the tail of F. This type of probability estimation comes with a confidence interval calculated using the asymptotic distribution of the estimator in question. This interval gives an idea of how precise the estimator is. If the probability of simultaneous extreme values is estimated over a grid of levels of extremeness, practitioners can then construct plots showing the probabilities of simultaneous extreme values, giving at a glance a visual idea of when the simultaneous probability exceeds certain limit probabilities chosen a priori.

As in the univariate case, it is possible to define the annual frequency of (now simultaneous) extreme events in several ways, including (i) the mean annual number of days where the simultaneous events occur, (ii) the number of episodes of simultaneous extreme events, and (iii) the probability that a simultaneous extreme event occurs at least once per year. Readers should examine Sect. 6.3.2 of this book for more precise details on the fundamental differences between these three definitions.

9.4.2 Inference Strategies

Sklar's theorem (Theorem 9.9) describes how a multivariate distribution is constructed starting from the marginal ones F_i and the dependence structure—the copula C_F. Thus, practitioners need to infer this function C_F as well as the set of univariate F_i. In the case of Model (9.9), obtained from the limit distribution of the maxima, inference is performed on the tails of the marginal distributions F_i, as well as on the tail dependence function ℓ.

In Ledford and Tawn's alternative model, determining the marginals F_i allows us to transform the data as in (9.2) and then model dependence in this transformed data using the model (9.10).

The question is then: in which order should inference on the marginal distributions and inference on the dependence structure occur? Several strategies are possible, which impact in different ways the overall estimation of the multivariate model.

9.4.2.1 First Strategy

This strategy involves first estimating the marginal distribution, then the dependence structure, in two separate steps.

(i) Parametric inference is used on the marginal distributions because the marginals of the limit distribution of the maxima follow GEV distributions, and those of approximation (9.9) are written with the help of GPDs.
(ii) Following this, the dependence structure can be estimated using the data transformed to follow a Fréchet distribution. We distinguish between two ways of transforming the data:

- using the models established for the marginal distributions F_i, or
- using the empirical marginal cumulative distribution functions, defined by

$$F_i^{emp}(x_i) = \frac{1}{n}\sum_{k=1}^{n} 1_{\{X_i^k \le x_i\}} \tag{9.12}$$

for $x_i \in I\!R$, where n is the number of data and $(X_i^1, \ldots, X_i^n)$ the n measured values of the i-th component of the phenomenon.

This first strategy increases the risk of poorly evaluating the dependence structure since estimation of the joint model is performed on data transformed by a potentially badly fitted model. In contrast, the approach below totally decorrelates the estimation of the joint model from that of the marginal ones. It is therefore recommended [65, p. 318].

9.4.2.2 Second Strategy (Recommended)

This strategy consists of inferring the whole multivariate model in a single step, i.e., the marginal distributions and dependence structure are estimated at the same time. A complete parametric model can be used; this means jointly estimating the whole set of parameters defining the marginal models and the tail dependence function ℓ.

This strategy thus allows information transfer between the marginal distributions and the dependence function. It also has the advantage of allowing constraints to be placed between parameters from different marginal distributions, e.g., imposing the same coefficient ξ on several marginal distributions F_i.

The downside of this strategy is possible complexity in the optimization problem it leads to, which only gets worse as the number of marginal distributions (and thus parameters) increases. One trick is to first perform a univariate analysis and use its output as the starting point for the global optimization problem.

9.4.3 Inference

Inference of the limit distribution of multivariate maxima is performed on block maxima data extracted from the original data. Maximum likelihood estimators are used here. As the extracted data is assumed to be independent, classical estimator consistency and convergence theory applies (cf. Sect. 4.2.2.3).

9.4.3.1 Censored Maximum Likelihood

By definition, models for the tails of distributions are valid in the tails, i.e., when all components are simultaneously extreme valued. It is, therefore, tempting to infer such models using data in which all components are extreme values. However, this naive approach considerably reduces the number of data that can be used, and makes model inference impossible in practice.

For this reason, *censored* statistical likelihood is used, whereby information is gained from *all* of the data (see Example 4.9 for more details). The principle is simple: data for which certain components are not extreme values are interpreted as *censored* but not removed. Then, each data point contributes to the model likelihood in proportion to its number of extreme-valued components. The larger this number, the more it contributes. For instance, in the bivariate case, the likelihood $L(x_1, x_2)$ of a data point (x_1, x_2), for a given threshold (u_1, u_2), is given by

$$L(x_1, x_2) \propto \begin{cases} F(u_1, u_2) & \text{if } x_1 \le u_1, x_2 \le u_2 \\ \dfrac{\partial F}{\partial x_1}(x_1, u_2) & \text{if } x_1 > u_1, x_2 \le u_2 \\ \dfrac{\partial F}{\partial x_2}(u_1, x_2) & \text{if } x_1 \le u_1, x_2 > u_2 \\ \dfrac{\partial^2 F}{\partial x_1 \partial x_2}(x_1, x_2) & \text{if } x_1 > u_1, x_2 > u_2. \end{cases} \tag{9.13}$$

A list of parametric models that allow the tail dependence function ℓ to be estimated can be found in [65]. Nonparametric estimators of ℓ also exist, but most have difficulty respecting the constraints on the function. Constructing estimators that do so is an open problem, even in the bivariate case.

Similarly, several parametric and nonparametric models for the function $\tilde{\mathscr{L}}$ from the Ledford and Tawn model can be found in the literature; these are defined starting from the spectral measure [243], an alternative way to represent distributions of multivariate extreme values.

Definition 9.24 (*Spectral measure*) Let G be a multivariate extreme value distribution. For all $\mathbf{z} \in [\mathbf{0}, \infty]$, the exponential measure is defined by

$$\mu_*(\mathbf{z}) = -\log G(\mathbf{z}). \tag{9.14}$$

The spectral measure H is the (pseudo-)polar representation of μ_*, defined by

$$H(B) = \mu_* \left(\mathbf{z} \in [\mathbf{0}, \infty] : \|\mathbf{z}\| \ge 1, \frac{\mathbf{z}}{\|\mathbf{z}\|} \in B \right) \tag{9.15}$$

for any Borelian B of the unit sphere in $I\!R^d$.

9.4.3.2 Example

In this section, we illustrate inference of the model obtained from the limit distribution of the maxima for the conjunction of flooding of the Vienne and Loire Rivers (case 1 in Sect. 9.1.3). Earlier in the chapter, we showed that these two phenomena had real dependence in their extreme values.

The two inference strategies were put into practice, with little difference appearing between them. The dependence function ℓ was fitted to a logistic model defined by

$$\ell(v_1, v_2) = \left(v_1^{\frac{1}{\alpha}} + v_2^{\frac{1}{\alpha}} \right)^{\alpha}, \quad \forall (v_1, v_2) \in (\mathbb{R}^+)^2, \quad 0 < \alpha \le 1. \tag{9.16}$$

Table 9.1 Parameters of the marginal distributions of the bivariate distribution of extreme flow measurements (Vienne and Loire Rivers)

Vienne	Estimation	St. dev.	**Loire**	Estimation	St. dev.
$u_1\ (m^3.s^{-1})$	1050	–	$u_2\ (m^3.s^{-1})$	2150	–
$\sigma_1\ (m^3.s^{-1})$	335.29	35.75	$\sigma_2\ (m^3.s^{-1})$	571.14	49.50
ξ_1	−0.126	0.08	ξ_2	−0.259	0.05
λ_1	0.012	–	λ_2	0.0119	–

Table 9.2 Parameter of the logistic model for dependence in the bivariate distribution of extreme flow measurements (Vienne and Loire Rivers)

Dependence	Estimation	St. dev.
α	0.879	0.02

The parameter α measures the strength of dependence between the extreme flows of the two rivers. In particular, total dependence corresponds to $\alpha \to 0^+$ and total independence to $\alpha = 1$.

As for F, it is given by, for each $(x_1, x_2) > (u_1, u_2)$,

$$F(x_1, x_2) = \exp\left[-\left\{(-\ln F_1(x))^{\frac{1}{\alpha}} + (-\ln F_2(x))^{\frac{1}{\alpha}}\right\}^{\alpha}\right],$$

where the tails of the marginal distributions F_i are modeled with the help of GPDs:

$$F_i(x) = 1 - \lambda_i\left[1 + \xi_i \frac{x - u_i}{\sigma_i}\right]^{-\frac{1}{\xi_i}}, \quad \forall x_i > u_i,$$

with $\lambda_i = \mathbb{P}(X_i > u_i)$.

Tables 9.1 and 9.2 summarize the parameter estimation for the whole model. The standard deviations of the estimators can be used to obtain confidence intervals constructed with the help of standard normal distributions. Figure 9.11 plots Pickands' dependence function $A(t)$ of the model selected and several iso-quantiles defined by $F(x_1, x_2) = \text{constant}$.

9.4.3.3 Inferring the Tail Dependence Coefficient

The previous sections have shown the importance of Ledford and Tawn's tail dependence coefficient η, due to it being able to help distinguish between different types of convergence.

We now have a look at some of the many estimators of η available. Several others can be found in [65], along with their asymptotic properties: those of Peng [605], Hill [389], Draisma [226], and Beirlant and Vandewalle [68].

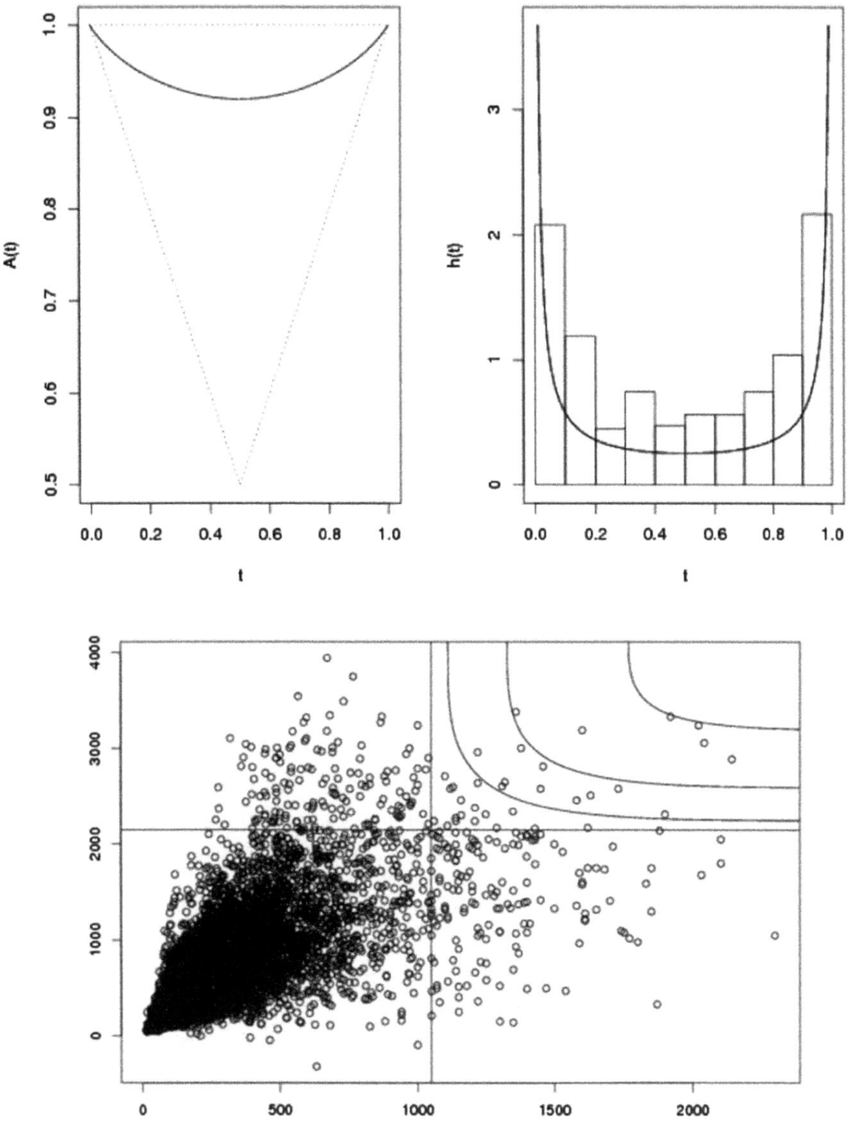

Fig. 9.11 Bivariate flow distribution: plot of Pickands' function $A(t)$ (top-left), spectral density (top-right), and iso-quantile lines (bottom). Bottom: the $x-$axis indicates the discharge values (m^3/s) of river Vienne, while the $y-$axis indicates the corresponding discharge values for river Loire

Ledford and Tawn's estimator is based on the following result: the tail of the distribution T, a scalar random variable defined by

$$T = \min(Z_1, Z_2),$$

where the variables (Z_1, Z_2) are the data transformed according to (9.2), follows a GPD with extremal index $\xi = \eta$. Thus, we can estimate η by maximizing the censored likelihood of a GPD model over a sample of data from T.

However, construction of a sample from T beginning with the observed data relies on the modeling of the marginal distribution functions F_i; the latter can be approximated by their empirical cumulative distribution functions—thus giving a sample denoted T_{emp}, or with the help of the Pareto model estimated from the data. The latter approach leads to a sample which we call T_{pot}. By mixing together various estimators of η based on samples from T, and the multitude of ways of forming this sample, we can construct a whole range of estimators of η.

Here are four examples:

1. Hill's estimator $\hat{\eta}_{he}$, based on T_{emp};
2. the estimator $\hat{\eta}_{me}$, obtained by maximizing the likelihood of the GPD model based on T_{emp};
3. Hill's estimator $\hat{\eta}_{hp}$, based on T_{pot};
4. the estimator $\hat{\eta}_{mp}$, obtained by maximizing the likelihood of the GPD model based on T_{pot}.

Figure 9.12 for case 4 (cf. Sect. 9.1.3) shows the behavior of these four estimators as a function of the *thresholding parameter* p_η. This parameter is set at the largest value for which estimation of η is stable and all of the estimators converge to the same value. We retain $p_\eta = 0.96$ and $\hat{\eta} = 0.51$. This value is very close to 0.5, which characterizes independence, factually true in this case.

9.4.3.4 Estimating the Probability of Simultaneous Extreme Values

This section studies several examples of estimators of the probability of simultaneous extreme values, defined as exceedances of extreme thresholds. These estimators, constructed differently, have different statistical properties. Practitioners need to compare and contrast them in order to decide how to proceed.

We proceed based on the method of Draisma et al. [226], which involves Ledford and Tawn's model (9.10). These estimators are shown here in two dimensions but can be generalized to an arbitrary number.

Let us consider the probability $p = \mathbb{P}(X_1 > x_1, X_2 > x_2)$, where x_1 and x_2 are marginal extreme values. Denoting $p_1 = \mathbb{P}(X_1 > x_1)$ and $p_2 = \mathbb{P}(X_2 > x_2)$, it is possible to show that Ledford and Tawn's model corresponds to approximating p by

$$p \approx \left(\frac{c}{q_0}\right)^{1/\eta} \mathbb{P}\left(F_1(X_1) < q_0\frac{p_1}{c}, F_2(X_2) < q_0\frac{p_2}{c}\right), \tag{9.17}$$

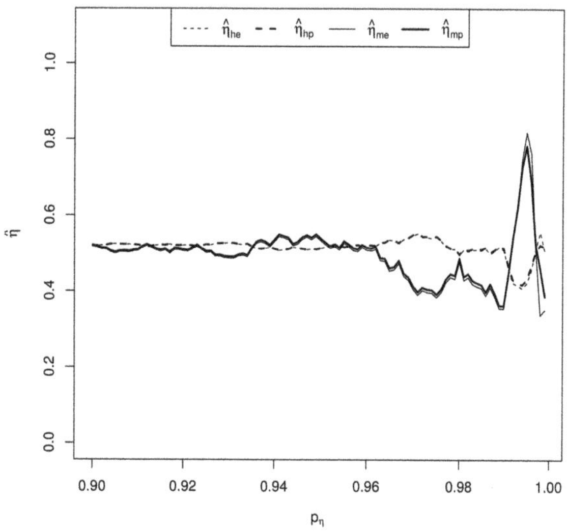

Fig. 9.12 Case 4: estimating η as a function of the parameter p_η

where q_0 is a probability for which $q_0 \to 0$, and

$$c = \sqrt{p_1^2 + p_2^2}.$$

The estimators detailed here are all derived from (9.17) and involve various estimates of the quantity

$$\tilde{p} = \mathbb{P}\left(F_1(X_1) < q_0\frac{p_1}{c},\ F_2(X_2) < q_0\frac{p_2}{c}\right),$$

alongside the tail coefficient η, estimated separately.

Estimator p_e uses the empirical estimator of the mean to approximate $\tilde{p}$ and model the marginal distributions F_1 and F_2 using their empirical estimators. If we take q_0 such that $nq_0 \to \infty$, then since $0 < p_1/c_0,\ p_2/c_0 < 1$, a certain number of observations X_1^i (resp. X_2^i) have a rank above $1 - q_0\frac{p_1}{c}$ (resp. above $1 - q_0\frac{p_2}{c}$). The use of the empirical cumulative distribution function is thus possible.

Estimator p_p uses the empirical estimator of the mean to approximate $\tilde{p}$, and models the approximate marginal distributions F_1 and F_2 with the help of Pareto models.

These first two estimators were studied in [226, 369], and their asymptotic distributions provided.

Estimator p_T (from [133]): set $r = p_2/p_1$ and let T_r be the variable defined by

$$T_r = \min\left(\frac{1}{1 - F_1(X_1)}, \frac{r}{1 - F_2(X_2)}\right).$$

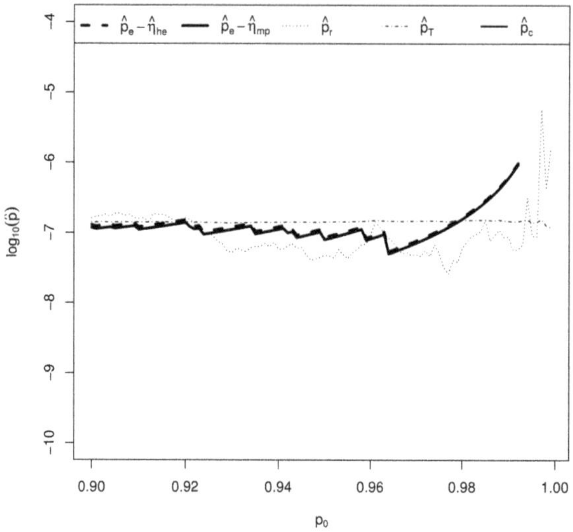

Fig. 9.13 Probability of simultaneous exceedances of extreme thresholds as a function of a parameter p_0: estimators p_e, p_p, p_T, and p_r

Then, [133] showed, under the Ledford and Tawn model's hypotheses, that $\tilde{p}$ can be estimated consistently by

$$p_T = \frac{k}{n}\left(p_1 T_{r,(n-k,n)}\right)^{1/\hat{\eta}}. \tag{9.18}$$

Estimator p_r takes temporal correlation in T_r into account. Under certain conditions on the correlation (which has to be short term), the limit distribution of the maxima of T_r is still a GEV distribution with parameters (ξ, σ, u). Let $\tilde{T}_r$ be the independent process with the same marginal distribution as T_r in the domain of attraction of $\tilde{G}$, a GEV with parameters $(\tilde{\xi}, \tilde{\sigma}, \tilde{u})$. Then it is possible to link (ξ, σ, u) and $(\tilde{\xi}, \tilde{\sigma}, \tilde{u})$ [65, p. 377] with the help of the extremal index θ. Details of this can be found in Chap. 6 (Sect. sec:Fund).

We can, therefore, deduce that the parts of $\tilde{T}_r$ exceeding the threshold $\tilde{u}$ follow a GPD distribution with parameters $(\tilde{\xi}, \tilde{\sigma})$. The estimator p_r is then defined by

$$p_r = \left(\frac{1}{n}\sum_{i=1}^{n} \mathbf{1}_{\{T_{r,i} > \tilde{u}\}}\right)\left(1 + \tilde{\xi}\frac{1/p_1 - \tilde{u}}{\tilde{\sigma}}\right)^{-1/\tilde{\xi}}. \tag{9.19}$$

Figure 9.13 shows the graphs obtained when estimating the probability of simultaneous exceedances of extreme thresholds for case 4, as a function of a parameter p_0. This parameter is set at the largest value for which the probability estimates are stable and converge.

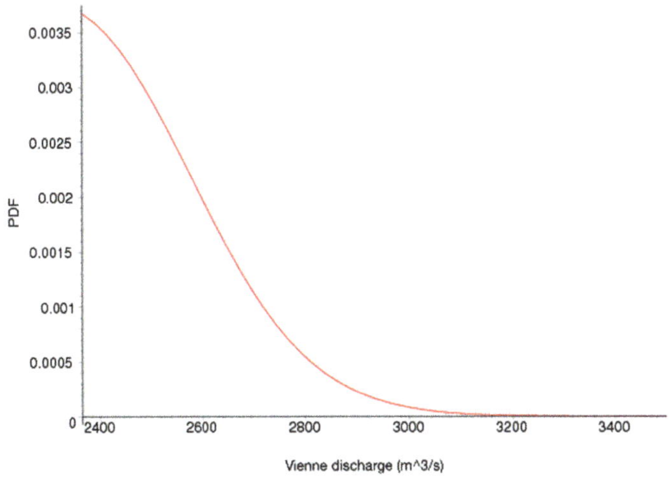

Fig. 9.14 Probability density of Vienne River flows when the Loire River is undergoing a 100-year flood

Table 9.3 Probabilistic characteristics of the flow distribution of the Vienne River in conjunction with a 100-year flood of the Loire River

Probabilistic quantities	Vienne flow rates ($m^3.s^{-1}$)
Mean	2513
St. dev.	141
Median	2517
Quantile of order 25%	2436
Quantile of order 75%	2629

9.4.4 Using the Assessed Models

We now describe several of the many ways to make use of results coming from a multivariate analysis.

When the analysis finds an expression for the tail of the multivariate distribution F of the phenomena, it is then simple to model various extreme-valued conditional scenarios. This is done by setting one of the phenomena to a particular scenario; the goal is then to regard the impact of this on the other phenomena.

For example, in Case 1 with the conjunction of two rivers flooding, we can look at the distribution of flows for the Vienne River when the Loire is in a 100-year flood situation. Figure 9.14 plots the probability density and the cumulative distribution function of the flow distribution in the latter setting. Table 9.3 lists several related probabilistic features.

The expression for F can also be used to help calculate iso-quantile surfaces, defined for levels $0 < q < 1$ by

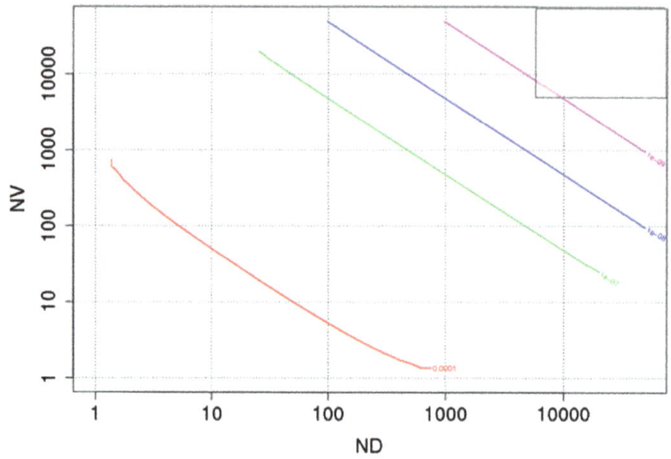

Fig. 9.15 Annual iso-frequency contours (log scale) with levels 10^{-9}, 10^{-8}, 10^{-7}, and 10^{-4}, in the return period space for flow and wind (N_D, N_V): simultaneous extreme values with the same annual frequency

$$\mathcal{S}_q = \left\{ \mathbf{x} \in \mathbb{R}^d \mid F(\mathbf{x}) = q \right\}.$$

If we associate the notion of quantile with that of criticality (or extremeness), then these surfaces correspond to the set of *iso-risk* combinations. For example, Fig. 9.11 shows the iso-quantile lines of order $q = 0.99, 0.995$, and 0.999 of the bivariate flow distribution.

Estimators for the probability of simultaneous exceedances of extreme-valued thresholds can be used to plot *probability maps*. For instance, for Case 4, Fig. 9.15 shows annual iso-frequency contours, defined in terms of the mean number of days per year where simultaneous storms and floods occurred. The level of extremeness of each marginal phenomenon is given by its return level. Chapter 17 also shows a probability map in a case of positively associated asymptotic independence.

Finally, let us mention an original use of the function $u \longrightarrow \bar{\chi}(u)$ in the asymptotic independence case: bounding $\bar{\chi}(u)$ for $u > u_s$ makes it possible to bound the probability of simultaneous extreme events occurring. Though less precise that an actual estimation of the probability, an upper bound on the latter probability may sometimes be sufficient in showing that the simultaneous extreme events in question are so unlikely to occur together that they are not worth studying in more detail. This type of reasoning assumes that the practitioner has provided extremely low probabilities of occurrence below which they will not study these events in more detail.

For instance, in Case 3, we found that the low-temperature and *river baseflow* phenomena had positively associated asymptotically independence. The hypothesis of their independence would, therefore, lead to an underestimate of the probability of their simultaneous extreme-valued occurrence.

A study of the function $u \longrightarrow \bar{\chi}(u)$ in Fig. 9.8, estimated on the opposite values of the phenomena (do not forget that the theory provides tools for studying extremely high values!), shows that for any $u > u_s$, $\overline{\chi}(u) < \epsilon_1$, where $u_s = 0.9$ and $\epsilon_1 = 0.3$. Note that the end of the curve should not be taken into consideration because estimation of $\bar{\chi}$ takes place via the empirical estimator of $\bar{C}(u, u)$, which is imprecise when u tends to 1 for lack of data.

A little algebra allows us to link the quantile q of order u of $(-X)$ to the quantile of order $(1-u)$ of X: $\mathbb{P}(-X < q) = u$, which is equivalent to $\mathbb{P}(X < -q) = 1-u$. We can then return to the initial flow and temperature processes. Thus, an upper bound on $\overline{\chi}(u)$ implies that for any extremely low order v such that $v < 1 - u_s = 0.1$, we have

$$\mathbb{P}(Q < q_v, T < t_v) < v^{\frac{2}{1+\epsilon_1}} \simeq v^{1.6},$$

where q_v and t_v are the order v quantiles of the flow variable Q and air temperature T.

9.5 Conclusion

Extreme value multivariate analyses, like univariate ones, require a certain number of choices on the part of the analyst. There is not yet a completely automatic and general approach available, and one often has to take details of the actual environmental processes of interest into account when decision-making.

In addition, certain multivariate problems remain little-studied in the literature. Let us now mention a few such difficulties that we are often faced with.

Temporal correlation between the observed values of each natural process (e.g., temperature, flow, etc.) complicates estimation of the joint distribution of multivariate processes. For example, there are still very few simple rules for the statistical estimation of the extremal index coefficient in the multivariate setting. More generally, the processes we study are often non-stationary, e.g., a trend in temperature measurements due to global warming. In such cases, it becomes impossible to apply the results presented in the chapter.

The threshold exceedance approach—though well-developed in the univariate setting—does not easily follow through to the multivariate one. An extension of the unifying framework of colored point processes to the multivariate setting, initiated by [546], is an area of active research. Indeed, in the multivariate case, it is difficult to define what *clusters* are, i.e., sequences of successive exceedances of a threshold that can be seen as members of the same exceedance events. Little work has been attempted on this, though we can cite two multivariate declustering methods: one from Coles and Tawn [178] and the other from Nadarajah [542]. However, these methods do not work with all types of dependence and require careful choices for numerous parameters.

The joint distribution estimated by extreme value theory is only valid in the zone where all of the components are simultaneously extreme valued. However, zones where only certain components are extreme valued are also interesting. In such cases, the results we have presented here cannot be used, and new techniques need to be developed.

Finally, note that the notion of a multivariate return period is more complicated to define than in the univariate setting. One way it can be defined is as the mean waiting time between two successive occurrences of the process in a domain chosen by the practitioner. This domain can be that which corresponds to all (or some) of the processes' components exceeding some extreme-valued threshold.

Kendall has suggested an interesting and original definition of this domain: it is defined by events of the process corresponding to quantiles above some given level. This, therefore, links the notion of quantile with that of criticality or extremeness [328, 329, 334]. On the iso-quantile surface, it, therefore, corresponds to choosing a particular combination that a practitioner will retain when dimensioning their structure. This choice may be based on probabilistic (the most likely combination) or project-specific (the combination which would have the largest impact on the structure in question) considerations.

Chapter 10
Stochastic and Physics-Based Simulation of Extreme Situations

Sylvie Parey, Thi-Thu-Huong Hoang, and Nicolas Bousquet

Abstract This chapter addresses two alternative approaches to extreme situations, which may be useful when the lack of extreme observations severely limits the relevance of a purely statistical approach. The first methodology is based on stochastic modeling of the regular phenomenon, for example, via autoregressive processes. Capturing this phenomenon allows extrapolation to extreme values based on theoretical properties of stochastic processes. The second approach is based on the use, by Monte Carlo-based methods, of numerical simulation models implementing the physical equations representing the phenomenon under study. This second approach, which is originally used in structural reliability, requires the development of specific simulation techniques that focus on very low-probability events; it appears to be more and more appropriate as expert knowledge of the phenomena increases.

10.1 Introduction

The preceding chapters have shown that it is not straightforward to directly apply the statistical theory of extreme values to natural hazards. On the one hand, variables are not necessarily stationary nor independent, and on the other, there is sometimes simply not enough data available for the theory to be applicable or the results reliable. Indeed, the theory is based on asymptotic convergence toward known extreme value distributions. The actual amount of available data is always limited, but it is important that the number of values over which a maximum is extracted be sufficiently large—or the threshold in the POT approach sufficiently high—for convergence to be considered attained. By way of example, when we wish to estimate the return level of a *frost index* (defined as the sum of the absolute values of temperatures below 0 °C [754]), there are in general very few values per year to work with. In this specific case, it is preferable to use a daily mean temperature model to simulate a larger sample of frost indexes. It is therefore essential to have *stochastic models* that

S. Parey · T.-T.-H. Hoang · N. Bousquet (✉)
EDF R&D, 6 quai Watier, 78401 Palaiseau, France
e-mail: nicolas.bousquet@edf.fr

N. Bousquet and P. Bernardara (eds.), *Extreme Value Theory with Applications to Natural Hazards*, https://doi.org/10.1007/978-3-030-74942-2_10

correctly simulate the entire distribution, including its extreme values. This approach is considered in the first part of this chapter (Sect. 10.2).

When dangerous events are extreme-valued examples of physical phenomena that can be described by a series of physics equations—giving a *mechanistic model* of the phenomenon, and when this model can be implemented on a computer, the numerical model (i.e., *code*) thus created can also be used to simulate virtual observations of the hazard and calculate occurrence probabilities and return levels. This alternative approach to the use of extreme value statistical theory is deeply associated with the field of *structural reliability* [475], which studies the response of complex physical systems (e.g., buildings) to probabilistic *stimuli* (e.g., strong winds) that are rarely or never seen.

For instance, the flow dynamics of a river are well described by free surface fluid dynamics equations. A set of solvers and tools dedicated to solving problems arising from these are found in the TELEMAC-Mascaret[1] software, initially developed by EDF/LNHE and managed since 2010 by a consortium of six international research centers (including EDF R&D). This software permits simulation of downstream river flows, taking as input parameters a set of upstream flows, information on the river's geometry, and its friction parameters. Depending on the level of detail sought at the output, and knowledge available on the inputs, adding probability distributions to the latter makes it possible to simulate a distribution of flows, followed by water levels, and conduct a Monte Carlo-based analysis.

Two important problems associated with Monte Carlo-based approaches are discussed in the second part of the chapter (Sect. 10.3): difficulties due to the calculation time required to explore the input parameter space before reaching areas characterizing extreme-valued outputs (often cripplingly long), and the validity of these models for such extreme situations of interest. This second point is crucial because such situations are sometimes characterized—in the case of natural hazards—by a sudden change in the physics. For instance, a flood can be defined as a dike overflow, which can highly modify the flow by adding new physical processes to the scenario. The domain of validity of an implemented numerical model may be strongly affected by the influence of these new physics and the presence of so-called *edge effects* such as local losses of continuity or differentiability. The uncertainty surrounding the extrapolation of a numerical model beyond its domain of validity is therefore likely to severely limit its use in practice.

10.2 Stochastic Models

The scientific literature is rich with works on the stochastic modeling of environmental variables, especially meteorological ones. These models are generally characterized by their faithful reproduction of the average, median, quartiles, etc., of the

[1] See www.opentelemac.org.

observed statistical distribution of a variable. However, they have more difficulty in correctly reproducing variability and, in particular, extreme values of phenomena [296].

10.2.1 A First Example: Air Temperature

The environmental variable most associated with stochastic modeling is without doubt the air temperature at a given altitude. The general principle underlying stochastic models of the daily temperature (whether it be the mean, maximum, or minimum), denoted $X(t)$, is the combination of a deterministic part $\{\Lambda(t), \Phi(t)\}$ and a stochastic process $Y(t)$ for relevant random fluctuations—aiming to encapsulate natural variation in the phenomenon (see Sect. 3.1.2):

$$X(t) = \Lambda(t) + \Phi(t)Y(t).$$

- The component $\Lambda(t)$ contains at least one seasonal cycle $S(t)$ [136], such that (typically):

  $$S(t+T) = S(t),$$

  and fairly often a trend $T(t)$ modeled by linear regression [136]:

  $$T(t) = \alpha t. \tag{10.1}$$

- The stochastic part is generally given by a stationary *autoregressive* (AR) function of order p of varying levels of sophistication. Typically this is in the form of an AR(1) ($p = 1$) up to AR(3) ($p = 3$) model:

  $$Y(t) = \alpha_0 + \alpha_1 Y(t-1) + \alpha_2 Y(t-2) + \ldots + \alpha_p Y(t-p) + \varepsilon_t,$$

  where $(\varepsilon_t)_t$ may be either:

(a) a *white noise*, i.e., a sequence of centered uncorrelated variables with constant variance:

$$\mathbb{E}[\varepsilon_t] = 0, \quad \mathbb{V}[\varepsilon_t] = \sigma^2, \quad \mathbb{C}ov[\varepsilon_t, \varepsilon'_t] = 0 \ \text{for}\ t' \neq t,$$

(b) or a *GARCH*(p, q) *process* (*generalized autoregressive conditional heteroskedasticity*), a standard choice in time series econometrics [100], defined by

$$\begin{aligned}
\mathbb{E}[\varepsilon_t|\mathscr{F}_{t-1}] &= 0, \\
\mathbb{V}[\varepsilon_t|\mathscr{F}_{t-1}] = \sigma_t^2 &= \omega + \sum_{i=1}^{q} \beta_i \varepsilon_{t_i}^2 + \sum_{j=1}^{p} \gamma_j \sigma_{t_j}^2,
\end{aligned}$$

where $\mathscr{F}_{t-1}$ is the filtration (or σ-algebra) representing the past of the process $(\varepsilon_t)_t$. This simply is called an ARCH process when $p = 0$ [252].

For some models, the variance can be seasonal; a yearly cycle seen in the variance can be modeled using seasonal coefficients [649].

However, the behavior of extreme values (their occurrence, heatwaves, cold snaps, etc.) cannot be well reproduced by analyzing the large deviations of these models, in particular, due to poor estimation of the conditional variance $\mathbb{V}[\varepsilon_t|\mathscr{F}_{t-1}]$ [296]. In reality, all studies based on the temperatures of the last 60 years indicate that the temperature distribution is bounded, meaning a negative shape parameter [427]—this independent of the method used to study the extreme values: GEV, POT, Hill's estimator, on either the original or normalized data. For this reason, linear Gaussian models or GARCH processes with unbounded tails generally model the extreme values poorly.

10.2.2 Second Example: Rainfall

Precipitation is a more complex variable, with high variability (both temporally and spatially [187, 740]), but is in particular characterized by numerous zeros: days when it does not rain. This random variable is therefore both continuous and discrete. Because of this, most rainfall generators involve two separate simulators:

1. A *rainfall simulator*, generally constructed as a discrete Markov process (cf. Definition 4.2), whose states (*classes*) may represent:
   (a) a rain or not rain event [513, 786] (two classes);
   (b) a dry day, light rain, heavy rain [743] (three classes);
   (c) *hidden* states of the regional climate [24].
2. A second simulator for the amount of precipitation, conditional on the occurrence of a rain event, founded on one of the following distributions:
   (a) a Gaussian distribution truncated to be always positive, and taken to a certain power to recreate the heavy tails of precipitation distributions [24];
   (b) a standard exponential distribution, gamma distribution, or a mixture of exponentials. The latter is used notably by the DTG (General Technical Division), EDF's team of hydraulics experts.

In parallel, *episode models* can also be used. These no longer model pointwise precipitation at given time steps, instead modeling rainfall episodes spread out over several [377, 486, 662]. These models have the advantage of reproducing the main features of precipitation observed over different time steps: intra-day, daily, and multi-day.

10.2.3 Third Example: Wind Speed

Numerous stochastic models make it possible to represent wind speed—sometimes coupled with other climatic and spatial variables to form generators [110]—but these models remain questionable with respect to the behavior of extreme values. In reality, wind is a naturally unstable variable; it is intermittent and its properties (including how its extreme values behave) can be different depending on the time step used.

For the wind speed variable, scientists sometimes admit that extreme wind speed values are Gumbel distributed. More generally, the Weibull distribution is accepted as correctly representing wind speed distributions, and its extreme values belong to the Gumbel's domain of attraction [31, 83, 198]. It does not seem, moreover, that there is an *observed* upper limit for wind speed, which is also consistent with a Gumbel distribution.

However, some reject the hypothesis of "Gumbel" behavior for extreme wind speeds. The authors of [424, 772] show that the wind speed is better represented using a bounded distribution. According to [757], "intuitively and physically", the wind speed should have an upper limit and thus be bounded. Many other distributions have been proposed to fit wind speed distributions—which are non-negative and right-asymmetric. Examples include the Weibull, gamma, Rayleigh, truncated normal, log-normal, square root normal, Skew-t, and the mixture of a Weibull and a normal distribution. For more details, see [142]. The Markovian models proposed by [269], which adopt a POT approach for the behavior of the extreme values, propose for their part an explicit representation of the temporal dependence, and are in this sense close to the non-stationary statistical models proposed in Chap. 8, and the multivariate ones in Chap. 9.

10.2.4 Fourth Example: Solar Radiation

The most popular stochastic models for daily, monthly, or annual solar radiation accumulation are based on linear or quadratic regression linking components of radiation (direct, diffuse, albedo*), duration, cloudiness, temperature, and other meteorological variables [28, 571]. Markov chain approaches have also been used, such as the one in [747], which models radiation using a continuous-time Markov chain.

Solar radiation is, in essence, a bounded variable: with a clear sky and no clouds, the earth receives direct solar radiation. It is therefore possible to estimate this upper bound by dividing the maximum observed irradiance by the *clear-sky index*. The latter is the ratio between the total irradiance observed on Earth and the (theoretical) irradiance that would be observed with clear skies [93, 535, 631]. Nevertheless, it would seem that no up-to-date study has specifically discussed the tail behavior of solar radiation.

10.2.4.1 Discretizing Continuous Stochastic Processes

Note that environmental variables are continuous over time. Though being measured at discrete time steps, it would thus appear appropriate to model their evolution using continuous stochastic processes such as *diffusion processes*—of which *Brownian motion* and *Wiener processes* are the most well-known.

Definition 10.1 (*Diffusion process*) A diffusion process $\{X_t\}_t^*$ is a stochastic process which is the solution to a stochastic differential equation of the form

$$\frac{dX_t}{dt} = \mathscr{M}(X_t) + \sum_{k=1}^{n} g_k(X_t)\xi_k(t), \tag{10.2}$$

where X_t is a differentiable random process, $\mathscr{M}$ a term representing the deterministic part of the evolution of X_t, and $\{g_k\}_k$ a set of functions coupling the process's evolution with a vector of white noise $\{\xi_k\}_k$ (usually Gaussian). The solution $\{X_t\}_t^*$, when it exists, is a continuous-time Markov process such that there is *almost surely* a *sample path* from each state of X_t to each other state.

This stochastic differential equation represents the noisy dynamics of a physical phenomenon. Though (10.2) does not have only one interpretation and can lead to several types of solution (Itô or Stratonovich formulation), it is unambiguously linked to the Fokker–Planck equation, which characterizes the temporal evolution of the probability density of the speed of a particle under the influence of drag and random forces. The latter can in turn be linked with the Langevin equation, which approximates the behavior of a system of particles. The choice of a solution process for modeling certain macroscopic variables thus implies, in theory, a change of scale and a relaxation of constraints in the description of underlying physical phenomena. This is however supported by the observed behavior of certain environmental variables (e.g., seismic activity [504]), and the possibility to perform approximation (of the Gauss–Galerkin type) under boundary conditions on the phenomena using sequences of discretized measures [248].

The discretization of a continuous process comes, however, with dangers since it is generally not guaranteed that theoretical results applying to the continuous setting—such as the tail behavior of distributions—are transferable to the world of discrete processes. By way of example, if only wind speed measurements at discrete time steps are taken into account, certain extreme events may not be recorded. In such cases, the characteristics of the extreme values may change depending on the time step chosen. Tail behavior, however, is preserved in the case of diffusion processes, which makes them of particular interest: the tail behavior of iid discrete variables derived from the marginal distribution of the process remains the same as that of the continuous-time diffusion process.

This justifies the pertinence of an approach in which a discretized version of a diffusion process is used as a simulation model for the discrete variables, and extreme value theory specific to diffusion processes [85, 200] is used to correctly reproduce discrete extreme values.

The use of stationary diffusion processes defined with Gaussian noise has proved particularly useful for modeling the (residual) air temperature. The following section describes this further. These same tools can be applied to radiation, combined with the use of a hidden Markov model [747]. For other variables such as wind and rain, physical constraints pose numerous problems limiting the relevance of using diffusion processes. Wind is unstable and has an asymmetric distribution, requiring a step transforming (e.g., log or *logit*) the discretized data to get back to the Gaussian setting. As for rainfall, the presence of a huge mass at the value zero makes this variable less continuous, but we could consider using a diffusion model alongside a Markov chain, as in [747].

10.2.5 Stochastic Modeling Using Bounded Diffusion Processes

This section summarizes an example of stochastic modeling applied to air temperatures, as detailed in [191]. The approach is quite general: first, remove the deterministic parts (trends, seasonality) to make the variable more stationary, then model what remains as a discretized diffusion process.

10.2.5.1 Stationarity

The direct use of a stochastic model (e.g., a diffusion one) whose coefficients vary as a function of time is not feasible. First, we must deal with the presence of a trend and seasonality in temperature time series. Indeed, such time series with time steps of one day exhibit clear seasonal cycles and trends (Fig. 10.1). The idea is to make the given temperature time series $\{X(t)\}_t$ stationary by removing trends and seasonality estimated based on the following model: $\forall t \in [1, T] \subset I\!N$,

$$X_t = m_t + u_t + s_t v_t Y_t, \tag{10.3}$$

where $\{u_t, v_t^2\}$ and $\{m_t, s_t^2\}$, respectively, correspond to seasonality and trends present in the mean and variance. $\{Y_t\}_t$ is called a *residual time series*. Model (10.3) is identifiable if the following conditions are satisfied [191]:

$$\sum_{t=1}^{T} m(t) = 0, \qquad \sum_{t=1}^{T} s(t)^2 = 1. \tag{10.4}$$

Trends are usually modeled using linear functions as in (10.1). However, this is not of use here because the trend depends on the considered period. It is preferable to

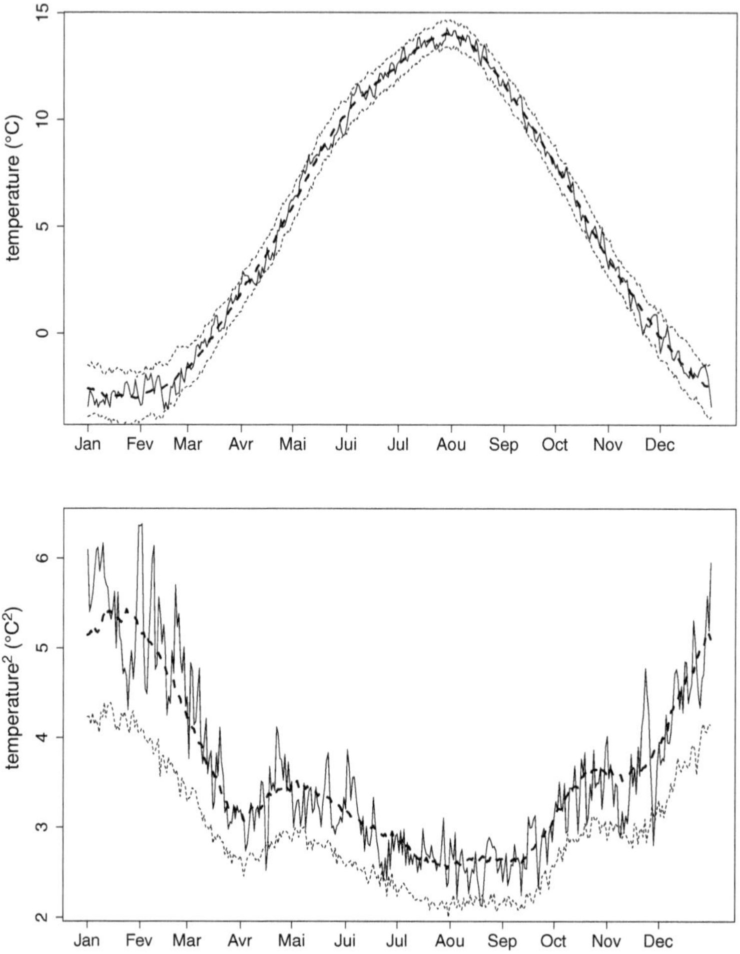

Fig. 10.1 Comparisons between the daily means (top) and variances (bottom) of observed data and simulations (with a 95% confidence interval)

use local regression modeling of the LOESS type [169], which is more flexible. As in most nonparametric approaches, LOESS requires an optimal estimation window (in years) to be determined, inside of which groups of data are considered homogeneous. However, the temperature is a strongly autocorrelated variable, meaning that standard selection criteria such as cross-validation are difficult to apply usefully. A new criterion, proposed in [391], can help determine the optimal window size. In most temperature studies, it turns out that for estimating low-frequency trends, a window of 10–15 years is generally used.

More generally, estimation of the other parameters in Model (10.3) proceeds as follows (technical details can be found in [191, 391]).

(a) **Estimating seasonality** u_t. Additive seasonal effects u can be adequately represented in terms of trigonometric polynomials:

$$P_p(t) = \psi_0 + \sum_{k=1}^{p} \psi_{k,1} \cos \frac{2\pi tk}{365} + \psi_{k,2} \sin \frac{2\pi tk}{365}, \tag{10.5}$$

and estimated using the series X_t. The degree p is chosen using AIC [25] or small sample corrected versions of it (see, for example, [400]). This approach leads to quite similar results to those obtained using nonparametric approaches of the STL type (*seasonal-trend decomposition based on LOESS smoothing* [167]).

(b) **Estimating the trend** m_t. Once u_t has been estimated by $\hat{u}_t$, consider the model

$$\hat{X}_t^1 = m_t + \tilde{Y}_t, \tag{10.6}$$

where $\hat{X}_t^1 = X_t - \hat{u}_t$. The series $\tilde{Y}_t$ is not supposed stationary. Estimation of m_t is performed using LOESS regression, with the help of the following theorem proved in [391], extending a result from [675]. This theorem results in quadratic control of the difference between the estimator and the unknown function m_t under a condition on the covariance of the $\tilde{Y}_t$. The use of a LOESS estimator (see Sect. 4.3.2) implies the choice of a kernel function K with compact support and smoothing (*bandwidth*) parameter h_T for which $h_T \to 0$ and $Th_T \to \infty$ as $T \to \infty$.

Theorem 10.1 *Suppose that m_t is a twice-differentiable function for $t \in I\!N$ (i.e., $m_t \in C^2$), that $\sup_{[1,T]} E[\tilde{Y}_t^2] < \infty$, and that $\tilde{Y}_t$ satisfies the following condition:*

$$\sup_{[1,T]} C_T(\tilde{Y}_t) = C(\tilde{Y}_t) < \infty, \tag{10.7}$$

where for any square-integrable stochastic process W,

$$C_T(W) = \frac{1}{T} \sum_{i=1}^{T} \sum_{j=1}^{T} \mid Cov(W_i, W_j) \mid. \tag{10.8}$$

Let $\hat{m}_t$ be the LOWESS estimator for m_t. Then

$$E(\hat{m}_t - m_t)^2 = \left(\frac{1}{2} h_T^2\, m^{(2)}(x) \mu_2^1\right)^2 + \frac{1}{Th_T} \mu_0^2 C_Y(T) + o\left(\frac{1}{Th_T} + h_T^4\right), \tag{10.9}$$

where μ_j^r is the j-th order moment of $K^{(r)}$.

For a stationary process, Condition (10.8) implies weak dependence (L^2), while for a non-stationary one, it controls the level of non-stationarity, which means that certain properties of iid series can be applied to dependent and non-stationary data.

Using (10.8), we can get asymptotic results which are close to those for ordinary regression.

(c) **Estimation of the seasonality** v_t^2. Like s_t, multiplicative seasonal effects v_t^2 can be estimated using trigonometric polynomials using the series

$$(X_t - \hat{u}_t - \hat{m}_t)^2 = \hat{X}_t^{(2)}.$$

(d) **Estimating the trend** s^2 **using the series** $\frac{(X_t-\hat{u}_t-\hat{m}_t)^2}{\hat{v}_t^2} = \hat{X}_t^{(3)}$. Consider the model:

$$\begin{aligned}\hat{X}_t^2 &= \frac{(X_t - \hat{m}_t - \hat{u}_t)^2}{\hat{v}_t} \\ &= s_t^2 + s_t^2(Y_t^2 - 1) \\ &= s_t^2 + R_t.\end{aligned}$$

We suppose that the stochastic process R_t satisfies Condition (10.8) (which is equivalent to a condition on the 4th-order cumulants of Y_t). Here again a *control theorem* proved in [391] allows a LOESS estimator $\widehat{s}_t^{\,2}$ for s_t^2.

Theorem 10.2 *Let k_T be the bandwidth parameter for the kernel K used to estimate s. Suppose that $k_T^4 + (Tk_T)^{-1} = o(h_T^2)$. Then,*

$$\mathbb{E}(\widehat{s}^{\,2}(x) - s^2(x)) = \frac{k_T^2}{2}\mu_2^1(s^2)''(x) + o(k_T^2), \tag{10.10}$$

$$\mathbb{V}(\widehat{s}^{\,2}(x)) \leq \frac{1}{Tk_T^2}\mu_0^1 C_T(\eta) + o\left(\frac{1}{Tk_T}\right). \tag{10.11}$$

Finally the *estimated* reduced series is defined by, $\forall t \in [1, T]$,

$$\hat{Y}_t = \frac{X_t - \hat{u}_t - \hat{m}_t}{\hat{v}_t \hat{s}_t}. \tag{10.12}$$

Let us terminate this section with several important remarks.

1. The rate of convergence of the estimator of s_t^2 is independent of m_t. An optimal choice for h_T based on Theorem 10.1 is impossible to obtain since the quantities $m_t^{(2)}$ and $C_Y(t)$ are unknown. Note also that standard cross-validation (CV) criteria cannot be applied to correlated and non-stationary data, and do not help select the bandwidths h_T and k_T. In [391], it is suggested to perform this selection using a *modified partitioned cross-validation* (MPCV) algorithm, a variant of the PCV approach proposed by [503]. Simulation studies in [391] have shown the improved behavior of MPCV over PCV, as well as over other bandwidth calibration algorithms in non-stationary settings, such as MCV [163], *plug-in* CV [289], and *block bootstrap* [370]. The published study reused here chose to set $k_T = h_T$.

2. A block bootstrap method [370] can be used to calculate confidence intervals. For the daily temperature, 10-day blocks suffice for taking into account correlation between temperatures on consecutive days.

10.2.5.2 Simulation Model for the Reduced Series and Taking Extreme Values into Account

The reduced series $\{Y_t\}_t$ for which $\mathbb{E}[Y_t] = 0$ and $\mathbb{E}[Y_t^2] = 1$ (centered and normalized) in (10.3) is potentially stationary. In reality, there may still remain some seasonality in the correlation and in its 3rd and 4th-order moments (this is generally the case for temperature data). Consequently, the residual series is not stationary but *cyclo-stationary*, i.e., its distribution does not change over certain time periods. Specific statistical tests are provided in [391].

We thus suppose that Y_t is a stationary diffusion process satisfying a stochastic differential equation of the type

$$dY_t = b(Y_t)dt + a(Y_t)dW_t, \tag{10.13}$$

where b is the *drift* term in the model, $a \geq 0$ a diffusion coefficient, and W_t the Brownian motion. The terms a and b depend on the time t (day of the year), which helps take seasonality remaining in the residual series into account. According to (10.13), the marginal density of the process Y_t is given by

$$\nu(y) = \frac{1}{a^2(y)} \exp\left(\int^y 2\frac{b(v)}{a^2(v)}dv\right).$$

The use of the theory of extreme values of bounded diffusion processes, studied by [85] and reformulated by [200], requires that the diffusion process (10.13) satisfies certain boundedness properties, which are not constraining in practice since the original process X_t represents a physical phenomenon with a limited amplitude (even if its maximum is not necessarily known a priori). By necessity, the Brownian motion W_t is supposed truncated. We thus denote $[r_1, r_2]$ the support of Y_t and make the following hypotheses (validated in [391] for the temperature):

(i) a and b are defined and continuous on $[r_1, r_2]$;
(ii) $b(r_1)b(r_2) \neq 0$.

Theorem 10.3, taken from [191], shows that under these hypotheses, if there exists a limit distribution for the standardized maxima $M_T = \{\max Y_t, 0 \leq t \leq T\}$, then one also exists for an iid sample from the distribution (10.14), and these two limit distributions are GEVs with the same shape parameter ξ. This result is valid for bounded processes where $\xi < 0$, but an extension to cases where $\xi \geq 0$ is also possible.

Theorem 10.3 (Maxima of a diffusion process)

Under Hypotheses **(i)** *and* **(ii)***:*

1. *If the distribution of the maximum of the diffusion process Y_t is in the domain of attraction of a GEV distribution with $\xi < 0$, then the marginal distribution ν of the diffusion process belongs to the same* max-domain *of attraction.*
2. *Furthermore,*

$$a^2(y) = -2b(r_2)\xi'(r_2 - y) + o(r_2 - x) \quad \textit{when } y \to r_2,$$

where $1/\xi + 1/\xi' = 1$.

To fit the model to the dynamics and other intrinsic characteristics of the residual temperature, a discrete version of the diffusion needs to be used [191]. The continuous-time residual temperature Y_t is thus represented by a seasonal and heteroscedastic* autoregressive model

$$Y_t = c(Y_{t-1}) + a(Y_{t-1})\eta_t(Y_{t-1}), \tag{10.14}$$

with

- $c(Y_{t-1}) = P_p(t)Y(t-1)$, where $P_p(t)$ is a p-th order trigonometric function with a period of $A = 365$ days,
- η_t an iid truncated Gaussian noise.

Seasonality and the existence of upper and low bound constraints requires the conditional variance $a^2(Y_{t-1})$ to be defined as follows:

$$a^2(Y_{t-1}) = (\hat{r}_2 - t)(t - \hat{r}_1)\sum_{k=0}^{5} P_{p',k}(t)Z_{t-1}^k,$$
$$\frac{da^2}{dy}(r_1) = \frac{2(c(r_1) - r_1)}{1 - 1/\xi_1},$$
$$\frac{da^2}{dy}(r_2) = \frac{2(c(r_2) - r_2)}{1 - 1/\xi_1}.$$

Finally, consistent statistical estimation of the model (10.14) can be obtained in the following way. For technical details, refer to [191].

(a) **Estimating the autoregressive part of** Y_t. The number p of cos and sin terms in the trigonometric function $P_p(t)$ is chosen using AIC (see Sect. 8.1.2.3), and the parameters of P_p estimated using a least squares method (cf. Eq. 10.17) on the data $\hat{Y}_t$ produced by (10.12).
(b) **Choosing the number of dimensions for** $P_{p'}$. The number of cos and sin terms p' in $P_{p'}$ is also chosen using AIC, based on the model without boundedness constraints:

$$\sum_{k=0}^{5} P_{p',k}(t) Y_{t-1}^{k} = (Y_t - \hat{c}(Y_{t-1}))^2.$$

(c) **Estimating the bounds** r_1 and r_2 by applying a GEV model (μ, σ, ξ) to the maxima of Y_t and a GEV model (μ', σ', ξ') to the *minima* of Y_t:

$$\hat{r_1} = \frac{\mu' - \sigma'}{\xi'}, \quad \hat{r_2} = \frac{\mu - \sigma}{\xi}.$$

(d) **Estimating the parameters of** $P_{p',k}$ on the interval $[r_1, r_2]$ using the series $(Y_t - \hat{c}(Y_{t-1}))^2$ and maximum likelihood with boundedness constraints.

10.2.6 Case Study 1: Extreme Air Temperatures

The modeling methodology described in the previous paragraphs was applied to temperature times series taken at three locations: Berlin (German), Fruholmen (Norway), and Death Valley (USA). Here, we illustrate putting simulation and validation procedures into practice to test model performance.

To validate the model, 100 trajectories of the same length as the actual sample are simulated according to the model using the procedure described below. A range of relevant functions can then be used to compare the simulations and actual data: moments, daily distributions, quantiles, GEV parameters for the maxima of Y_t, cold snaps, and heat waves.

Simulation procedure.

- Preprocessing: estimate the trends (m_t, s_t) and seasonal effects (u_t, v_t) in the mean and standard deviation of the original time series X_t.
- Calculate $\hat{Y}_t = (X - \hat{u} - \hat{m})/(\hat{s}\hat{v})$ and apply model (10.14) to Y_t.
- Estimate $a(y)$ and $c(y)$.
- Use the estimators $\hat{a}(y)$ and $\hat{c}(y)$ to construct 100 simulated trajectories $Y*_t$.
- Obtain 100 simulated trajectories of X using the model:

$$\hat{X}_t = \hat{s}_t \hat{v}_t Z *_t + \hat{m}_t + \hat{u}_t.$$

With regard to daily temperature distributions, 365 Kolmogorov–Smirnov tests made it possible to test the homogeneity of the observed and simulated distributions. This led to, for both measurement stations, not rejecting homogeneity at a 5% confidence level. Moreover, the simulations of the daily means and variances represent well the characteristics of the observations; the results in Fig. 10.1 are for the minimum daily temperature in Fruholmen: mean values and the 2.5% and 97.5% quantiles of the estimators provide the lines for the mean estimator and its 95% confidence interval.

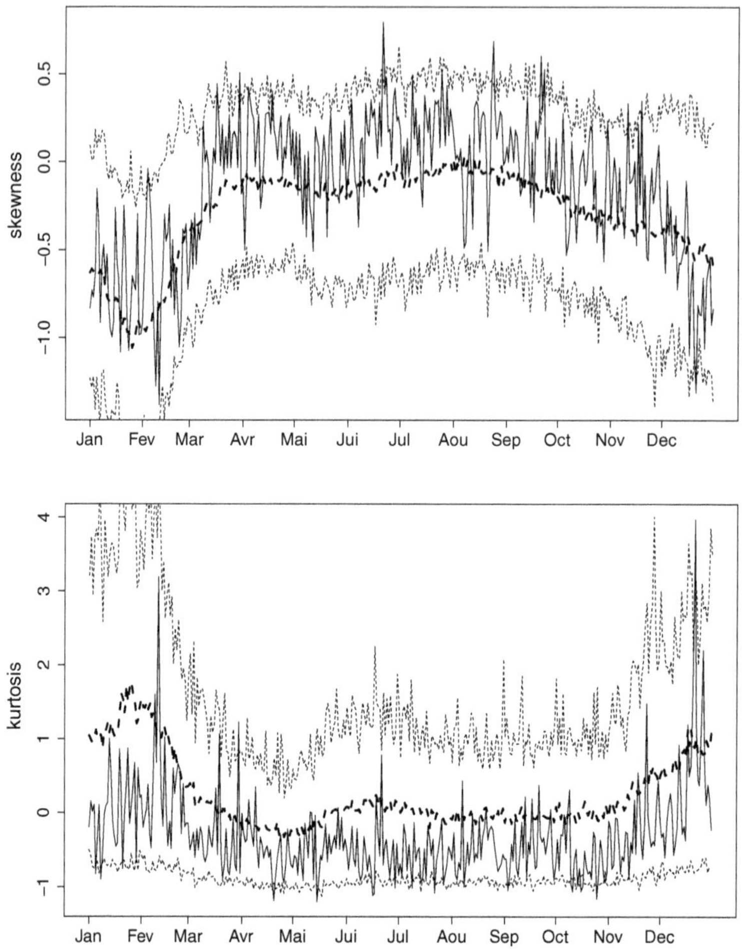

Fig. 10.2 Comparison between the daily skewness (top) and kurtosis (bottom) of the data recorded at Fruholmen and simulated data (95% confidence interval). The actual data are plotted as a solid line

As for the *skewness** and *kurtosis** of the daily distributions, the results are not as good as for the daily means and variances. We see in Fig. 10.2 (minimum temperatures in Fruholmen) that the indicators for the 42nd and 355th days are quite low or high. For these days, indeed, extremely low temperatures were recorded: −21.6 °C and −24.5 °C.

In order to confirm the usefulness of the model, simulations from it can be compared to those from simpler models, e.g., ones with a constant function $a(y)$ or a trigonometric function $a^2(t)$ which depends only on time and not the state y_{t-1}.

The example simulating the minimum daily temperature in Berlin (Fig. 10.5) presents the model's performance for most of the quantiles with respect to the other models. The extreme quantiles seem to be well represented by our model since the observed quantiles are located within the confidence interval of the simulations defined for a corresponding level.

Compared with other temperature generators in the literature, this model differs in that it is bounded. This a priori feature leads to better reproduction of the extreme values. To see whether the obtained trajectories have good extreme value characteristics, the parameters of the GEV distribution can be estimated from the maxima and *minima* of the residual series of simulated trajectories. The results (illustrated in Fig. 10.3) show that the shape parameter ξ is better estimated than the position μ and dispersion σ, not forgetting that the estimation of extreme value parameters is extremely sensitive to the block size.

We can also evaluate the capacity of these models to reproduce cold snaps and heat waves. The cold snaps considered here are defined as a set of consecutive days during which the daily minimal temperature is less than the 2% quantile of the daily minimal temperature distribution, and heat waves as sets of consecutive days whose maximal daily temperature is above the 98% quantile of the daily maximal temperature distribution. The length of such episodes essentially varies from 1 to 15 days, with a small number lasting longer than 15 days. In practice, the 2% and 98% quantiles are calculated using the actual time series data, and the frequency of periods of all lengths are compared to the minimal, maximal, and mean frequencies (for each length) obtained from the 100 simulated trajectories. The results are good in general; while the model tends to overestimate the frequency of periods lasting one day only, it is able to reproduce longer periods with reasonable frequencies with respect to the observed ones. Figure 10.4 illustrates this on cold snaps in Berlin and heat waves in Death Valley. Return levels can also be used as validation criteria, but we do not detail this further here—see [595] for a discussion about this (Fig. 10.5).

10.2.7 Case Study 2: Extreme Values of a Frost Index

The *frost index* characterizes the strength of a freezing episode. Such episodes are defined as sets of consecutive days on which the daily mean temperature remains below 0 °C. The frost index then corresponds to the absolute value of the temperature gap with respect to 0 °C over an episode [754].

Here, we look at the daily mean temperature recorded in Colmar, France from 1957 to 2012 (56 years). During this time, 283 frost indices were recorded—on average 5.1 per year, which makes applying extreme value theory unsuitable. The largest observed value of the frost index was 418.2 during the winter of 1962–1963, and the second largest was 189.6 during the 1985–1986 winter. In order to estimate return levels for frost indices, it may therefore be useful to stochastically simulate the daily mean temperature.

By simulating the daily temperatures over a 56-year period 1000 times with the model calibrated to the observed data, we obtain a total of 291,446 frost indices, i.e., 5.2 per year on average, which is consistent with the 5.1 observed in the real data.

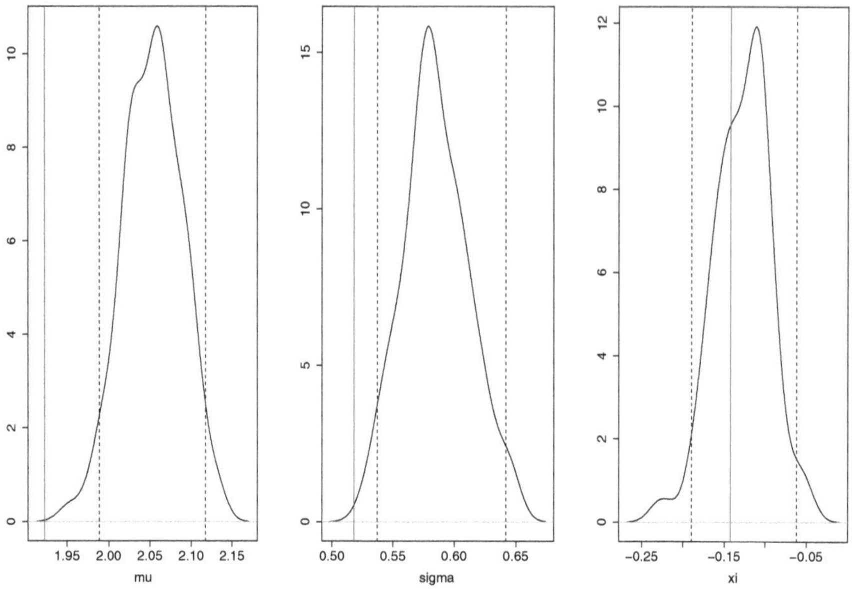

Fig. 10.3 Extreme value parameters estimated using the observed residual time series (vertical lines) and their empirical distributions constructed using the minimal daily temperature simulations for Berlin

Among these indices, 11,449 are above 100, i.e., a proportion of 0.039, in comparison to the observed proportion of 0.042 (12 in 56 years). A comparison of the numbers of simulated indices per year and the annual maxima in Fig. 10.6 also supports the relevance of the model, with 3 observed indices found outside the 90% confidence interval of the simulated ones, compared to the number expected: 2.8 (5% above and 5% below).

It is then possible to estimate a return level for frost indices in two ways:

- Estimate the corresponding quantile of the distribution of 291,446 simulated indices; i.e., we look for the level exceeded on average once every N years, which corresponds to the quantile $1 - 1/(n_b \cdot N)$, where n_b is the mean number of indices per year of this distribution. For a return level of 1000 years, this approach leads to a value of 391 and the 95% confidence interval [376.2, 404.5].
- Apply extreme value statistics theory by fitting a GEV distribution to the 1000 maxima of each 56-year long simulation (the maximum is then taken on average over $56 \times 5 = 280$ values, which seems more reasonable). For a return level of 1000 years, this approach gives a value of 386.2 and a 95% confidence interval of [376.7, 395.7].

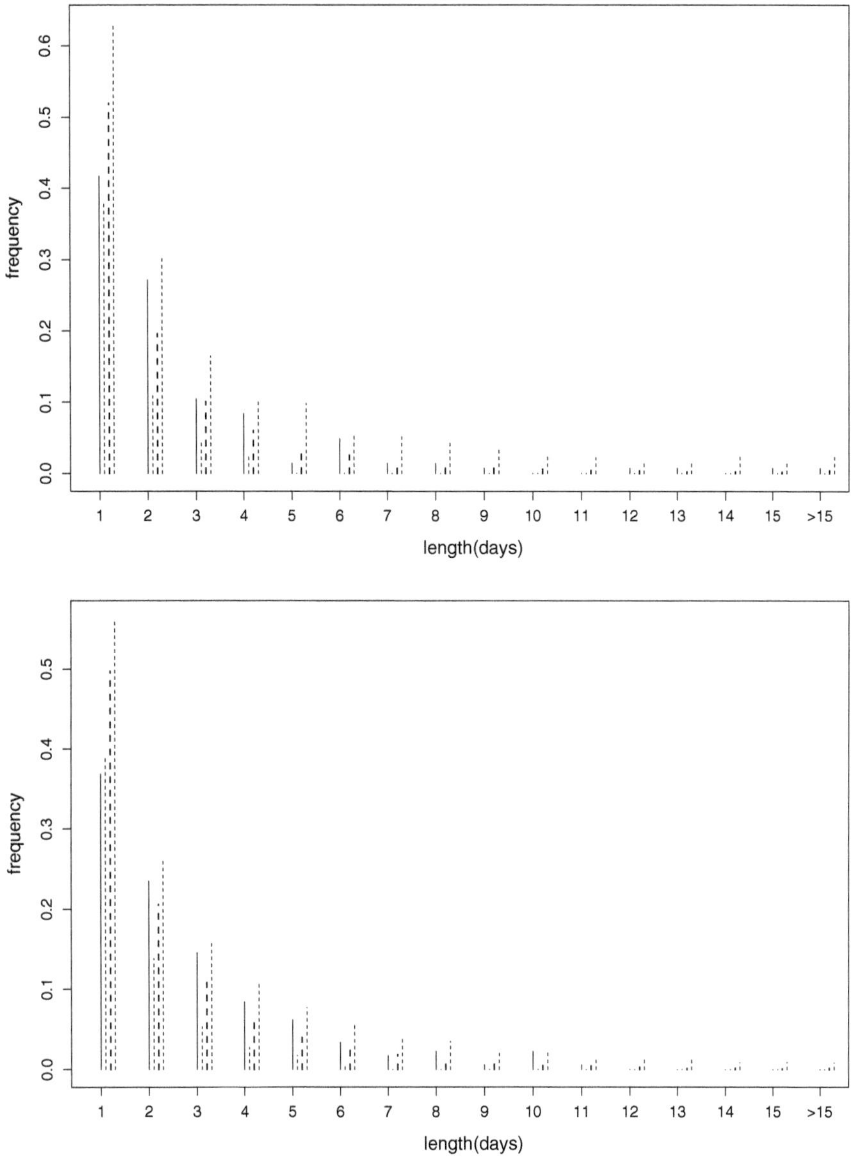

Fig. 10.4 Observed frequencies (solid line) and maximal, minimal, and mean frequencies (dotted lines) obtained from simulating cold snaps in Berlin (top) and heat waves in Death Valley (bottom)

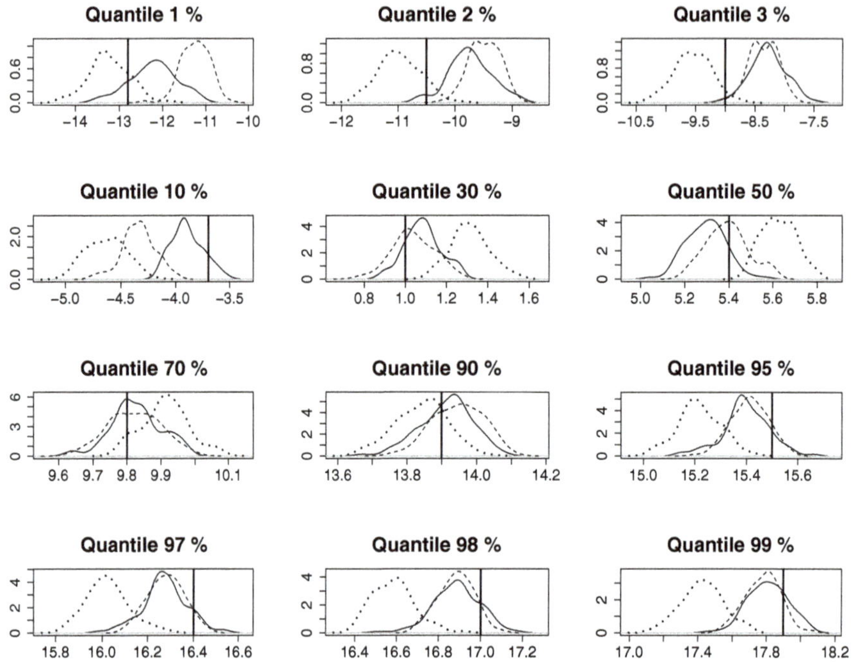

Fig. 10.5 Observed quantiles (vertical lines) of the daily minimal temperature in Berlin and their respective empirical distribution constructed using different models: constant $a(y)$ (dotted lines), trigonometric $a^2 = f(t)$ (dashed lines), and Model (10.14) (solid lines)

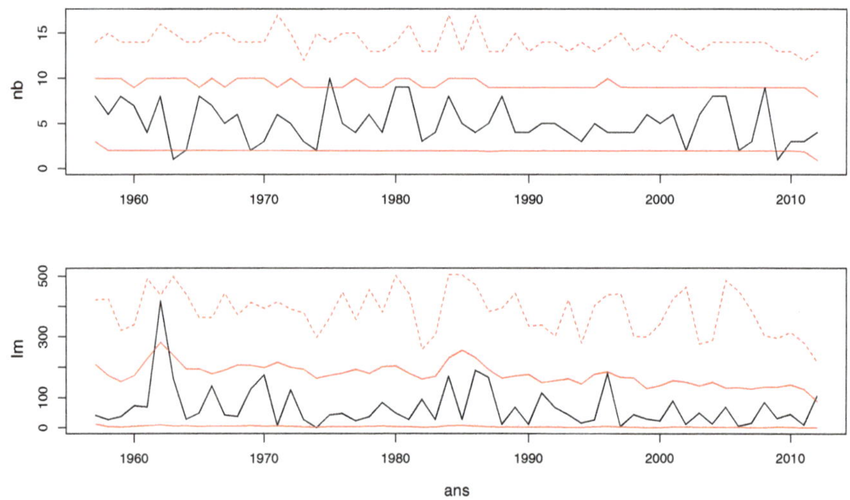

Fig. 10.6 Number of frost indices (top) and maximal index (bottom) per year recorded in Colmar (solid line) with a 90% confidence interval provided from simulations (solid line—center), and the simulated maxima (dotted line)

10.3 Physics-Based Simulation of Natural Hazards

10.3.1 Computer Models

The obvious lack of data on natural hazards, and sometimes the difficulty to get a fine description of them through stochastic processes, has pushed the search for an alternative way to both statistical and stochastic direct modelings. An indirect approach is the following: stochastic modeling of the most uncertain yet influential parameters ψ (*input parameters*) in the dynamic equations leading to the appearance of extreme events, if such dynamics equations can be simulated. These equations correspond to deterministic, discrete, or continuous modeling of events [331], and we call computer implementation of such models *numerical modeling*, *computer modeling*, or *computer codes*. An important review of dynamic models, illustrated by examples from climatology, seismology, and socio-economics, can be found in the article [331]. The use of such models to characterize potential causes and consequences of extreme natural events such as earthquakes, tsunamis, severe storms, avalanches, and even meteorite strikes is discussed in [13] (Sect. 5.10). In particular, hydrological (rainfall) forecasts generally benefit greatly from such tools [323, 447, 536].

Let us return to the example in the introduction of this chapter. The flow of a watercourse can be understood as a *calculable* variable whose values are produced by an algorithm that implements a structured set of equations $X = g(\phi)$ from free surface fluid dynamics [27, 37]. The upstream flow, surface roughness of the submerged earth in a section of the watercourse [120], and the parameters describing the geometry of the river along the section in question ([785], Sect. 2.17) form part of the explanatory parameters ϕ for the downstream flow. The approximate resolution of fluid dynamics equations—with which the watercourse is approximated as a curvilinear abscissa [120] or as a more precise multidimensional mesh [213]—necessarily implies model uncertainty with respect to g, to which is added uncertainty in the value of influential parameters [559]. For instance, the measured flow generally comes with a non-negligible measurement error [218]; also, the geometry is summarized using a small number of parameters [548]; finally, there are no friction measures that can help to assess the corresponding input parameters. More generally, flow and friction are variables subject to a mixture of *epistemic* and *random* uncertainty (see [393] and Sect. 3.1.2). The epistemic part is linked to the fact that our knowledge of physics—corresponding to the level of model sophistication—improves over time. The intrinsic variability in the value of the parameters at any physical point is a signal of random uncertainty. Note that the theoretical model g is itself subject to a epistemic-type error with respect to the reality of the phenomenon X, while its numerical version inevitably adds approximations and (supposedly under control) rounding errors.

The representation of epistemic and random uncertainty with the same tools (stochastic processes or probability distributions) remains controversial [278, 392, 568, 671]. A rapid review of methodological principles alternative to probability-based ones can be found in [807]. For practical reasons [658] but also epistemological ones ([240, 712] and Sect. 3.1.2), the joint representation of uncertainties through probability tools remains the predominant approaches in risk analyses. This is evidenced by the most recent attempts to formalize procedures for the exploration and use of complex computer codes $\boldsymbol{g}$ that model physical phenomena and whose entries ϕ contain uncertainty (so-called *verification, validation, and uncertainty quantification* approaches [13]). This predominance concerns in particular most of the studies taking place under regulatory frameworks [531]. The probabilistic modeling of influential parameters ϕ thus makes it possible to simulate the variable of interest X, find values for ϕ which lead to extreme values in X, then calculate risk indices.

Studies seeking to treat epistemic uncertainty differently, through the use of modeling derived from non-probability-based theories,[2] are still essentially research topics and not fully accepted methodologies that can become predominant today in regulatory settings. The main reason is computational complexity, which can rapidly become prohibitive with respect to the dimension of ϕ [753, 769]. Let us however mention the article [663] which studies the feasibility of the *possibilistic* approach [231] to represent a lack of knowledge of the explanatory parameters of earthquakes, [458] which estimates extreme sea levels with the same approach, and also [604] which considers a case study similar to that of Example 10.2 (further in the text) and proposes to dimension dikes with the joint use of probabilistic and possibilistic tools. This type of approach has been gaining ground in risk studies since the turn of the XXIst century [61, 368]. Nevertheless, formally speaking, these methods do not make it possible to define typical notions in natural hazard management like return levels and periods. Rather, they provide bounds in extreme value situations, given as solutions to deterministic problems—often in a quite conservative manner [238].

Example 10.1 (*Predicting tsunamis caused by meteorite impact*)

Simplistic, pseudo-physical models can also be used to evaluate the probability (or return period) of an extreme-valued event. The prediction of characteristics of a tsunami caused by a meteorite impact is a prime example. The methods proposed in [583, 775] involve products of probabilities of occurrence

$$\mathbb{P}(\text{ meteorite impact during } [t_0, t_1]) \times \mathbb{P}(\text{ meteorite size } > 100 \text{ m}) \times \ldots \quad (10.15)$$

with physics-based impact models. For example, [583] has proposed modifying (conservatively) a model that correctly predicted the consequences of the impact of the Shoemaker–Levy 9 comet on Jupiter in 1994. The two components of this type of study are subject to extremely high epistemic uncertainty. Thus, approaches like (10.15) rely on mechanisms similar to *probability safety assessments* (PSA, [775]) which are used to rank extreme accident scenarios in industry (nuclear in particular), rather than quantifying them precisely [237]. Furthermore, the error in modeling such impacts obviously remains very diffi-

[2] Often called *extra-probabilistic* theories.

cult to estimate in the (thankful) absence of observed data. Nevertheless, such models are the only ones available today, and at minimum a stochastic exploration of them can help compare this type of extreme event with other more frequent ones, in terms of *risk* (cost × probability).

10.3.2 Monte Carlo-Based Exploration and Extreme Risk Calculation

The most general approach available for estimating extreme event occurrence probabilities or the corresponding return levels—using black box numerical models—is based on Monte Carlo sampling [673]. The following example introduces this exploration principle (see also Fig. 10.8) (Fig. 10.7).

Example 10.2 (*Flood probability and return period*)

Suppose that X represents a water level (typically obtained through a transformation of the flow using a *calibration curve*—see Example 4.4), and that there exists a deterministic hydraulic model $\boldsymbol{g}$ such that $X = \boldsymbol{g}(\phi)$, where ϕ is random with density f on Φ, and $\dim \Phi = d$. Let x_0 be the height of a dike. If we suppose that $\boldsymbol{g}$ is valid for representing X in some domain $\chi = \boldsymbol{g}(\Phi)$, and that $x_0 \in \chi$, then the flood

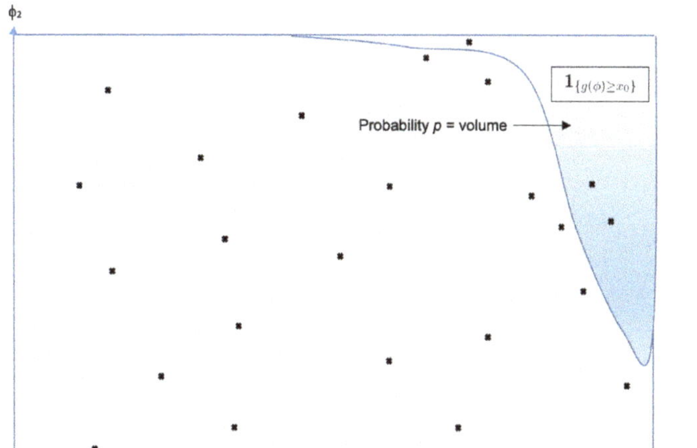

Fig. 10.7 Two-dimensional illustration: the probability p of an extreme event can be interpreted as the volume (i.e., the area in two dimensions) of a small grey zone. A numerical experimental design (black dots) requires to surround the surface of this zone in order to correctly estimate its volume

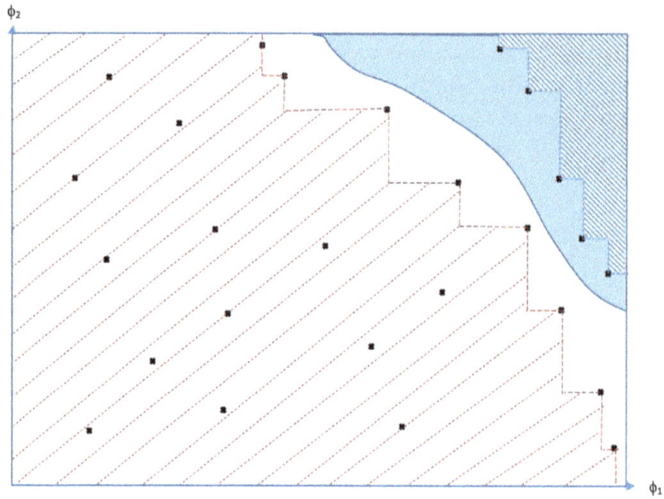

Fig. 10.8 Under monotonic constraints on g, it is possible to bound the boundary $g(\phi) = x_0$ by two hyper-stairs [115, 659]. The two hatched volumes then allow two sure bounds on the probability p (grey area) to be determined. The same reasoning can be used to bound a quantile z [117]

probability (overflow) is given by:

$$p = \mathbb{P}(X \geq x_0) = \int_\Phi \mathbf{1}_{\{g(\phi) \geq x_0\}} f(\phi)\, d\phi, \tag{10.16}$$

and can be estimated using a Monte Carlo method: $\hat{p}_m \xrightarrow{a.s.} p$, where

$$\hat{p}_m = \frac{1}{n} \sum_{k=1}^{m} \mathbf{1}_{\{g(\phi_k) \geq x_0\}}, \tag{10.17}$$

and $\{\phi_1, \ldots, \phi_m\} \overset{iid}{\sim} f$, with $m \gg 1$. Estimator (10.17) is unbiased and is such that

$$\frac{\sqrt{m}}{\sigma} \left(\hat{p}_m - p\right) \underset{m \to +\infty}{\xrightarrow{\mathcal{L}}} \mathcal{N}(0, 1), \tag{10.18}$$

where the variance $\sigma^2 = p(1-p)/n$ is estimated by $\hat{\sigma}_m^2 = \hat{p}_m(1-\hat{p}_m)/(m-1)$. An asymptotic confidence interval for (10.17) is then

$$\left[\hat{p}_m - \Phi^{-1}(1-\alpha/2)\frac{\hat{\sigma}_m}{\sqrt{m}},\ \hat{p}_m + \Phi^{-1}(1-\alpha/2)\frac{\hat{\sigma}_m}{\sqrt{m}}\right].$$

If conversely we are looking for the $1/p_0$ return level for X, denoted z, then

$$z = \inf \{x \in \chi \;\; \mathbb{P}\left(g(\phi) \leq x\right) \geq 1 - p_0\} \tag{10.19}$$

can be consistently estimated by

$$\hat{z}_m = X^*_{\lfloor m(1-p_0)\rfloor+1}, \tag{10.20}$$

where $X_k = g(\phi_k)$, and $X^*_1, \ldots, X^*_m$ is the order statistic of $X_1, \ldots, X_m$ [44]. When the distribution f_X of $X = g(\phi)$ is differentiable at $\hat{z}_m$, the bias of the estimator $\hat{z}_m$ is:

$$\mathbb{E}[\hat{z}_m] - z = -\frac{p_0(1-p_0)f'_X(z)}{2(m+2)f^3_X(z)} + \mathcal{O}(1/m^2),$$

which becomes negligible when $m \to \infty$, and is asymptotically Gaussian [199]:

$$\sqrt{m}\left(\hat{z}_m - z\right) \underset{m\to+\infty}{\overset{\mathcal{L}}{\longrightarrow}} \mathcal{N}\left(0, \frac{p_0(1-p_0)}{2(m+2)f^2_X(z)}\right). \tag{10.21}$$

As the distribution f_X remains unknown, it is necessary to estimate it using sampling in order to estimate the variance of $\hat{z}_m$, and in particular to estimate it well in its tail $x \sim z$, which requires a large sample size.

An alternative approach to (10.20) is *quantile regression* [126, 441], which in general supposes that z is a linear function of the vector of variables ϕ, with coefficients β. We define, theoretically:

$$\beta = \arg\min_b \mathbb{E}\left[\rho(X - \phi^T b)\right], \tag{10.22}$$

where $\rho(y) = (1 - p_0 - \mathbf{1}_{\{y\leq 0\}})$. An estimator of the coefficients β can be obtained by minimizing the empirical estimator of the *objective function* in (10.22) produced by Monte Carlo sampling. However, this approach has for the most part been reserved for *real experimental designs* (not numerical ones) and the evolution of quantiles along trajectories measured in situ [794]. We also cite [134], which suggests applying this approach to calculating extreme sea levels.

10.3.3 Robustness and Calculation Time

Exploration of the model g (or rather its encoded version) by simulating the parameter vector $\phi \sim f$ is the most-used approach to estimate the quantities of interest in Example 10.2. The aim is to have a minimum of hypotheses on g to ensure as general as possible results (*non-intrusive approaches*). In the absence of any regularity hypothesis or even the assumption of continuity, g can be seen as a "black box", and the Monte Carlo approach described in Example 10.2 gives powerful results whose precision is independent of the number of dimensions d. This wish for a minimum of hypotheses results most often from the fact that $\boldsymbol{g}$ models a chain of several physical processes, increasing the risk of edge effects (cf. Sect. 10.1).

However, this standard approach comes with a major weakness, in practice, when $p \ll 1$. In such cases a Monte Carlo approach is not *robust* [341], in the sense that the estimation error on $\hat{p}_m$, given in terms of its coefficient of variation:

$$\mathrm{CV}[\hat{p}_m] = \sqrt{\frac{1-p}{m \cdot p}},$$

tends to infinity when $p \to 0$ for fixed m. Obtaining a typical precision of $p \sim 10^{-q}$ requires at least $m = 10^{q+2}$ calls to $\boldsymbol{g}$. This need for a large number of calls also affects the estimation of the quantile (10.20). However, continual improvement in the precision of numerical models leads to an increased calculation cost for each run of $\boldsymbol{g}$, which can make it prohibitive to calculate the estimators $\hat{p}_m$ and $\hat{z}_m$. This is particularly true for climate models [633].

Computing time and memory cost problems can be greatly diminished by performing parallel calculations (in particular for climate models [495, 514]) or using supercomputers; see, for example, [756] for applications involving heavy rainfall due to planetary oscillations, earthquakes, and tsunamis. Moreover, a number of *accelerated Monte Carlo* methods [672] can be used to significantly decrease calculation costs by numerically "zooming" into a zone $\Phi' \subsetneq \Phi$ where $\mathbb{P}\left(\boldsymbol{g}(\phi) \geq x_0 | \phi \in \Phi'\right)$ is a non-extreme probability. A very brief summary of such approaches usable by engineers is presented later in this section. As this area of research has been rapidly evolving since the year 2000, readers interested in a more comprehensive take on the subject should consult [672].

10.3.3.1 Conservative Estimates of Probabilities and Return Levels

Starting from a Monte Carlo sample, Wilks [787], later extended by [773], proposed an empirical estimate of a return level z, described in (10.19) as an order $1 - p_0$ quantile, though with high computation costs. This is a conservative approach to calculating the quantile (10.19) in the sense that, with high probability $1 - \beta \gg 0$, it does not underestimate the risk when the model $\boldsymbol{g}$ has to be explored using numerical trials. It is therefore an important element in certain safety demonstrations in the structural reliability field [564, 806], and can be used to help mitigate certain extreme-valued natural phenomena.

From (10.21), we can deduce that asymptotically the estimator $\hat{z}_m$ satisfies $\mathbb{P}(\hat{z}_m \geq) \simeq 1/2$. To increase this probability to $1 - \beta$, we can replace the calculation (10.20) by

$$\hat{z}_m = X^*_{\lfloor m(1-p_0) \rfloor + r}, \tag{10.23}$$

where the integer $r \geq 1$ can be selected optimally as the smallest positive integer r_0 for which [564, 787]

$$C_{p_0}(m,r) = \sum_{k=0}^{mp_0 - r} \binom{k}{m} p_0^k (1-p_0)^{m-k} \leq \beta.$$

The estimator (10.23) defined by $r = r_0$ is called the *Wilks estimator*, and satisfies

$$\mathbb{P}\left(\hat{z}_m \geq z\right) \geq 1 - \beta.$$

This estimator is defined for any sample size m above or equal to some minimum sample size m_c defined by the limit case:

$$m_c = 1 + \left\lfloor \frac{\log \beta}{\log(1-p_0)} \right\rfloor.$$

To give an example, for $p_0 = 1 - \beta = 5\%$ a minimum value of $m_c = 59$ runs is required to obtain an upper bound for the quantile or return period z.

Several improvements have been made to this approach to help bound quantiles and increase the "strength" $1 - \beta$ of the conservatism; these can be found notably in [271, 722]. Note also that it can be shown that a conservative estimator of a quantile z obtained using an empirical bootstrap is similar to that of the Wilks estimator. Practical tips for using this approach are provided in [20] (Sect. 4.3.22). However, remember that this conservatism is not certain, and the approach can also strongly overestimate the risk when the size of the sampling design is highly constrained, since this estimator suffers from high dispersion [405].

A more sophisticated approach is the *monotonic reliability method* (MRM), which provides both statistical estimation (of the accelerated Monte Carlo type—see Sect. 10.3.4) and bounding of a certain probability or quantile, and works by taking into account a *monotonic* constraint on $\boldsymbol{g}$. To continue with Example 10.2, it may be realistic, or conservative, to suppose that the true water level (and its model $\boldsymbol{g}$) increases when the flow and/or friction does. If so, the regions Φ close to the limit state surface $\{\phi \in \Phi,\ \boldsymbol{g}(\phi) \simeq x_0\}$ can be quickly contoured using two numerical experiment designs. The measure of the space enclosed by these two designs can be used to bound a probability or quantile. This approach is quite well developed for estimating and bounding a given probability p [115, 119, 533, 659], and much more robust and rapid than Monte Carlo—though restricted to low dimensions (less than 10). Adapting it to be used to estimate and bound quantiles is an active area of research [117].

10.3.4 Accelerated Monte Carlo Methods

Accelerated Monte Carlo methods broadly encompass all probability (10.16) and quantile (10.19) estimation methods based on *stochastic* or *deterministic* sampling of the space Φ which are more efficient than classical Monte Carlo ones. What "efficiency" means depends on the context. When sampling is stochastic, it principally refers to the rate of convergence of the estimator to a Gaussian distribution (i.e., the inverse of its asymptotic variance) and its robustness as defined previously. When sampling is deterministic, the notion of *discrepancy* plays the role the rate of convergence does in the probabilistic setting. In general, efficiency is linked to the number of calls to g required to obtain a certain estimation precision. In this section, we provide a rapid overview of the major accelerated Monte Carlo methods that can be used when dealing with extreme probabilities and quantiles. Readers interested in more technical details should turn to the book [672] and review article [525].

10.3.4.1 Improving Numerical Experiment Designs

A first type of acceleration involves modifying the sample (or *numerical experiment design*) $u_1, \ldots, u_m \overset{iid}{\sim} \mathscr{U}_{[0,1]^d}$, whereby $\{\phi_1, \ldots, \phi_m\} = \{F^{-1}(u_1, \ldots, u_m)\}$, where $F(., \ldots, .)$ is the joint cumulative distribution function of ϕ. Figure 10.9 (left) illustrates a well-known phenomenon in standard Monte Carlo sampling: due to the necessarily finite sample size, certain zones are not filled—unlike with a regular grid (right), though the latter comes with an exponential computational cost in g that scales with the number of dimensions d. Several classes of methods have been developed to carry out a more complete exploration of a space Φ (or $[0, 1]^d$) under the constraint of at most m runs of g. Such methods are generally called *space filling* designs [193], and can also be separated into deterministic and stochastic approaches.

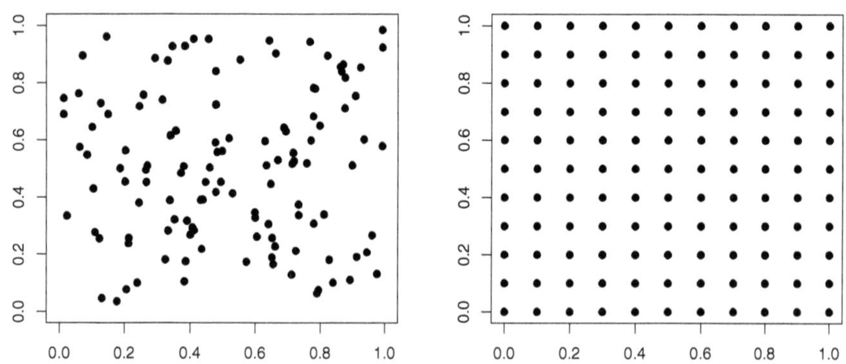

Fig. 10.9 Standard Monte Carlo sampling on $[0, 1]^2$ (left) and a regular 2-dimensional grid (right)

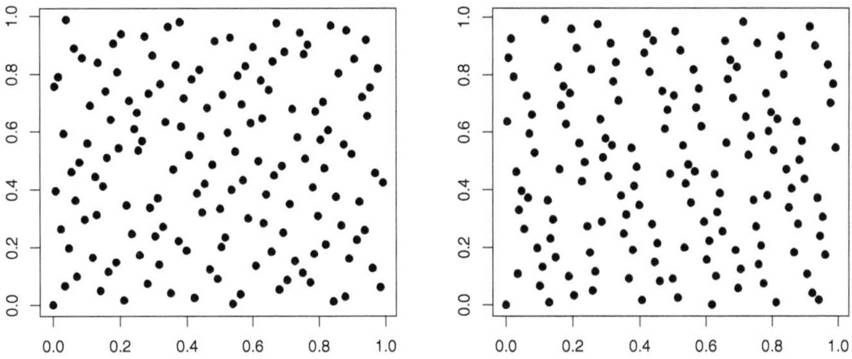

Fig. 10.10 Example of two Halton sequences in dimension 2

Deterministic approaches. These approaches involve constructing a deterministic sampling of the space Φ (or $[0, 1]^d$) using sequences with uniform properties, then calculating pseudo-statistical estimators using classical formulas (10.16) and (10.19), and deterministic error bounds [476]. Going under the name of *quasi-Monte Carlo methods,* these sequences $u_1, \ldots, u_m$ (in the space $[0, 1]^d$) can be generated in a way that optimizes a deviation from uniformity criterion called L^2-norm *discrepancy*, which is of the form:

$$D(u_1, \ldots, u_m) = \sqrt{\int_{[0,1]^d} \left\{ \frac{1}{m} \sum_{i=1}^{m} \mathbf{1}_{\{u_i \in [\mathbf{0}, y]\}} - \text{Volume}([\mathbf{0}, y]) \right\}^2 dy},$$

where $\mathbf{0}$ is the origin $(0, \ldots, 0)$ of the space $[0, 1]^d$, and $[\mathbf{0}, y] = [0, y_1] \times \ldots \times [0, y_d]$. This criterion compares the volume of all of the sub-intervals (anchored at the origin) of the domain of the u_i with the proportion of points contained in each. Robust choices for discrepancy and anchoring with respect to inputs as the number of dimensions decreases were studied in [193]. To obtain faster estimators (less costly in the number of calls to $\boldsymbol{g}$) than Monte Carlo, *low-discrepancy sequences* are of fundamental importance. In effect, for such sequences, the rate of convergence when estimating a probability (10.16) is $(\log m)^d/m$, better than the Monte Carlo one of $1/\sqrt{m}$. In practice, Halton [373], Hammersley [415], Faure [268], Sobol [711], or Niederreiter [558] sequences are usually used, the latter two generally considered the best in terms of robustness with respect to increasing the number of dimensions (Sobol) or asymptotic properties (Niederreiter), especially when they are associated with what are known as *scrambling* techniques [582]. An illustration of two Halton sequences is provided on Fig. 10.10.

Concepts other than deviation from uniformity using the discrepancy have been subsequently considered; these define useful properties for numerical experiment design. For instance, [416] has proposed the construction of *space filling* experiment designs which satisfy one or the other of the following two properties:

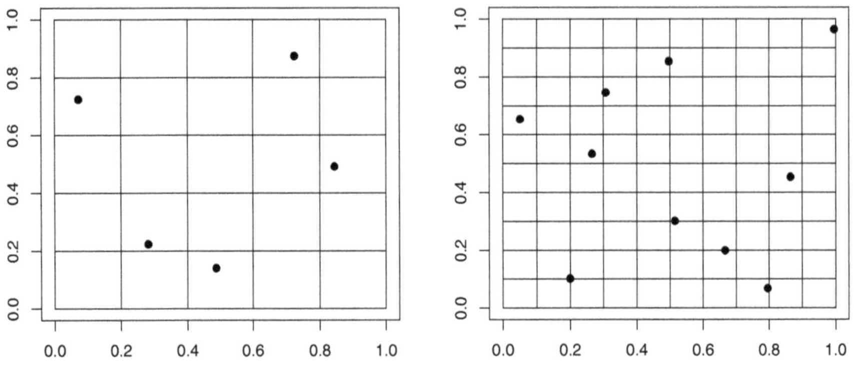

Fig. 10.11 Latin hypercube sampling with $m = 5$ (left) and $m = 10$ (right) on $[0, 1]^2$

- *maximin designs*: the minimal distance between two points of the design is maximized, and the number of points exactly the minimum distance apart is minimized,
- *minimax designs*: the maximal distance between two points of the design is minimized, and the number of points exactly the maximal distance apart is minimized.

Such designs are generally non-unique and non-deterministic, and the selection of points needs to be performed numerically, e.g., using a *simulated annealing* algorithm. *Minimax* designs should generally be used in practice as they tend to better fill spaces, whereas *maximin* ones generally fill the boundaries better than the center, especially in high dimensions. However, numerous technical difficulties are involved in constructing *minimax* designs [629], and *maximin* ones are still frequently used. A recent overview of the construction and properties of these types of experiment design is provided in [628].

Stochastic approaches. The *Latin Hypercube sampling* (LHS) method was historically one of the first stochastic simulation methods to improve the filling of the space Φ (or $[0, 1]^d$) with respect to classical uniform simulation. The idea is to partition the interval for each input variable X_k into N segments of equal probability, then uniformly draw from each of these intervals (see Fig. 10.11 for an illustration in 2 dimensions). Sampling is then produced by combining the $m = N^d$ draws while supposing that the input variables are independent [509, 721]. LHS has since been extended to treat dependent cases [404, 611], work with non-hypercubic domains, and incorporate distance constraints between points generated using *maximin* approaches [49, 511]. In summary, this method ensures uniform coverage of the domain of each input, and leads to faster overall convergence of classical statistical estimators [385].

The good uniform coverage properties of LHS points are nevertheless *marginal*, in the sense that correlations between the dimensions of X are not generally taken into account. To do so, *orthogonal tables* [381] can be used, which allow points to be distributed in subspaces of Φ (or $[0, 1]^d$). A combination of the best space exploration features of both is also possible [476].

The statistical estimation of probabilities q or quantiles z by a simple *static* improvement to numerical experimental designs—using a better uniform distribution of points in the space $[0, 1]^d$ and before inverting using $F(., \dots, .)$—is not, however, entirely satisfying. Indeed, these experimental designs are constructed using dependent values, which is of little (or no) help in formally proving consistency and convergence of the statistical estimators $\hat{p}_m$ and $\hat{z}_m$ to their true values p and z since asymptotic results like Theorem 4.5 and bootstrap techniques are no longer valid. The work of Nakayama [224, 544] has nevertheless led to significant progress in adapting these theorems when estimating quantiles using optimal sampling. However, in general (and in particular in extreme value settings [336]) it is necessary to construct new statistical estimators associated with sampling designs that are both stochastic and *progressive* (also called *sequential* or *iterative*), that improve an original experimental design, for hoping to improve on the Monte Carlo approach [352].

10.3.4.2 Adaptated Statistical Estimators

A full overview of techniques for constructing statistical estimators with less variance than those using Monte Carlo sampling is given in [476, 673]. From the 1990s on, research at EDF [91, 115, 336, 533, 537], CEA [137], INRIA [148, 149, 367], ONERA [524, 527], the École Supérieure d'Électricité (*Supélec*) [63, 64],[3] and the University of Sheffield [566], have led to several non-intrusive approaches well-suited to attaining small probabilities and extreme quantiles.

These methods generally share certain useful ideas such as *control variables, stratification, directional* simulation, *preferential* (or *importance*) sampling, *multi-level* sampling, and the selection of criteria to optimize when choosing the next point in progressive sampling. In the following paragraphs, we rapidly summarize these ideas. Other classical variance-reduction techniques like conditional Monte Carlo sampling and the antithetic variables method are not considered to be sufficient for attaining small probabilities and extreme quantiles [336].

Meta-modeling. We begin by introducing the idea that sampling can be efficiently guided using *meta-models*, also known as *response surfaces* [20], the name given to any function $\tilde{g}$ which "mimics" the behavior of g over a domain $\Phi_I \subseteq \Phi$ of interest but with much lower (negligible) computational time. Constructed (estimated) from an initial experimental design obtained using tools from Sect. 10.3.4.1, and possibly refined after each improvement to it, this may be, in general:

- a *deterministic interpolator*: linear, quadratic, polynomial (e.g., spline), or logistic regression model, neural network, etc.
- a *stochastic* (or *kernel*) *interpolator*: e.g., Gaussian-process based kriging [680] derived from geostatistics, or *chaos polynomials* (or a combination of both [436]). Such interpolators are based on the hypothesis that the value $x = g(\phi)$ can be

[3] More generally, work performed by the CNRS Mascot-Num research group (http://www.gdr-mascotnum.fr/).

considered as the output of a stochastic process, with the goal of better taking into account, when using $\tilde{g}$ in place of g, the meta-modeling error, and deduce zones of Φ_I to preferentially explore when calculating the quantities of interest q and z.

Readers interested in knowing more about the extensive world of meta-models can refer to numerous recent publications; in particular we recommend [439] and the references given in [20].

Control variates. For calculating the probability (10.16), the control variates method [483] consists of introducing an auxiliary variable $\xi(\phi)$ such that $\mathbb{E}[\xi(\phi)] = \xi_0$:

$$p = \mathbb{E}\left[\mathbf{1}_{\{g(\phi)\leq x_0\}}\right] = \mathbb{E}\left[\mathbf{1}_{\{g(\phi)\leq x_0\}} - \xi(\phi)\right] + \xi_0,$$

estimated via Monte Carlo sampling using:

$$\tilde{p}_{m,\xi} = \frac{1}{m}\sum_{k=1}^{m}\left\{\mathbf{1}_{\{g(\phi_k)\leq x_0\}} - \xi(\phi_k)\right\} + \xi_0,$$

where

$$\mathbb{V}[\tilde{p}_{m,\xi}] = \frac{1}{m}\left[p(1-p) + \mathbb{V}[\xi(\phi)] - 2\mathbb{C}ov\left\{\mathbf{1}_{\{g(\phi)\leq x_0\}}, \xi(\phi)\right\}\right].$$

If $\xi(\phi)$ is strongly and positively correlated with the indicator function $\mathbf{1}_{\{g(\phi)\leq x_0\}}$, the covariance term becomes negative and leads to a decrease in the variance σ^2 of the usual Monte Carlo estimator. This approach can also be modified to calculate quantiles, using a meta-model as the control variate [388].

Stratification.
The idea behind stratified sampling is to partition the domain Φ into disjoint sub-domains known as *strata* (or layers):

$$\Phi = \bigcup_{i=1}^{M}\Phi_i \quad \text{such that} \quad \Phi_i \bigcap \Phi_j = \varnothing \text{ when } i \neq j.$$

The integral in (10.16) is then calculated in each stratum, and weighted by volume $\omega_i = \int_{\Phi_i} f(\phi)\, d\phi$:

$$\begin{aligned}\int_{\Phi}\mathbf{1}_{\{g(\phi)\geq x_0\}} f(\phi)\, d\phi &= \sum_{i=1}^{M}\omega_i\int_{\Phi_i}\frac{\mathbf{1}_{\{g(\phi)\geq x_0\}}}{\omega_i} f(\phi)\, d\phi \\ &= \sum_{i=1}^{M}\omega_i \lim_{M_i\to\infty}\frac{1}{M_i}\sum_{k=1}^{M_i}\mathbf{1}_{\{g(\phi_{k,i})\geq x_0\}},\end{aligned}$$

where M_i is the number of Monte Carlo draws $\phi_{1,i}, \ldots, \phi_{M_i,i}$ in stratum i, with $\sum_{i=1}^{M} M_i = M$. This method helps, for example, to ensure that draws occur in the tails of distributions. The number of draws M_i in each stratum, as well as the number of strata, can be optimized with respect to the desired precision of the estimator

$$\tilde{p}_M = \sum_{i=1}^{M} \frac{\omega_i}{N_i} \sum_{k=1}^{M_i} \mathbf{1}_{\{g(\phi_{k,i}) \geq x_0\}},$$

and the postulated order of magnitude of p. This technique, which does not guarantee a reduction in variance [336], is generally combined with directional simulation (see below) to produce powerful estimators of the probability p [537]. Coupled with the use of a meta-model to guide the choice of partition, it can also be used to estimate extreme quantiles [137].

Directional simulation. This method involves transforming the physical space Φ into a *standard Gaussian* one using an isoprobabilistic transform $y = \Psi \circ F^{-1}(\phi)$ (where Ψ is the joint cumulative distribution function of a standardized multivariate Gaussian distribution), randomly generating directions in this space, estimating probabilities p conditional on these directions, then averaging these estimates to give an unconditional estimator of p. Though limited in the number of dimensions it can handle, this approach is useful for performing an initial exploration of a space and detecting influential variables using *sensitivity analysis* [407]. The idea of transforming the space was original used in older reliability methods (FORM-SORM[4] methods), which turn the probability estimation problem into an optimization one but offer little or no information on the estimation error involved [475]. Requiring very little computation time, but not very robust with respect to the number of dimensions, these methods can also be adapted to iteratively look for quantiles of interest [215, 610].

Preferential or importance sampling (IS). The principle of IS is to change the draw density $f(\phi)$ to $f_{IS}(\phi)$ and introduce a ratio of densities into the statistical estimator of the quantity of interest. The idea is to concentrate the random draws in an area of interest, i.e., one that corresponds to extreme situations X. Under non-restrictive conditions on the support, the estimator (10.17) is replaced by

$$\tilde{p}_{IS} = \frac{1}{m} \sum_{k=1}^{m} \frac{f(\phi_k)}{f_{IS}(\phi_k)} \mathbf{1}_{\{g(\phi_k) \geq x_0\}}. \tag{10.24}$$

An optimal choice for f_{IS} is theoretically possible [672]:

$$f_{IS}^{*}(\phi) = \frac{\mathbf{1}_{\{g(\phi) \geq x_0\}} f(\phi)}{\int_{\Phi} \mathbf{1}_{\{g(\phi) \geq x_0\}} f(\phi)}, \tag{10.25}$$

[4] *First and second order reliability methods* [475].

characterized by a null variance, but this is unattainable in practice as it corresponds to knowing how to explicitly solve the problem (i.e., knowing the domain of ϕ where $g(\phi) \geq x_0$). Nevertheless, this result is a fundamental one because almost all modern approaches based on IS look to dynamically construct[5] a succession of approximations to the ideal distribution (10.25). This can typically be done:

- by learning about g, via a meta-model [51, 56, 232], while reducing the dimensions of the problem to the most influential components of ϕ [538];
- by roughly yet rapidly "zooming" in to the zone of Φ close to the border $g(\phi) \geq x_0$, using, for example, FORM-SORM approaches [295, 409];
- by learning directly about f_{IS}, e.g., using a mixture of kernels whose weights are updated [139, 183], or in an entirely nonparametric way [599].

The inversion of the cumulative distribution function $\mathscr{F}$ of X required in (10.19) can also benefit from (10.24) by defining the estimator

$$\hat{\mathscr{F}}_{IS}(x) = \frac{1}{m} \sum_{k=1}^{m} \frac{f(\phi_k)}{f_{IS}(\phi_k)} \mathbf{1}_{\{g(\phi_k) \geq 0\}},$$

and the estimation of a quantile z can be achieved using classical [340, 388] or adaptative [246, 523] IS. In particular, several strategies involving adaptative importance functions were tested by [599]. The most recent results use a mix of meta-modeling and adaptative IS [532, 691]. The incremental part of the adaptative methods can notably be performed by taking advantage of statistical learning tools (it is then called *active learning* [691]).

Example 10.3 (*Prediction of climate extremes*)

A promising use of adaptative importance sampling applied to temperature trajectories derived from climate models, with random initial conditions, is based on iterative resampling of importance weights in order to select periods with extreme temperatures [633]. In doing so, these authors are capable of parallel simulation of rare (or never observed) heatwaves over periods of 90 days or more with the help of the *principle of large deviations* [745] which specifies the way in which the extreme values of an unknown probability distribution behave, in order to estimate extreme return levels (associated with probabilities of between 10^{-5} and 10^{-7} per year). The "zoom" enabled by importance sampling drastically decreases the required number of calls to the climate model (*Planet Simulator* [287]) by a factor of 100–1000.

---

[5] Namely, during the numerical exploration of g, through a progressively enriching experimental design.

Multilevel sampling. Multilevel sampling, often called *subset simulation* [48] or *importance splitting* [147, 149, 472, 526], is based on the principle of expressing the quantity to be estimated as a function of a decreasing (in size) sequence of nested subsets containing the desired domain. The most illuminating example of this involves the estimation of the exceedance probability $p = \mathbb{P}(X \geq x_0)$. This can be rewritten as

$$p = \prod_{i=1}^{M} \mathbb{P}(X \in D_i | X \in D_{i-1}),$$

where $D_M = \{x \in \chi,\ x \geq x_0\}$, $D_0 = \chi$, and $\forall i \in \{1, \ldots, M-1\}$,

$$D_i = \{x \in \chi,\ x \geq \varepsilon_i\},$$

with $\varepsilon_1 \leq \ldots \varepsilon_{M-1} \leq x_0$. If the size of the subsets D_i (i.e., the values of ε_i) are well chosen, each conditional probability $\mathbb{P}(X \in D_i | X \in D_{i-1})$ can be estimated using a (perhaps accelerated) Monte Carlo method, simulating according to the density

$$f_{X_{D_i}}(x) = \frac{f_X(x)\mathbf{1}_{\{x \in D_i\}}}{\mathbb{P}(X \in D_i)}. \tag{10.26}$$

This simulation can be performed using Markov chains (Sect. 4.1.5): *Markov chain Monte Carlo* (MCMC). The most common implementation of this method is based on the acceptance-rejection Metropolis-Hastings algorithm, which is recalled in Chap. 11 (in a Bayesian context). This type of algorithm was implemented by [48] and has undergone numerous improvements since (see, for example, [367, 524] for a recent history).

These improvements are essentially based on the construction of the Markov chains themselves in ways which guarantee accelerated convergence to each target distribution (10.26), as well as the construction of better estimators of each probability $\mathbb{P}(X \in D_i | X \in D_{i-1})$. However, the main difficulty posed by the use of MCMC is weak control over the number of calls required to $\boldsymbol{g}$. For this reason, several authors like [64] have proposed a modification to these methods based on meta-models which have low computation costs, under the name *Bayesian subset simulation*. Multilevel sampling methods perform optimization by favoring an adaptative selection of thresholds, choosing them as fixed-order quantiles. The number of steps M then becomes random.

Finally, these techniques can be modified in order to estimate quantiles z with the help of an efficient choice of subsets D_i, also constructed with adaptative thresholding [527, 599].

10.3.4.3 Conclusions on the Use of Methods and Tools

The use of calculation codes in place of insufficient real or reconstructed data is an engineering subject all to itself, more often known as *treatment of uncertainties in numerical models*. The search for physically extreme situations requires numerical "zooming" tools that take into account different aspects of the problem: the mathematical dimension itself (i.e., the number of model inputs that are considered to be stochastic), the computational time (which generally increases as the algorithmic model implemented in the code becomes more complex as knowledge of the physical phenomenon improves), the *level of extremeness required* (i.e., the *a priori* order of magnitude of an exceedance probability, the range of values considered for a return level—given as a quantile), and lastly, prior knowledge of part of the phenomenon's physics, notably whether is it smooth or not.

Lacks of smoothness generally come in the form of *edge effects*. A *validated* code (cf. Sect. 10.3.6) must be able to reproduce any loss of continuity in the physical phenomenon. A loss of *differentiability,* weaker than that of continuity, also needs to be taken into account. An example of non-locally differentiable physics is that of the water level of a watercourse, which when reaching the level of a protective dike, overflows into the flood plain* and thus radically changes in value, due to spreading out over a much larger volume than the river bed* (where it usually flows).

The presence of edge effects is much in evidence at the boundary of a domain of validity, precisely in situations bordering a safety domains. The Monte Carlo method—by the fact that it requires no hypotheses on the numerical model—remains the preferred method here, on the condition that we can overcome computational time difficulties, very generally correlated with a good recovery of critical situations. In other situations, an additional hypothesis on the physics, possibly a conservative one (e.g., monotone outputs with respect to inputs), may allow accelerated methods to be used.

However, the most common hypothesis is *regularity* or smoothness, which corresponds both a set of differentiability classes and a certain stationarity of the phenomenon. In particular, the use of meta-models with continuity and differentiability features in the place of code requires this hypothesis in practical settings.[6] This is true notably for Gaussian processes (meta-modeling using kriging), which, since the start of the century, have become commonly used tools in studies dealing with uncertainty. The use of classification tools coupled with continuous meta-models can help to free ourselves from some regularity hypotheses (e.g., stationarity—see [353]), but this area of research is still in its infancy. An important and unresolved problem with the increasing use of meta-models is the question of whether they can produce confidence intervals which do not underestimate the estimation error. Indeed, true confidence intervals must take into account both the error related to the choice of an estimator and that of the meta-model.

---

[6] Not necessarily from a theoretical point of view since the validity of a statistical estimator can remain independent of that of a meta-model, which generally helps to guide the numerical design.

Table 10.1 Summary of the uses and properties of rare event evaluation methods (from [406]). Y/N (yes/no) means that the method can be used (or not) for the given goal or requires (or not) hypotheses on the code. The notation ($-$, $\sim$, $+$) is a scale representing increasing quality. An empty entry means that the goal is attainable but the methods and tools required are still being developed or unavailable

	MC	QMC	Wilks	AIS	SS	F(S)ORM	MRM	GP	BSS
Probability p	Y	Y	N	Y	Y	Y	Y	Y	Y
Level of z return quantile	Y	Y	Y	Y	Y	Y	Y	Y	
Hyp. of $\mathscr{G}$	N	Y	N	N	N	Y	Y	N	N
Adaptive	N	N	N	Y	Y	Y	Y	Y	Y
Dimension	+	$\sim$	+	$-$	+	$\sim$	$-$	$\sim$	$-$
Rarity of the event	$-$	$-$	$-$	$\sim$	+	+	$\sim$	$\sim$	+
Computing cost	$-$	$-$	+	+	$\sim$	+	$\sim$	+	++
Error	+	$-$	+	$-$	$-$	$-$	+	$\sim$	$-$
Ease of running	+	+	+	$-$	$-$	+	+	$-$	$-$

Inspired by the work of B. Iooss [406], Table 10.1 provides a summary, in terms of relative advantages and disadvantages, of the best methods mentioned above for calculating exceedance probabilities, which are more or less comparable when estimating quantiles. The abbreviations used for the methods are as follows:

- MC: Monte Carlo,
- QMC: Quasi-Monte Carlo,
- Wilks: Wilks estimator,
- AIS: Adaptative importance sampling,
- F(S)ORM: First (Second) order reliability methods,
- MRM: Monotonic reliability method,
- GP: Meta-modeling using Gaussian kriging,
- BSS: Multilevel sampling coupled with meta-modeling using Gaussian kriging (*Bayesian subset sampling*).

Furthermore, Fig. 10.12 shows the settings in which each of the main methods can be used in order to help engineers who wish to quickly test one or several of the approaches.

10.3.5 Practical Implementation of Methods

Monte Carlo, quasi-Monte Carlo, and space filling algorithms have long been found in standard statistical software packages (e.g., `randtoolbox` and `lhs` in `R`) and in software for analyzing uncertainty (Open TURNS [20]: http://www.openturns.org). The research about and industrialization of the great majority of

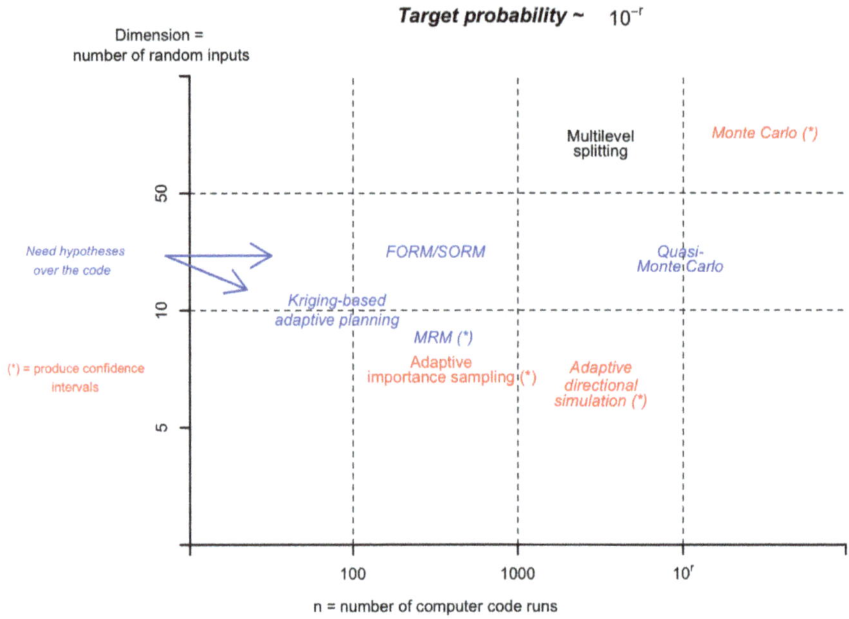

Fig. 10.12 A classification of rare event methods (adapted from [406])

the more advanced methods presented above benefit greatly from the dynamism of the highly active MASCOT-NUM (http://www.gdr-mascotnum.fr) and SIAM-UQ communities. This dynamism is reflected in the rapid availability (principally in `R` and `Python`) of tools for the construction of refined experimental designs and advanced meta-modeling using kriging, accompanied by calculation optimization methods and even estimation methods for quantities of interest; we cite in particular: `DiceDesign`, `DiceKriging`, and `DiceOptim` [670], resulting from work undertaken by the DICE (http://dice.emse.fr/) and then ReDICE (http://www.redice-project.org/) consortiums—which are close to MASCOT-NUM. Certain tools from the `mistral` package run the Wilks and MRM approaches for calculating probabilities and conservative quantiles.

Chaos polynomials are available for use in the `GPC`, `polychaosbasics`, and `plspolychaos` packages associated with sensitivity and regression analysis tools. Finally, a few *hybrid* uncertainty analysis tools, combining probabilities and tools for extra-probabilistic theories can be found in the recent `HYRISK` package.

10.3.6 Validation in Extreme Settings

Validation corresponds to conferring on the model g and its implementation[7] a certain objectivity in its *predictive* description of the phenomenon linking ϕ to X. This objectivity is derived from a *Popperian consensus* [450, 622], i.e., results from a *consensual decision by the protagonists* (the scientific community, and beyond that, civil society), supported by the publication of repeated testing of the model. A consensual definition for the validation of a computational model has been suggested by the US National Research Council [13] in 2006:

Validation is the process of determining the degree to which a model is an accurate representation of the real world from the perspective of the intended uses of the model.

This is a response to a concern which appears in numerous reference documents in this discipline [38, 39, 539, 540] and internationally shared under the general framework of *verification, validation and uncertainty quantification* (VV&UQ). In principle, such repeated trials must allow simulations of X to be compared with real observations, and help delimit a *domain of validity* $\Phi_v \subseteq \Phi$ of the numerical model, i.e., the domain covering the ranges of the input parameter values ϕ that leads to simulations of X that correspond to real physical situations, with an acceptable margin of error. As stated by Chang and Hanna (2004) who built a software dedicated to validating computational models [153]:

(...) according to Oreskes et al. [578], *verification and validation of numerical models of natural systems are impossible, because natural systems are never closed and because model solutions are always non-unique. The random nature of the process leads to a certain irreductible inherent uncertainty. Oreskes et al.* [578] *suggest that models can only be confirmed or evaluated by the demonstration of good agreement between several sets of observations and predictions.* [152].

The notion of an acceptable error thus presupposes knowledge—or even perfect control—of the experimental conditions and model, and that it is possible to define metrics $\mathscr{D}(\mathbf{x}, \mathbf{x}^*)$ that can compare the simulation results $\mathbf{x}$ with observations $\mathbf{x}^*$ [567]. Such calculations of metrics correspond to indicators which must be

- calculable with a reasonable calculation cost (and thus can be produced for large volumes of data coming from simulations);
- interpretable;
- always well defined and robust (in particular: able to deal with missing data, e.g., resulting from interrupted numerical calculations).

In practice, such indicators may

1. be *visual,* typically in the form of QQ-plots or *scatter plots*, cf. Fig. 10.13;
2. more formally, correspond to the result of a statistical test [569], which also implies being able to set limit values d such that under the (null) hypothesis of agreement between the model and reality, we have $\mathscr{D}(\mathbf{x}, \mathbf{x}^*) \leq d$ with high

[7] Supposed *verified* in the sense that the implemented algorithm correctly translates the model up to a controlled error [530].

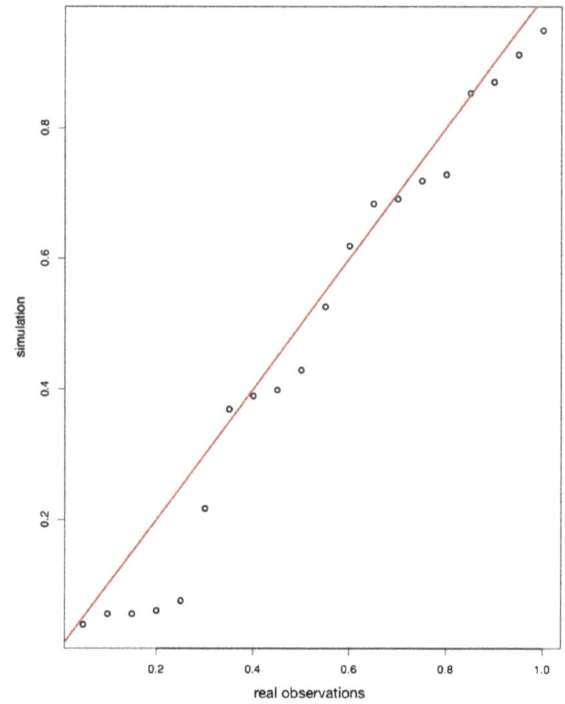

Fig. 10.13 Illustration of a QQ-plot highlighting the possible disagreement between real and simulated data

probability $1 - \alpha$. However, no test statistic is universally good for validation [569], nor is there a universally good way to choose threshold values α. It is thus necessary to use several tests and a number of classical thresholds recommended by [417]: p-values of 1% or even 0.1% are now considered more acceptable than the traditional choice of 5% as they provide a reasonable agreement with Bayesian tests (see Chap. 11).

According to [152, 153], such indicators can be separated into two groups, summarized on Table 10.2:

1. *weak indicators*, which provide a *local* vision of the validation (e.g., on average, the real phenomenon is reproduced, or its direction of variation is coherent with that proposed by the numerical model, etc.);
2. *strong indicators*, which suggest a better and more general match. A mixture of weak indices can produce a strong indicator.

One of the most robust validation metrics is the *area metric* d_{AM} proposed by Ferson et al. [274], defined as the difference in ℓ^1 norm between the cumulative distribution function F_{x^*} of the real data and F_x of the simulations of X produced by $g(\phi)$:

$$d_{AM} = \int_{\chi} |F_x(u) - F_{x^*}(u)| \, du,$$

Table 10.2 The main strong and weak validation indices in the sense of [152, 153]. We denote $\mathbf{x} = (x_1, \ldots, x_n)$ a vectorial output of the model g and $\mathbf{x}^* = (x_1^*, \ldots, x_n^*)$ a real observation. Here, $y_i^{(*)}$ is the rank of y_i in its sample ($i = 1, \ldots, n$). The notation $\bar{\mathbf{y}}$ means the arithmetic mean of the vector $\mathbf{y}$, and $\sigma_{\mathbf{y}} = (\overline{(\mathbf{y} - \bar{\mathbf{y}})^2})^{1/2}$

Weak indicators	
Bias	$\overline{\mathbf{x}^* - \mathbf{x}}$
Bias factor	$\overline{\mathbf{x}^*/\mathbf{x}}$
Fractional bias	$2(\bar{\mathbf{x}}^* - \bar{\mathbf{x}})/(\bar{\mathbf{x}}^* + \bar{\mathbf{x}})$
Geometric bias	$\exp\left(\overline{\log(\mathbf{x}^*)} - \overline{\log(\mathbf{x})}\right)$
FAC$-\alpha$ (*fraction of data*)	$\overline{\mathbf{1}_{\{\alpha \leq \mathbf{x}/\mathbf{x}^* \leq 1/\alpha\}}}$
Pearson's linear correlation	$\dfrac{\overline{(\mathbf{x} - \bar{\mathbf{x}})(\mathbf{x}^* - \bar{\mathbf{x}}^*)}}{\sigma_{\mathbf{x}}\sigma_{\mathbf{x}^*}}$
Spearman's rank correlation	$\dfrac{\overline{(\mathbf{x}^{(*)} - \overline{\mathbf{x}^{(*)}})(\mathbf{x}^{*(*)} - \overline{\mathbf{x}^{*(*)}})}}{\sigma_{\mathbf{x}^{(*)}}\sigma_{\mathbf{x}^{*(*)}}}$
Strong indicators	
Mean quadratic error (ℓ^2 norm)	$\sqrt{\overline{(\mathbf{x} - \mathbf{x}^*)^2}}$
Normalized mean quadratic error	$\dfrac{\overline{(\mathbf{x} - \mathbf{x}^*)^2}}{\bar{\mathbf{x}}\bar{\mathbf{x}}^*}$
R^2 prediction coefficient	$1 - \dfrac{\overline{(\mathbf{x} - \mathbf{x}^*)^2}}{\overline{(\bar{\mathbf{x}} - \bar{\mathbf{x}}^*)^2}}$
Geometric variance	$\exp\left\{\overline{(\log(\mathbf{x}) - \log(\mathbf{x}^*))^2}\right\}$
Maximal absolute error (ℓ^∞ norm)	$\max_{i=1\ldots,n} \lvert y_i - y_i^* \rvert$

which can easily be estimated as the area between the empirical cumulative distribution functions (Fig. 10.14). This metric has the advantages of being always defined, fairly robust, and producing results in the same measurement units as $X \in \chi$. The same authors, faced with difficulties in dealing with multivariate outputs X, have proposed an aggregation technique called *u-pooling* which, though it has the advantage of providing a visual diagnostic tool, supposes that the comparison indices between empirical and simulated distributions have equivalent weights per dimension. We instead prefer (for example) recent statistical tests matching observed and simulated data which are based on a set of multivariate statistics (known as *energy* statistics) involving Euclidean distances between data [729, 730]. Such tests are currently some of the most powerful available when the dimension of X becomes large [732] and are often used by the machine learning community.

Lastly, we can expect that validation methods will be increasingly influenced by the spectacular advances in automatic classification seen in recent years. The starting assumption is simple: if a simulation model producing a great quantity of data similar

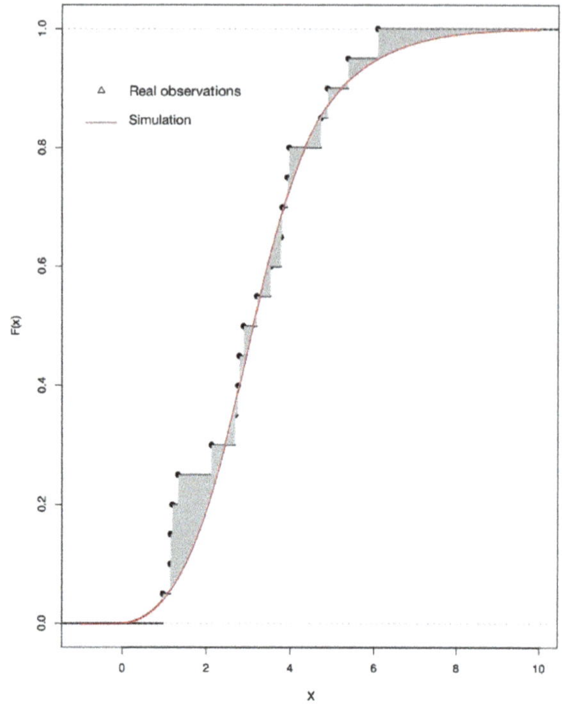

Fig. 10.14 Illustration of the empirical estimation of the *area metric* [274] based on the difference between cumulative distribution functions: area of the grey zone

to real data is "valid", then any classification of the union of these two types of data should not be able to distinguish between them. This suggests that to validate a simulation model, it should be possible to confront it with a *detection model* that can test whether the data shown to it are generated or in fact correspond to real data, with balanced test results. This is exactly what is known in classification as the *adversarial* principle, which originated in game theory and has become well known due to the rise of generative neural networks (*generative adversarial networks* or GANs) [42, 347]).

The difficulty of validation in extreme-value settings is linked to the unlikeliness of being able to run a direct comparison between simulation results, known as *system response quantities* in [569], and observed data, since there may be very few of the latter available. Automatic classification approaches, requiring usually large quantities of data, seem to be of little relevance here. In fact, it appears easier to confer simulation results de facto validity using an *information fusion* procedure with the real data, than look for a binary response confirming or rejecting calculation outputs. This type of procedure, generally based on the use of Bayes rule (see the following chapter)—for instance, *Bayesian melding* [170, 242]—implies a comparison metric and the recalibration of one or several uncertain inputs to the numerical code, possibly with the help of meta-models [294, 637]. Though Bayesian inference seems to be the most relevant and used methodology for calibrating computer models,[8] the difference between (re)calibration and validation in this setting is not always clear [192].

Remark 10.1 It is also important to pay close attention to the following phenomenon, which occurs often: certain key parameters of numerical models "pushed" into extreme settings are calibrated using data coming for the most part (or entirely) from a *regular* domain, i.e., a non-extreme one. This is the case in particular for *limit conditions*, as illustrated in Example 10.4, which act as hidden, non-probabilistic parameters, therefore severely limiting the pertinence of the model when it is used close to the boundaries of its domain of validity.

Example 10.4 (*Hydraulic codes*)

Hydraulic codes are typically calculated over a well-defined territory by providing a hypothesis on the link between the water height and the downstream flow, through a rating curve [223]. This corresponds to a limit condition on the fluid mechanics equations modeling the flow [301]. "Downstream" is a sufficiently distant place (ideally, the ocean in the case of a river) that we can consider this link little affected by changes in the flow domain upstream. Nevertheless, gauging performed to construction of this curve occurs in standard flow conditions.

Examples of such studies, however, remain rare in extreme value statistics. The works [181, 182], which focus on fusing information in order to estimate the frequency of extreme rainfall in a climate change context, are relevant here. Alternative approaches focus on validating the distributions of stochastic inputs of these computational models, in relation with worst-case scenarios most often defined as the values located at the borders of the models' domain of validity. For instance, [488] has proposed using concentration inequalities to bound the probability of rare events output from numerical models, whatever the statistical characteristics of the input distributions are. However, such approaches are still a matter of debate, as they are susceptible to leading to highly conservative return level values.

At present, the validation of numerical models in extreme settings remains a methodological "bet", fueled by the gradual refinement of the various parts of these models[9] and, in the precise case of climate models, by the increasing attention paid to problems involving changes in scale. Validation of extreme situations produced by mesoscale climate models (typically square grids with 200 km sides), then a transfer by *downscaling* of these situations to regional extreme events using extreme value statistics models, is a current area of research [292] whose main constraint is

[8] See, for example, [294, 664] for an illustration on hydrological models for flood forecasting purposes.

[9] Structural equations, domains of variation of parameters, numerical calculation precision, etc.

how to deal with the high levels of uncertainty that permeate climate models [689]. Methodological advances have been made in recent years to help render *ensemble* climate projections probabilistic [90, 632] and then provide local projections of the evolution of extreme climate values with the help of specific statistical models [728].

Chapter 11
Bayesian Extreme Value Theory

Nicolas Bousquet

Abstract This chapter provides an introduction to Bayesian statistical theory and a broad review of its application to extreme values. The Bayesian methodology differs greatly from the traditional approaches considered in the other chapters. Indeed, it requires the construction of so-called prior measures for the parameters of extreme value models and defines estimation through the minimization of a cost function adapted to the event. While Bayesian calculations (as those based on Monte Carlo Markov Chains) remain superficially discussed, this chapter focuses on modeling features, which allow expert opinion and historical knowledge to be integrated into the estimation of quantities of interest. In this respect, the Bayesian approach constitutes a methodology of increasing use, allowing the mixed treatment of aleatoric and epidemic uncertainties and adapted to the specific needs of engineers.

11.1 Methodological Principles

The great majority of the methods and tools presented in the earlier chapters treat the problems of inference, prediction, and extrapolation in the *classical* statistical framework, which supposes that the parameters ψ of extreme value models F_ψ take unknown but unique fixed values. This is a natural consequence of a strict application of the Fisher–Tippett [283] and Pickands [615] theorems, the supporting pillars of extreme value theory. This acknowledged, it is necessary to estimate these unknown parameters using available observations $\mathbf{x_n} = (x_1, \ldots, x_n)$, which leads to the difficulty mentioned previously in Sect. 4.2.2: rules are required to choose the best estimator $\hat{\psi}_n$. The quality of a given estimator $\hat{\psi}_n$ corresponds in part to the level of certainty that the statistical confidence region it defines contains the true parameter.

The classical statistical framework is not the only one available to us, however. The *Bayesian* framework—just as important—can help us overcome certain limits of the classical setting, without disrupting the applicability of standard theorems.

N. Bousquet (✉)
EDF R & D, 6 quai Watier, 78401 Chatou, France
e-mail: nicolas.bousquet@edf.fr

N. Bousquet and P. Bernardara (eds.), *Extreme Value Theory with Applications to Natural Hazards*, https://doi.org/10.1007/978-3-030-74942-2_11

This chapter reviews the advantages and disadvantages of using a Bayesian approach for handling extreme value statistics models. Examples will focus on one-dimensional and stationary extreme value problems, though there are increasing numbers of applications and improvements to such methods in the literature for the spatial and temporal multivariate cases. For a more in-depth look at important concepts, methods, and results in Bayesian statistics, we suggest the article [715] as an excellent starting point.

11.1.1 Credibility Rather Than Confidence

The Bayesian paradigm criticizes the notion of *classical statistical confidence* which is, in reality, not so efficient at decision-making from available observed data, since it is based on infinitely repeated virtual observations, dependent on the estimated model. Arguments leading to the construction of confidence regions are indeed often based on the convergence in distribution of estimators $\hat{\psi}_n$ (cf. Sect. 4.2.1), which is by essence asymptotic (i.e., when $n \to \infty$). This is the case for estimators for extreme value statistical models invoked earlier. However, the only information we have—without modeling hypotheses—is the data $\mathbf{x_n}$ itself, and the data size n may not be large enough to ensure that we are close enough to the asymptotic setting.

Bayesian theory replaces this notion with the idea of *credibility*. Here, **conditional on observed data**, an estimated credibility region contains ψ with a certain probability.

Conditioning ψ on the observations implies that it is no longer considered as a vector of fixed and unknown parameters, but rather as a random latent variable, whose distribution is denoted generically by Π (and its density π). The existence of a probability measure on ψ is assured by the De Finetti representation theorem and extensions thereof [577] when observations are considered to be *exchangeable* (and thus potentially dependent) and no longer iid. The Bayesian framework thus encompasses the classical framework. The book [626] (Chap. I.5) provides numerous details concerning the existence of Π.

The inference procedure consists of performing an *update of the distribution* Π, conditional on the observations $\mathbf{x_n}$, using Bayes' theorem [651]: the *prior* density $\pi(\psi)$ is modified adding information coming from the statistical likelihood $\ell(\mathbf{x_n}|\psi)$ into a *posterior* density $\pi(\psi|\mathbf{x_n})$, written as

$$\pi(\psi|\mathbf{x_n}) = \frac{\pi(\psi) f(\mathbf{x_n}|\psi)}{\int_\Psi \pi(\psi) f(\mathbf{x_n}|\psi)\, d\psi}. \tag{11.1}$$

Hence, while classical statistics calculates an estimator $\hat{\psi}_n$ (often maximizing the likelihood $f(\mathbf{x_n}|\psi)$) which has a known asymptotic distribution, Bayesian statistics postulate that the true nature of ψ is better described by another statistical distribution defined for n fixed, and described by Formula (11.1). Strong connections exist

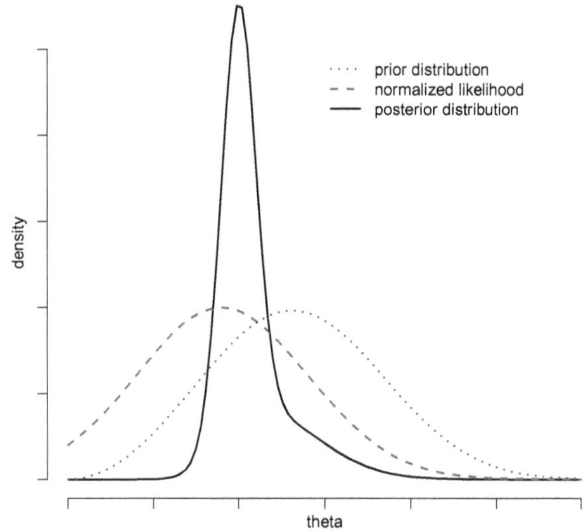

Fig. 11.1 Illustration using density functions of the improved information on ψ from the prior distribution and the data likelihood (seen as a function of ψ and renormalized here)

between the two approaches. In particular, when $n \to \infty$ and under mild conditions, the respective distributions of $\hat{\psi}_n$ and $\pi(\psi|\mathbf{x_n})$ become similar. These connections are described in greater detail in [651].

Figure 11.1 shows the classical form of the statistical likelihood $f(\mathbf{x_n}|\psi)$ seen as a function of ψ, as compared with the densities $\pi(\psi)$ and $\pi(\psi|\mathbf{x_n})$. The pooling of information on ψ leads logically to a posterior distribution with a sharper—and thus more informative—peak than the prior distribution.[1]

Nature of the prior distribution. What does the prior distribution $\pi(\psi)$ express? It corresponds to the knowledge we may have about the possible values of ψ, independent of the observations $\mathbf{x_n}$. Encoded in probabilistic form, this type of information is therefore likely to enable the addition of real and serious knowledge of the phenomenon expressed in other ways through recorded observations: expertise, predictions made by physical models, etc. It should be noted that prior information on ψ always exists; indeed, ψ is a choice of parametrization of the selected extreme value statistics model (MAXB or POT), and the structure of this model is known. This gives rise to constraints on the correlation structure of $\pi(\psi)$ (cf. Sect. 11.2). We recommend, to readers interested further in possible interpretations of $\pi(\psi)$, the book [591] and article [598].

[1] Except for cases where the two sources of information disagree; see [259].

11.1.2 Dealing with Parametric Uncertainty

The second criticism that can be made of the classical approach is its inability to take into account all parametric uncertainties. If ψ is estimated by $\hat{\psi}_n$, whose distribution is denoted by G_n, how should we predict a new observation x_{n+1} of the extreme-valued natural hazard? This should in theory be produced using the conditional distribution

$$x_{n+1} \sim f(x_{n+1}|\psi, \mathbf{x_n}). \tag{11.2}$$

1. By supposing that the next random event x_{n+1} is independent of the previous ones $\mathbf{x_n} = (x_1, \ldots, x_n)$, the term x_{n+1} can simply be estimated by

   $$x_{n+1} \overset{\text{approx.}}{\sim} f(x|\hat{\psi}_n),$$

   but it remains necessary to include the uncertainty in $\hat{\psi}_n$, represented by G_n, to lower the approximation error. However, G_n also needs to be estimated, generally using $\hat{\psi}_n$ itself (e.g., with the help of the delta method, cf. Theorem 4.4). It is thus necessary to set a practical limit which allows us to work with a G_n with "sufficient" uncertainty to make reasonable predictions of x_{n+1}. This practical limit is linked to the precision of the estimation of G_n, which depends on asymptotic theoretical results that are often unverifiable in practice. Here again, asymptotism is a limiting factor, all the more so in a context where the number of real data is often low.
2. If we suppose that the next random event x_{n+1} is dependent on certain previous ones in $x_1, \ldots, x_n$, the distribution (11.2) has to be estimated by

   $$f(x_{n+1}|\hat{\psi}_n, \mathbf{x_n}).$$

   The difficulties presented in the first point above still remain, and moreover, there is redundant information, arriving both from the data $\mathbf{x_n}$ and $\hat{\psi}_n$. This redundancy artificially inflates the available information and threatens to lead to an underestimation of the true uncertainty in the value of x_{n+1}, making it even more important to resolve the earlier problems.

In the Bayesian framework, the expression (11.2) can be rewritten uniquely, without theoretical approximations or untoward redundancy, as follows:

$$f(x_{n+1}|\psi, \mathbf{x_n}) = \int_{\Psi} f(x_{n+1}|\psi, \mathbf{x_n})\pi(\psi|\mathbf{x_n})\, d\psi, \tag{11.3}$$

and in the case where conditional on ψ, the future observation x_{n+1} is independent of $\mathbf{x_n}$:

$$f(x_{n+1}|\psi, \mathbf{x_n}) = \int_{\Psi} f(x_{n+1}|\psi)\pi(\psi|\mathbf{x_n})\, d\psi. \tag{11.4}$$

It turns out that Bayesian predictions are generally better than classical ones, in particular when looking at extreme quantiles [708]. According to Coles [173] (Chap. 9), *Bayesian techniques offer an alternative that is often preferable.* He also notes that as the principal use of extreme value distributions is for prognosis and anticipation of future events, the most natural framework to estimate them should be *predictive* [173, 177]. The framework needs to transfer as completely as possible uncertainty in model parameter estimation into the functions of interest (such as return periods).

Example 11.31 (*Forecasting floods*)

Numerous authors have warned about the unfortunate consequences of poor (or non-existent) taking into account the uncertainty in extreme value hydraulics problems, whether in terms of extrapolation and forecasting [454, 534, 763] or in fitting the Manning–Strickler friction coefficient [223] for strong flow situations (see also Remark 10.14).

Remark 11.15 Standard Monte Carlo techniques [652], requiring simulation of the posterior distribution $\pi(\psi|\mathbf{x_n})$, can be used to calculate the integrals in Formulas (11.3) and (11.4).

11.1.3 Model Selection

Several other criticisms of the classical frequentist framework are examined in detail in the reference work [651]. In particular, classical statistical tests do not allow to compare symmetrically two hypotheses: rejecting one hypothesis is not similar to give credibility to the other one. (cf. Sect. 4.1.3). Besides, the arbitrary choice of extreme value thresholds in classical tests is a clear weak-point [258]. The Bayesian framework offers a clear alternative, placing several prior models on an equal footing and then selecting the *most likely* one conditional on the data $\mathbf{x_n} = (x_1, \ldots, x_n)$. Let us consider the case where the classical statistical test involves testing two Bayesian models

$$\mathscr{M}_i : X|\psi_i \sim f_i(X|\psi_i) \text{ and } \psi_i \sim \pi_i(\psi_i), \quad \text{for } i \in \{1, 2\}.$$

Conditional on the data $\mathbf{x_n}$, the *Bayes factor*

$$B_{12}(\mathbf{x_n}) = \frac{\mathbb{P}(\mathbf{x_n}|\mathscr{M}_1)}{\mathbb{P}(\mathbf{x_n}|\mathscr{M}_2)} = \frac{\int_{\Psi_1} f_1(\mathbf{x_n}|\psi_1)\pi(\psi_1)\,d\psi_1}{\int_{\Psi_2} f_2(\mathbf{x_n}|\psi_2)\pi(\psi_2)\,d\psi_2}, \tag{11.5}$$

or ratio of integrated likelihoods, gives the ratio of the probability that the data matches with model $\mathscr{M}_1$ over the probability it matches with model $\mathscr{M}_2$. This model selection problem can be more broadly defined as part of a decision-making framework (see Sect. 11.1.4) which provides a scale for the ratio's value (after what value of B_{12} can we consider that model 1 is better than model 2?). To this end, we also include a penalty term ρ_{12} related to the fact that one of the competing models may be more flexible (or even over-parametrized) than the other, which could lead to a selection bias. Let

$$\rho_{12} = \frac{P_{\mathscr{M}}(\mathscr{M}_1)}{P_{\mathscr{M}}(\mathscr{M}_2)},$$

where $P_{\mathscr{M}}$ is a *prior* probability measure over the discrete and bounded Bayesian model space $\mathscr{M}_i$. $P(\mathscr{M}_1)$ and $P(\mathscr{M}_2)$ represent, in practical terms, the *prior* probabilities that the Bayesian models $\mathscr{M}_1$ and $\mathscr{M}_2$ correctly model the process producing the data X, before observing any. This probability measure is difficult to handle (except for cases where the models are supposed equitable a priori) and the ratio ρ_{12} helps to more easily make a choice. In this case, the *posterior* probability that the model $\mathscr{M}_1$ could have produced the data $\mathbf{x_n}$ is defined by Bayes' rule:

$$\begin{aligned} P_{\mathscr{M}}(\mathscr{M}_1|\mathbf{x_n}) &= \frac{\mathbb{P}(\mathbf{x_n}|\mathscr{M}_1)P_{\mathscr{M}}(\mathscr{M}_1)}{\mathbb{P}(\mathbf{x_n}|\mathscr{M}_1)P_{\mathscr{M}}(\mathscr{M}_1) + \mathbb{P}(\mathbf{x_n}|\mathscr{M}_2)P_{\mathscr{M}}(\mathscr{M}_2)}, \\ &= \{1 + 1/(\rho_{12}B_{12}(\mathbf{x_n}))\}^{-1}, \\ &= \{1 + \rho_{21}B_{21}(\mathbf{x_n})\}^{-1}, \end{aligned}$$

and the "averaged" predictive posterior distribution—which helps us better simulate new data—is given by

$$\mathbb{P}(x|\mathbf{x_n}) = \mathbb{P}(x|\mathscr{M}_1)P_{\mathscr{M}}(\mathscr{M}_1|_b f x_n) + \mathbb{P}(x|\mathscr{M}_2)P_{\mathscr{M}}(\mathscr{M}_2|_b f x_n).$$

These model selection tools, which strongly differentiate Bayesian from classical statistics, can notably thus help us to refine our knowledge of the behavior of the parameter ξ, which is very important when characterizing the nature of the differences between centennial, millennial, and decamillennial orders of magnitude (see Example 11.32). Also, further pleasing properties of the Bayes factor include that it does not need to be weighted as a function of a difference in dimension between competing models,[2] and is sure to select—in a finite list of models—as $n \to \infty$, the true model if it exists, or the best one according to an information-theoretic loss [89]. We recommend the articles [34, 384, 425, 776] for a more detailed introduction to understanding and calculating with Bayesian model selection tools. A readable

[2] As do, for example, classical asymptotic criteria like AIC [25] and BIC [693].

comparison of the Bayesian and frequentist approaches for hypothesis testing is provided in [715, 763] performs one such comparison for extreme and non-extreme value models in natural hazard management.

Remark 11.16 We strongly discourage using standard Monte Carlo techniques like

$$\hat{B}_{12}(\mathbf{x_n}) = \frac{M_1^{-1} \sum_{i=1}^{M_1} f_1(\mathbf{x_n}|\psi_{1i})}{M_2^{-1} \sum_{k=1}^{M_2} f_2(\mathbf{x_n}|\psi_{2i})},$$

where $(M_1, M_2) \gg 1$, $\psi_{1i} \overset{iid}{\sim} \pi_1(\psi_1)$, and $\psi_{2i} \overset{iid}{\sim} \pi_2(\psi_2)$ to estimate a Bayes factor (11.5). Doing so leads to high instability in calculations [140]. More robust estimation techniques are based on posterior simulation of ψ_1 and ψ_2 [778]. Note that a promising model selection methodology based on mixtures, making it possible to dispense with the sometimes tricky estimation of the Bayes factor (notably when the prior measures are improper, cf. Sect. 11.2.1), was recently proposed by [422].

Example 11.32 (*Torrential rain*)

A model selection exercise for choosing between Gumbel, Fréchet, and Weibull distributions was performed by [590] in the POT setting on snowfall data and by [118] in the MAXB setting on the rainfall data in Example 11.34. In the latter, an alternative method for calculating the Bayes factor, inspired by [422], is used. Lastly, in a regional study by [634], a Bayesian selection method for the parameter ξ was proposed.

11.1.4 Decision-Making for Risk Mitigation

More generally, the Bayesian framework is well-adapted at formalizing and making mitigation decisions under uncertainty [337, 452, 708]. Hypothesis testing and model selection can also be defined under this framework [651]. Decision-making is conditional on the choice of:

1. A *function of interest* $h(\psi)$. In the extreme hazards context, this may typically be

   - the identity $h(\psi) = \psi$, if we want to estimate the parameter vector ψ;
   - a return level, calculated in the MAXB or POT setting;
   - the probability $1 - F(x_0|\psi)$ that a future hazard will exceed a fixed level x_0.

As ψ is treated as random in the Bayesian setting, $h(\psi)$ is random too. We therefore denote by $\pi(h(\psi)|\mathbf{x_n})$ its posterior density.

2. A *cost function* $L(z, h(\psi))$ which measures the *cost* or error resulting from the pointwise choice of z for estimating the random variable $h(\psi)$; indeed, a Bayesian-type decision means choosing a unique representative value which minimizes *on average* this cost function over the uncertainty carried by $h(\psi)$, expressed by $\pi(\psi|\mathbf{x_n})$:

$$\widetilde{h(\psi)} = \arg \min_{z \in h(\Psi)} \int_{\Psi} L(z, h(\psi))\, \pi(h(\psi)|\mathbf{x_n})\, dh(\psi). \tag{11.6}$$

The choice of a cost function integrated over the posterior distribution corresponds to the choice of an *expected gain*. Generally speaking, the cost function depends on the study context as it possesses an *economic interpretation* [591]. Below are two classical examples of cost functions, which may be seen as first-order expansions of more complicated cost functions, that remain usually unknown to the statistician.

The **quadratic cost function** L^{quad} defined by

$$L^{\text{quad}}(z, h(\psi)) = (z - h(\psi))^2, \tag{11.7}$$

applied to a function of interest $h(\psi)$, provides as its "natural" estimator the posterior expectation (or *mean*)

$$\widetilde{h(\psi)} = \mathbb{E}[h(\psi)|\mathbf{x_n}] = \int_{\Psi} h(\psi)\, \pi\left(h(\psi)|\mathbf{x_n}\right)\, d\psi.$$

This symmetric and convex cost function penalizes—strongly and similarly—large discrepancies between the pointwise estimator $\widetilde{h(\psi)}$ and the random variable $h(\psi)$. It is therefore useful for estimating position and scale parameters in extreme value models.

The **piecewise linear cost function** L^c is defined as follows for (11.6):

$$L^c(z, h(\psi)) = C_1|z - h(\psi)|\mathbf{1}_{\{z \leq h(\psi)\}} + C_2|z - h(\psi)|\mathbf{1}_{\{z \geq h(\psi)\}}.$$

It involves asymmetric costs: $C_1(z - h(\psi))$ of overestimating $h(\psi)$ if it is less than z, and $C_2(h(\psi) - z)$ of underestimating $h(\psi)$ if it is greater than z. The estimator $\widetilde{h(\psi)}$ that comes out of this is the quantile of order $\alpha = C_2/(C_1 + C_2)$ of the posterior distribution of $h(\psi)$ [591]:

$$\alpha = \mathbb{P}\left(h(\psi) \leq \widetilde{h(\psi)}\right) = \int_{-\infty}^{\widetilde{h(\psi)}} \pi\left(h(\psi)|\mathbf{x_n}\right)\, dh(\psi).$$

Clearly, this type of convex cost function is good for one-dimensional functions of interest, like return levels, for which overestimation corresponds to a conservative approach in terms of risk. Indeed, underestimating the return level is intuitively more costly than overestimating it ($C_2 > C_1$). The cost of overestimation is the cost of needlessly strong mitigation, while that of underestimation is essentially the damage caused by the natural hazard.

An engineer can then easily communicate the risk level (or corresponding costs ratio) associated with their choice of the estimator; the posterior probability of having poorly evaluated the return level in the non-conservative direction can be estimated by $C_1/(C_1 + C_2) = 1 - \alpha$.

Example 11.33 (*Flood protection*)

In [591, 592] can be found more detailed examples of the development of cost functions well-adapted to dike dimensioning for mitigating the consequences of flooding. The link between decision-making support in dimensioning and the selection of an extreme value statistical model,is also described in [452], which deals with cases of accidental chemical release, seen as random events.

It is common, if no cost function stands out, to examine the stability of a Bayesian estimator by testing a variety of them. A practical way to do so is to use, in addition to the functions mentioned above, the *linex* family (*linear exponential*). This allows us to pool certain good properties of the previously mentioned functions, which is particularly of interest in finance when estimating quantiles (or *value-at-risk*) [805]. This family is defined by

$$L_c(z, h(\psi)) = \exp\{c(z - h(\psi))\} - c(z - h(\psi)) - 1$$

for $c \in I\!R$. When $c > 0$, the function is asymmetric around 0; overestimation is penalized more than underestimation. When $|z - h(\psi)| \to \infty$, it increases almost exponentially when $z > h(\psi)$ and almost linearly when $z < h(\psi)$. When $c < 0$, the two speeds are swapped. Lastly, when $|z - h(\psi)|$ is close to 0, the cost function is close to $c(z - h(\psi))^2/2$.

Remark 11.17 Note that when using a sample $h(\psi_1), \ldots, h(\psi_M)$, where the ψ_i are drawn from the distribution $\pi(\psi|\mathbf{x_n})$, the right-hand side expression in Eq. (11.6) can be easily approximated using standard Monte Carlo sampling if M is large:

$$\int_\Psi L(z, h(\psi))\pi(h(\psi)|\mathbf{x_n})\, dh(\psi) \simeq \frac{1}{M}\sum_{i=1}^{M} L(z, h(\psi_i)).$$

11.1.5 A Practical Review of the Bayesian Framework

Why not work exclusively in the Bayesian setting then? It gives us more profound and systematic insights into managing uncertainty, while also incorporating knowledge other than observed data when doing statistical modeling. In theory, the methods and tools mentioned in earlier chapters can be carried over into this framework and provide equivalent if not better results, as explained by numerous authors (see for example [293] for forecasting extreme wind gusts). In the usual extreme value study context, the capacity of the Bayesian approach to take information in addition to observed data into account is welcome. In this sense, the usefulness of the Bayesian setting for industrials confronted with extreme natural hazards and regulation authorities was highlighted as early as 1993 by Lambert and his coauthors [452].

Nevertheless, using Bayesian statistics in an industrial setting is not entirely straightforward, for the following reasons:

1. construction/choice of a prior distribution $\pi(\psi)$ is not necessarily obvious and can lead to much criticism [715]; this can be seen as introducing unwanted subjectivity to the problem;
2. calculating an estimator $\hat{\psi}_n$ is replaced by estimating a posterior distribution $\pi(\psi|\mathbf{x_n})$, which can require specific, time-intensive algorithms for which convergence is uneasy to prove in practice;
3. a gain in terms of precision in the estimation of a quantity of interest (like a return level, for example) cannot be systematically ensured;
4. engineers or researchers in charge of a study do not have all of the elements required to run a full Bayesian analysis. For instance, the cost function L, indispensable for determining an estimator of quantities of interest, is not available to them (only to decision-makers). Furthermore, there are no guarantees that Equation (11.6), based on the expected gain, is adequate for dealing with problems involving extreme hazard mitigation (see Sect. 11.3.2.1).

Such arguments can be partially countered. First of all, it is possible to make an *objective* choice of prior distribution in numerous situations (see Sect. 11.2), even if the search for a perfect statistic free of all subjectivity seems hopeless (if not counterproductive) according to philosophers of science [225]. In the above-mentioned situations, the algorithms required to estimate the posterior distribution are well understood, and methods being developed for more complex cases continue to benefit from the growing power of computers (see Sect. 11.4).

The search for an improvement in estimation is a problem related to the comparison between a simulated model (for which the parameters are known and fixed) and an estimated model; de facto, such a comparison is biased because the comparison tools presuppose that the classical model is the norm. A fair comparison would involve comparing the location of actual observations within statistical confidence or credibility regions. Thereupon, Bayesian statistics prove to be highly competitive in extreme value modeling [175–177].

Lastly, the search for cost functions can be simplified: either results can be provided to decision-makers in the form of the posterior distribution itself—leaving them to make use of it—or estimates of ψ and functions thereof can be provided by the engineer through a choice of cost functions which approximate (at first-orders) more complex and/or unknown ones [651], such as those presented in Sect. 11.1.4.

However, the use of Bayesian methods in extreme value statistics remains essentially, for the time being, in the academic domain. The main reasons for reluctance in using them further are probably due to the following:

(a) The difficulty in defining an *prior* modeling methodology that is as universal and general as that which consists of maximizing a likelihood. Practical strategies are often perceived as being case-specific;
(b) A certain degree of statistical conservatism from applied research engineers, a behavior found in numerous applied settings [471] and the vast majority of public policy-making worldwide [276].

As stated earlier, this chapter aims above all to provide an up-to-date view of the field to engineers and researchers who are perhaps interested in refining initial studies performed with the help of classical statistics, as well as looking at how to adapt this framework to real-world case studies.

11.2 Prior Modeling

The choice of a measure $\pi(\psi)$ is no simple task. While there is still no comprehensive prior modeling methodology with as much universality as extreme value theory itself, solid principles—summarized for example in [82, 789]—can be implemented and used by practitioners.

11.2.1 Choosing a Reference Measure

If no prior information other than the parametric model $X|\psi \sim f(x|\psi)$ is available, the chosen measure $\pi(\psi)$ has to be considered a benchmark prior, usually called an "noninformative prior"[3] which acts to regularize the likelihood [79]. The choice and construction of this measure obey certain formal rules based for the most part on the invariance of prior information and a weak sensitivity of the posterior distribution $\pi(\psi|\mathbf{x_n})$ to the choice of $\pi(\psi)$. The main formal rules are given and discussed in [426]. Several common measures are described below.

- The *Jeffreys prior* [412] is defined by

$$\pi^J(\psi) \propto \sqrt{\det I_\psi},$$

[3] Even though this term is overused, precisely due to our knowledge of $X|\psi \sim f(x|\psi)$ [408].

where I_ψ is the Fisher information associated with the sampling model $f(x|\psi)$. This measure has the property of being invariant with respect to any bijective reparametrization $\psi \to h(\psi)$, which is desirable when proposing a formal definition of "noninformative".

Remark 11.18 (*Uniform measure (Laplace's prior)*)
The above property is not valid for the classical uniform measure derived from Laplace's principle of indifference (see for example [230]). It can therefore not be considered as noninformative, and we should beware of its use in practical situations as it can lead to paradoxes [426].

- The *Berger–Bernardo reference prior* $\pi^{BB}(\psi)$ [80] is constructed in such a way that its posterior distribution $\pi^{BB}(\psi|\mathbf{x})$ is *maximally deformed* or *enriched* by each dataset $\mathbf{x} \in \chi_\otimes$ produced by the Bayesian model

$$\mathbf{X} \sim f(\mathbf{x}) = \int_\Psi f(\mathbf{x}|\psi)\pi(\psi)\, d\psi.$$

The measure $\pi^{BB}(\psi)$ can thus be interpreted as the one that has the least influence on the posterior distribution, on average, over the set of possible datasets $\mathbf{x}$. The idea of deformation of a probability distribution $\pi(\psi|\mathbf{x_n})$ with respect to $\pi(\psi)$ is defined here with the help of the Kullback–Leibler divergence[4] between $\pi(\psi|\mathbf{x_n})$ and $\pi(\psi)$:

$$KL(\pi, \mathbf{x}) = \int_\Psi \pi(\psi|\mathbf{x}) \log \frac{\pi(\psi|\mathbf{x})}{\pi(\psi)}\, d\psi.$$

The mean of $KL(\pi, \mathbf{x})$ over $f(\mathbf{x})$ defines the mutual information between ψ and $\mathbf{X}$ [185]:

$$I_m(\psi, \mathbf{X}) = \int_{\chi_\otimes} KL(\pi, \mathbf{x}) f(\mathbf{x})\, d\mathbf{x}.$$

Thus,

$$\pi^{BB}(\psi) = \arg \max_{\pi(\psi) \geq 0} I_m(\psi, \mathbf{X}),$$

where the size of the set of random variables $\mathbf{X}$ tends to infinity. Invariance of mutual information guarantees that the reparametrization invariance property holds for $\pi^{BB}(\psi)$.

Remark 11.19 In multidimensional settings, when constructing $\pi^{BB}(\psi)$, it is possible to distinguish between two types of parameter for each component of ψ.

[4] Or the negative (relative) entropy between $\pi(\psi|\mathbf{x})$ and $\pi(\psi)$.

Parameters *of interest* are typically those that play a part in prediction. *Nuisance* parameters, on the other hand, are generally those linked to the observation process (typically bias or variance in observations) whose influence we would like to curb when making predictions [60]. We do not go down this path in the rest of the chapter; further details can be found in [80].

- The *Maximal Data Information Prior* (MDIP) [795, 797] is constructed so as to maximize the average information $G(\pi)$ contained in the density $f(x|\psi)$ relative to that found in the prior measure. If we denote by $H(\psi) = -\mathbb{E}_X[\log f(X|\psi)]$ the entropy* of $f(x|\psi)$, and

$$H_\pi = -\mathbb{E}_\psi \left[\log \frac{\pi(\psi)}{\pi_0(\psi)} \right]$$

the entropy of $\pi(\psi)$ relative to a measure $\pi_0(\psi)$ (chosen by default), then $G(\pi) = \mathbb{E}_\psi[H_\pi - H(\psi)]$ and the MDIP is defined by

$$\begin{aligned}\pi^{MDI}(\psi) &= \arg \max_{\pi(\psi)\geq 0} G(\pi), \\ &= \arg \max_{\pi(\psi)\geq 0} \left\{ \int_\Psi \pi(\psi)\mathbb{E}_X[\log f(X|\psi)]\, d\psi - \int_\Psi \pi(\psi) \log \frac{\pi(\psi)}{\pi_0(\psi)}\, d\psi \right\},\end{aligned}$$

which gives

$$\pi^{MDI}(\psi) \propto \pi_0(\psi) \exp\left(\mathbb{E}_X[\log f(X|\psi)]\right).$$

Though $\pi^{MDI}(\psi)$ is no longer systematically reparametrization invariant, it can be made so for a finite number of useful reparametrizations of ψ [426]. This class of prior measures often performs well with respect to other approaches. It allows us to more frequently obtain informative prior measures when extra constraints are added [798], by virtue of the principle of maximum entropy (see 11.2.2).

Equalities like $\pi^J(\psi) = \pi^{BB}(\psi)$ in unidimensional situations, the study of desirable theoretical behavior [164], and the quality of consistency properties of Bayesian estimators [426] can be used to support the choice or rejection of a prior measure. Other properties can be used to construct a hierarchy, such as for instance *posterior matching frequentist coverage* [718], which measures the closeness between maximum likelihood and noninformative Bayesian results. These prior measures are generally *improper*, i.e., non-integrable:

$$\int_\Psi \pi(\psi)\, d\psi = \infty,$$

and are therefore not probability measures. Recent work has allowed us to see them as limiting cases of proper (integrable) probability measures for a certain topology [95]. The lack of prior integrability is not prohibitive as long as the density of the posterior distribution $\pi(\psi|\mathbf{x_n})$ remains integrable, as it is this which allows estimation and statistical prediction to be possible.

As for the use of these priors in the extreme value distribution framework, Tables 11.1 and 11.2 summarize known noninformative priors for GEV and GPD distributions. While usually the Berger–Bernardo reference prior works better than the Jeffreys one in multidimensional settings [426], the former remains difficult to construct [80]. Indeed, it has not yet been published for the GEV case. The Berger–Bernardo reference prior is appreciated for its good posterior frequentist matching coverage and is known explicitely[5] for Fréchet sub-model [21]:

$$\pi^{BB}(\mu, \sigma, \xi) \propto \frac{1}{\sigma \xi}, \tag{11.8}$$

and Weibull sub-model too [725]:

$$\pi^{BB}(\mu, \sigma, \xi) \propto \frac{1}{\sigma \xi^2}. \tag{11.9}$$

As for the Gumbel distribution, only the Jeffreys prior has been obtained, by [752]:

$$\pi^{J}(\mu, \sigma) \propto \frac{1}{\sigma^2}.$$

In practice, it is recommended—when several measures are in competition for a specific applied setting with a fixed number of data n—to run studies simulating the posterior distribution in order to select the prior reference measure, looking in particular at the proximity of a classical Bayesian estimator (typically the posterior expectation) to the parameter values chosen in the simulation, as well as the frequentist coverage matching property.

An alternative approach to using improper measures is to use prior probability measures that are "close" to them in a certain sense (in general, by letting hyperparameters of these distributions tend toward specific values). For instance, when the domain of attraction of an extreme value distribution is Weibull [790], the use of distributions constructed in this way shows that biases or inconsistencies linked to the use of multidimensional Jeffreys posterior estimators can disappear.

[5] Results (11.8) and (11.9) can be obtained by reparametrizing the choices made in [21, 725].

Table 11.1 Characteristics of prior reference measures and their posterior distributions for n data points under the MAXB (GEV distribution) approach: $\psi = (\xi, \mu, \sigma)$. Here, γ refers to Euler's constant ($\gamma \simeq 1.57722$)

Prior measure		Posterior measure	Reference(s)
Jeffreys	defined if $\xi > -1/2$	improper $\forall n > 0$	[443, 563]
MDIP (truncated)	$\pi^{MDI}(\psi) \propto \exp\{-\gamma(1+\xi)\}/\sigma$ with $\xi \geq -1$	proper $\forall n \geq 2$	[66, 563]
Log-uniform	$\pi(\psi) \propto \mathbf{1}_{\{\sigma>0\}}/\sigma$	proper if $n \geq 4$	[563]
Northrop & Attalides	$\pi(\psi) \propto \pi(\xi)/\sigma$ with proper $\pi(\xi)$	proper $\forall n \geq 2$	[563]

Table 11.2 Characteristics of prior reference measures and their posterior distributions for n data points under the POT (generalized Pareto distribution) approach: ψ is restricted to the parameters (ξ, σ) (the threshold u is considered known or estimated)

Prior measure		Posterior measure	Reference(s)
Jeffreys	$\pi^J(\psi) \propto \frac{1}{\sigma(1+\xi)\sqrt{1+2\xi}}$	Proper	[257]
MDIP (truncated)	$\pi^{MDI}(\psi) \propto \exp(-1-\xi)/\sigma$ with $\xi \geq -1$	Proper $\forall n > 0$	[66, 563]
Pickands (log-uniform)	$\pi(\psi) \propto \mathbf{1}_{\{\sigma>0\}}/\sigma$	Proper if $n \geq 3$	[563, 617]
Northrop & Attalides	$\pi(\psi) \propto \pi(\xi)/\sigma$ with proper $\pi(\xi)$	Proper $\forall n > 0$	[563]

Example 11.34 (*Torrential rain in Corsica*)

We consider the set of 29 yearly precipitation maxima (Fig. 11.2) measured or reconstructed in the village of Penta-di-Casinca (Corsica), which were analyzed by [674]. In the MAXB setting, under the stationarity hypothesis, the plots in Fig. 11.3 show a comparison, for each of the GEV parameters, between the distribution of the Maximum Likelihood Estimator (MLE) and the posterior distributions of the MDIP and log-uniform priors (Table 11.1). We see that the domain values are similar for the three parameters, though the MLE distribution is generally more informative than the two posterior distributions, which incidentally are asymmetric for σ.

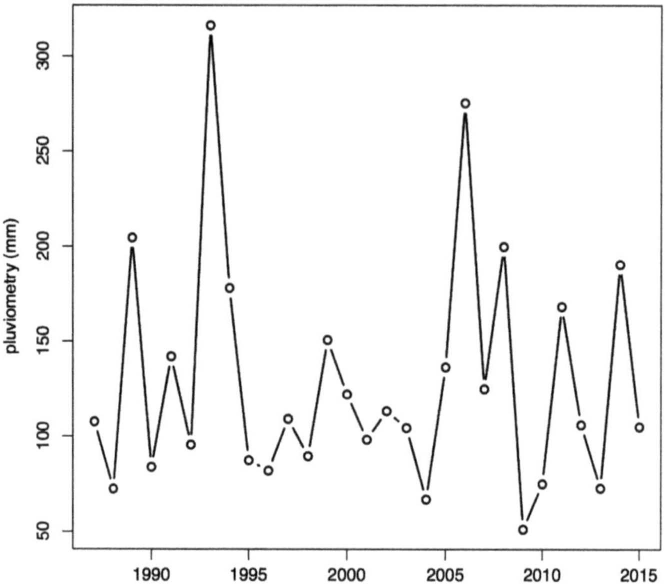

Fig. 11.2 Maximum annual precipitation per day (in mm) in Penta-di-Casinca, Corsica. *Source* http://penta.meteomac.com

11.2.2 Selecting a Prior Algebraic Structure

If supplementary information is available, one or several parametric choices must be made for $\pi(\psi)$. One simple idea [481] for guiding the algebraic structure of $\pi(\psi)$ is to consider that

(a) prior information is expressed in a "hidden" way in the form of a virtual iid sample $\tilde{\mathbf{x}}_{\mathbf{m}}$;
(b) truly available information provides us with (direct or indirect) knowledge of certain features of this sample;
(c) the prior distribution $\pi(\psi)$ should ideally be defined as the posterior distribution of ψ given $\tilde{\mathbf{x}}_{\mathbf{m}}$ and an noninformative prior π^{NI}:

$$\pi(\psi) = \pi^{NI}(\psi|\tilde{\mathbf{x}}_{\mathbf{m}}). \tag{11.10}$$

Besides guiding the choice of the algebraic form of the prior distribution, this idealized vision has several merits [789]. First, the correlations expressed in $\pi(\psi)$ between the coordinates of ψ are coherent with the nature of supposed or observed data. Second, the number m of virtual data is an intuitive calibration tool which can be directly compared with the number n of real data $\mathbf{x_n}$.[6] The combination of several

[6] Or rather with the effective level of information carried by the data $\mathbf{x_n}$, some of which may be only partially informative (e.g., *censored*); see Sect. 5.3.3 of [626, 651].

prior distributions—a well-known situation to specialists [572]—is then equivalent to aggregating several virtual datasets and operates directly via Bayes' rule if potential correlations between information sources can be removed. Other benefits are highlighted in [480, 528, 549].

The following definition introduces a practical vocabulary element for handling parametric Bayesian laws: hyperparameters.

Definition 11.26 (*Hyperparameters*)
In the rest of the chapter, we will concentrate on the use of parametric prior distributions and call *hyperparameters* the parameters in $\pi(\psi)$ and $\pi(\psi|\mathbf{x_n})$ to distinguish them from those in ψ which contains the extreme value distribution's parameters.

Constraining the form via meaningful hyperparameters—in terms of information quantity relative to that brought by the real data—is a necessity in the extreme value Bayesian framework; Coles [177] has indeed shown that a purely subjective prior distribution can uncontrollably take precedence over information contained in the data when estimating parameters and, in particular, over the fundamental parameter ξ that controls the heaviness of the distribution's tails.

When the parametric model $f(x|\psi)$ belongs to the *natural exponential family*[7] [651], applying Formula (11.10) allows us to construct the major part of the *conjugate prior* family of distributions, the form of which being explicit and remaining stable when conditioning on data (for more details on the extent and properties of this family, see [279]). For such models, the virtual size m and the mean of virtual data define natural hyperparameters [789].

However, extreme value statistical distributions do not accept conjugate prior measures over the totality of their parameter vector, except for the Gumbel distribution, so only *semi-conjugate* approaches can be proposed. Currently, we distinguish between three principles of prior modeling.

(a) Chose the form of the prior based on an approximation of the posterior distribution (11.10) or based on semi-conjugation properties. Examples can be found in Sect. 11.2.3 for the POT and MAXB approaches.

(b) Justify another formal rule allowing us to select the algebraic form from specifications other than a virtual sample. In this vein, an important approach is the principle of *maximum entropy* [411]. This provides a measure $\pi^*(\psi)$ by maximizing its entropy relative to an noninformative measure π_0:

$$\pi^*(\psi) = \arg \max_{\pi(\psi)\geq 0} \left\{ - \int_\Psi \pi(\psi) \log \frac{\pi(\psi)}{\pi_0(\psi)} \, d\psi \right\}$$

---

[7] Which includes the normal, gamma, multinomial, etc.

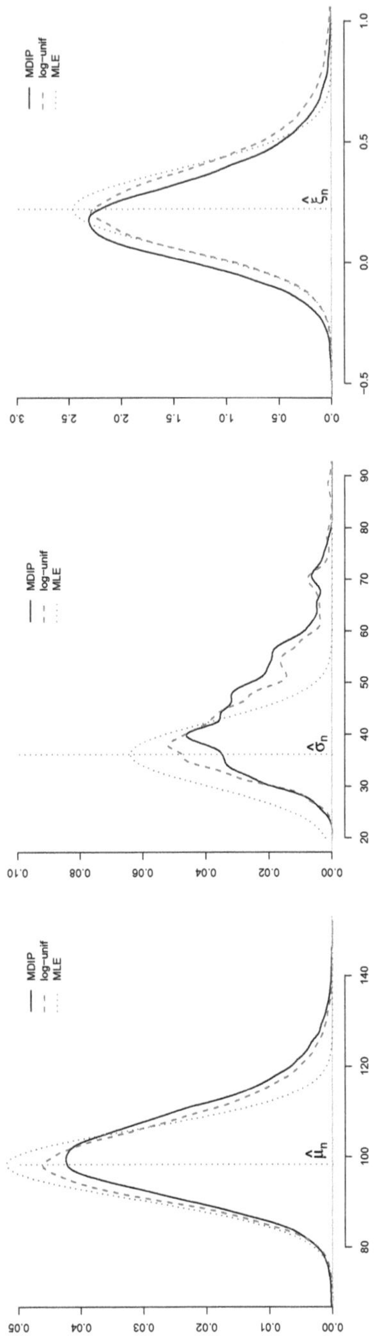

Fig. 11.3 Posterior densities of noninformative prior distributions on (μ, σ, ξ) and the density of the maximum likelihood estimator obtained using the delta method, for a GEV model of maximal annual daily rainfall in Corsica (Casinca). The Bayesian calculations were run using a Metropolis-Hastings-within-Gibbs algorithm with 6 parallel chains over 50,000 iterations (cf. Sect. 11.4)

under linear constraints in π: $\forall i \in \{1, \ldots, q\}$, $\int_\Psi h_i(\psi)\pi(\psi)\, d\psi = c_i$. Solving this type of functional optimization problem using Lagrange multipliers [626] gives the explicit result:

$$\pi^*(\psi) \propto \pi_0(\psi) \exp\left(\sum_{i=1}^{q} \lambda_i h_i(\psi)\right), \tag{11.11}$$

where the Lagrange multipliers $\lambda_i \in I\!R$ must be estimated using the linear constraints. This strategy suffers nevertheless from several weak points, including:

- the reference measure π_0 is typically chosen to be uniform for practical reasons, whereas it should in reality have good properties like reparametrization invariance (π_0 thus should intrinsically depend on the algebraic structure of $f(x|\psi)$);
- the normalization constraint $\int_\Psi \pi(\psi)\, d\psi < \infty$, which encodes the fact that we want a *proper* prior, is not automatically satisfied.[8] However, bringing together the maximum entropy approach and MDIP construction rules nevertheless can allow to improve the chance of obtaining a integrable $\pi^*(\psi)$ [114];
- the algebraic form of the prior distribution can fluctuate wildly as a function of the linear constraints, and the Lagrange multipliers (which define a family of hyperparameters) are not easily interpretable.

(c) Arbitrarily choose a common distributional form and check that its features (the tails of the distribution in particular) have little influence on the posterior results. This general approach has been pushed by Coles [173, 175, 177, 179] who often uses independent truncated normal (or gamma) prior distributions on the parameters of GEV distributions. The reasoning here is practical: this method simplifies the computations necessary to obtain the posterior distribution of these parameters (see Sect. 11.4).

In particular, in [173] Coles shows the feasibility of the Bayesian approach on yearly seal level maxima data at Port Pirie (Australia) by putting independent normal priors with high variance on GEV parameters. This approach—also chosen by [706] (Chap. 3) for modeling rainfall—corresponds to a crude approximation of a noninformative approach. The choice of priors remains tricky in practice since it can lead to physically implausible models (for instance in the case of regional analyses [634]).

Remark 11.20 Though this chapter looks at extreme value distributions coming from the asymptotic theorems presented in Chap. 6, certain authors equally employ Bayesian approaches for other distributions, justifying this with the use of statistical tests. For instance, [491, 763] propose using generalized exponential distributions

---

[8] Indeed, the linear constraint here is the identity ($h_i(\psi) = 1$), which plays no role in Formula (11.11) as it is folded into the underlying proportionality constant.

[366] for modeling extreme river flows and high precipitation, respectively. These are defined by the cumulative distribution function

$$\mathbb{P}(X \leq x|\psi) = \{1 - \exp(-\lambda x)\}^{\alpha} \quad \text{with } \psi = (\alpha, \lambda) \in I\!R_*^+.$$

They then perform calculations in a very similar way to Coles and his coauthors using weakly informative gamma priors. An essentially identical study has been performed in [490] with the Pareto distribution for modeling temperature maxima.

11.2.3 Conjugate and Semi-conjugate Prior Modeling

Semi-conjugate priors $\pi(\psi)$ are those whose posterior structure is partially stable. A common situation is to have—for certain coordinates of the parameter ψ—*conditional conjugation* on the other coordinates of ψ. Here, some hyperparameters in $\pi(\psi)$ can still be viewed as the size and statistics of a virtual sample, which can help calibration.

11.2.3.1 The POT Setting

First approach. Recall that the *conditional* POT approach consists of estimating a model defined by the cumulative distribution function of the Generalized Pareto Distribution (GPD)

$$\mathbb{P}(X \leq x|X \geq u) = \begin{cases} 1 - \left\{1 + \xi \frac{(x-u)}{\sigma}\right\}^{-1/\xi} & \text{for } \xi \neq 0, \\ 1 - \exp\left(-\frac{x-u}{\sigma}\right) & \text{for } \xi = 0, \end{cases}$$

using data $x_1, \ldots, x_n$ with values greater than u. The unconditional approach, examined by [589, 770], consists in setting the value of u and adopting a POT approach, which brings into the study the counting process of excesses $\mathbf{1}_{\{X_i \geq u\}}$ over the sampling time period T. Under a classical independence hypothesis between exceedance events, this is a homogenous Poisson process with intensity λ:

$$\mathbb{P}(N_T = n) = \frac{(\lambda T)^n}{n!} \exp(-\lambda), \tag{11.12}$$

where N_T is the (random) number of times u is exceeded during the time period T. The complete likelihood of the data $x_1, \ldots, x_n$ is then

$$\begin{aligned} &\mathbb{P}(X_1 = x_1, \ldots, X_n = x_n, X_1 \geq u, \ldots, X_n \geq u|T) \\ &= \mathbb{P}(X_1 = x_1, \ldots, X_n = x_n|X_1 \geq u, \ldots, X_n \geq u)P(X_1 \geq u, \ldots, X_n \geq u|T) \quad (11.13) \\ &= \quad f(\mathbf{x_n}|\psi)\mathbb{P}(N_T = n), \quad (11.14) \end{aligned}$$

where $f(\mathbf{x_n}|\psi)$ is the generalized Pareto likelihood, which can be written as [589]

$$f(\mathbf{x_n}|\psi) = \rho^n \exp\{(\rho - \beta)s_n(\mathbf{x_n}, \beta)\},$$

using the reparametrization $\beta = -\xi/\sigma$ and $\rho = 1/\sigma$, with

$$s_n(\mathbf{x_n}, \beta) = \frac{1}{\beta}\sum_{k=1}^{n} \log(1 - \beta(x_k - u)). \tag{11.15}$$

Remark 11.21 When β (or ξ) equals 0, the statistic (11.15) simplifies by continuity to $s_n(\mathbf{x_n}, 0) = -\sum_{k=1}^{n}(x_k - u)$.

The parameter vector ψ to be estimated then becomes $\{\mu, \rho, \beta\}$ (as u is constant). A semi-conjugate prior modeling from [589] is given in Proposition 11.7, which can be partly interpreted as a posterior distribution given a virtual sample. Note that [589] proposes numerous calibration choices for the hyperparameters without necessarily linking them with this interpretation. For a different parametrization, [770] provides a similar methodology.

Proposition 11.7 (GPD prior [589]) *Let $\pi(\mu, \rho, \beta)$ be a prior distribution defined by*

$$\begin{aligned}\mu &\sim \mathcal{G}(m, t),\\ \rho|\beta &\sim \mathcal{G}(m, s(\beta)),\\ \beta &\sim \pi(\beta),\end{aligned}$$

where $\pi(\beta)$ is any density on $\beta \in I\!R$. Then, $\pi(\mu, \rho, \beta)$ is conjugate for the likelihood (11.14) conditionally on β and can be interpreted as the posterior distribution of a sample of m data values $\tilde{\mathbf{x}}_{\mathbf{m}}$ greater than u observed over time period t and such that $s(\beta) = -s_m(\tilde{\mathbf{x}}_{\mathbf{m}}, \beta)$.

Indeed, given $\{\mathbf{x_n}, T\}$, the conditional posterior distributions are given by

$$\left.\begin{aligned}\mu|\mathbf{x_n}, T &\sim \mathcal{G}(m+n, t+T),\\ \rho|\beta, \mathbf{x_n}, T &\sim \mathcal{G}(m+n, s(\beta) + s_n(\mathbf{x_n}, \beta)),\\ \pi(\beta|\mathbf{x_n}, T) &\propto \frac{\Gamma(m+n)}{\Gamma(m)}\frac{\pi(\beta)\exp\{-\beta s_n(\mathbf{x_n}, \beta)\}}{[1 - s_n(\mathbf{x_n}, \beta)/s(\beta)]^{m+n}}.\end{aligned}\right\} \tag{11.16}$$

A Metropolis-Hastings-within-Gibbs algorithm then allows us to sample from the posterior distribution (Sect. 11.4).

Second approach. Restricting ourselves to the common case $\xi \geq 0$, an original semi-conjugate approach for the distribution of the data $y_i = x_i - \mu$ (with μ supposed known), which follows a GPD reparametrized by (α, β) with $\alpha = 1/\xi$ and $\beta = \sigma/\xi$, was proposed by [220]. The density $f(y|\alpha, \beta)$ of this distribution has the property of being able to be expressed as

$$f(y|\alpha, \beta) = \int_0^{\infty} z \exp(-yz) g(z|\alpha, \beta)\, dz, \tag{11.17}$$

where $g(z|\alpha, \beta)$ is the density of the gamma distribution $\mathcal{G}(\alpha, \beta)$ [639]. By introducing auxiliary variables $z_i \sim \mathcal{G}(\alpha, \beta)$, it is possible to define the statistical likelihood of the data y_i with respect to these z_i:

$$f(\mathbf{y_m}|\mathbf{z_m}) \propto \exp\left(-\sum_{i=1}^{m} z_i y_i\right),$$

which corresponds to a likelihood for exponential distributions. The full Bayesian model is then

$$\left.\begin{array}{ll} y_i & \sim \mathcal{E}(z_i), \\ z_i & \sim \mathcal{G}(\alpha, \beta), \\ \alpha & \sim \mathcal{G}_{II}\left(\eta/\lambda, \tilde{m}\right), \\ \beta|\alpha & \sim \mathcal{G}\left(\tilde{m}\alpha + 1, \tilde{m}\eta\right), \end{array}\right) \tag{11.18}$$

where $\mathcal{G}_{II}(c, d)$ is the Gamcon II distribution [194] with density

$$\pi(\alpha|c, d) \propto \frac{\Gamma(d\alpha + 1)}{\Gamma^d(\alpha)} (cd)^{-d\alpha}.$$

The data z_i play the role of parameters (common in Bayesian modeling). The choice of $\pi(\alpha, \beta)$ is such that if we have reconstructed data $z_1, \ldots, z_m$,

$$\begin{aligned} \alpha|z_1, \ldots, z_m &\sim \mathcal{G}_{II}\left(\eta'/\lambda', \tilde{m} + m\right), \\ \beta|\alpha, z_1, \ldots, z_m &\sim \mathcal{G}\left([\tilde{m} + m]\alpha + 1, [\tilde{m} + m]\eta'\right), \end{aligned}$$

with $\eta' = (\tilde{m}\eta + \sum_{i=1}^{m} z_i)/(\tilde{m} + m)$ and

$$\lambda' = \lambda^{\frac{\tilde{m}}{\tilde{m}+m}} \left(\prod_{i=1}^{m} z_i\right)^{1/(\tilde{m}+m)}.$$

Hence, the prior information can be interpreted as coming from a virtual gamma distribution of size m, arithmetic mean η, and geometric mean λ. It is then easy to construct a Gibbs algorithm for the Bayesian calculation (Sect. 11.4) of the distribution of (α, β) given $\mathbf{x_m}$.

11.2.3.2 The MAXB Case

The GEV distribution (uni- or multivariate) does not allow for any conjugate priors. Semi-conjugate ones have been proposed by [251]. As an aside, it is possible to define practical prior models for each of the three extreme value distributions: Fréchet, Weibull, and Gumbel (cf. Sect. 6.2.1), which seems coherent with respect to the POT results. This is of particular advantage for the Gumbel with its cumulative distribution function:

$$\mathbb{P}(X < x|\psi) = \exp\left\{-\exp\left(-\frac{x-\mu}{\sigma}\right)\right\} \quad \text{with } \sigma > 0, \mu \in I\!R \text{ and } x \in I\!R,$$

very often used in hydrology in a Bayesian framework, and for a long time, the most used distribution for modeling extreme rainfall [444] despite its drawbacks (cf. Sect. 6.5). Indeed, it accepts the conjugate prior [159]:

$$\pi(\mu,\sigma) \propto \sigma^{-m} \exp\left(m\frac{(\mu-\bar{\tilde{\mathbf{x}}}_m)}{\sigma} - \sum_{i=1}^{m} \exp\left\{-\frac{\tilde{x}_i-\mu}{\sigma}\right\}\right),$$

where the hyperparameters $(m, \bar{\tilde{\mathbf{x}}}_m, \tilde{x}_1, \ldots, \tilde{x}_m)$, respectively, correspond to the size of the prior dataset, its mean, and the data itself. Given real data $\mathbf{x_n}$, the posterior distribution is written as

$$\pi(\mu,\sigma|\mathbf{x_n}) \propto \sigma^{-m-n} \exp\left(\{m+n\}\frac{\left(\mu - \frac{m\bar{\tilde{\mathbf{x}}}_m + n\bar{\mathbf{x}}_\mathbf{n}}{m+n}\right)}{\sigma} - \sum_{i=1}^{m} \exp\left\{-\frac{\tilde{x}_i-\mu}{\sigma}\right\} - \sum_{k=1}^{n} \exp\left\{-\frac{x_k-\mu}{\sigma}\right\}\right).$$

This prior modeling is thus relatively useable for dealing with historical data (see Sect. 11.3.1.1). When m is small, we could imagine using a relationship like (11.32) to look for the values $\tilde{x}_i$ [118]. Note that $\pi(\mu, \sigma)$ is proper only if $m \geq 3$.

An alternative (semi-conjugate) approach, proposed by [515], uses the reparametrization $\psi = \exp(\mu/\sigma)$. Conditional on σ, a gamma distribution is conjugated. However, it remains necessary—in a completely Bayesian approach—to specify a prior on σ.

The Fréchet distribution $\mathscr{F}(\psi)$ accepts a semi-conjugate prior [118]. Reparametrizing $\psi = (\mu, \nu, \xi)$ with $\nu = \sigma^{1/\xi} > 0$, the Fréchet cumulative distribution function is written as

$$\mathbb{P}(X < x|\psi) = \exp\left\{-\nu\,(x-\mu)^{-1/\xi}\right\} \quad \text{with } \xi > 0, \mu \in I\!R \text{ and } x \geq \mu.$$

Proposition 11.8 Fréchet prior [118] *Let $\pi(\mu, \nu, \xi)$ be the prior distribution defined by*

$$\begin{aligned}
\nu|\mu, \xi &\sim \mathscr{G}\left(m, s_1(\mu, \xi)\right), \\
\xi|\mu &\sim \mathscr{IG}\left(m, s_2(\mu)\right), \\
\pi(\mu) &\propto \frac{\mathbf{1}_{\{\mu \leq x_{e_1}\}}}{(x_{e_2}-\mu)^m s_2^m(\mu)},
\end{aligned}$$

where $x_{e_1} < x_{e_2}$ and

$$\begin{aligned}
s_1(\mu, \xi) &= m(x_{e_1}-\mu)^{-1/\xi}, \\
s_2(\mu) &= m \log\left(\frac{x_{e_2}-\mu}{x_{e_1}-\mu}\right).
\end{aligned}$$

Then, $\pi(\mu, \nu, \xi)$ is conditionally conjugate on (ξ, μ) for the Fréchet likelihood and interpretable as the reference posterior distribution of a sample $\tilde{\mathbf{x}}_{\mathbf{m}}$ of m Fréchet data points with shifted geometric mean

$$x_{e_2} = \mu + \prod_{i=1}^{m} (\tilde{x}_i - \mu)^{1/m}.$$

In effect, given $\mathbf{x_n} \sim \mathscr{F}(\psi)$, the conditional posterior distributions present a kind of balance between the real data's and virtual data's statistics:

$$\begin{aligned}
\nu|\mu, \xi, \mathbf{x_n} \quad &\sim \mathscr{G}\left(m+n, s_1(\mu,\xi) + \sum_{i=1}^{n} (x_i-\mu)^{-1/\xi}\right), \\
\pi(\xi|\mu, \mathbf{x_n}) \quad &\propto \frac{\xi^{-n-m-1} \exp\left\{-\frac{1}{\xi}\left(m\log(x_{e_2}-\mu) + n\log(\bar{\bar{x}}-\mu)\right)\right\}}{\left(\sum_{i=1}^{m}(x_{e_1}-\mu)^{-1/\xi} + \sum_{i=1}^{n}(x_i-\mu)^{-1/\xi}\right)^{m+n}}, \\
\pi(\mu|\nu, \xi, \mathbf{x_n}) &\propto \frac{\exp\left(-\frac{1}{\xi}\left[m\log(x_{e_2}-\mu) + n\log(\bar{\bar{x}}-\mu)\right]\right)}{(\bar{\bar{x}}-\mu)^n (x_{e_2}-\mu)^m} \\
&\quad \times \exp\left(-\nu\left[\sum_{i=1}^{m}(x_{e_1}-\mu)^{-1/\xi} + \sum_{k=1}^{m}(x_k-\mu)^{-1/\xi}\right]\right),
\end{aligned} \tag{11.19}$$

with

$$\bar{\bar{x}} = \mu + \prod_{i=1}^{n}(x_i-\mu)^{1/n}.$$

A similar result for the Weibull distribution $\mathscr{W}(\psi)$ can be easily deduced by symmetry from Proposition 11.8. Its cumulative distribution function is written, with the same reparametrization, as

$$\mathbb{P}(X < x|\psi) = \exp\left\{-\nu\,(\mu - x)^{1/\xi}\right\} \quad \text{with } \nu > 0, \xi > 0, \mu \in I\!R \text{ and } x \leq \mu.$$

Proposition 11.9 Weibull prior [118] *Let $\pi(\mu, \nu, \xi)$ be a prior distribution defined by*

$$\begin{aligned}
\nu|\mu, \xi &\sim \mathscr{G}\,(m, s_3(\mu, \xi))\,, \\
\xi|\mu &\sim \mathscr{IG}\,(m, s_4(\mu))\,, \\
\pi(\mu) &\propto \frac{\mathbf{1}_{\{\mu \geq x_{e_4}\}}}{(\mu - x_{e_4})^m s_4^m(\mu)},
\end{aligned}$$

where $x_{e_3} < x_{e_4}$ and

$$\begin{aligned}
s_3(\mu, \xi) &= m(\mu - x_{e_3})^{-1/\xi}, \\
s_4(\mu) &= m \log\left(\frac{\mu - x_{e_3}}{\mu - x_{e_4}}\right).
\end{aligned}$$

Then, $\pi(\mu, \nu, \xi)$ is conditionally conjugate on (ξ, μ) for the Weibull likelihood and is interpretable as the posterior distribution of a sample $\tilde{\mathbf{x}}_m$ of m Weibull data points whose shifted geometric mean is equal to x_{e_4}:

$$x_{e_4} = \mu - \prod_{i=1}^{m} (\mu - \tilde{x}_i)^{1/m}.$$

Remark 11.22 The Weibull distribution used in extreme value statistics is a kind of "reverse" version of the one usually seen in survival analysis or when studying renewal processes [650]

$$\mathbb{P}(X < x|\mu, \eta, \beta) = \exp\left\{-\left(\frac{x - \mu}{\eta}\right)^{\beta}\right\} \quad \text{with } \eta > 0, \beta > 0, \mu \in I\!R \text{ and } x \geq \mu.$$

Numerous prior Bayesian models have been suggested in these domains, providing alternatives to the semi-conjugate choice and inspiring the construction of Bayesian models for the Fréchet distribution. In particular, the works [112, 255, 423] suggest ways to include engineering expertise in modeling. Nevertheless, remember that questioning an expert in an industrial reliability setting is not the same thing as doing so on the subject of extreme hazards. The former is probably less *risk averse* than the latter (see Sect. 11.3.2.1).

11.2.4 Hierarchical Modeling

Physical processes generating the occurrence and intensity of extreme hazards can often be partially understood. Physical models, implemented using numerical simulations (Sect. 10.3), are sometimes capable of reproducing certain features of events. For instance, the global climate models used by the IPCC[9] since 1988 are built using simulations of the evolution of major climate variables (temperature, rainfall, relative humidity, etc.) over mesoscale (several hundred kilometers) spatial grids, as a function of measurable or predictable parameters (notably greenhouse gas concentration). As a function of the spatial and temporal precision, the uncertainty affecting these simulations can vary to a considerable extent. Nevertheless, the use of physical models allows us, a minima, to get a sense of the influence of correlations between hazards from the past and with locations near to a given site. This knowledge, if it exists, can add to *statistical correlation* analyses between sites.

The interest in exhibiting a partial knowledge of the underlying physical processes, by means of contextual data and/or models, is that it makes it possible to place the extreme statistical model in a *hierarchical modeling* framework. The state of nature represented by ψ which generates the random variable X is dependent on (*conditioned on*) a hidden variable (called *latent*) denoted by ζ, on which knowledge is available. As this knowledge is by its very nature uncertain, ζ can practically be considered as a random variable with density $\pi(\zeta)$. In the simple MAXB and POT settings, we can then define a hierarchical model:

$$\begin{aligned} X &\sim \mathcal{G}_{EV}(\psi) \text{ or } \mathcal{G}_{PD}(\psi) && \textit{(modeling the data)} \\ \psi &\sim \pi(\psi|\zeta) && \textit{(modeling the underlying process)} \\ \zeta &\sim \pi(\zeta) && \textit{(modeling process parameters)} \end{aligned} \tag{11.20}$$

and the prior distribution $\pi(\psi)$ is thus defined by $\int \pi(\psi|\zeta)\pi(\zeta)\,d\zeta$. Clearly, we can extend this schema further by imagining that, for instance,

$$\begin{aligned} \zeta &\sim \pi(\zeta|\kappa), \\ \kappa &\sim \pi(\kappa). \end{aligned}$$

Reasoning in this way often leads to more robust posterior results with respect to uncertainty in prior information [651]. A concrete example is where several prior information sources provide values $\zeta_1, \ldots, \zeta_q$. A practical strategy is then to consider this as a sample which can be used to calibrate the distribution $\pi(\zeta)$.

---

[9] Intergovernmental Panel on Climate Change.

Example 11.35 (*Multiple information sources for the parameter* ξ)

The sign and value of ξ, whether in the MAXB or POT setting, greatly influence the nature of the extreme value distribution's tail and estimation of return levels. Suppose for the sake of convenience that a priori,

$$\xi \sim \mathcal{N}\left(\mu_{xi}, \sigma_\xi^2\right),$$

and that we have several pairs $\{\mu_{\xi,1}, \sigma_{\xi,1}^2\}, \ldots, \left\{\mu_{\xi,S}, \sigma_{\xi,S}^2\right\}$ resulting from the estimation of ξ at sites similar to the one of interest. Writing $\zeta = \{\mu_\xi, \sigma_\xi^2\}$, we can run statistical tests on the sample of ζ_i for testing hypotheses of the form

$$\mu_\xi \sim \mathcal{N}\left(\bar{\mu}_\xi, \sigma_{\mu_{x_i}}\right), \tag{11.21}$$

$$\sigma_\xi^2 \sim \mathcal{IG}(a_\xi, b_\xi). \tag{11.22}$$

When different pieces of prior information on ψ are correlated—for example, when similar sites are geographic neighbors of the site of interest—this correlation can be taken into account by supposing (for example) that the vector of ζ_i was generated by a multivariate model in dimension $2q$ and no longer by (11.21–11.22). In this spirit, [30] provided an example of a hierarchical approach allowing the combination of several correlated information sources coming from experts (Sect. 11.3.2) who influence each other. Also, note that the prior modeling (11.18) proposed in Sect. 11.2.3.1 for the POT framework, which uses auxiliary variables (or data) z_i, is by nature hierarchical.

11.2.4.1 Spatio-Temporal Covariates and Dependency

In numerous situations, hierarchical modeling is the result of a link between the parameters of extreme value distributions and *explanatory covariates*: at a given site, the extreme hazard depends on factors $\mathbf{Z} = (Z_1, \ldots, Z_q)$ for which information is available. See [695] for a (frequentist) comparison of the most well-known extreme value spatial hierarchical models. Generally speaking, we can characterize the parameter vector ψ (or a transformed version thereof) of an extreme value distribution in the following way:

$$\psi = g(\beta_0, \beta_1, \ldots, \beta_q, Z_1, \ldots, Z_q), \tag{11.23}$$

where $\boldsymbol{\beta} = (\beta_i)_i$ is a vector of coefficients to be estimated, which represents the influence of the variable Z_i on ψ (β_0 being an effect independent of the Z_i), and g a usually monotonic function, linear or otherwise. In Formula (11.23), the values Z_i are generally known (measured in situ, for example) or modeled using statistical distributions. The latent vector $\zeta = \boldsymbol{\beta}$ and the stochastic setting (11.20) then simplifies to a deterministic one.

Example 11.36 (*Extreme rainfall*)

The distribution of yearly precipitation maxima (or exceedance values) at site i is assumed, in [332], to follow a GEV (or GPD) distribution whose location parameter μ_i is defined by

$$\mu_i = \boldsymbol{\beta}^T \mathbf{Z}_i m, \tag{11.24}$$

where $\mathbf{Z}_i$ is a set of measurable covariates at site i, including in particular the geographic coordinates (latitude, longitude). Without taking into account temporal variation in $\boldsymbol{\beta}$ (see Example 11.37 below), the prior distribution on $\boldsymbol{\beta}$ is a multivariate Gaussian one:

$$\boldsymbol{\beta} \sim \mathcal{N}_q\left(\boldsymbol{\beta}_0, \Sigma_{\boldsymbol{\beta}}\right). \tag{11.25}$$

A priori, we know little about the distribution of the factors of influence β_i. It therefore seems like a good idea to take a hierarchical approach and put weakly informative distributions on $(\boldsymbol{\beta}_0, \Sigma_{\boldsymbol{\beta}})$, making sure that the joint posterior distribution of $\psi_i = (\mu_i, \sigma_i, \xi_i)$ remains well-defined. Gelman [325] has proposed a family of distributions that adapts well to the specific choice of $\pi(\Sigma_{\boldsymbol{\beta}})$.

Remark 11.23 Bayesian regionalization approaches (Sect. 11.3.1.2), which allow for the introduction of spatial correlation into extreme value distributions, can be interpreted as approaches where the explanatory covariates Z_i are the values of ψ (or alternatively of the hazard X) itself in regions nearby the site of interest.

Temporal dependency in extreme value models (Chap. 8) can also be inserted at the level of the latent variable ζ. For example, the vector $\zeta_t, \zeta_{t+1}, \ldots, \zeta_T$ allows us to model the impact of seasonality over an entire region. More precisely, according to [176], it can help model non-stationary effects via seasonality ones when the temporal variation is expected but undetectable in small datasets. The dimension-reduction techniques used in Chap. 8 apply here; thus, ζ_t generally reduces to an autoregressive process.

Example 11.37 (*Extreme rainfall*)

The modeling in Example 11.36 (from [332]) can be expanded to take into account the temporal trend seen in μ. In (11.24), the Z_i have fixed values, independent of time, and an autoregressive process is placed on the components of $\boldsymbol{\beta}$:

$$\beta_{k,t} = \alpha_k \beta_{k,t-1} + \varepsilon_t \quad \text{for } k \in \{1, \ldots, q\}, \tag{11.26}$$

with $\varepsilon_t \sim \mathcal{N}(0, \sigma_\beta^2)$. The prior modeling (11.25) has therefore to be turned into a prior modeling of $(\alpha_1, \ldots, \alpha_q, \boldsymbol{\beta}_0, \sigma_\beta^2)$. Following the suggestion of [325], we place a weakly informative inverse gamma distribution $\mathcal{IG}(0.1, 100)$ on σ_b^2. Though [332] proposed an ad hoc pointwise estimation approach for $\boldsymbol{\beta}_0$, a hierarchical choice is made for each α_k by introducing a new hyperparameter σ_α^2:

$$\begin{aligned} \alpha_k | \sigma_\alpha^2 &\sim \mathcal{N}\left(0, \sigma_\alpha^2\right), \\ \sigma_\alpha^2 &\sim \mathcal{IG}(0.1, 100). \end{aligned}$$

In doing so, however, the prior information on the α_i is made quite vague, and the posterior distribution is nearly exclusively dependent on data information.

Remark 11.24 The special case of Bayesian linear regression is the subject of a vast quantity of literature. In particular, the use of Zeller's prior modeling (the so-called *G−prior* [796]) on the vector of coefficients $\boldsymbol{\beta}$ is generally recommended. For more information, see [651] (Sect. 4.3.3).

11.2.4.2 Model Complexity

The use of correlation models between ψ and other variables is fairly standard when dealing with extreme natural hazards problems because we look to take advantage of the spatial and temporal dependence of extreme values to increase the information in the dataset—or make up for its absence. Hierarchical prior modeling thus appears to be a useful tool for engineers, on the condition that the stack of hierarchical layers still allows the production of a posterior distribution that is more informative than the prior. Indeed, it might be tempting to build a complex model that tries to take into account spatial and/or temporal random effects, and then be confronted on the one hand by complicated posterior calculations (Sect. 11.4), and on the other by the problem of *nonlearning* or an improper posterior distribution.

A nonlearning problem can appear when the posterior distributions correspond to the prior ones for some or all of the parameters. This means that the data is not informative for these parameters. An improper posterior distribution is generally due to the stacking of improper (noninformative) priors on the parameters in the upper levels of the hierarchy (those furthest from the likelihood; for instance, Σ_{β} in Example 11.36). An added difficulty is that computational techniques used to produce posterior distributions (Sect. 11.4) can incorrectly give the impression that the distribution being looked for is well-defined. Warnings to this effect are described in [652] (Sect. 10.4.3).

Therefore, it is important to algebraically confirm that the posterior distribution is indeed proper.[10] Besides it is strongly advised to progressively increase the level of hierarchical complexity, for example by setting some parameters as constants when testing Bayesian calculation tools, rather than putting poorly informative prior measures on them. The software tool described in Sect. 11.4.4 is able to quickly and easily modulate the level of complexity. So as to limit such complexity, results on taking into account trends like (11.26) in hierarchical models suppose in general that only one parameter is concerned (usually the location parameter μ, as ξ is tricky to estimate and often left constant).

Example 11.38 (*Extreme rainfall in Spain*)

A spatio-temporal Bayesian GEV model was recently proposed by [305] for modeling the occurrence and intensity of extreme rainfall in Extrémadure, Spain. Non-stationarity was taken into account through a linear temporal trend placed in the location parameter at site $i \in \{1, \ldots, S\}$:

$$\mu_i(t) = \mu_0(i) + \mu_1(i)t + \varepsilon_i(t),$$

where (μ_0, μ_1) are terms of the type $\boldsymbol{\beta}^T Z$ (Example 11.36) and $\varepsilon_i(t)$ is model noise not taken into account in the spatio-temporal part. As in [332], autoregressive models were used for the regression coefficients $\boldsymbol{\beta}$ and weakly informative prior measures put on the coefficients α of the model. Though complex, this model can still be coded using (for instance) the BUGS language (from the `OpenBUGS` software, cf. Sect. 11.4.4). The authors were then able to run a progressive series of tests, including notably whether incorporating a temporal trend affected the estimation quality.

---

[10] If the prior is proper in Ψ and the likelihood has a maximum in the interior of the same space, the posterior distribution is automatically proper.

11.3 Calibration

Methods for calibrating a prior distribution $\pi(\psi)$, i.e., estimating its hyperparameters, vary widely depending on the nature of available information, of which there are two main types:

1. Prior information can be carried by other data obtained from studying the past (*historical data*) or from studying the region around the site of interest (*regional data*). The empirical Bayesian approach (Sect. 11.3.1.3) is a special case—generally disapproved from a theoretical point of view—where the actual data $\mathbf{x_n}$ is used, thus bringing a troublesome redundancy when using the posterior results. We nevertheless mention it here as it is of practical interest.
2. Prior information can come from an *elicitation*[11] study of human expertise. Specifying who an expert is eminently related to the applied setting being considered. Along with [572], we consider that this typically means engineers, researchers, project managers, and specialists in one or several domains (hydrology, oceanography, climatology, etc.) with plenty of "field experience".

Calibration techniques are therefore based on a direct or indirect statistical interpretation of prior information, and the infinite number of possible combinations of such sources leads to just as many possible methods. However, we focus on only a small number here. Their strengths include being easily explainable and able to clearly signal the choices made due to the subjectivity of analysts.

11.3.1 Data-Dependent Calibration

11.3.1.1 Calibration from Historical Data

In cases where the information is directly provided by *historical data* $\tilde{\mathbf{x}}_\mathbf{m} \in \chi_I$, several authors have proposed to use it in a Bayesian framework to calibrate prior distributions, especially in hydrology [588, 602, 638]. Three representative studies are [759], which looked at extreme sea levels for dimensioning dikes in the Netherlands (Example 11.39), [588] which considered flood forecasting for the Garonne river, and [128] which looked at coastal defences in France in the context of Cyclone Xynthia (February 2010) in La Rochelle.

[11] From the Latin word *eliciere*, meaning to *extract*, *obtain*, *bring to mind*.

Example 11.39 (*Marine flooding*)

To study dike dimensioning in the Netherlands, [759] proposed a Bayesian approach on a Gumbel distribution $\mathcal{M}(\psi)$ by constructing a prior distribution based on a sorted sample of the $m = 10$ highest sea levels $\tilde{x}_1^* < \ldots < \tilde{x}_m^*$ observed in the past. The distribution of each ith order statistic x_i^* is known explicitly given $\mathcal{M}(\psi)$:

$$f(x|\psi) = \frac{m!}{(i-1)!(n-i)!} \{F_{\mathcal{M}}(x|\psi)\}^{i-1} \{1 - F_{\mathcal{M}}(x|\psi)\}^{n-i} f_{\mathcal{M}}(x|\psi).$$

A statistical likelihood can therefore be determined, as well as its maximum likelihood estimator $\hat{\psi}_m$. The bootstrap distribution of this estimator can be used to define the distribution $\pi(\psi)$. This kind of method is relatively simple to implement, but the quality of the bootstrap distribution is limited by the small sample size, leading perhaps to overly informative prior distributions. The POT approach (limited to the $\xi = 1$ case) can also be used to check that the results are coherent in a noninformative Bayesian setting.

Remark 11.25 A similar approach based on sorting the smallest recorded values (without an environmental application in mind) has been proposed by [497]. Using the Gumbel distribution, the authors provided a full Bayesian framework for predicting new values. Based on a Jeffreys prior, this resulted in defining a new semi-conjugate prior (Sect. 11.2.3) with the help of the recorded sample.

More generally, the point of view given in Equation (11.10), with $\pi(\psi)$ defined as a posterior distribution of historical data, is meaningful: the conjugate Gumbel prior given in Sect. 11.2.3.2 falls exactly into this framework. Note that by expanding (11.10), there therefore exists a likelihood $\tilde{f}(\tilde{\mathbf{x}}_\mathbf{m}|\psi)$ such that

$$\pi(\psi) \propto \tilde{f}(\mathbf{x}_\mathbf{m}|\psi)\pi^{NI}(\psi). \tag{11.27}$$

If the historical data $\tilde{\mathbf{x}}_\mathbf{m}$ is in the same form as the data $\mathbf{x}_\mathbf{n}$ (block maxima or POT), $\tilde{f}$ is defined using the same densities as f. Thus, [128] used the uniform measure $\pi^{NI}(\psi) \propto 1$ proposed by [602] for data with a GPD distribution $\tilde{f}(\psi) = f(\psi)$, despite the difficulties mentioned in Remark 11.18 on the subject of the uniform distribution and the possible prior choices provided in Table 11.2.

However, historical data can potentially be a mixture of maxima, exceedance, or simply Poisson count data [374, 759], which moreover may be noisy or censored. A consequence of this is that we need to take particular care[12] in defining $\tilde{f}$ (and

---

[12] Especially in the choice of the noninformative prior π^{NI}, which should ideally be able to adapt to the presence of censored historical data; see for example [206].

$\pi^{NI}(\psi)$). See [128] for obtaining such likelihoods, as well as Example 4.9. Recall that ψ can define a parametrization containing both GEV and GPD parameters through (asymptotic) theoretical results between parameter triples [428, 487]:

$$\begin{aligned}
\xi_{GEV} &= \xi_{GPD} = \xi, \\
\sigma_{GPD} &= \sigma_{GEV} + \xi\left(\mu_{GPD} - \mu_{GEV}\right), \\
\mu_{GPD} &= \mu_{GEV} + \frac{\sigma_{GPD}}{\xi}\left(\lambda^{-\xi} - 1\right), \\
\lambda &= \left(\frac{\sigma_{GEV}}{\sigma_{GPD}}\right)^{1/\xi}.
\end{aligned}$$

Remark 11.26 In fact, most authors treat historical information like a choice of statistical likelihood, guided by the type of data (censored, truncated, noisy, etc.), and describe their Bayesian framework as *objective* in that they rely on noninformative measures π^{NI} [588]. There is thus no *elicitation* of prior distributions, strictly speaking. Nevertheless, a prior defined as $\pi^{NI}(\psi|\mathbf{x_m})$ is similar to the one defined as $\pi^{NI}(\psi)$ plus a historical data likelihood.

Giving weight to historical information. Another problem is weighting historical information with respect to the data $\mathbf{x_n}$. This is because "historical data" may be informative but sparse (not measured/recorded systematically over time), poorly calibrated with respect to today's standards, or less representative of current and future hazards than recently obtained data $\mathbf{x_n}$. For instance, very old data on a certain river breaking its banks may have been measured at a time when the river's flow (its riverbed, etc.) was different or the measurements less precise (see the first chapter of this book and Sect. 5.2.1 for more details).

It is thus necessary to decide on an *equivalent data quantity* $\hat{m} \leq m$, "flattening" the historical data's likelihood $\tilde{f}(\tilde{\mathbf{x}}_\mathbf{m}|\psi)$ (Fig. 11.4); that is, we replace $\tilde{f}(\mathbf{x_m}|\psi)$ in (11.27) with

$$\hat{f}_{\hat{m}}(\tilde{\mathbf{x}}_\mathbf{m}|\psi) = \frac{\tilde{f}^{\hat{m}/m}(\tilde{\mathbf{x}}_\mathbf{m}|\psi)}{\int_{\chi_I} \tilde{f}^{\hat{m}/m}(\tilde{\mathbf{x}}_\mathbf{m}|\psi)\, d\tilde{\mathbf{x}}_\mathbf{m}}. \tag{11.28}$$

This flattening rule is also called *fractional* [94] and indirectly defines what numerous authors call *power priors* [160]. The reason for doing this is as follows: let $\tilde{f}_i(.|\psi)$ be the density of each historical data event $\tilde{x}_i$. Supposing the $\tilde{x}_i$ are independent, we have

$$\tilde{f}(\tilde{\mathbf{x}}_\mathbf{m}|\psi) = \prod_{i=1}^{m} \tilde{f}_i(\tilde{x}_i|\psi)$$

and

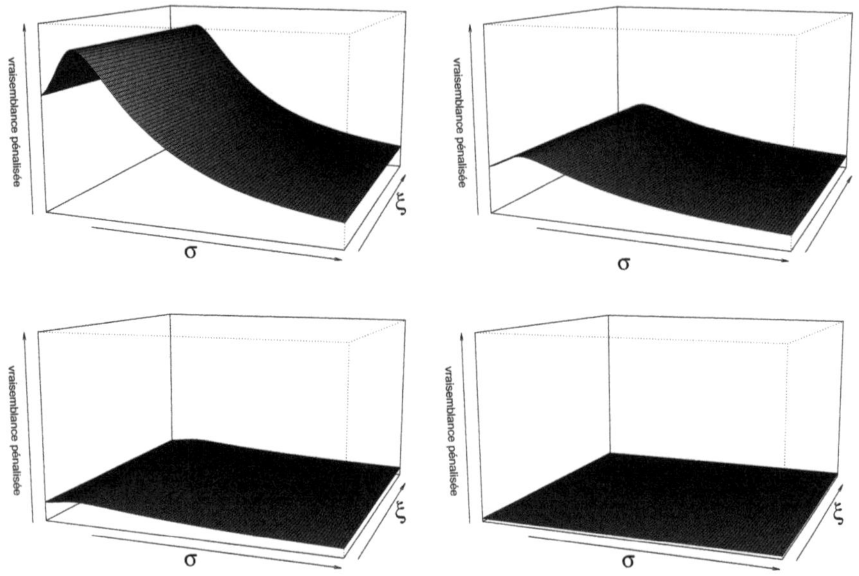

Fig. 11.4 Illustration of the flattening of a GEV likelihood with $m = 40$ historical data values, with μ constant, for $\hat{m} \in \{m, 2m/3\}$ (right-hand column) and $\hat{m} \in \{m/2, 1\}$ (left-hand column)

$$\hat{f}_{\hat{m}}(\tilde{\mathbf{x}}_{\mathbf{m}}|\psi) = \arg\min_f \sum_{i=1}^{m} \omega_i KL\left(f, \tilde{f}_i(.|\psi)\right),$$

where

- each weight $\omega_i = \omega = \hat{m}/m$ is an estimate of the fraction of information found in the historical data point $\tilde{x}_i$ with respect to a normal (non historical) data point. This so-called *weighted likelihood* approach was initially proposed by [398].
- the Kullback–Leibler divergence $KL(f, \tilde{f}_i(.|\psi))$ is a measure of the information loss occurring when choosing $\tilde{f}_i$ in the place of f to represent historical information on ψ.

Essentially, the geometric mean $\hat{f}_{\hat{m}}$ corresponds to an estimate of an unknown optimum f "conveying" in an exhaustive manner historical information on the parameter vector.

If we suppose we are able to more precisely quantify the percentage of information $\delta_i \in [0, 1]$ contained in historical data points $\tilde{x}_i$ with respect to current-day ones, the rule (11.28) can be refined into

$$\hat{f}_{\hat{m}}(\mathbf{x}_{\mathbf{m}}|\psi) = \frac{\prod_{i=1}^{m} \tilde{f}_i^{\delta_i}(\tilde{x}_i|\psi)}{\int_{\chi_I} \left[\prod_{i=1}^{m} \tilde{f}_i^{\delta_i}(\tilde{x}_i|\psi)\right] d\tilde{\mathbf{x}}_{\mathbf{m}}}. \tag{11.29}$$

The approaches proposed in [128, 588] correspond to choosing $\hat{m} = m$, flirting the risk of over-weighting the final posterior distribution with the prior. Note that other rules could also be considered for choosing $\tilde{m}$. For example, we could by default give the entire historical data record the same weight as a single regular data point ($\tilde{m} = 1$), representing a summary of the state of prior knowledge. This approach was considered by [220], among others. We could also provide a calibration rule of the form

$$\tilde{m}^* = \arg\max_{\tilde{m} \leq m} \mathscr{C}_{oh} \left(\pi(.|\tilde{m}) \| \mathbf{x_n}\right),$$

where $\mathscr{C}_{oh}$ is a coherency rule between the prior $\pi(\psi|\tilde{m})$ and regular data $\mathbf{x_n}$ (see for example [113]).

Remark 11.27 The flattening procedure (11.28) is similar to the one used to calculate *fractional Bayes factors* [573], concurrents to *intrinsic Bayes factors* [81], which allow for model selection when priors $\pi(\psi)$ are improper. Building a posterior distribution given a likelihood weighted by a power term is also referred to as a so-called generalized Bayesian inference [359].

11.3.1.2 Calibration from Regional Data

It would seem natural to construct a prior $\pi(\psi)$ on the parameters characterizing the distribution of a natural hazard at a given site of interest i with the help of information from neighboring sites $j \in \{1, \ldots, S\}$, if these are subject to similar natural hazards and have the same physical characteristics (similar land type, same relationship to a water feature, etc.) The region containing these sites is then called *spatially homogenous*, corresponding to a similar notion to the stationarity of temporal processes [322]. A typical approach is the one proposed by [177]: calibrate a multidimensional normal prior $\pi(\psi)$ starting from a range of frequentist estimations (maximum likelihoods) at various neighboring sites.

More generally, Bayesian regionalization methods aim to provide a sample $\mathbf{x}$ of extreme values characterizing the behavior of hazards across the supposed homogenous regional spatial grid. Another possibility is to produce a sample of estimates $\boldsymbol{\theta} = \{\psi_1, \ldots, \psi_S\}$ of the parameter vector ψ of the regionalized extreme value distribution. Samples $\mathbf{x}$ or $\boldsymbol{\theta}$ can then be used to calibrate a prior $\pi(\psi)$ on the parameter vector ψ at site of interest i. This is only allowed if $\mathbf{x}$ does not incorporate observations (potentially) made at site i or equivalently, if $\psi_i \notin \boldsymbol{\theta}$.

As for standard regional approaches (Chap. 7), these methods require a homogenization metric for data (or estimators of ψ) to be put in place between sites,[13] as well as a method for delimiting the homogeneous region [45]. The choice

[13] More generally, the terms *upscaling* or *downscaling* are used to describe this kind of homogenization.

of metric may be directly inspired by a standard rule of thumb, as in the following example.

Example 11.40 (*Regional flooding analysis*)
In order to predict flooding at a site $i \in \{1, \ldots, S\}$ using data from the same region, which is supposed homogenous in terms of hydrological characteristics, hydrologists define what is known as a *index-flood* C_i, where the rate of flow Q_i at site i is such that $Q_i = C_i Q_R$, where Q_R is a regionalized rate of flow (see Chap. 7). The strategy suggested by [562] and implemented by [648] is to consider that the extreme values at each site $j \neq i$ should be homogenized via C_i. Considering a GPD model (POT approach), [648] proposed a direct homogenization of the estimated parameters (μ_j, σ_j, ξ_j) at site j. The new estimated homogenized parameters $(\tilde{\mu}_j, \tilde{\sigma}_j, \tilde{\xi}_j)$ are calculated using

$$\tilde{\mu}_j = C_i \mu_j, \ \tilde{\sigma}_j = C_i \sigma_j, \ \tilde{\xi}_j = C_i \xi_j,$$

and the scaling factor C_i is estimated independently of the measurements taken at site i. The regionalized sample of $(\tilde{\mu}_j, \tilde{\sigma}_j, \tilde{\xi}_j)$ then allows an estimation of the prior hyperparameters to be made on the vector $\psi = (\mu, \sigma, \xi)$ characterizing the extreme value distribution at site i. In the case study [648], the choice of types of prior (log-normal distributions) was essentially justified by the positivity constraint on (μ, σ). In the same POT setting and for the same application—but in a non-stationary context—[642] proposed a prior construction method using gamma distributions appreciated by Coles and Tawn [179] for their flexibility, summarizing *prior* regional information in terms of estimated quantiles at neighboring sites to i.

The study [322], working in a GEV framework for the prediction of extreme floods, joined a supposed noninformative[14] uniform prior $\pi^{NI}(\psi)$ with a statistical likelihood defined as the product of likelihoods, each characterizing the observations at one site, as a function of ψ (including censored data at ungauged sites, cf. Example 4.9).

Consequently, ψ does not refer to one site in particular, but a whole region, and the authors supposed that the data was independent, conditional on ψ. Removing from this likelihood the part $L_i(\psi)$ corresponding to the site i of interest, this approach leads to a posterior distribution which can play the role of an informative prior to combining to $L_i(\psi)$, following similar rules to those given for dealing with historical data (cf. Sect. 11.3.1.1).

More recent approaches are based on metrics defining the proximity of sites with the help of geostatistics tools. Prior objects constructed in this way are generally *hierarchical* (cf. Sect. 11.2.4), as they aim to model the hypothesis of the existence of subjacent physical processes that characterize the data.

Using such a strategy for a study of regionalized rainfall prediction under the MAXB framework, [634] suggested using a hierarchical approach to model M simultaneous parameter values of the form ξ over the set of M regions surrounding the

[14] See Remark 11.18.

one of interest. These values came from statistical estimates and made it possible to define a "hidden" regional sample. To take into account correlation between values in ξ—a necessity due to the physical nature of the problem—it would seem meaningful to put a prior on the vector $(\xi_1, \ldots, \xi_M)$, such as for instance a multidimensional Gaussian:

$$(\xi_1, \ldots, \xi_M) \sim \mathcal{N}_M \left(\begin{bmatrix} \nu_1 \\ \cdots \\ \nu_M \end{bmatrix}, \Sigma_\xi \right),$$

where Σ_ξ is a covariance matrix defined by

$$\Sigma_\xi = \sigma_\xi^2 \boldsymbol{\rho}_\xi,$$

with $\boldsymbol{\rho}_\xi$ a correlation matrix and σ_ξ^2 the variance of ξ. Here, a hierarchical approach corresponds to proposing a method for producing the ν_i and putting priors on σ_ξ^2 and $\boldsymbol{\rho}_\xi$ which improve the posterior separation between the estimates of these two components of the covariance of $(\xi_1, \ldots, \xi_M)$. We could for example consider an approach where $\pi(\sigma_\xi^2)$ is chosen to be noninformative and a correlation model based on the distance $r_{i,j}$ between measuring stations (or regions) i and j, such that

$$\boldsymbol{\rho}_\xi(i, j) = \exp\left(-r_{i,j}/R\right) \quad \text{for } i \neq j, (i, j) \in \{1, \ldots, M\}^2,$$

where R is a (hyper)parameter having the sense of a representative correlation distance (which could itself be subject to supplementary information and merit its own prior distribution).

11.3.1.3 The Empirical Bayes Approach

The *empirical Bayes* approach [144], which consists of estimating the prior hyperparameters using the data,[15] is conceptually hard to accept [643] because it involves artificially duplicating all or part of the information provided by the likelihood. Posterior credibility regions are therefore potentially under-estimated. For such reasons, the regional hydrological analyses performed by [277, 493], based on an empirical calibration that uses data from the site being studied, have been firmly criticized by [648].

Nevertheless, if a sensitivity study demonstrates the robustness of the posterior distribution to this empirical calibration, this type of approach continues to be used, particularly in engineering, where it is perceived as a quick and often efficient approximation to the full Bayesian methodology (see for example [655]). Indeed, this was

[15] As opposed to the *semi-empirical* approach which estimates some of the hyperparameters using the data and the rest using real complementary information.

the approach adopted by [220] for the quasi-conjugate Bayesian model for a GPD described in Sect. 11.2.3.1, and by [515] for a Gumbel distribution.

Example 11.41 (*Extreme rainfall and temperature*)

Work by [162] has shown using simulation in the bivariate case that this approach, originally proposed by [430] for predicting floods, can be efficient when there are fifty data points or more for extreme value models conditioned on non-extreme-valued data. The method was applied to both rainfall and temperature data.

11.3.2 Calibration from Expert Judgment

11.3.2.1 Elicitation, Probabilistic Interpretation, and a Warning

Expert judgment plays an important role in decision-making in many domains. It can give a guiding hand to studies and help to put scientific results in order of importance [180]. It can also considerably improve economic [464] and actuarial [748] studies on the financial impact of risks and associated flows. It may even be a deciding factor in a court case, public policy-making [522], or environmental governance [228, 516]. Its influence on technological, economic, political, societal, and even personal predictions and especially in establishing strategies for obtaining a profit has been analyzed in the epistemological literature for a long time [241, 280]. Recent studies conducted in the field of climate change use structured expert judgment to tackle the difficulties of deterministic modeling approaches, and to improve the range of predictions. Especially, the impact of the ice sheet on future sea-rise, threatening the viability of coastal communities, was assessed in such a way [55]. In the case of extreme value statistics applied to natural hazards, [177, 179]—among others—have highlighted the possibility of questioning experts on the upper quantiles of these types of distribution. In a similar manner, [706] (Sect. 3.4.2) used a regional analysis to build a prior model for a GEV distribution at a site of interest using expert knowledge on neighboring sites.

Because it is a decision-making theory, the Bayesian theory allows the integration of expert opinions into a statistical problem, by interpreting such an expertise with a bet, namely a subjacent decision-making.

This subjective point of view, progressively developed by Ramsey, Savage, De Finetti, and Koopman [281], simultaneously acts as one of the most attractive (based on "true-life" experience) and most criticized (how should the subjectivity be controlled?) features of the Bayesian approach.

This kind of criticism is even more justified when dealing with extreme events. As Coles and Powell have noticed [177], a fundamental question is what importance should we accord to advice coming from an expert who can be submitted to the stress of assuming some part of the responsibility for a possible serious accident. In this field, it would seem difficult to guarantee that expertise be *risk neutral**, with little emotional bias, conflicts of interest, or other sources of bias that typically appear when asking for expertise [77, 755]. Court rulings, such as the one condemning six Italian experts from the "Major Risks" commission to seven years prison (October 22nd, 2012) for having under-estimated the destructive power of the L'Aquila earthquake of April 6th, 2009 (decision finally reversed in 2014), can strongly limit the willingness of experts to provide honest opinions about the imminence or extent of catastrophic events [470]. Indeed, it is quite possible that an expert will cede to the idea of *non-compensation*, described in the economics setting by Bernard Cazes. It means that "the successes of a forecaster cannot compensate for their failures, since supposing that if one given method has been used several times, it should lead to comparable results, whereas the presence of errors leads us to the question if the successes were simply fortuitous". [702].

However, it is also a good idea to highlight the conclusion of Franklin and Sisson [290] from their in-depth bibliography on the application of extreme value distributions:

It is reasonable to give human intuition the 'last word' in risk assessment, while at the same time trying to use formal statistical methods as a kind of prosthesis to supplement its known weaknesses.

Subjectivity—which is also expressed in the choice of probability distributions whose validity depends on the difficulty to verify theoretical conditions—will always taint the quantification of extreme hazards, whether Bayesian or not. In much broader applied settings, authors such as Gelman and Hennig [326] and Sprenger [716, 717] have emphasized the scientifically relevant and acceptable nature of the subjective Bayesian approach. For these authors, the use of available, moderate, and reliable expert judgment is a useful way to improve forecasts and extrapolation; to ignore it could be damaging or even nonsensical.

To bolster confidence in posterior results, many formal elicitation techniques have been developed over the last 50 years, aiming to train experts and thus improve the accuracy of their assertions. Confronted with the permanent need to ensure that knowledge is retained, industry, the health sector, and financial industries have often provided the required funding to do so. In [572], the history and development of such methods are examined in detail, and useful advices are provided to limit the impact of biases coming from experts. The seminal article [755] and the works [335, 421] illustrate such biases and provide *decision heuristic** allowing expert advice to be "captured" by deciding whether *risk aversion** (or the opposite: *risk propensity*) has corrupted the use of probabilities in the case being examined. Indeed, recent work—notably in neuroscience—tends to show that human reasoning, confronted with situations where knowledge is incomplete, leads to probabilistic outputs following a Bayes rule type the Bayes rule type [151, 209, 346, 624, 683, 684]. The difficulty appears, however, when it is time to *explain this inference process*, using interpretative steps we typically call "providing expertise". In a famous article, Allais

[33] showed that an explanation of certain decisions could not respect the maximum expected gain principle formulated in Eq. (11.6).

The probabilistic encoding of expert judgment should therefore, above all, be seen as a useful (but certainly approximative) modeling tool that nevertheless brings formal coherence to judgments. Applying the rules of probability theory structures such judgments according to *axioms of rationality*, which provide a logical framework to uncertain information. The important Cox–Jaynes theorem [186, 411] from 1946, constantly improved since then [597, 710, 739, 762], provides such axioms and shows that probability theory is isomorphic to the logical system built based on these axioms. Certain axioms have been sharply criticized as incapable of taking into account the reality of how knowledge is explicited [372]. Relaxing them has led to theorems based on belief functions [699, 705] and possibility functions [231], according to which two dimensions are required to correctly represent the plausibility of a proposition. The addition of axioms founded on a formal definition of risk aversion tends to lead to *non-standard* probability theorems (see for example [354]).

However, these axioms are supposed to represent *objectivised information*, not knowledge suffering from possible internal inconsistencies[16] in which expressions depend on arbitrary symbolic systems (for example, the choice of a language). Recent results have shown that the conclusions of the Cox–Jaynes theorem remain robust even when the axioms of rationality are slightly modified [345]. It would thus seem legitimate to use probability theory to take into account uncertainty about a subject examined by a cognitive system (human or machine) that could suffer from inconsistencies.[17]

For these reasons, this theorem remains one of the foundations of artificial intelligence, as well as one of the main justifications for the use of probability distributions in representing prior information derived from expert knowledge.

11.3.2.2 Expert Judgment Based on Observed Variables

For natural events, it would seem more simple to express prior information in terms of particular values of the observed variable X [320, 419, 420, 608]—considered as a natural *anchoring variable** [421]. For extreme values, useful information means information on distribution tails, and we may wish to provide a return level x_α, for a given α, such that:

$$\mathbb{P}\left(X \leq x_\alpha\right) = \alpha, \tag{11.30}$$

[16] For example, the knowledge E on X can be reformulated as knowledge on the transform $\psi(X)$ without knowing $D = \psi(E)$, since ψ may not be easy or entirely possible for an expert to work with.

[17] Note that under the current state of knowledge, probability theory provides a pertinent *description* of information related to an incompletely understood phenomenon, rather than an objective characterization of the true variability found in the phenomenon which would be accessible to an omniscient being [50].

or, inversely, for an observed level x_α, provide an estimator of α. Being able to give one or several predictive prior quantiles $(x_{\alpha_i}, \alpha_i)_{1\leq i\leq N}$, such that $\alpha_i < \alpha_{i+1}$ and $x_{\alpha_i} < x_{\alpha_{i+1}}$, defined by

$$\begin{aligned} \mathbb{P}\left(X \leq x_{\alpha_i}\right) &= \int_\Psi \mathbb{P}\left(X \leq x_{\alpha_i}|\psi\right)\pi(\psi)\,d\psi \\ &= \alpha_i, \end{aligned} \tag{11.31}$$

is considered by numerous Bayesian practitioners to be the best way to encode expert knowledge [78, 420]. Several arguments are in favor of this:

- On the one hand, expertise provided by an expert can be considered to be made up of correlated information; that is, by giving a value of X, other values of X can be suggested (typically, two probabilistic bounds on the most probable values of X). Also, the joint distribution of prior information is invariant to any permutation of this information (basically, the order in which the quantiles are proposed should not matter). In other words, available information is considered *non-independent* but *exchangeable*, which characterizes data output by a prior predictive distribution [626].
- On the other hand, it is relatively simple to link the pair of variables (x_{α_i}, α_i) to the statistical interpretation (11.30) with the help of a piecewise linear cost function (cf. Sect. 11.1.4). We should however keep in mind that in doing so, the expert providing (11.30) may be considered neutral in terms of the risk generated by his or her own expertise.

Example 11.42 (*Predicting sea levels*)

For the reason mentioned above, methods involving questioning a group of experts—who formalize the choice of a probabilistic model over a set of options—look to provide weights for pooling often contrary opinions across experts, coherently, while attempting to limit bias related to the correlation between experts and possible conflicts of interest (see [572] for a qualitative review). This type of procedure, proposed in [180], was put into practice in [54] in order to give a prior evaluation of future sea levels. Readers interested in this still largely open problem can turn to the articles [30, 166, 330] for more information.

The estimation of quantities viewed as moments or combinations thereof (predictive expected value and variance of X, etc.) is trickier as it presupposes statistical knowledge (a common example is the frequent confusion between the median and mean of the intensity of a phenomenon). Also, unlike quantiles, it is not always true that certain moments of a probability distribution exist. For this reason, the numerous methods listed in [572] suggest *training* experts to better understand the questions raised by analysts and then better formulate responses.

Let us assume a prior form[18] $\pi(\psi) = \pi(\psi|\omega)$ hyperparametrized by ω. We can then write

$$\tilde{\alpha}_i(\omega) = \mathbb{P}\left(X \leq x_{\alpha_i}\right)$$

as defined by Eq. (11.31). These quantiles are related to the marginal prior density

$$f(x|\omega) = \int_{\Psi} f(x|\psi)\pi(\psi|\omega)\, d\psi.$$

The goodness of fit between the prior information as represented by the pairs (x_{α_i}, α_i) and that which it is possible to model—the pairs $(x_{\alpha_i}, \tilde{\alpha}_i(\omega))$—can be characterized by a loss function minimized with respect to ω. Cooke [180] for instance has proposed a loss function originating in a discretized version of the Kullback–Leibler divergence between the unknown distribution on X which gives the quantiles x_{α_i} (which we could call the "expert" distribution), and the distribution $f(x|\omega)$:

$$\omega^* = \arg\min_{\omega} \sum_{i=0}^{M} (\alpha_{i+1} - \alpha_i) \log \frac{(\alpha_{i+1} - \alpha_i)}{(\tilde{\alpha}_{i+1}(\omega) - \tilde{\alpha}_i(\omega))}, \tag{11.32}$$

with $\alpha_0 = \tilde{\alpha}_0 = 0$ and $\alpha_{M+1} = \tilde{\alpha}_{M+1} = 1$, which ensures similar relative importance is given to the quantiles in the calibration. This expression could be assigned weights if we want to highlight the goodness of fit of certain prior predictive quantiles. However, it is by no means certain that the function of ω defined by the right-hand side expression in (11.32) is globally convex, which should allow for a unique optimizer. Decreasing as much as possible the dimensionality of ω (for instance by making the other hyperparameters constant) will increase the possibility of the existence of a unique optimizer. As an aside, it can always be useful to test several loss functions (e.g., a relative least squares loss function).

Estimating ω^* necessarily requires a numerical estimate of the right-hand side term in (11.32), which has to be sufficiently smooth to allow a gradient descent algorithm to work. This can be achieved using quadrature methods (like the Runge–Kutta method) or importance sampling [483]. See Example 11.43 for an illustration of this. Stochastic optimization algorithms like the Kiefer–Wolfowitz [437] one also perform well here.

[18] By "form", we define the algebraic structure of the probability distribution.

Example 11.43 (*Torrential rainfall*)

Here, we pursue Example 11.34 and look at a Fréchet distribution $\mathscr{F}(\psi)$ for modeling rainfall maxima in Corsica. An expert from Météo-France, schooled in statistics, gave in [118] three rainfall values described as predictive prior quartiles (11.30). These values x_α and their credibility thresholds α are given in Table 11.3. Using the semi-conjugate modeling defined in Sect. 11.2.3.2 for the Fréchet distribution, we have

$$\mathbb{P}\left(X \leq x_{\alpha_i}\right) = \iint \left(1 + \frac{(x_\alpha - \mu)^{-1/\xi}}{s_1(\mu, \xi)}\right)^{-m} \pi(\xi|\mu)\pi(\mu)\, d\xi d\mu. \quad (11.33)$$

For fixed m, we therefore have to calibrate the hyperparameters $\omega = (x_{e_1}, x_{e_2})$ of the functions s_1 and s_2. Minimizing Cooke's criteria (11.32) involves calculating numerous times (11.33) for different values of ω. Also, the normalization constant in $\pi(\mu)$ is unknown. For these reasons, the choice was made in [118] to use a bi-dimensional grid over $[50, 200] \times [100, 200]$, disallowing $x_{e_1} \geq x_{e_2}$ and computing each estimation of (11.33) using a unique importance sampling

$$(\xi_k, \mu_k)_{k\in\{1,\ldots,M\}} \overset{iid}{\sim} \pi^I(\xi, \mu)$$

produced by

$$\mu \sim \mathscr{N}(0, 50),$$
$$\xi \sim \mathscr{IG}(m, m\log 2).$$

In effect, $\mathbb{P}\left(X \leq x_{\alpha_i}\right)$ can be estimated consistently by

Table 11.3 Statistics-based specification of expert knowledge on torrential rain in Corsica (*Example* 11.34 *cont.*) given as a triple of predictive prior quartiles

$\alpha-$confidence level	Rainfall x_α (mm)
25%	75
50%	100
75%	150

Table 11.4 Calibration of the hyperparameters (x_{e_1}, x_{e_2}) of the Fréchet prior for Example 11.43, in agreement with the available expertise on torrential rainfall in Corsica

Virtual size	x_{e_1}	x_{e_2}	Effective quantile thresholds		
m			(75 mm)	(100 mm)	(150 mm)
1	100.00	127.27	25%	50%	75%
2	95.45	136.36	23%	49%	73%
3	90.90	133.33	23%	51%	73%
5	87.91	133.31	23%	50%	74%
7	86.36	133.28	24%	52%	73%
10	84.84	131.81	25%	52%	74%
15	84.78	131.24	25%	52%	74%

$$M \frac{\sum_{k=1}^{M} \beta_k \left(1 + \frac{(x_\alpha - \mu_k)^{-1/\xi_k}}{s_1(\mu_k, \xi_k)}\right)^{-m}}{\sum_{k=1}^{M} \beta_k},$$

where

$$\beta_k \propto \frac{\pi(\xi_k|\mu_k)\pi(\mu_k)}{\pi^I(\xi_k, \mu_k)}$$

are importance weights. The choice of $\pi^I(\xi, \mu)$ here is specifically fitted to the case study in order to obtain relatively balanced weights (see [652] for more details): we simulate μ around a plausible value of the expectation of the Fréchet distribution and ξ by supposing a plausible ratio of 2 between $x_{e_2} - \mu$ and $x_{e_1} - \mu$.

In doing so, all of the estimates of (11.33) are correlated, but the search for the optimum (11.32) becomes insensitive to the sampling. The calibration results produced are slightly biased due to correlation, but globally satisfactory as they reproduce the expert-defined constraints quite well (Table 11.4). An approach based on an estimation of the integrals in (11.33) using quadrature was slower and did not give better results [618].

Remark 11.28 As noticed by Berger [78], chap. 3, we should not expect a perfect fit between the α_i and the $\tilde{\alpha}_i$ because the former are provided without knowledge of the sampling model for extreme values used in the problem, whereas the parametric distributions have limited flexibility. The choice of a prior coherent with the sampling model[19] implies that certain constraints on the quantiles cannot be satis-

[19] It is not therefore a *relaxation* of the sampling model, which could be interpreted as injecting an error into the model.

fied. Hence, it makes sense to prioritize a prior modeling that can satisfy one or two specifications—typically those for which our confidence is highest—and calibrate the hyperparameters using a rule like (11.32).

Prior return levels and "hidden" historical data. It is of interest to note that return levels $(x_{\alpha_i})_i$ specified by an expert can also be interpreted as an information source corresponding to a virtual historical dataset of size m, the latter defined by the highest value of the precision α. A value of $\alpha = 0.9$ (centennial level) requires that an expert "is aware of" at least 9 historical values below x_α and one above (i.e., $m \geq 10$). Upon accepting this hypothesis and not allowing m to change, it becomes meaningful to associate with the provided expertise the censored likelihood defined by

$$\ell(x_\alpha|\psi) = [F(x_\alpha|\psi)]^{\frac{9m}{10}} [1 - F(x_\alpha|\psi)]^{m/10},$$

and to define the prior as

$$\pi^E(\psi) \propto \ell(x_\alpha|\psi)\pi^{NI}(\psi),$$

where $\pi^{NI}(\psi)$ is a noninformative prior. An alternative to the method (11.32) for calibrating a vector of hyperparameters ω can therefore be proposed for a given choice of prior distribution $\pi(\psi|\omega)$ by minimizing a distance or divergence $\mathscr{D}$ [129] between the distributions:

$$\omega^* = \arg\min_{\omega} \mathscr{D}\left(\pi^E(\psi) \mid \pi(\psi|\omega)\right).$$

To our knowledge, however, this approach has not yet been attempted.

11.3.2.3 Expert Knowledge (Direct or Indirect) on Parameter Values

To suppose that prior information is available on the values of ψ (by simple algebraic relationships) is by nature risky. The information provider is expected to know the parametric form and properties of the extreme value statistical model, and thus provide predictions that are perfectly coherent with it. In the approach described in Sect. 11.3.2.2, information is available on the prior marginal probability (11.30), whereas the parametric approach carries information on $\mathbb{P}(X \leq x_\alpha|\psi)$ directly. Though the expert advice still seems (in general) to be based on the observed variable X, the latter is no longer considered *unconditionally* on ψ.

Calibration of the hyperparameters ω of a prior distribution $\pi(\psi|\omega)$ with a known form is generally performed by supposing that information is transmitted in the form of pairs (x_{α_i}, α_i), such that

$$\mathbb{P}\left(X \leq x_{\alpha_i}|\psi\right) = \alpha_i. \tag{11.34}$$

A set of values $(\psi_k)_{k\in J}$ for ψ can be obtained by using basic algebra. Many authors propose calibrating features of $\pi(\psi)$ (moments, etc.) with the help of the ψ_k.

For instance, interested in a Poisson distribution for severe earthquake events, [135] has linked expert knowledge on the frequency of occurrence directly to parameters of a prior distribution. In the MAXB setting, Coles and Tawn [179] have proposed to use expert statements, interpreted as parametric quantiles, to construct gamma priors on the parameters of a GEV distribution. In [706] (Sect. 3.4), numerous technical details on how to calibrate gamma distributions are provided, based on algebraic relationships and approximations to the behavior of these distributions. [706] has also looked at the extension of these methods to the bivariate case while supposing that information from experts exists on the *range parameter* of a dependency relationship between two GEV distributions. This corresponds to the maximal geographic distance under which we suppose that catastrophic events at the two sites remain correlated. Nevertheless, this approach remains highly focused on the direct case study in mind.

Moving back to the univariate MAXB setting, Gaioni and his coauthors [298] have constructed samples of (ψ_k) and conducted inference on the means and variances of a GEV distribution $\pi(\psi)$. They apply this method to estimate the occurrence of flash floods in watercourses. A necessary precondition is to know three distinct quantiles (11.34), which is similar to the requirements in [179].

In the POT framework, [220] (Appendix 2) proposed to calibrate the modal value of $\pi(\psi)$ from constraints placed on the quantiles. When predicting heavy rainfall [177] and flooding [589], these authors considered that experts interrogated on X were capable of providing two types of information. The first was on the mean annual number of events exceeding the extreme limit μ, while the second was on differences between extreme value quantiles (return levels); for example:

$$\begin{aligned} x_{0.99} - x_{0.9} &= \text{difference between centennial and decadal levels,} \\ x_{0.999} - x_{0.99} &= \text{difference between millennial and centennial levels,} \end{aligned}$$

but also on the ratio of these differences, such as:

$$\frac{x_{0.999} - x_{0.99}}{x_{0.99} - x_{0.9}}.$$

Here again, sampling work on the prior values of ψ output from algebraic relationships allows calibration of the prior distributions.

Though this approach is simpler than the one in Sect. 11.3.2.2 (which explains why it is used so often), this type of calibration is arbitrary because it requires us to define estimators of features of $\pi(\psi)$ using a very small number of prior values, with the help of formulas like (11.34). In our opinion, preferring one estimator over the others based on rules from classical (non-Bayesian) statistics does not make a lot of sense. For this reason, we recommend the *marginal* vision of prior information given in Sect. 11.3.2.2 and to invoke the parametric point of view only if we have access to a large quantity of prior information.

11.4 Bayesian Computational Methods

11.4.1 General Principle

In the simplest case (one-dimensional, iid data), the posterior density $\pi(\psi|\mathbf{x_n})$ of the parameters of extreme value models is known up to a constant: the (integral) normalizing constant in Eq. (11.1). Numerical integration techniques (e.g., the Runge–Kutta method) can be used to calculate it. However, it is necessary to know the whole posterior distribution, whose tails influence the determination of credibility regions and Bayesian predictions.

A general approach is to *simulate* a sample $\psi_{\mathbf{M}} = (\psi_1, \ldots, \psi_M)$ from $\pi(\psi|\mathbf{x_n})$, which then allows Monte Carlo estimation of quantities of interest like (11.6). This simulation is necessarily *indirect*, i.e., produces a more simple distribution $\rho(\psi)$ (called *instrumental*), using the following *accept–reject* algorithm (see [652], Sects. 2.3, and 2.4 for technical details and proofs):

Accept–reject method

- *Initialization* $\psi_{\mathbf{M}} = \varnothing$.
- While Card $(\psi_{\mathbf{M}}) < M$

1. Simulate $\psi_k \sim \rho(\psi)$;
2. Test whether ψ_k could have been output by $\pi(\psi|\mathbf{x_n})$;
3. If yes, update: $\psi_{\mathbf{M}} = \psi_{\mathbf{M}} \cup \{\psi_k\}$.

This test accepts or rejects the candidate ψ_k at each step. If the choice of $\rho(\psi)$ is far from $\pi(\psi|\mathbf{x_n})$, the reject frequency is high and the algorithm works poorly in practice (Fig. 11.5). An efficient approximation to a sample with distribution $\pi(\psi|\mathbf{x_n})$ can then be output iteratively by:

1. producing a first-order Markov chain on Ψ with stationary distribution $\pi(\psi|\mathbf{x_n})$ (cf. Sect. 4.2),
2. the use of an instrumental distribution $\psi_k \sim \rho_k(\psi|\mathscr{F}_{k-1})$ which evolves over iterations k of the chain and can be constructed dependent on the whole history $\mathscr{F}_{k-1}$ of the chain,
3. the selection and decorrelation of a simulated sample under the stationary distribution.

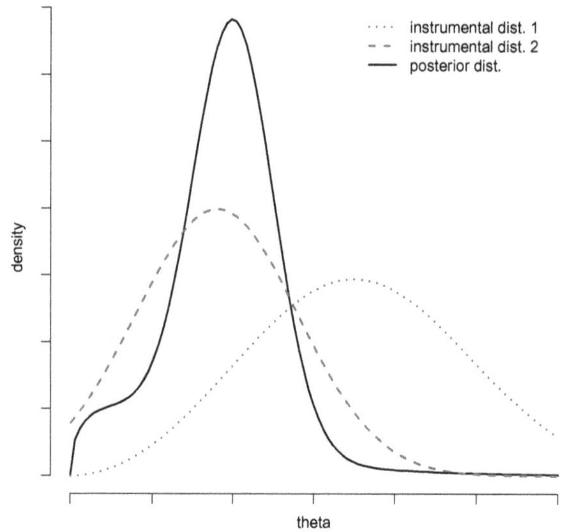

Fig. 11.5 Two instrumental densities and a target posterior density. The second of the two instrumental densities allows us to simulate data that is more likely to have come from the posterior distribution. The first distribution is significantly shifted with respect to the posterior distribution, so a large number of draws will be required to simulate in its left tail, which are the most acceptable values for any compatibility test

11.4.2 Markov Chain Monte Carlo Methods

Monte Carlo sampling algorithms based on the use of Markov chains are usually named MCMC for *Markov Chain Monte Carlo* algorithms. These have become the go-to tool in Bayesian calculation [652], growing in importance in parallel with the rise in computing power. The most advanced such inference methods, working in high-dimensional[20] settings and for complex statistical likelihoods and integration constants in (11.1), continually generalize and improve on the underlying *Metropolis–Hastings* algorithm, which we describe below. Readers interested in a more theoretical approach should turn to the book [652] and to [501] for practical implementation details. Also, the article [638] gives a useful methodology for the extreme value statistical setting. Accelerated MCMC methods previously evoked are inspired by physics (Langevin dynamics) and link up with stochastic gradient descent techniques. See 11.4.4 for some additional details, and the interested reader is advised to read the article [552] for a recent review of these approaches.

[20] Including the number of latent variables in hierarchical models (Sect. 11.2.4).

Hastings–Metropolis algorithm.

- Let $\{\psi_1, \ldots, \psi_{k-1}\}$ be a chain *(initialized for example by simulating from the prior distribution)*.
- Select a new value ψ_k:

1. Generate a new candidate $\tilde{\psi}_k \sim \rho_k(\psi|\psi_{k-1})$
2. Calculate the Metropolis ratio

$$\alpha_k = \min\left\{1, \frac{f(\mathbf{x_n}|\tilde{\psi}_k)\pi(\tilde{\psi}_k)\rho_k(\psi_{k-1}|\tilde{\psi}_k)}{f(\mathbf{x_n}|\psi_{k-1})\pi(\psi_{k-1})\rho_k(\tilde{\psi}_k|\psi_{k-1})}\right\};$$

3. Statistical test: let $U_k \sim \mathscr{U}[0, 1]$;

$$\text{choose } \psi_k = \begin{cases} \tilde{\psi}_k & \text{if } U_k \leq \alpha_k, \\ \psi_{k-1} & \text{otherwise.} \end{cases}$$

The Metropolis ratio involves the ratio of the posterior distributions (which allows us to remove the integration constant): if $\tilde{\psi}_k$ is in a higher-density zone of $\pi(\psi|\mathbf{x_n})$ than ψ_{k-1}, the ratio will be greater than one. The inverse ratio of these instrumental distributions blocks us from automatically accepting the new point, thus helping the Markov chain to better explore the whole space Ψ. A candidate point is accepted with probability α_k. Here again, for more details on the convergence of MCMC chains produced in this way, we recommend turning to [652]. In our situation, a typical choice of instrumental distribution is a Gaussian, which implies that the sequence of candidates is a *random walk* [501]:

$$\tilde{\psi}_k = \psi_{k-1} + \varepsilon_k \quad \text{with } \varepsilon_k \sim \mathscr{N}(0, \tau_k^2),$$

and the sequence $(\tau_k^2)_k$ converges[21] to a limit value $\tau^2 > 0$. See [41, 654] for practical examples and methodological suggestions (Fig. 11.6).

If the parameter $\psi = (\psi_1, \ldots, \psi_d)$ is multidimensional (notably for hierarchical models), it is recommended to use *Gibbs sampling* [652]. Suppose that we have conditional posterior distributions

$$\begin{aligned} \psi_1|\psi_2, \ldots, \psi_d, \mathbf{x_n} &\sim \pi(\psi_1|\psi_2, \ldots, \psi_d, \mathbf{x_n}), \\ \psi_2|\psi_1, \psi_3, \ldots, \psi_d, \mathbf{x_n} &\sim \pi(\psi_2|\psi_1, \psi_3, \ldots, \psi_d, \mathbf{x_n}), \\ \ldots \quad & \ldots . \end{aligned}$$

Then the Markov chain produced by the iterative conditional simulations also converges to the posterior distribution under quite general hypotheses [652]. When conditional posterior distributions are not themselves completely explicit, a hybrid strategy

[21] Such MCMCs are called *adaptive* [46].

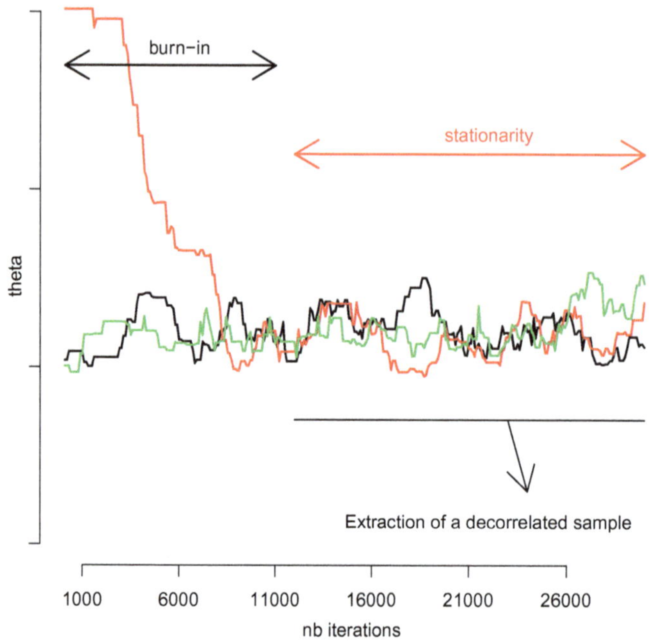

Fig. 11.6 Three MCMC chains converging in parallel to the same posterior target distribution. The *burn-in* period corresponds to the number of iterations necessary for the chains to mix, which is an indication that stationarity has been attained

called *Metropolis-Hastings-within-Gibbs* (MHwG) [414] can be used, consisting in running a Metropolis test in each dimension.

To illustrate how this algorithm works, let us take the conditional structure of the posterior distribution (11.19) on the Fréchet parameters, applied to Example 11.43 on torrential rainfall in Corsica. A typical MHwG algorithm for simulating this distribution takes the following form.

Example 11.44 (*MHwG algorithm for Fréchet (Example* 11.43 *cont.)*)

The Fréchet likelihood is written, for $\psi = (\mu, \nu, \xi)$, with $\nu = \sigma^{1/\xi} > 0$,

$$f(\mathbf{x_n}|\nu, \mu, \xi) = \frac{\nu^n}{\xi^n} \left(\prod_{i=1}^{n} (x_i - \mu) \right)^{-1/\xi - 1} \exp\left\{ -\nu \sum_{i=1}^{n} (x_i - \mu)^{-1/\xi} \right\}.$$

The MHwG algorithm takes the following form.

Step 0 Initialize (ν_0, μ_0, ξ_0).

Step $k \geq 0 \rightarrow k + 1$

1. Simulate $\nu_{k+1} \sim \mathcal{G}\left(m+n, s_1(\mu_k, \xi_k) + \sum_{i=1}^{n}(x_i - \mu_k)^{-1/\xi_k}\right)$.
2. Simulate a candidate $\tilde{\xi}_{k+1} \sim \rho_\xi(\xi|\xi_k)$;
   - calculate $\alpha_{1,k+1}$ as the minimum of 1 and
   $$\frac{f(\mathbf{x_n}|\nu_{k+1}, \mu_k, \tilde{\xi}_{k+1})\pi(\nu_{k+1}|\tilde{\xi}_{k+1}, \mu_k)\pi(\tilde{\xi}_{k+1}|\mu_k)\rho_\xi(\xi_k|\tilde{\xi}_{k+1})}{f(\mathbf{x_n}|\nu_{k+1}, \mu_k, \xi_k)\pi(\nu_{k+1}|\xi_k, \mu_k)\pi(\xi_k|\mu_k)\rho_\xi(\tilde{\xi}_{k+1}|\xi_k)};$$
   - accept $\xi_{k+1} = \tilde{\xi}_{k+1}$ with probability $\alpha_{1,k+1}$, otherwise choose $\xi_{k+1} = \xi_k$.
3. Simulate a candidate $\tilde{\mu}_{k+1} \sim \rho_\mu(\mu|\mu_k)$;
   - calculate $\alpha_{2,k+1}$ as the minimum between 1 and
   $$\frac{f(\mathbf{x_n}|\nu_{k+1}, \tilde{\mu}_{k+1}, \xi_{k+1})\pi(\nu_{k+1}|\xi_{k+1}, \tilde{\mu}_{k+1})\pi(\xi_{k+1}|\tilde{\mu}_{k+1})\pi(\tilde{\mu}_{k+1})\rho_\mu(\mu_k|\tilde{\mu}_{k+1})}{f(\mathbf{x_n}|\nu_{k+1}, \mu_k, \xi_{k+1})\pi(\nu_{k+1}|\xi_{k+1}, \mu_k)\pi(\xi_{k+1}|\mu_k)\pi(\mu_k)\rho_\mu(\tilde{\mu}_{k+1}|\mu_k)};$$
   - accept $\mu_{k+1} = \tilde{\mu}_{k+1}$ with probability $\alpha_{2,k+1}$, otherwise choose $\mu_{k+1} = \mu_k$.

Simulations were run choosing for $\rho_\xi(.|\xi_k)$ a normal distribution truncated below zero, with mean ξ_k and coefficient of variation 20%, and for $\rho_\mu(.|\mu_k)$ a normal distribution truncated above x_{e_1}, with mean μ_k and the same coefficient of variation. Convergence of the MCMC chains is illustrated for the parameter ξ in Fig. 11.7 for the case where the prior distribution is basically noninformative. Several thousand iterations are required before starting to mix the chains. Calculated after a decorrelation step, the posterior characteristics of the parameters and several return levels are summarized in Table 11.5.

Predictive posterior densities are also plotted, along with the histogram of observed rainfall, in Fig. 11.8. The influence of the choice of prior is visible in the form of the density function: the bigger the m, the larger the weight given to the expert's advice. The dissymmetry in the data distribution with respect to the value of 100 mm (which is the empirical median) is slightly pessimistic regarding the expert's advice. This explains the shift to the left of the predictive distribution as the weight of the expert's advice increases. This progressive increase in the force of the expert's knowledge has little influence on the decadal return level (around 200 mm, in agreement with the advice of the Météo-France expert), but much more on the centennial return level, where the uncertainty decreases dramatically. Despite the failure to take into account a trend linked to climate change, a centennial level slightly above 400 mm appeared quite plausible to the expert, who was a specialist of Mediterranean weather patterns and Corsican micro-climates.

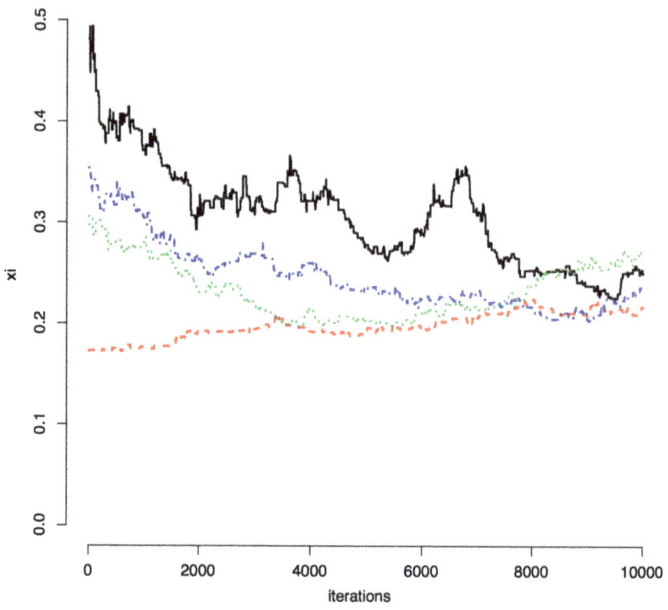

Fig. 11.7 Paths of four MCMC chains towards the posterior distribution of ξ for a very small virtual data size of $m = 0.1$ (Fréchet distribution)

Table 11.5 Posterior results as a function of the size of the virtual sample m accorded to prior expertise, for the Fréchet model and conditional on rainfall data in Corsica. The results shown are the posterior expectations; standard deviations are in brackets. Calculations were performed after 50,000 MCMC iterations

Virtual size m	μ	σ	ξ	Return level (mm)	
				10 years (decennial rainfall)	100 years (centennial rainfall)
1	−8 (19)	110 (21)	0.30 (0.04)	210 (27)	465 (190)
5	−2 (17)	98 (22)	0.32 (0.07)	208 (25)	449 (109)
10	3 (6)	90 (10)	0.34 (0.03)	202 (22)	420 (96)
15	5 (6)	86 (7)	0.35 (0.03)	200 (23)	410 (87)

11.4.3 Reconstruction of Incomplete Data

Under the Bayesian framework, no distinction is made between parameters and variables in terms of probabilistic nature. Incomplete data, which may come from reaching a technical limit or a bias of some measuring device or historical missing data censored by one or several historical point values (cf. Examples 4.8 and 4.9), may potentially be reconstructed by conditioning on these historical values and a current estimate of the parameters. Other types of incomplete data involve mixture models of extreme value distributions (Sect. 12.2.2.3); in such cases, the "data miss-

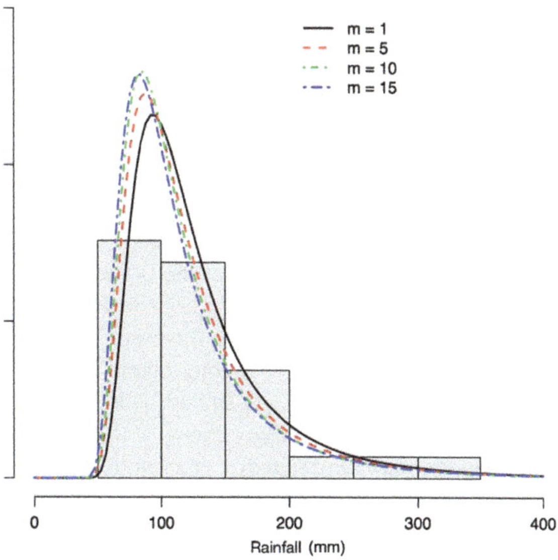

Fig. 11.8 Predictive posterior densities (Table 11.5) and histogram of observed data

ing part" is its attribution to one of the models in the mixture. Latent variables in hierarchical models (Sect. 11.2.4) are yet another example.

Reconstruction of such types of incomplete data is called *data augmentation*, which can be done iteratively and naturally falls under the Gibbs framework. This method is the Bayesian counterpart to the *Expectation–Maximization* (EM) algorithm and its generalizations [510], which iteratively reconstruct a complete dataset (or a related statistic) in order to run maximum likelihood estimation without being bothered by complicated integrals in the definition of the data likelihood [690].

Curiously, though this type of approach is well-known in the reliability setting—examined in numerous missing data situations—it remains little used in extreme value statistics. [686] has looked only at the use of EM algorithms in the mixture model setting. In the Bayesian case, only [676, 677, 679] seem to have proposed the Gibbs algorithms for dealing with data augmentation in the extreme value statistics setting, in order to run a regional study with historical data and resolve multivariate inference problems.

Experiments conducted at EDF for developing EM-type algorithms for GEV distributions have shown that this approach can highly computationally intensive since three-dimensional functionals have to be optimized at each iteration. Stochastic approaches, either SEM [146] or Gibbs (in the Bayesian setting), which *simulate* the complete data, appear to be better candidates for improving the estimation of extreme value distributions using incomplete data. Formalizing and scaling up such approaches to the industrial setting—if they prove useful—is an interesting research avenue.

11.4.4 Other Computational Approaches

The use of MCMC as applied to extreme value models (in simple cases) can be greatly simplified with the help of the `OpenBUGS` software [726], which provides an implementation of the hybrid Gibbs techniques.[22] The major constraint of this software is that it requires input Bayesian models to have proper priors. Indeed, these measures are required, among others initializing the MCMC chains and the instrumental simulation to run smoothly. Recent MCMC methods [550] are based on a comparison of the Markov chains with deterministic simulations of molecules following *Hamiltonian dynamics*; an evolution-gradient approach in this setting leads to much less correlation between elements of the chain than for standard MCMCs, as well as being more efficient (in terms of run time) [92]. This approach is thus particularly useful for dealing with high-dimensional spaces in hierarchical and/or multivariate problems, for instance with the help of the `Stan` software [141]. Most of these software packages can link up with `R`. See [446] for a well-presented look at how to implement them.

This computational setting is undergoing incessant development. For instance, *variational Bayes* methods can be used to quickly approximate the main characteristics of $\pi(\psi|\mathbf{x_n})$ when the posterior distribution is difficult to get using MCMC (see for example [435]). In an application involving heavy rainfall in Australia (in the MAXB setting), [556] highlighted how variational Bayes methods can be useful when looking to evaluate the impact of continuous covariates on the occurrence and intensity of extreme hazards.[23]

The promising *bridge sampling* approach can now use these variational methods to accelerate the Gibbs algorithm iterations and help the algorithm function efficiently in high-dimensional settings [358]. In conclusion, we therefore recommend that readers interested in using extreme value models in larger and larger dimensions (spatial, non-stationary, etc.) first inform themselves as to the current state-of-the-art, with the help of review articles and seminal works as [552].

11.5 Conclusions

This chapter has given a (necessarily incomplete) overview of the use of Bayesian methods for statistically dealing with modeling, predicting, and extrapolating extreme

---

[22] In early 2018, extreme value distributions are still not encoded in its competitor `JAGS` [621]; users interested in the power and flexibility of `JAGS` must encode them using programming tricks.

[23] The authors approximated a GEV distribution, under weak conditions, by a Gaussian mixture.

hazards. A deeper look into the theory can be made with the help of the works [591, 651], especially in terms of case studies. Fundamental aspects of the validation of Bayesian results are discussed in detail in [768]. The Bayesian framework is still little used in industrial studies and civil engineering, even though it provides a large number of opportunities for engineers working in industrial and environmental risk. In particular, it facilitates the inclusion of historical and/or regional data, as well as expert opinion, and helps to formalize essential questions on the issue of risk mitigation. From the academic point of view, this research area is extremely fast-moving, especially as computational tools required for Bayesian computation are more and more powerful. Therefore why this chapter has mainly focused on modeling practice, which remains one of the trickiest difficulty in Bayesian practice.

We are willing to bet that an increasingly important task in the industrial and environmental risk engineering community will be to improve the ease of understanding methods for eliciting prior information in the extreme hazards setting, stating precisely what was chosen arbitrarily, and justifying parameter choices to "moderate" this. Obviously, any Bayesian analyst confronted with the problem of calibrating prior distributions must conduct a sensitivity study on quantities of interest (return levels, etc.) with respect to prior choices. These choices naturally include those relating to the algebraic structure of prior distributions. To decrease the influence of such choices, it may be possible to take advantage of recent work looking to directly provide *robust* Bayesian inference [82]. For instance, [674] considered an extreme value setting where no formal rule allows for a simple approximation of a prior form, and attempts are made to avoid having to choose a particular form. The inference is performed in a framework including all possible prior distributions. This type of approach lies at the confluence of Bayesian theory and possibility theory [231].

Other methodological alternatives have also been proposed for integrating expert knowledge into the estimation of extreme events, for return periods in particular. This is notably true for the imprecise probabilities and the Info-Gap theory [71], which deal with the problem of robust decision-making vis-à-vis inherent uncertainty in the handling of extreme value distributions. Several developments and applications are summarized in [290] (Sect. 5.5) and should probably be compared with methods currently used in engineering.

Furthermore, the likelihood-regularization mechanism permitted by the use of prior measures, combined with the ability of the Gibbs algorithms to reconstruct observed data and take advantage of the complete likelihood, already makes it possible to significantly improve frequentist estimations in multivariate settings [222]. These extremely recent results from academia, for the moment essentially tested on simulated data, should lead to problems in more than three dimensions now becoming accessible. The question of adapting them to environmental data, in a world where the relationships between different natural hazards are increasingly of interest, appears to be of immediate importance.

Chapter 12
Perspectives

Marc Andreewsky, Pietro Bernardara, Nicolas Bousquet, Anne Dutfoy, and Sylvie Parey

Abstract This chapter provides a brief summary of the latest results obtained in extreme value theory, and offers many suggestions for the reader interested in using these tools in a context where, in particular, statisticians and engineers cannot ignore the Big Data paradigm and the industrialization of machine learning tools, now essential components of modern artificial intelligence. Parallels are also made with other disciplines interested in extreme statistics.

12.1 Conclusions and Recommendations

The philosophy behind extreme event modeling can be described as an attempt to reduce the impact of strong cognitive biases caused by easy access to non-extreme observations, and the loss of historical memory of past extreme events. Some such biases are in play in events called *black swans* (economic crashes, wars, scientific discoveries, etc.), popularized in 2009 by [734]. Arriving unexpectedly and with huge societal impact, these types of events are rationalized afterwards, as if they could have been predicted. This is an illusion, as [736] states openly. In effect, the complexity of the systems around us continues to increase, and it is already difficult in practice to predict in a more or less near future, non-extreme events. According to [735, 736], this error is the worst possible epistemological one that can be made in risk management. Due to increasing complexity, it is important to not systematically and indiscriminately rely on past data within mathematical models that claim to provide predictions of extreme events. Such data may become rapidly obsolete, and are no help in looking for new sources of risk. Risk management generated by those particular extreme events should instead be based on *non-inductive** decision logic, or on the smallest possible number of axioms. This would imply not using mathematical modeling, which requires a clear delineation of the most influential factors, modeling theorems, and validation techniques based on simulation.

M. Andreewsky · P. Bernardara · N. Bousquet (✉) · A. Dutfoy · S. Parey
EDF R&D, Chatou, France
e-mail: nicolas.bousquet@edf.fr

N. Bousquet and P. Bernardara (eds.), *Extreme Value Theory with Applications to Natural Hazards*, https://doi.org/10.1007/978-3-030-74942-2_12

The class of events considered in this book does not belong, by *scientific and societal consensus*, to the *black swans* category [116]. This means that the relationship between the nature of the objects looked at and their mathematical description based on probability theory results from a long and ongoing process of discussion and testing. Arguments justifying the stochastic treatment of natural hazards were provided in Chap. 1 (Sect. 3.1.2), but this justification remains relative to the state of science and technology, and its relevance should be re-examined on a regular basis.

On the basis of this consensus, the theory of extreme values has for many years been a common and unique reference point for the estimation of extreme values output from natural phenomena. This theoretical framework has allowed engineers to overcome subjective extrapolation techniques and converge toward a common approach. It thus plays a leading role in guiding choices in natural risk mitigation, whether for the industrial world or society as a whole. We hope that readers are now fully convinced of this.

Nevertheless, practical applications of the theory—which we hope to have illustrated well in this book—are fraught with problems that are difficult to solve, and that sometimes put the credibility of extrapolation results into question. The estimation of the form parameter ξ, for example, is often the subject of long discussions, since a small change in its assessment (ordinarily a difficult task) has significant consequences for extrapolation to rare events. Controversy surrounds the occasionally very extreme values that are produced in estimates. Are they physically plausible?

Often, debate ignites around the existence (or inexistence) of physical limits to the intensity of phenomena. Since the theory of extreme values provides for the existence of distributions unbounded above, what does this say about the acceptability of hypotheses involving limits? And if we admit the existence of an upper limit for the intensity of a phenomenon, how should it be defined? Physically speaking, the total rain capacity is limited by the (finite) amount of water in the atmosphere, but to imagine a phenomenon reaching this limit is obviously not realistic. In this case, can deterministic approaches, such as *Probable Maximum Precipitation* (PMP; [741]) or *Probable Maximum Flood* (PMF; [456]) serve as references? Would we be able to introduce this information into the estimation of statistical distributions if it were available and reliable? And how should we proceed for phenomena which logically should be bounded above, if analysis of the data leads us to estimate unbounded statistical distributions? And lastly, if we are content to extrapolate from distributions advocated by the theory, how can we integrate knowledge on physical phenomena that trigger extreme events? Here we touch on the limits of the application of statistical theory, which has justified, in this book, a deeper look at computational approaches (Chap. 8).

Some adherents to the theory mention the fact that sometimes major assumptions (such as data independence and convergence of estimates) do not hold, which could explain statistical results that do not conform to expectations and to physical knowledge of a phenomenon. In this context, they work on how to give advice to ensure that methods are applied in a reasonable manner in accordance with the assumptions. Others—who are more skeptical—point to the fact that extrapolations of statistical distributions are made without taking into account underlying physical phenomena, and are therefore unreliable by definition. In particular, it is common to refuse to extrapolate up to the frequency 10^{-m} if we do not have a sample of at least 10^m data points allowing us to validate it. By the same token, it is the fundamental asymptotic hypothesis, which underlies the entirety of the theory, which is considered unrealistic.

Thus, the question is can this theory still be used, taking into account these limits? Some experts advise stopping at estimates that are close to the length of time over which data is available. This hypothesis is certainly very reasonable, but quickly prevents the use of the theory in a large number of practical cases where industrials, insurers or legislators wish to estimate frequencies lower than those which can be empirically provided by the given data. Our experience in using the data and models described in this book, and our knowledge of industry, leads us to suggest the following.

First, it is essential to make several estimates, via a number of different methods, under the two frameworks (MAXB and POT). Confidence intervals around parameters and functions of interest should always be calculated, provided, and discussed. Their definition depends on the models used, and their calculation is strongly influenced by the number of data. The calculation hypotheses must therefore be stated clearly, and the validity of these intervals, which are conditional on the choice of model, should be tested using simulation techniques. Moreover, it is always useful to compare the results produced with deterministic approaches with the probabilistic distributions resulting from statistical calculations. Discussions of these results must include input from one or more experts, which requires that statisticians produce meaningful orders of magnitude, called *anchoring values**. Keep in mind that the use of extreme value distributions is a simplified—sometimes oversimplified—response to strong social pressure on decision makers and risk quantifiers [107]. It is therefore necessary to consult and consult again before approving a result.

Second, important work needs to be conducted on the validity of the data being used, including the statistical interpretation of historical and regional data. Statisticians must call upon historians and physicists to agree on one or several interpretations, and any strong sensitivity of the estimates to the keeping or removing of a certain data point must be discussed from a physical point of view. Is it an *outlier*? Is it due to the influence of an incorrect measure? The ongoing collection of new data and a critical analysis thereof is necessary, and this "novelty" should be connected with the forgetting of past extreme events that characterizes modern society (as described in the first chapter of this book).

Third, the meaning of the concepts of confidence and credibility must be kept in mind. Very often, dimensioning rules are produced in terms of upper limits of confidence intervals, which offer a certain level of conservatism. However, this conservatism must be justified, and its nature depends on the notion of underlying cost in dimensioning problems. The more information we have on this cost, the more the probabilistic interpretation of the limits can be clarified. Note that information on cost is part of the prior knowledge we have on a problem, and though imperfect, it would be a pity to deprive ourselves of it, as it helps us to place the theory of extreme values into a (simplified) decision-support framework, which then allows statisticians to conduct sensitivity studies that can be comprehended by engineers.

The increasing desire of civil society, including industry, to mitigate the effects of extreme natural hazards, requires a constant increase in the arsenal of techniques dedicated to risk quantification. While we believe that despite its shortcomings, extreme value theory is a fundamental building block of this arsenal, we also believe that it must continue to evolve in relation to recent scientific advances. This concerns, for example, data collection and dropping calculation costs, which are starting to permit the analysis of high-dimensional models. The following paragraphs list some future paths in this direction. Our long-term vision is that the decision-support allowed by the theory leads to the proposition of *anti-fragile* mitigation rules [735]. That is, decisions that become more and more solid when new knowledge is added, at a level above and beyond *robustness* (if a hazard is powerful, mitigation is effective) and *resilience* (if a hazard is strong, mitigation may occasionally not be 100% effective, but still relatively so).

From experience gathered in our industry working group, we are deeply convinced that approaches leading to evolution and innovation in these methods must be multidisciplinary. This includes the skills of various scientific communities, people working at the forefront of applied settings—dedicated to sharing the experience of the industrial sector—and that of other sectors of society, such as civil engineering. We hope that this book marks a significant step in this direction.

12.2 Perspectives

12.2.1 Analyzing Large Databases

Without changing the theory itself, we have today hope of increased access to larger and larger and/or spatially distributed databases. The scientific community as a whole has new methods of observing the Earth (i.e., a massive deployment of sensors and satellites whose cost has decreased significantly since around 2010), and is benefitting from an increasing effort to reconstruct data via historical analyses, and the use of increasingly sophisticated simulation models. This bringing together of data types is a major scientific advance and a formidable asset, provided that models are developed that know how to exploit such large quantities of information, especially in the spatial sense.

The detection of useful *patterns*, or variables (*feature selection*), by *machine learning* techniques is a fundamental issue for companies and institutions facing massive increases in available data (the *Big Data* setting). Challenges include not only the increasing number of sensors but also the diversity of measured variables, the range of validity of these sensors, their *accuracy* (which influences data quality), the increasing number of supposedly explanatory variables, as well as the high *rate of acquisition*.[1] The first attempts to use probabilistic modeling for extreme values from the real world uncovered in extremely large databases have been published quite recently [338, 701, 802], and seem promising. These are counterparts to methods aiming to identify systemic weaknesses in industrial infrastructure by the joint use of fault trees (classical tools in probabilistic safety assessment (PSA) studies) and statistical relationships between probabilities of occurrence and covariates [361].

12.2.2 Making Distributions More Robust

The statistical approach to extreme values is still criticized, within the scientific community, for a certain lack of robustness (which implies we should not stick to a unique estimate in practice). This includes restrictions on types of data, the choice of estimators, and the choice of distributions used. As stated by [107], the theory applies only when the phenomenon is continuous, purely random, and the rate of convergence of the non-stationary (or *minima*) toward the limit distribution strongly depends on the nature of the distribution of non-extreme values. One potential path forward is to construct modeling theorems that work for a greater variety of data.

12.2.2.1 Robustness of Statistical Estimates and Limits to Extrapolation

The robustness of statistical estimates will clearly continue to improve due to regionalization when natural phenomena are not too "local", and the Bayesian framework can be used effectively if—very generally speaking—additional information on top of local data is available (see Sect. 11.5). Also, the use of the entire extreme value data sample, including correlated data *clusters*, can also help strengthen statistical estimates of return levels or periods (or any other quantity of interest). It is therefore necessary to take into account *similar but correlated over time extreme values* in analyses. The copula-based approach described in Chap. 9 would then need to be modified, as it requires events to occur at simultaneous times.

A promising approach not requiring *declustering*, proposed by [647] and illustrated on flood forecasting, relies, for example, on Markovian modeling using dependency structures between exceedance extreme values, and is situated somewhere between a POT approach and time series analysis. The number of dimensions of the underlying copula determines the order of the Markov process. This methodology may likely complement the functional approaches described in Chap. 8. In the GEV framework, a similar approach, Bayesian but restricted to Gaussian copulas, was recently proposed by [560]. As the use of extreme copulas improves, it is becoming

[1] That is, the spacing used between measurement times, which can limit or even prohibit the data being loaded into computer memory, and force the detection of extreme values to be conducted in situ.

essential to focus on how to use them in high-dimensional settings (using *vine* copulas for instance [76]), in order to capture information related to inertia in extreme phenomena (e.g., aftershocks of earthquakes, repeated floods, etc.), and to refine probabilistic forecasts by considering sequences of extreme events that happen in a very short time period. Indeed, it is not obvious that a dimensioning produced from a single hazard, even if bounded above, is also robust to a sequence of slightly less powerful hazards over a short time period. Beyond this, it appears essential to develop statistical models that consider several variables at the same time, and are also able to characterize the spatial distribution of phenomena and interactions between them. Other approaches based on mixing non-extreme data with extreme data, using conditioning, also constitute possible research directions, and have barely been looked at until now. The articles [162, 383] give an outline of such studies, in a multivariate context, for the frequentist and Bayesian frameworks.

Lastly, the robustness of an extrapolation (such as a return level) itself strongly depends on the robustness of the statistical estimation. Establishing formally (and not just empirically) limits to extrapolation is a major roadblock, which has only recently begun to be attempted in a one-dimensional framework [29]. Here again, extensions to settings with multivariate, spatial, and trending data, are important research paths of the future.

12.2.2.2 Distributional Robustness

The difficulty of verifying theoretical conditions leading to the use of distributions from extreme value theory has recently led some researchers to study so-called *distributional* robustness in the context of extreme natural hazards studies, in order to severely limit the impact of modeling errors resulting from the theory. These theoretical limits have for example been studied by [496] for financial indicator time series. In response, the study [98] proposes a methodology for estimating quantiles (return levels) in a *nonparametric* way, by looking at a class of heavy-tailed statistical distributions located in the neighborhood of an extreme value distribution, rather than using one distribution only. The neighborhood is defined via the choice of a class of divergences between statistical distributions, and neighborhoods can be bounded by taking advantage of the convex geometry underlying the problem. As a result, it becomes possible to define bounds on the tails of the statistical distribution representing the phenomenon, and deduce robust estimates of return levels.

Modeling errors resulting from the choice of a correlation structure between multivariate hazards are also problematic. How to produce risk measures which are robust against errors caused by not completely certain choices? Recently, bounds on Pickands' function and excess probabilities have been obtained by [250] as a partial answer to this question. Tested on financial data, this approach—again based on neighborhoods of measures (defined by divergences)—provides a conservative view of the co-occurrence of extreme natural hazards, in high-dimensional settings.

The practical implementation of the results of this initial research, and extensions to more general frameworks, remains to be done.

12.2.2.3 Extending Results to Other Distributions

Extensions of distributions described in this book are numerous. As an example, spatio-temporal characteristics of multivariate extreme values are now currently being studied through the prism of peaks-over-threshold analysis applied to functions, that allow to define complex extreme events as special types of exceedances, and then obtain their limit tail distribution, the so-called *generalized r-Pareto process* [284], a generalization of the approach proposed by [222].

To limit the risk of error due to the use of extreme value models, whose construction assumptions are difficult to verify, and to relax constraints on the nature of the data, an additional possibility is to include, in the class of distributions mentioned above, some with new properties. For instance, the instability of extreme value distributions by summing extreme variables may prevent modeling a coalescence phenomenon. Levy's distributions, that are sum-stable and correspond to the asymptotic behavior of sums of random variables that don't fall under the central limit theorem [36, 498], could be useful.

Another possibility is to produce classes based on mixtures of extreme value distributions. Little-used in engineering, this point of view is currently being investigated by [286, 489, 580, 678].

12.2.3 Interactions Between Statistical and Numerical Models

The coupling of statistical models to data from numerical models, used in a stochastic framework, represents a kind of parallel tool that attempts to decide how to implement encompassing hierarchical models [695], with the help of ever-increasing computing power. The most significant attempts [637] in this direction still require strong assumptions on the invariance of parameters in the extreme value distributions. Additional difficulties posed by the calculation time required of computational models can be partly countered using meta-models (evoked in Chap. 10)—possibly based on classical *machine learning* tools—which are capable of simulating large amounts of data. Recent approaches are reviewed in [455]. Developing meta-models that are well adapted to extreme value situations is therefore an important goal in linking these two major information sources.

12.2.4 Sustaining a Multidisciplinary Approach

Though this book focuses on the use of extreme value statistical theory for characterizing natural hazards, the theoretical and practical insights of the theory are part of a much broader scientific field. In particular, keeping an eye on methods developed in finance, insurance, and fraud detection is an ongoing necessity. Though these areas have long been characterized by large quantities of data, where noise is low or non-existent, the economic stakes are such that many developments from these domains will become useful in the study of natural hazards.

Let us briefly look at three such examples of important work, dealing with *risk perception*, the selection of explanatory variables, and the definition of new measures of extreme risk.

12.2.4.1 Risk Perception

The perception of risk by people working in financial markets can be partially covered by insurance contracts; the case of extreme risks associated with large levels of uncertainty remains, however, problematic, in the sense that no clear methodology exists for determining the price of insurance premiums [235]. The inability of economic actors to solve this kind of problem testifies to the need to adopt, with a certain dose of humility, multiple approaches for calculating extreme risks, which is in line with the recommendations in the preceding section.

12.2.4.2 Selecting Explanatory Variables

In the field of insurance, hierarchical Bayesian analysis is often used to produce a model that can deal with *heterogeneity* problems in extreme value data, when these originate from different sources [102]. Even though applying this to extreme value data coming from the natural world (which have normally been arranged to form an as homogeneous as possible dataset) is not immediately obvious, the scarcity of certain events could lead us to pool data from events sharing explanatory phenomena, whether correlated or not (for example, floods due to heavy rains and/or melting ice), or measured by different methods, and to differentiate the specific characteristics of the shared features in the statistical analysis.

12.2.4.3 New Measures of Extreme Risk

In the specific context of risk mitigation associated with natural events, exceedance probabilities, levels (quantiles), and return periods are not the only subjects of possible dialogue with regulatory authorities [452]. Other risk measures that are well adapted to extreme events have been studied by various communities—in particular financial and actuarial—and have better coherence properties.[2] This is true in particular for *conditional values-at-risk* [656], also called *Bregman super-quantiles* [449], which characterize how tails of statistical distributions behave beyond extreme quantiles.

---

[2] Especially, the *sub-additivity* of a risk measure $\mathscr{R}(X+X') \leq \mathscr{R}(X)+\mathscr{R}(X')$. See [656] for more details.

Part III
Detailed Case Studies on Natural Hazards

Chapter 13
Predicting Extreme Ocean Swells

Pietro Bernardara

Abstract This chapter illustrates the conduct of a classic univariate extreme statistical study on severe ocean swell forecasting.

13.1 Context of the Study

13.1.1 The Variable of Interest and Probability Levels

Following the procedure described in Sect. 3.1.3, the variable of interest here is the *significant spectral height* (see below) of the sea state, which describes the surface of the sea when under the influence of wind (which generates wave systems) and swell. The significant spectral height is an often-used quantity for characterizing the sea state. It is often denoted H_s, H_{m0}, or SWH (*significant wave height*). This variable can be obtained via the spectro-angular distribution of the sea state variance $F(f, y)$ (usually expressed in the units $\mathrm{m}^2 \cdot \mathrm{H}_z^{-1} \cdot \mathrm{rad}^{-1}$), also known as the directional spectrum of the sea state variance. This function characterizes the division of the sea state's energy across frequencies f and the directions y of origin. The significant spectral height is calculated starting from the zero-order moment M_0 of the directional spectrum of the variance:

$$H_{M_0} = 4\sqrt{M_0},$$

with

$$M_0 = \int_0^\infty \left[\int_0^{2\pi} F(f, y)\, dy \right] df.$$

P. Bernardara (✉)
EDF R&D, Chatou, France
e-mail: pietro.bernardara@edf.fr

N. Bousquet and P. Bernardara (eds.), *Extreme Value Theory with Applications to Natural Hazards*, https://doi.org/10.1007/978-3-030-74942-2_13

Under the so-called *deep sea* hypothesis that swell heights are distributed according to a Rayleigh distribution, the significant spectral height is equal to the *statistically significant height* $H_{1/3}$, which corresponds to the mean height (measured between peak and trough) of the highest third of wave heights.

The probability level required for this study corresponds to a centennial return period, i.e., $p = 0.01$. The quantity of interest to estimate is, therefore, the centennial significant spectral height. Following recommendations similar to those expressed by the ASN (the French Nuclear Safety Authority) in [15], we would also like to estimate the upper limit to the 70% confidence interval associated with this quantity of interest (see Sect. 5.3).

13.1.2 Data Collecting

Our case study is about high swell events off the coast of Yeu Island (in the Vendée department of France—see Fig. 0.1). We do not, however, have directly observed significant swell height data (e.g., from wave buoys or high-frequency tide gauges); instead, the data used has been generated using sea state and swell height models.

These (reanalyzed) simulations of spectral heights were run using the ANEMOC database (a numerical atlas of oceanic and coastal sea states), which contains simulated sea state data created by LNHE (National Hydraulics and Environment Laboratory) of EDF in collaboration with CETMEF (the Center for Maritime and Fluvial Technical Studies) [74]. The ANEMOC database was constructed using the TOMAWAC software [75] to perform 23-year simulations with a temporal resolution of 240 seconds over the period 1979–2002. The simulated data are available at a time resolution of 1 hour. TOMAWAC has been developed by LNHE to model the temporal and spatial evolution of the directional spectrum of the sea state.

These reanalyses produced with the help of ANEMOC were calibrated and validated using comparisons with wave buoy measurements found in the CANDHIS database from CETMEF for the years 1999 and 2000 [74].

13.2 Sampling Extreme Values

13.2.1 Visualizing the Data

The aim is to extract from the available data a sample of independent extreme values, sufficiently numerous and stationary for classical theorems (Chaps. 4 and 6) to be applied. This process leads to a sample of 23 annual significant sea heights, as shown in Fig. 13.1.

This sample seems rather small for conducting a study of extreme value statistics, so we instead use a POT threshold exceedance approach to increase the number of

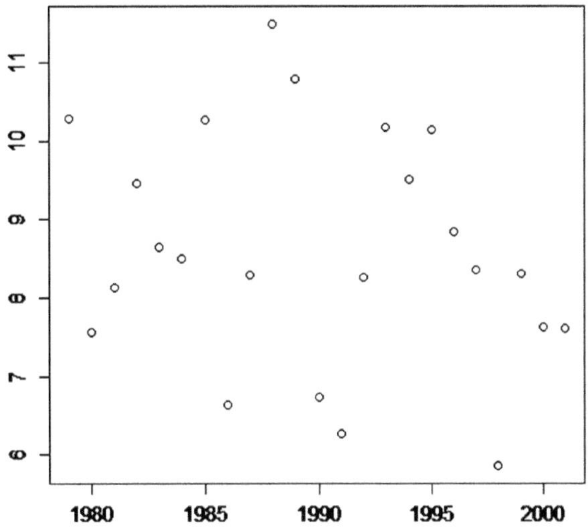

Fig. 13.1 Sample of annual significant heights (in m) of the sea state, extracted from the ANEMOC database for 1979–2002

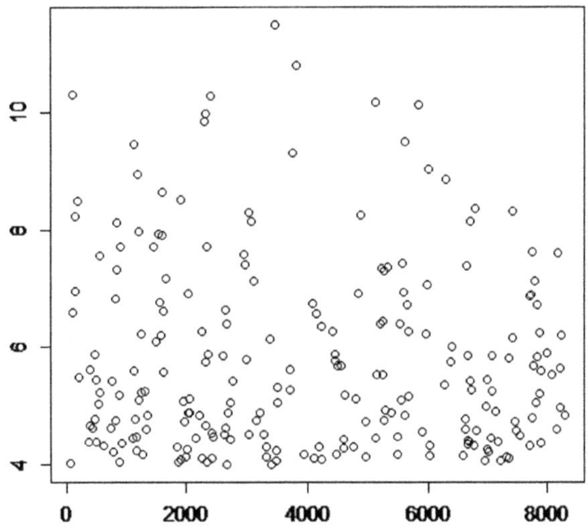

Fig. 13.2 Sample of significant heights (in m) of sea states over 4 m extracted from the ANEMOC database from 1979 to 2002. The x-axis corresponds to the number of days between two dates

useful data points. For this second sample, the threshold is fixed at $u = 4$ meters, and a minimal separation of 2 days between consecutive values above this value is used, leading to a set of 226 data points. These values (2 days, 4 meters) were chosen after consulting with experts on the duration of typical storms in the region and what they consider an extreme swell to be. This sample is shown in Fig. 13.2.

It turns out that the number of occurrences per year can be represented by a Poisson distribution, whose parameter λ is estimated by $\hat{\lambda} = 9.83$ using maximum likelihood (equal here to the mean number of events exceeding the threshold per year). The quality of this fit is shown in Fig. 13.3.

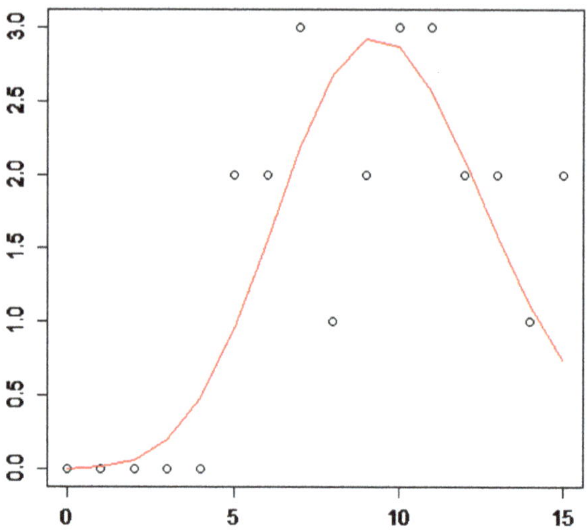

Fig. 13.3 Fitting a Poisson distribution to the process of the number of occurrences per year exceeding the threshold (y-axis) in the POT sample of significant heights over 4 m. The x-axis corresponds to the number of threshold exceedances

13.2.2 Testing for Stationarity

Stationarity is a fundamental hypothesis when working with extreme value statistics. In the present situation, stationarity of the sample of annual non-stationar is not rejected by the Mann–Kendall test (cf. Sect. 5.4.3.2), which was run using the `ManKendall` function from the `Kendall` package in `R`. For a type I error of 5%, the p-value is 0.245. This large value indicates that the observed test statistic is not found in the range of extreme values of the distribution of this statistic under the stationarity hypothesis, meaning that it cannot be rejected. The same test run on the sample made up of peaks over 4 m confirms the result, with a p-value of 0.689.

13.2.3 Analyzing Data Independence

Here we wish to check whether the sample is composed of iid values and therefore consistent with the hypotheses involved in the main extreme value statistics methods. To check for independence of the sample values, Fig. 13.4 shows the calculation of the autocorrelation coefficient (cf. Sect. 5.4.1.1). As the vertical lines remain within an interval—bounded by the dotted lines—indicating the predominance of a white noise, the observed annual non-stationar can therefore reasonably be considered decorrelated, if not independent (see Sect. 6.5.2.2, Fig. 6.8 for more details). Similarly, for the sample of POT values, the calculation of the autocorrelation does not lead to the rejection of the independent data hypothesis (Fig. 13.5).

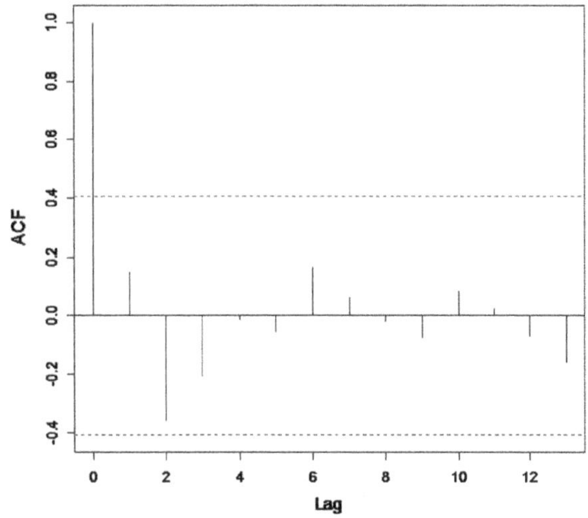

Fig. 13.4 Autocorrelation of the sample of annual non-stationar

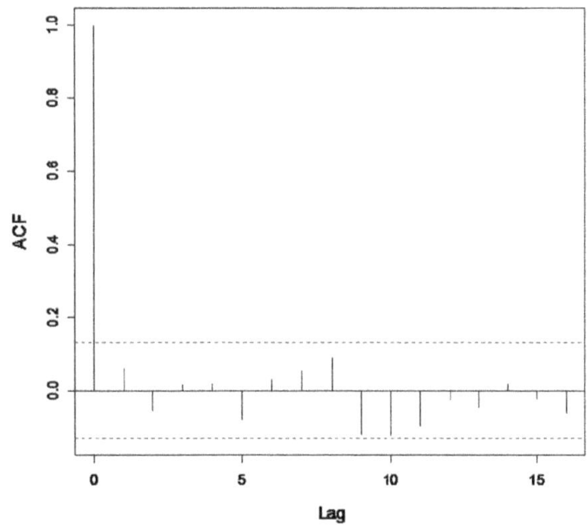

Fig. 13.5 Autocorrelation of the POT sample

13.2.4 Analyzing the Sample Size

The sample size of annual non-stationar is relatively small. This may lead to a high level of estimation uncertainty and thus to particularly wide confidence intervals. Here we recommend paying particular attention to the obtained widths. The size of the POT sample is, however, satisfactory in view of the criteria considered in Chap. 5.

13.3 Statistical Estimation

The two extreme value distributions induced by the sampling methods (GEV and GPD) are estimated here using maximum likelihood, and the confidence intervals on the estimators produced using the delta method (Theorem 4.4).

For the GEV distribution estimated using the sample of annual maxima, the estimated parameters have the following values: $\hat{\xi} = -0.31$, $\hat{\mu} = 8.10$, and $\hat{\sigma} = 1.46$. The value for the shape parameter $\hat{\xi}$ is strongly negative. Because of this, the distribution is bounded by $x_{\max} = 12.80$ m. It is impossible, however, with current knowledge, to justify the statistically estimated upper bound by the existence of a physical bound on the phenomenon. It is simply a feature of the tail of the best-fitting distribution for the data available to us.

This feature appears again when estimating a GPD distribution using the sample of values over the threshold $u = 4$ m. The estimated parameters have the values $\hat{\xi} = -0.15$ and $\sigma = 2.04$ m. Because of the negative value of the shape parameter, this distribution is bounded by $x_{\max} = 17.60$ m. The confidence intervals obtained in both situations are coherent with each other. However, the much larger number of data in the latter sample means that its estimate of ξ is more pertinent.

13.4 Tests and Graphical Validation

The statistical distribution chosen for the study needs to be validated using statistical tests and an empirical analysis of the results. We thus apply the Kolmogorov–Smirnov test to the sample of block non-stationar (cf. Example 4.4) using the `ks.test` function in `R`, in order to test the goodness of fit to a GEV distribution with parameters $\mu = 8.10$ m, $\sigma = 1.46$ m, and $\xi = -0.31$. For a type I error of 5%, the test statistic is $R_n = 0.112$ and the p-value is 0.902. The goodness-of-fit hypothesis is, therefore, not rejected at the 5% level. This result is corroborated by the same test applied to the threshold exceedance data for a GPD distribution with parameters $u_0 = 4$ m, $\sigma = 2.04$, and $\xi = -0.15$, giving

$$R_n = 0.049,$$
$$p\text{-value} = 0.650.$$

The plots shown in Fig. 13.6 help visualize the quality of the fit of the parametric distribution to that of the sample of annual *maxima*. The top two plots illustrate the fit with respect to a line of reference, indicating that the data are globally well modeled, including the most extreme ones. The return level plot (bottom left) shows modeled and observed return levels as a function of annual return periods. We see that all of the points are located inside the confidence interval. Finally, the last plot (bottom right) contrasts the theoretical distribution with the empirical one.

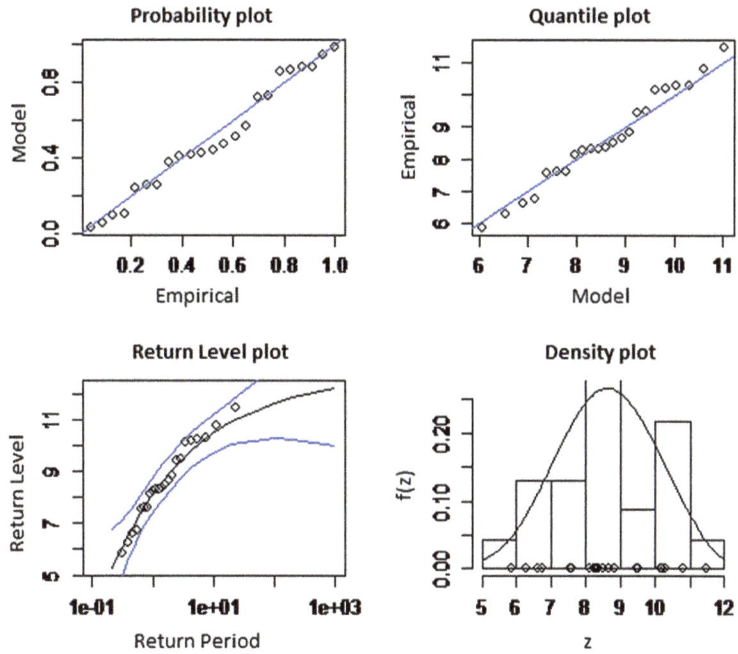

Fig. 13.6 Goodness of fit of the GEV distribution with respect to the distribution of the sample of annual non-stationar

Figure 13.7 gives the results for the sample of POT values. The fit of the GPD distribution to the data appears better than the GEV one. We remark also that the confidence intervals around the GPD distribution are not tighter than those around the GEV one. The effect of the small sample size of annual non-stationar is offset by the fact that the GPD distribution has a less negative shape parameter, meaning that the GEV distribution is more tightly bounded. Given the difficulty in fitting this parameter, and its particularly high absolute value, its value obtained for the GEV distribution is quite uncertain and probably not reliable.

13.5 Results and Critical Analysis

In this case study, the variable of interest is principally the significant height z associated with a required probability level, here for a centennial return period. According to the GEV fit, we obtain an estimate of the significant centennial height z_{100} of 11.67 m, with an upper bound to the 70% confidence interval of 12.40 m. As for the GPD fit, with an average of $\mu = 9.83$ values selected per year above the threshold $u = 4$ m, we obtain an estimate of the significant centennial height z_{100} of 12.67 m, and a upper bound to the 70% confidence interval of 13.90 m.

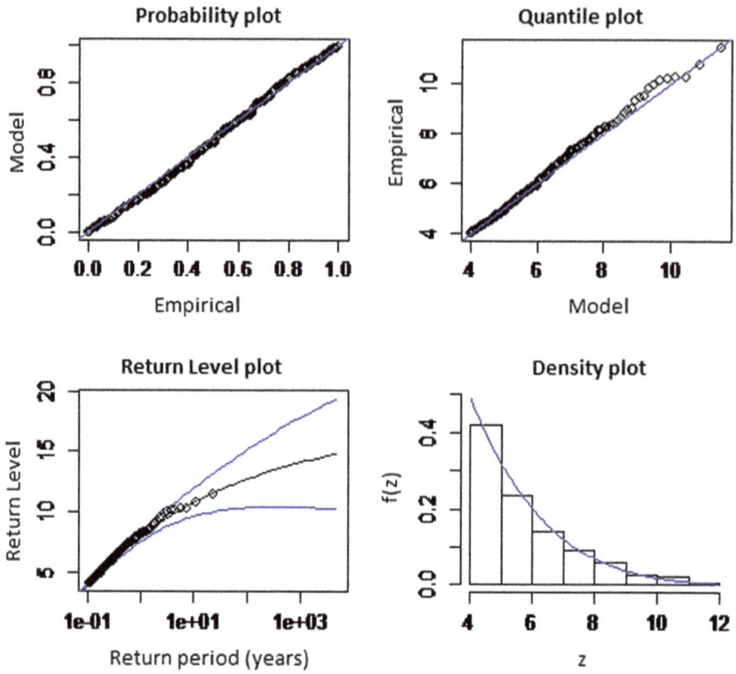

Fig. 13.7 Goodness of fit of the GPD distribution to the set of significant heights over 4 m

The generally good fits obtained mean that we can approve—in this initial analysis—the choice of the POT approach and the results associated with it. The main factor in this decision is the much larger dataset available. However, before definitively validating the results, it is advisable to carry out a more precise study of the use of the GPD distribution, in particular, by refining the level of significance (or type I error risk) of the statistical tests being used. In effect, the choice of a 5% level is now being challenged in a growing number of settings (cf. Sect. 4.1.3), and lowering it to 1% or even 0.1% can help better evaluate a phenomenon's properties based on the data (by testing the independence and stationarity hypotheses more comprehensively) and progressively move toward more realistic models. In this sense, we consider that the results from this initial study are an illustration of the "historical approach" to extreme value statistics, which is still nevertheless widely used in certain engineering communities.

Chapter 14
Predicting Storm Surges

Marc Andreewsky

Abstract In this chapter, we present a second example of statistical estimation of extreme quantiles: millennial quantiles for storm surges at Brest (France), based on hourly sea-level measurements. We run sensitivity tests on parameter values, on the choice of analytic models for distributions, and on the statistical estimation methods chosen. Uncertainty in estimates and associated confidence intervals are also calculated and compared.

14.1 Data Collection and Preprocessing

14.1.1 Source Data

The available data is an hourly record of the sea level at Brest (Figure 0.1) over the period 1846–2010, with an effective duration of observation of 150.25 years. This data is the property of the French Navy and SHOM*, and is accessible at the REFMAR website.[1] The quality of the data was studied in a recent Ph.D. thesis [625], among others. This study also takes advantage of *retro-predictions* of tide levels obtained using the software PREDIT, which originated from a collaboration between SHOM and EDF.

[1] http://refmar.shom.fr.

M. Andreewsky (✉)
EDF R&D, Palaiseau, France
e-mail: marc.andreevsky@edf.fr

N. Bousquet and P. Bernardara (eds.), *Extreme Value Theory with Applications to Natural Hazards*, https://doi.org/10.1007/978-3-030-74942-2_14

14.1.2 Constructing Storm Surges Samples

14.1.2.1 Calculating Open Sea Storm Skew Surges

Recall that high-tide (HT) exceedance (*storm skew surge*) is defined as the difference between the observed and predicted sea levels at open sea. As we are talking about the difference, it must be taken into account that, due to timing errors, the zero points of the phases of the two curves are not necessarily the same. This difficulty is illustrated in Fig. 5.4.

Calculation of the HT exceedance occurs as follows: from hourly tide measurements, a quadratic polynomial is constructed, and its maximum M_1 is recorded. Since the theoretical maximum M_2 is known, the HT exceedance is simply defined as $M_1 - M_2$. Remember that M_2 is calculated without taking into account eustatic change, which will also need to be dealt with (cf. Example 5.21).

The use of HT exceedances has the advantage of simplifying the dependence model between storm surges and the tide itself, and limits the risk of error due to the phase shift mentioned above. However, information on exceedances is only available at open sea. We therefore have less data that if we calculated the hourly exceedances. Nevertheless, those hourly exceedances can be calculated with high bias due to this phase shift.

At sites with a high *tidal range** like Brest, exceedance/tide dependence and the error linked with the phase shift cannot be ignored. One way to deal with these is to simply work with HT exceedances, which is the choice we make here. The HT exceedances and the tide levels themselves can be considered independent in this study because open sea levels are always relatively large (unlike in the Mediterranean Sea, for example). Furthermore, the risk of coastal flooding is a priori only high at open sea. The HT exceedance variable has therefore several benefits.

14.1.3 Accounting for Eustatic Change

The definition of the mean annual sea level (useful for calculating eustatic change) we adopt here comes from the *Permanent Service for Mean Sea Level*[2] (PSMSL). To calculate this, temporal means of measured levels are used, based on daily, monthly, and annual periods. These means are calculated while accounting for missing data.

Variation in the mean annual sea level (eustatic change) is principally linked to variations in the climate, movements in the Earth's crust, atmospheric pressure, and ocean currents. However, the tide retro-prediction software PREDIT does not take eustatic change into account: tides are retro-predicted around a constant (and current) mean annual level. Eustatic change must therefore be taken into account when calculating storm surges if we do not want to add bias to the calculations between theoretical and measured tide levels.

---

[2] http://www.psmsl.org/data.

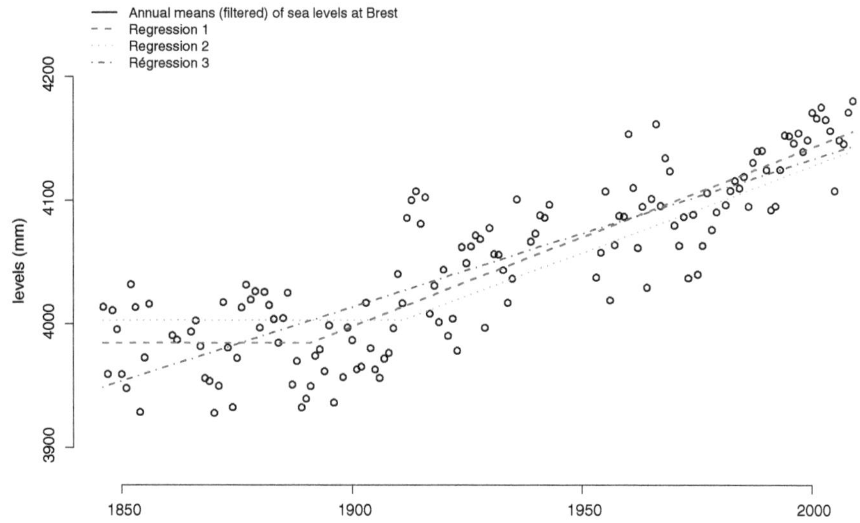

Fig. 14.1 Four models of eustatic change tested over storm surge Brest data. Calibration based on CAL1: annual means, CAL2: linear regression over the annual means, CAL3: a constant followed by linear regression, and CAL4: two optimized linear regressions

Data has been recorded at Brest over largely more than 50 years (actually, 150 years), which is a sufficiently long period to observe whether a clear eustatic trend emerges—which it does here. It is therefore possible to model variation in mean sea level in order to remove it from measured levels. Four calibration methods, illustrated in Fig. 14.1, have been tested for correcting the hourly levels of the eustatic change:

CAL1 : setting based on annual means;

CAL2 : setting based on regression over the annual means;

CAL3 : setting based on a continuous piecewise linear regression in which the first segment is constrained to have zero slope. The junction between the two segments is chosen visually; (this selects the year 1892)

CAL4 : a setting method similar to the previous one, except that the first segment is constrained to have a non-negative slope, and the junction between the two segments is found by minimizing the quadratic error (this selects the year 1915).

In the end, a maximal difference of only 4 cm is obtained between the values associated with the empirical quantiles of the storm surges corrected of eustatic change across the four methods. The choice of method therefore has little influence on the extreme values. We have chosen to use CAL3 in this study, for which the change in slope occurs in 1892. Its second linear segment (after 1892) has the equation:

$$Y = 1.45X + 1251.05 \quad \text{(mm)}.$$

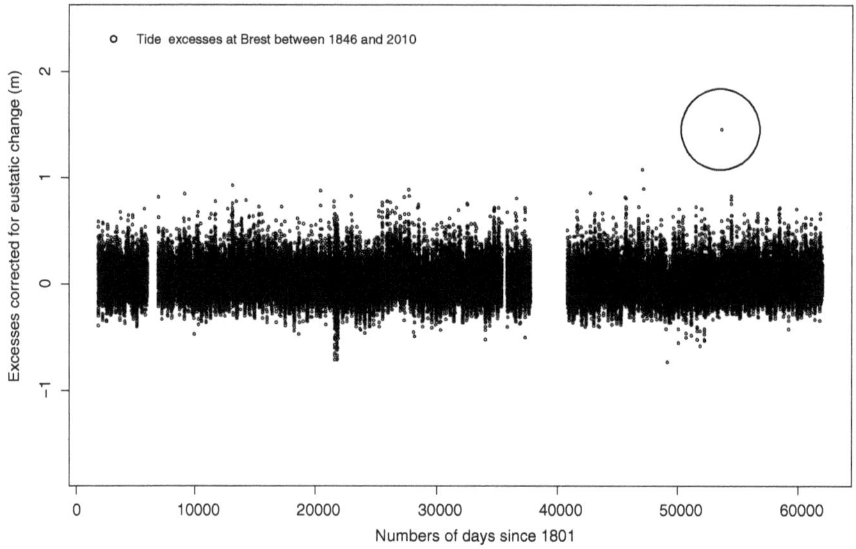

Fig. 14.2 Storm skew surges at Brest since 1846 after accounting for eustatic change using the CAL3 method. The circled exceedance is related to the severe storm that occurred on 16 October 1987

According to this model, an increase of 1.45 mm occurs each year after 1892, leading to a shift of about 15 cm over the 20th century. The sample of eustatic-corrected exceedances is displayed in Fig. 14.2.

14.2 Methods

14.2.1 Sampling Methods

As described earlier in this book, two sampling methods are advocated in extreme value theory: peaks over threshold (POT) and block maxima (MAXB, usually blocks of one year). The difference between the two is the way in which they select extreme values.

1. Under the POT approach, the sample used in the current study is constructed as follows:

   - only observed values greater than or equal to a predefined threshold u_0 are taken into account;
   - storms (characterized by one or several high tides above the chosen threshold) are delimited in time using some *redescent* criterion (cf. Example 5.20) on the

storm surges; a storm ends when the value of the HT exceedance drops below some threshold (e.g., $u_0/2$);
- if during a storm we observe several successive storm surges above the chosen threshold, we retain only the maximal value among them (*declustering*);
- the resulting sample studied is intimately linked to the predefined threshold. As this choice suffers from arbitrariness and uncertainty, we test several values, keeping in mind that a typical year has 2–10 storms. In the end, however, it is the total number of storm surges retained that is of most importance (for Brest, even though we have less than two values per year, the fact that the time series is quite long means that the final number of exceedances will be quite high).

2. Under the MAXB approach, the sample used here is constructed as follows:

- the original sample is divided up into *maritime years* (1 August–31 July) mainly so that the whole of the northern hemisphere winter is contained in the same year[3];
- years with too many missing data are removed (the criterion used here is: more than 20% missing data over a year and/or more than 10% missing data in winter), except when we are sure that a year's maximum has been recorded;
- the final sample to be fitted is then simply the set of annual maxima.

14.2.2 Statistical Distributions and Tests

In the context of this study, the distributions tested and fitted by maximum likelihood are shown in Tables 14.1 and 14.2.

A certain number of tests, described in Sects. 4.1.3 and 5.4.3.2, can be used to check data quality, help choose the threshold (in the POT case), and look closer at the relevance of the parameters estimated for the fitted distributions in each setting (POT and MAXB):

- *Testing stationarity*: we use a test for equal means and equal variance.
- *Uniform distribution test*: this tests whether the dates corresponding to peaks in the sample are uniformly distributed over the period of observation.
- *Testing for independence*: for testing the independence of successive peaks over the threshold, autocorrelation coefficients between these selected events are estimated (cf. Sect. 5.4.1.1). Recall that the criterion used to select storm surges in the POT setting is as follows: two exceedances close in time are both retained if and only if they are both over the threshold u_0 and, between them, there is another exceedance below $u_0/2$ (i.e., *redescent* criterion—see Example 5.20).

[3] Most storms that hit France occur in winter.

Table 14.1 Distributions used to fit the POT sample

Distribution	Cumulative distribution function	Comments
GPD	$F(x) = 1 - \left\{1 - \frac{\xi}{\sigma}(x - u_0)\right\}^{1/\xi}$	Advocated by extreme value theory
Exponential	$F(x) = 1 - \exp(-\rho\{x - u_0\})$	Special GPD case for $\xi = 0$
Weibull	$F(x) = 1 - \exp\left(-\rho\{x - u_0\}^{\beta}\right)$	β is a shape parameter

Table 14.2 Distributions used to fit the MAXB sample

Distribution	Cumulative distribution function	Comments
GEV	$F(x) = \exp\left(-\left\{1 - \frac{\xi}{\sigma}(x - \mu)\right\}^{1/\xi}\right)$	Advocated by extreme value theory
Gumbel	$F(x) = \exp\left(-\exp\left(-\frac{1}{\sigma}\{x - \mu\}\right)\right)$	Special GEV case for $\xi = 0$

- *Chi-2 test*: to run the POT method and look for the most appropriate threshold to use, a χ^2 test (among others) can be used, measuring the relevance of the estimated probability distribution. This test is in addition to a visual analysis of the fit.

14.2.3 Evaluating Data and Results in Practice

Below we list several criteria for deciding whether to retain or remove obtained samples, based on the tests listed above, in the MAXB and POT settings:

- the number of extreme values selected to be used in the fit must be greater than or equal to 20 at the very least;
- the mean number of extreme values per year needs generally to be 2–8, perhaps stretching to 1–20. However, in the end, it is the total number of events retained which is more important;
- the p-value associated with the occurrence of peaks with respect to a Poisson distribution should be at worst 5%, though, given the specific case, it could be set as small as 0.1%;
- tests for stationarity, independence of peaks, and uniform distribution of peak dates, are accepted with very small error risks $\alpha \ll 1$, ideally following the recommendations in [417] (in general, values of 5% and 10% are still commonly used in the literature).

In addition, several exclusion and acceptance criteria for estimated distributions can be used, which depend in particular on the confidence level chosen to define confidence intervals over estimated functions of interest (cf. Sect. 4.13). In this study, this is set at 70%, a typical choice for dimensioning (see for example, [15]). Depending on the project, it may be desirable to increase this value to 90% or even 95% [321]. The criteria are as follows.

- We set the risk associated with the χ^2 goodness-of-fit test to a low threshold β (e.g., if $\beta = 10\%$, 10 equiprobable classes are taken into account in the calculations). The table of χ^2 fractiles then tells us that the value of the test statistic needs to be less than 14.7 so that we have a one over ten chance of erroneously concluding that the distribution of the data and that of the proposed fit are not the same.
- One exclusion criterion is based on characteristics of the graphical representation of the fitted distribution: if it has an empirical tail distribution (the last 4 or more values) outside the 70% confidence interval and pointing in a radically different direction to that of the distribution being fitted, the estimated distribution is not retained.
- One criterion of admissibility is that the value of the χ^2 criterion (4.2) associated with the fit is small or minimal for the selected threshold.
- If in a neighborhood of the selected threshold, the plots with respect to the centennial (or millennial) threshold and the parameters of the fitted distributions are relatively stable, these add credibility to the choice of threshold.
- Finally, if the graphical representation of the fitted distribution shows that a large number of observations are contained within the 70% confidence interval—in particular for the extremities of the tail—it also adds credibility to the fit.

14.3 Estimation Using Annual Maxima

The MAXB approach consists of selecting the largest storm surge of each year and fitting a GEV or Gumbel distribution to the resulting sample. Here, the test for stationarity is accepted for risks of 10% and 20% (remember this is Student's t-test of equality of the means) for all time intervals. The fits obtained are shown in Fig. 14.3. Note that the fits associated with the GEV (top) and Gumbel (bottom) distributions are almost identical. The millennial return level estimation results are shown in Table 14.3.

In Table 14.3, we can see several differences in the point values, depending on the distribution. We also note that the differences seen between the two distributions tested for a given quantile are smaller than the lengths of the estimated confidence intervals. The fits were obtained using maximum likelihood, and the parameter estimates are shown in Table 14.4. We further note that the χ^2 values do not contradict these fits.

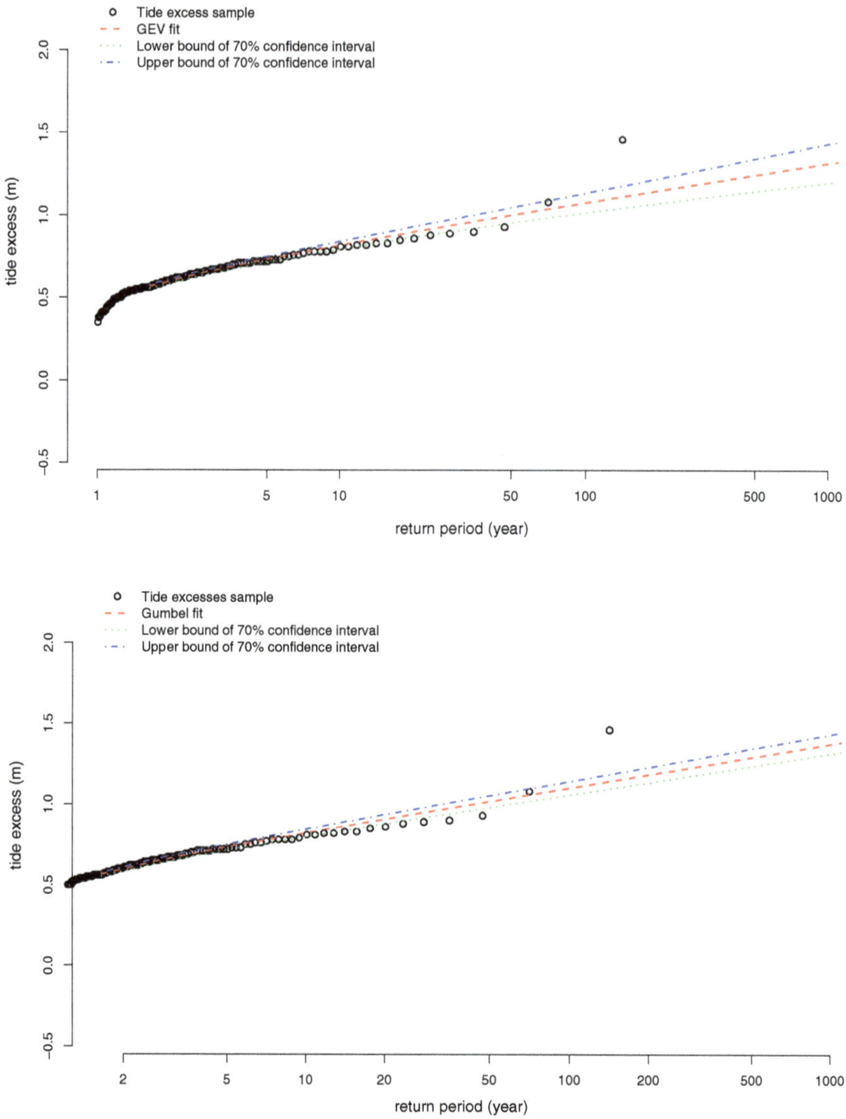

Fig. 14.3 GEV (top) and Gumbel (bottom) fits along with 70% confidence intervals

Table 14.3 Millennial return level and its 70% confidence interval, and length of this confidence interval (CI), for the GEV and Gumbel distributions under the MAXB approach

Distribution	70% lower bound (m)	Millennial level (m)	70% upper bound (m)	Length of CI (m)
GEV	1.2	1.32	1.44	0.24
Gumbel	1.32	1.38	1.44	0.12

Table 14.4 Parameter estimates for the GEV and Gumbel distributions along with p-values of the χ^2 test for a 10% significance level

Distribution	Parameter	Values
GEV	μ	0.56
	σ	0.12
	ξ	0.024
	p-value χ^2	10.0
	μ	0.56
Gumbel	σ	0.12
	p-value χ^2	9.71

14.4 Estimating Storm Surges Using POT Approach

14.4.1 Search for Optimal Thresholds

Three distributions are estimated from a sample of independent exceedances (storm surges) over some threshold (i.e., the POT approach): the theoretically based GPD distribution and two other distributions considered statistically reasonable based on the data histogram and the historical treatment of such data, even though they are at odds with the theory (cf. Sect. 3.2.1)—the Weibull and exponential distributions. Their historical use in maritime hydrology is reflected in their existence in traditional software tools like ASTEX [433], developed by LNHE. We conclude that the redescent criterion (introduced in Example 5.20) allows us to claim that the two selected exceedances come from different storms. For HT exceedances, we consider that successive storms are separated by at least 24 hours.
For each distribution tested, we look for an optimal threshold by minimizing the χ^2 criterion (4.2), by using an iterative approach. The candidate optimal thresholds for each are as follows:

- the best χ^2 for the exponential distribution is obtained for a threshold of 0.63 m, with 96 exceedances selected above it;
- the best χ^2 for the GPD distribution is obtained for a threshold of 0.60 m, with 129 exceedances selected above it;
- the best χ^2 for the Weibull distribution is obtained for a threshold of 0.64 m, with 85 exceedances selected above it.

In all three cases, stationarity, independence, and uniform distribution hypotheses are tested and validated, as the Poisson distribution. Moreover, the differences between the optimal thresholds are very small, and for each, the number of exceedances selected is relatively high (which is a positive point in general in terms of reliability of the fitting results). We also wish to check, in addition to a visually acceptable fit, whether the results are relatively stable across the board for the millennial quantiles, upper and lower bounds of the 70% confidence intervals, and parameters around the candidate optimal thresholds, for the three distributions tested (Figs. 14.4, 14.5, 14.6, 14.7, 14.8 and 14.9).

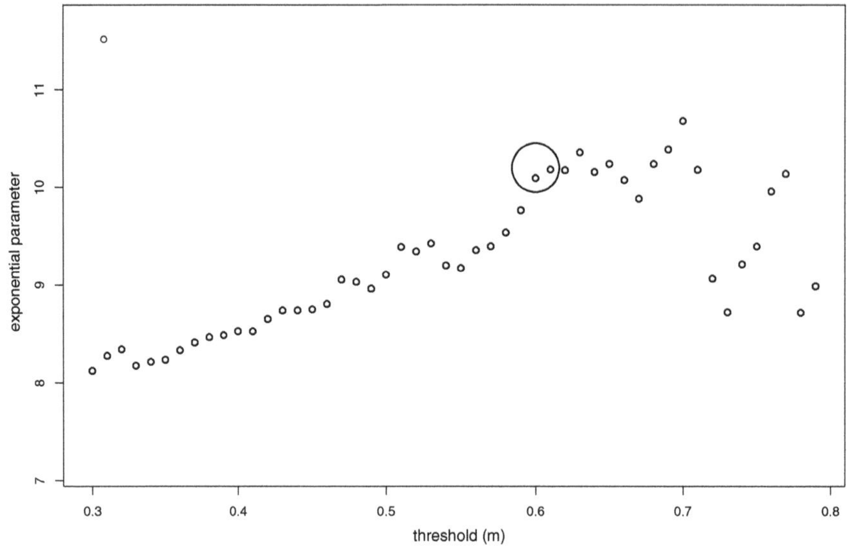

Fig. 14.4 Variations of the exponential parameter as a function of the threshold. The circle indicates the zone in which the threshold associated with the optimal χ^2 is found for this distribution

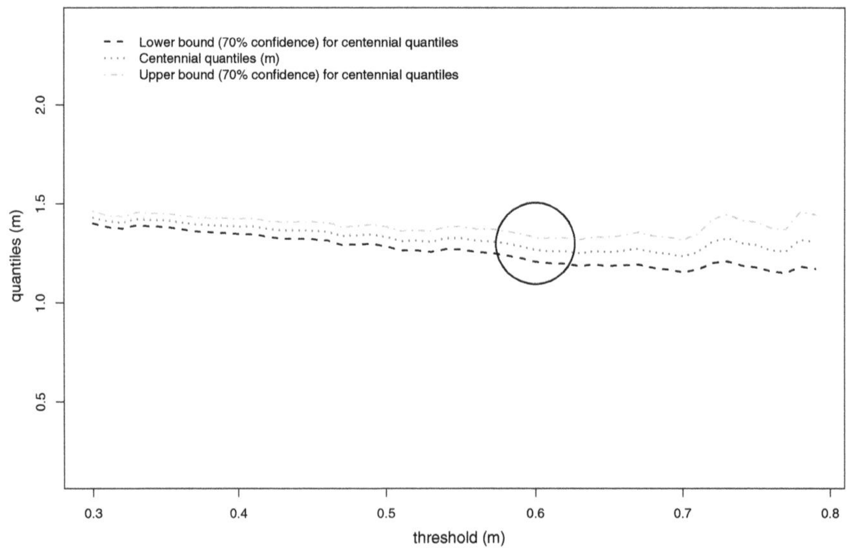

Fig. 14.5 Variations in the upper and lower bounds of the point estimate and its 70% confidence interval as a function of the threshold for the exponential distribution. The circle indicates the zone in which the threshold associated with the optimal χ^2 is found for this distribution

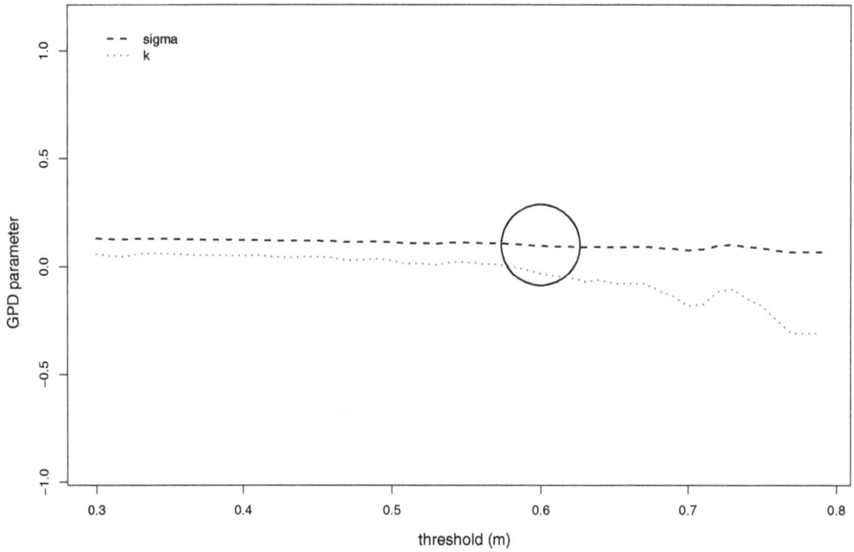

Fig. 14.6 Variation in the GPD distribution's parameters as a function of the threshold. The circle indicates the zone in which the threshold associated with the optimal χ^2 is found for this distribution

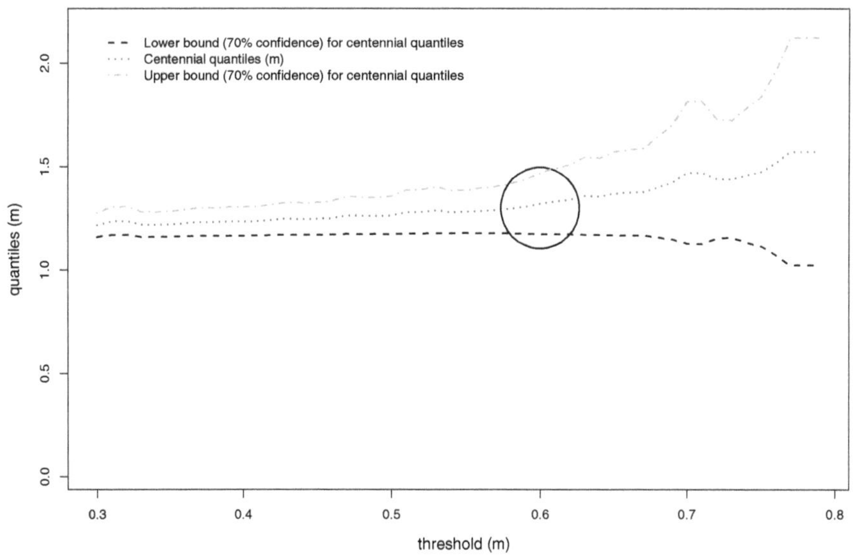

Fig. 14.7 Variation in the upper and lower bounds of the point estimate and its 70% confidence interval as a function of the threshold for the GPD distribution. The circle indicates the zone in which the threshold associated with the optimal χ^2 is found for this distribution

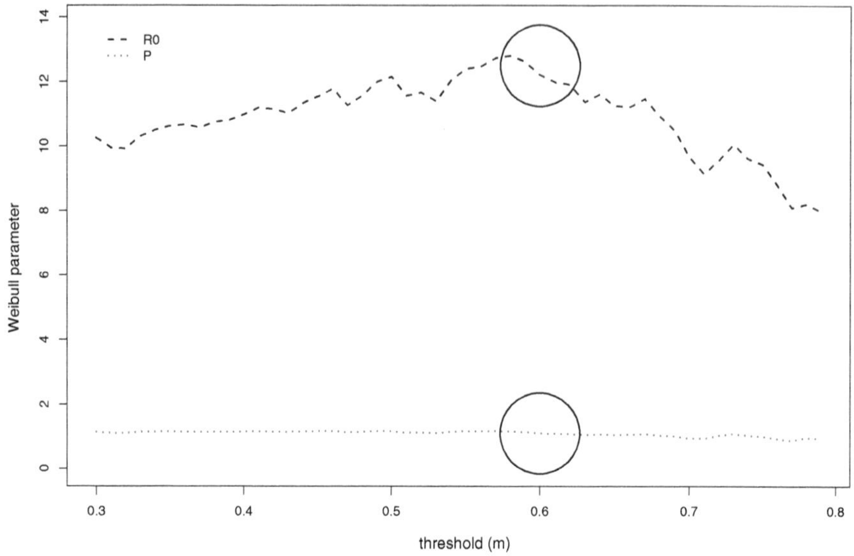

Fig. 14.8 Variation in the Weibull distribution's parameters as a function of the threshold. The circle indicates the zone in which the threshold (=0.64) associated with the optimal χ^2 is found for this distribution

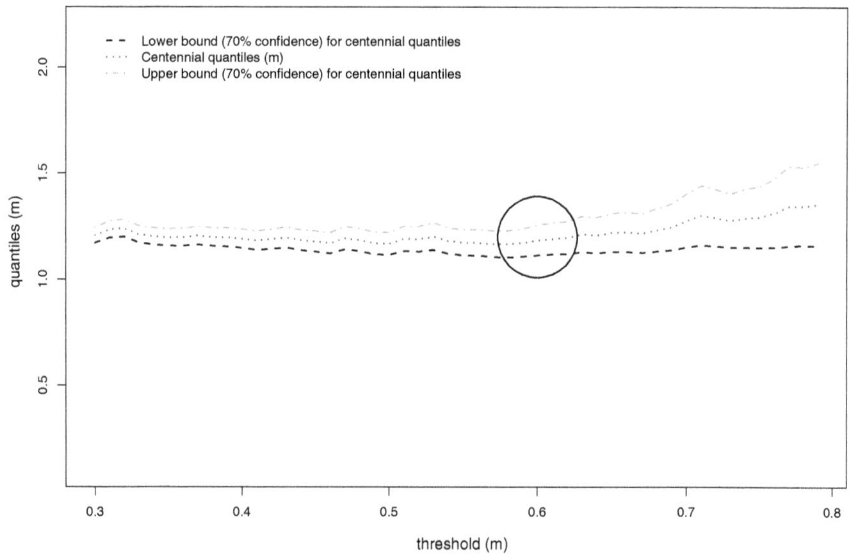

Fig. 14.9 Variation in the upper and lower bounds of the point estimate and its 70% confidence interval as a function of the threshold for the Weibull distribution. The circle indicates the zone in which the threshold associated with the optimal χ^2 is found for this distribution

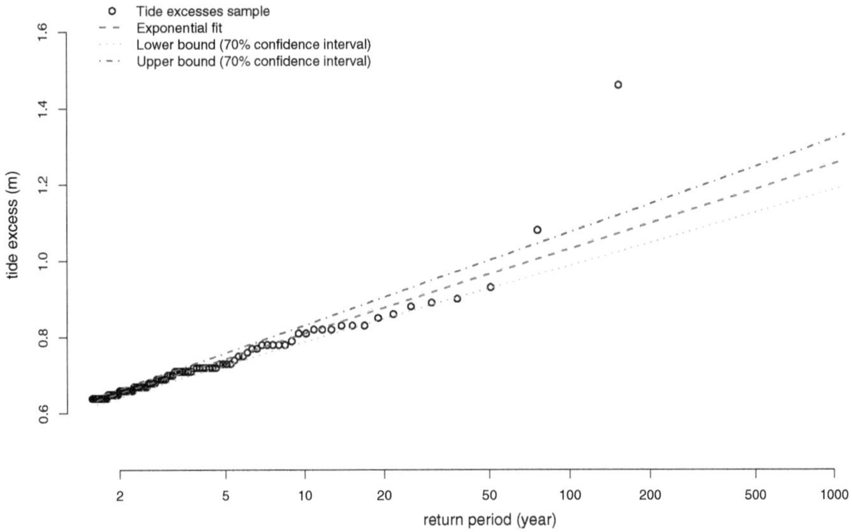

Fig. 14.10 The exponential distribution fit (threshold = 0.63 m)

In each case, there is no sign of high instability around the optimal threshold candidates for either the lower or upper bounds of the 70% confidence interval, quantile values, or parameter values (if this had not been the case, we could of course have considered other candidate thresholds). The fits obtained for the distributions tested are shown in Figs. 14.10, 14.11, and 14.12. We see in these one exceedance located far from the 70% confidence intervals, corresponding to the major storm from October 1987 mentioned earlier. Apart from this outlier, the fits are relatively good (the other points are inside the confidence intervals—themselves not overly wide).

Ultimately, the POT approach is broadly relevant here, as indicated by the robust estimates obtained. The optimal thresholds and parameter estimates for each distribution are shown in Table 14.5.

14.5 Conclusion and Discussion

In this example, the estimated quantiles (return levels) were essentially the same, independent of which fitting method we used, applied to either the distribution from the theory or from other "expert" distributions. But these results are not generic. The selection of a quantitative approach (the most conservative or otherwise) must always be conditioned on the importance of the installation requiring protection; a "naturally" conservative approach (powered by control authorities in France as in other countries) would be to choose the upper bound of the associated confidence interval in question.

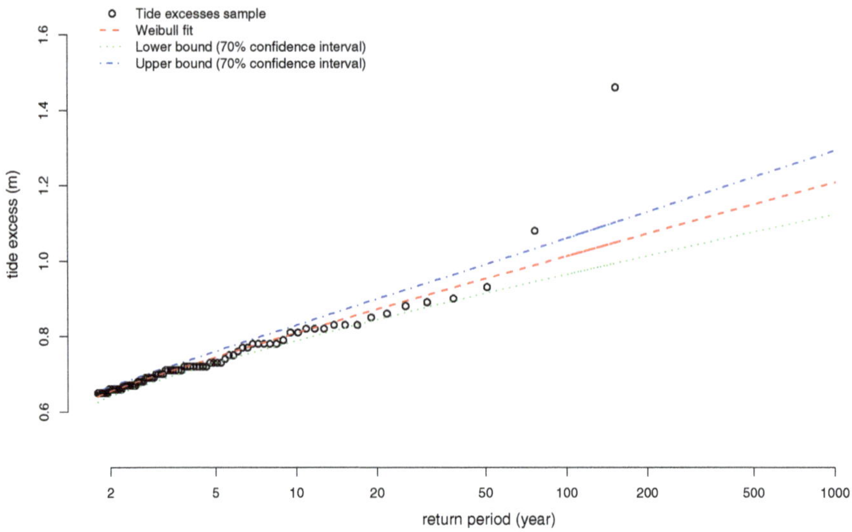

Fig. 14.11 The Weibull distribution fit (threshold = 0.64 m)

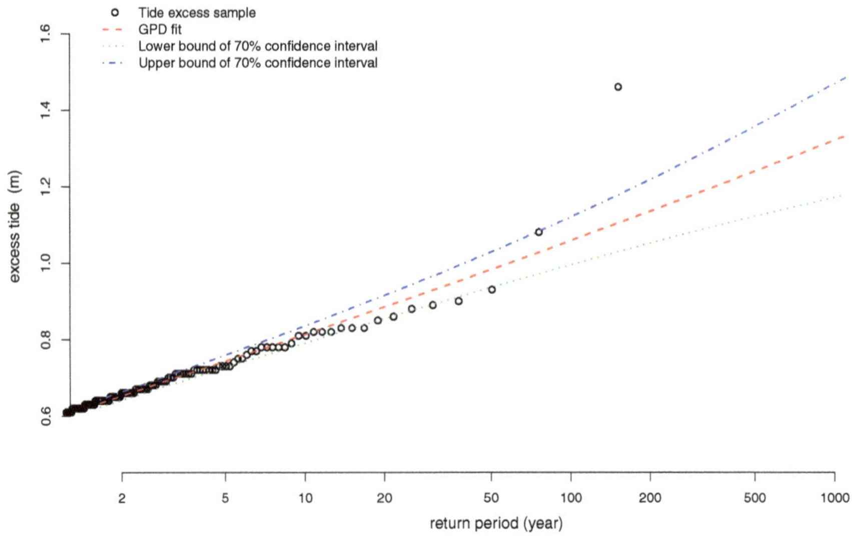

Fig. 14.12 The GPD distribution fit (threshold = 0.60 m)

Table 14.5 Optimal thresholds and parameter estimates for the GPD, Weibull, and exponential distributions

Distribution	Threshold (m)	Parameters		Lower limit 70% (m)	Millennial level (m)	Upper limit 70% (m)
GPD	0.6	σ	0.096	1.21	1.27	1.33
		ξ	−0.031			
		u_0	0.86			
Weibull	0.64	ρ	11.6	1.12	1.21	1.29
		β	1.07			
		u_0	0.57			
Exponential	0.63	ρ	10.36	1.19	1.25	1.32
		u_0	0.639			

Note also that the 1987 storm event, sitting outside the calculated confidence intervals, is a red flag to some experts, leading to doubts as to the pertinence of the model fitting. Regional analysis (see Chap. 7) can help provide answers in this type of situation.

Finally, note that the plots shown in this chapter, linked to statistical tests, are driving factors in decision-making (accepting or not an estimated statistical distribution), but can also be supplemented by further distribution selection tools, e.g., *QQ-plots* (cf. Fig. 6.6) and *mean exceedance plots* [333] which displays the evolution of $X - u|X \geq u$ as a function of u. As indicated in Proposition 6.4, the latter is a linear function of u when $X|X \geq u$ is a GPD distribution.

Chapter 15
Forecasting Extreme Winds

Sylvie Parey

Abstract This chapter considers the modeling and forecasting of extreme wind values. This one-dimensional problem is dealt with in depth and provides a third example of the deployment of the "classic" methodology for the treatment of natural extreme values.

15.1 Context of the Study

15.1.1 Study Variable and Probability Levels

The focus of this study is the maximal instantaneous wind speed in Nantes (Loire-Atlantique department of France, cf. Fig. 15.1) at an altitude of 10 m. We are looking to estimate the 50-year probability level, i.e., a return level of 50 years, along with the upper boundary of its 95% confidence interval. The variable of interest is exactly the same as the study variable, so it can be investigated directly.

15.1.2 Collecting Data

Data source. In general, wind studies in France involve data recorded at Météo-France weather stations, which are spread out across the country. The data includes the daily maximal wind speed and the hour in which it occurred. Our study will use this daily maximal wind speed data. The Nantes weather station data covers the period 1 January 1972–31 December 2003, i.e., 32 years or 11,688 days, and in particular contains the (non-aberrant) data for 26 December 1999: Cyclone Lothar. As this dataset is quite complete, there is no need to include other sources of information.

S. Parey (✉)
EDF R&D, Palaiseau, France
e-mail: sylvie.parey@edf.fr

N. Bousquet and P. Bernardara (eds.), *Extreme Value Theory with Applications to Natural Hazards*, https://doi.org/10.1007/978-3-030-74942-2_15

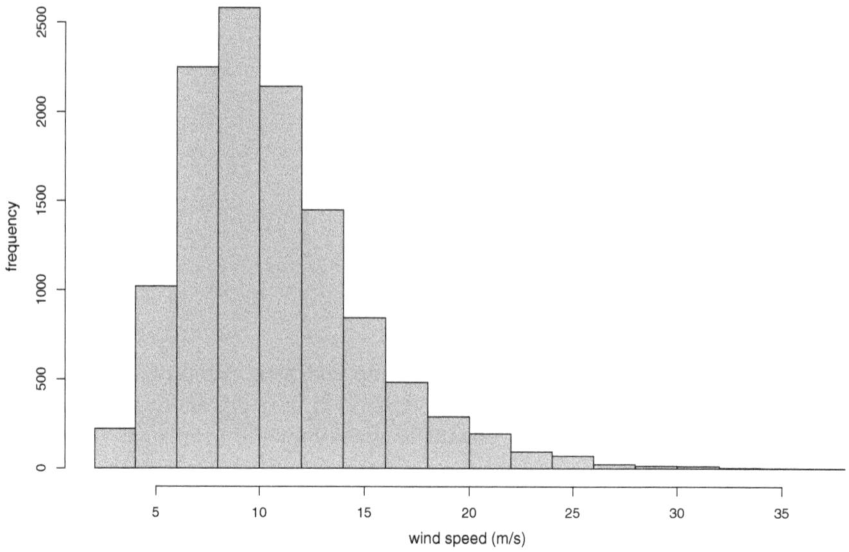

Fig. 15.1 Histogram of the daily maximal wind speed in Nantes

Data preprocessing. The Météo-France dataset has very few missing values—less than 10% across the whole set of weather stations. In fact, for the Nantes station, there are no missing data. Nevertheless, the following steps are for the general case, in which the mean in the data for a given date is assigned to a specific example of that date which has a missing value. Note that when there is missing data, it is important to make sure they do not correspond to dates where extreme events occurred.

The time series of daily maximal wind speeds in Nantes is made up of whole numbers between 3 and 37 m/s. As the statistical theory of extreme values is designed for continuous variables rather than discrete ones, a small continuous noise in the form of a random draw between $-1/2$ and $1/2$ has been added to the values. This is a heuristic known as *jittering* (described on Sect. 5.4.1.3), and it can be seen as a way to compensate for errors due to rounding by the measuring equipment.

In addition, to simplify the process (e.g., in terms of matrix operations), the data for each 29 February has been removed, such that each year has the same number of days.

We can then plot the histogram of observed daily maximal wind speeds after preprocessing, which is shown in Fig. 15.1.

15.2 Extreme Value Sampling

Recall again that two types of sampling are possible: block maxima (MAXB) and threshold exceedance (POT). An analysis of the properties of the subsequently obtained sub-samples (e.g., stationarity, independence, and sample size) must be conducted.

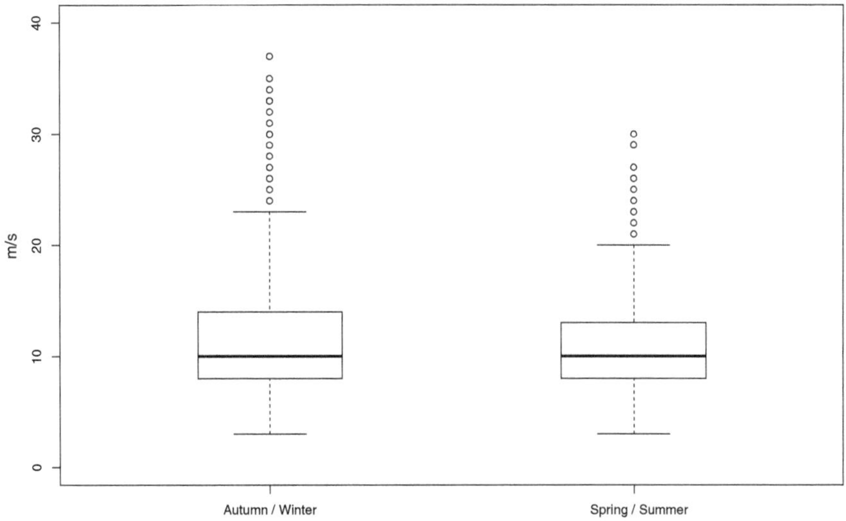

Fig. 15.2 Distribution of the instantaneous daily maximal wind speed per season

15.2.1 *Sub-sample Analysis*

Wind is a seasonal phenomenon, and so is strong wind. A seasonal analysis, shown in Fig. 15.2, confirms that the wind is strongest in autumn and winter. It would therefore be wrong to suppose that extreme wind values are evenly spread across the year; we will thus limit our study to autumn and winter. In order to precisely select the months to be retained, we can look at the distributions of 1–5 annual maxima per month of the year, compared with evenly spread distributions (Table 15.1). It seems reasonable to retain October–March as the months in which strong winds occur.

15.2.2 *Sampling Methods*

Each autumn–winter period straddles two years. In order to work with years which entirely contain each seasonal cluster of extreme values, we shift the definition of a year so that it starts on 1 April and ends on 31 March. Under the MAXB approach, we retain the maximum value obtained each year, which, given the length of the time series, means a total of 31 values. As for the POT approach, it begins by looking for a threshold which best satisfies Pareto distribution properties. In particular, this means linearity in the mean excess as a function of the threshold (see Proposition 6.4) and how constant the shape parameter is.

To study these, we plot (using the `R` package `extRemes`) the evolution of the mean excess and the shape parameter of the fitted distribution for various values of

Table 15.1 Comparison of the distribution of 1–5 maxima with evenly distributed data

Max. num.	Evenly distributed	Jan.	Feb.	Mar.	Apr.	May	June
1	2.67	6	8	5	0	0	0
2	5.33	14	11	8	3	2	1
3	8.00	23	13	10	5	2	1
4	10.67	32	16	10	6	4	1
5	13.33	38	23	13	7	4	1
Max. num.	Evenly distributed	July	Aug.	Sep.	Oct.	Nov.	Dec.
1	2.67	0	0	2	2	3	6
2	5.33	0	0	3	3	7	12
3	8.00	1	0	3	8	10	20
4	10.67	1	0	5	14	13	26
5	13.33	1	0	6	14	20	33

the threshold (Fig. 15.3). In addition, to ensure independence, if consecutive values exceed the threshold, only the largest is retained (*declustering*, cf. Sect. 6.5.2.1). The `mrlplot` function of the `extRemes` package plots the evolution of the excess above different thresholds. We can also recode these diagnoses by accessing and modifying the function's code; here we do this with the aim of zooming in on the highest thresholds. Figure 15.3 appears to indicate that a threshold of 23 m/s would appear reasonable.

15.2.3 Detecting Non-stationarity

Stationarity is a fundamental requirement for fitting extreme value distributions. It is necessary to ensure that the mean and standard deviation remain stable over time. Running the Mann–Kendall test (see Sect. 5.4.3.2) using the `Kendall` package in `R` on the wind data for the autumn/winter period given in terms of the maxima does not identify any significant trend: for a 5% risk, the test does not reject the hypothesis of an absence of trend in the sample, even though the associated plot (Fig. 15.4) seems to indicate a slight increase in the maximal speed over time.

15.2.4 Data Independence Analysis

We now test whether the resulting samples are composed of independent and identically distributed data. Figure 15.5 shows the calculation of the autocorrelation coef-

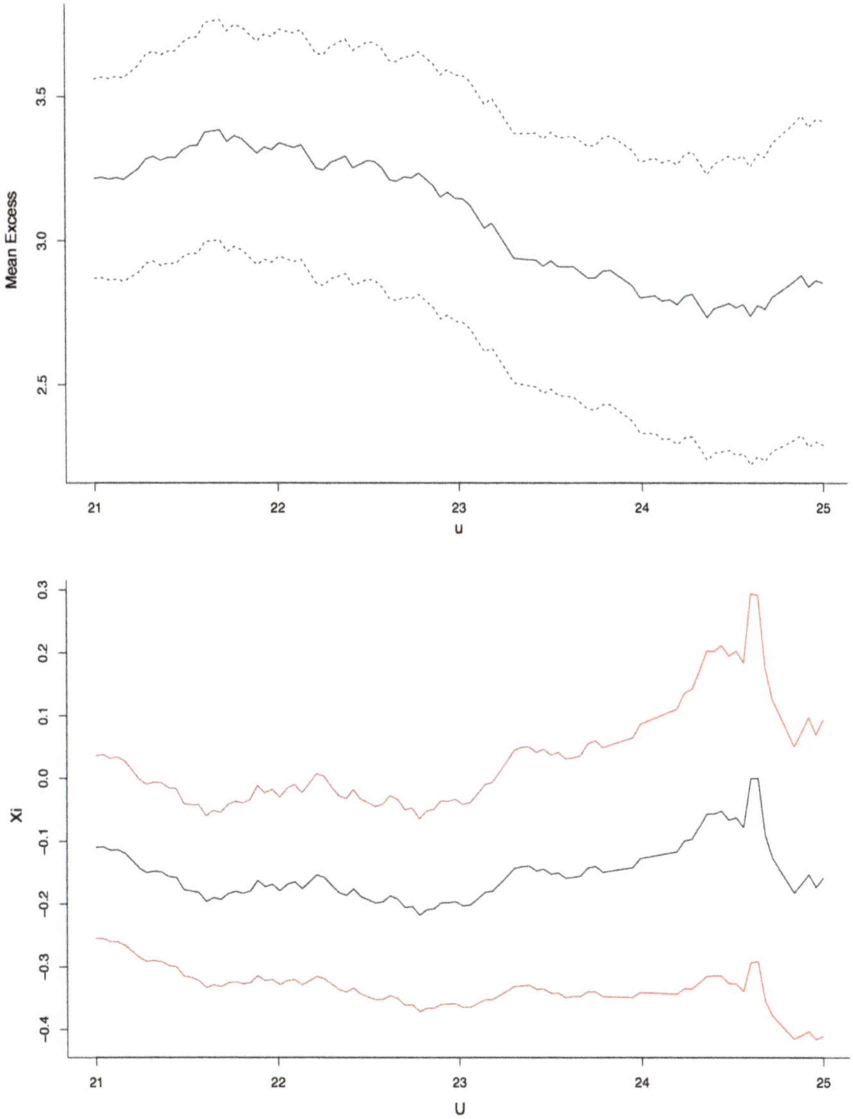

Fig. 15.3 Mean excess (top) and evolution of the shape parameter (bottom) as a function of the threshold

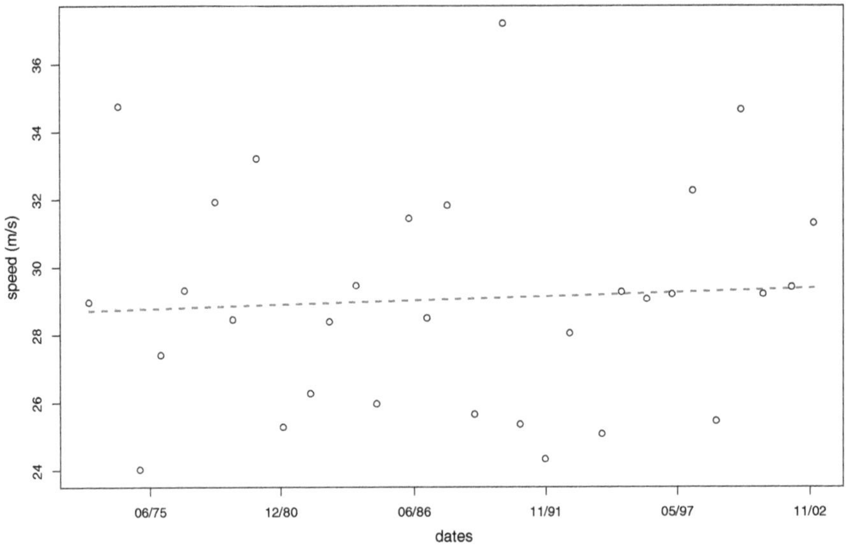

Fig. 15.4 Applying the Mann–Kendall test to the maximal annual wind speeds

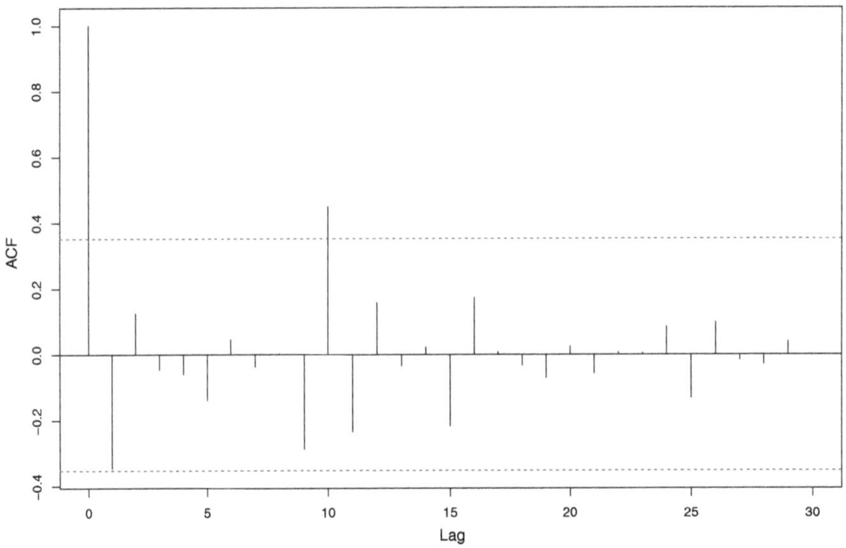

Fig. 15.5 Autocorrelation of the block maxima sample

ficient (Sect. 5.4.1.1) using the R function `acf`. As the vertical lines remain in the confidence interval (dotted lines), the independence hypothesis is not rejected. As for the POT approach with its 130 values over the threshold of 23 m/s, the estimated autocorrelation does not invalidate the independence hypothesis either (Fig. 15.6).

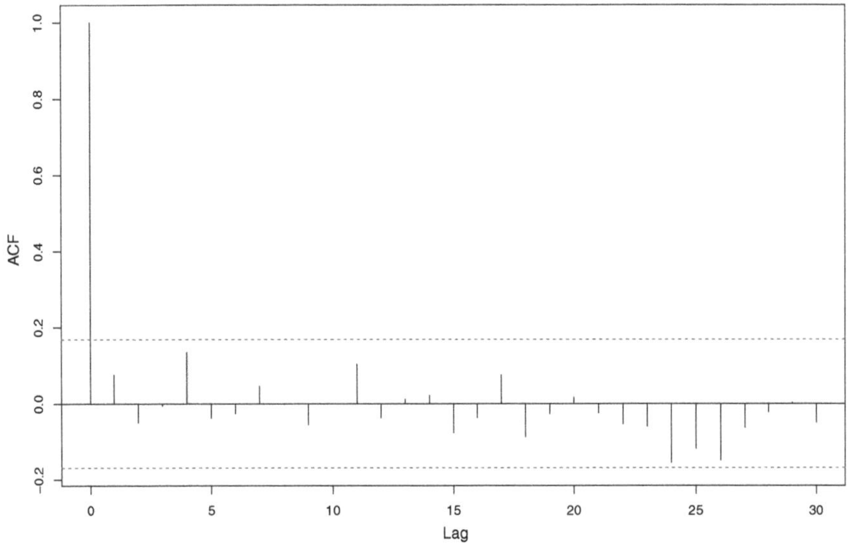

Fig. 15.6 Autocorrelation of the POT sample with a threshold of 23 m/s

15.3 Statistical Estimation

For each of the approaches (MAXB and POT), the parameters of the corresponding extreme value distribution can be estimated using maximum likelihood. As for the confidence intervals for extreme quantiles, we use the delta method (cf. Theorem 4.4); this can be run using the R function `return.level`.

15.3.1 MAXB Approach

For the first fit, the estimated parameters and their 95% confidence intervals are as follows:

$$\begin{aligned}\mu &= 27.4 \quad [26.3; 28.6],\\ \sigma &= 2.75 \quad [1.92; 3.58],\\ \xi &= -0.05 \quad [-0.36; 0.27].\end{aligned}$$

The confidence interval around the shape parameter ξ is quite wide, which is not surprising given the low number of maxima (=31). In addition, it covers 0, which suggests that a Gumbel distribution could be chosen. Indeed, the point estimate for ξ is itself quite close to 0.

15.3.2 POT Approach

As a reminder, according to extreme value theory, if the block maxima sample has a GEV distribution, then the sample of maxima above a threshold u has a GPD one. Estimation of this distribution's parameters can be done using the `fevd` function in the `R` package `extRemes` with $u = 23$ m/s. In the initial fit, the estimated parameters and their 95% confidence intervals are as follows:

$$\sigma = 4.07 \quad [3.10; 5.03],$$
$$\xi = -0.18 \quad [-0.34; -0.01].$$

The shape parameter ξ here is clearly negative, unlike in the previous case. However, remember that with a threshold of 23 m/s, we have 130 independent values above the threshold, whereas before, only 31 values were used. This second fit can therefore be considered more reliable.

15.4 Tests and Graphical Aids

15.4.1 MAXB Approach

The plots shown in Fig. 15.7 help us to visualize the goodness of fit of the parametric distribution to the obtained sample. The plots at the top show the fit with respect to lines of reference; the whole of the sample appears to be well-modeled, including points at the extremes. The bottom-right plot shows the modeled and observed return level as a function of the return period, given on a log scale. We see that all of the data points are contained within the confidence interval.

15.4.2 POT Approach

Here too the goodness of fit can be visualized in the same ways, as shown in Fig. 15.8. Overall, the fit is reasonable, and broadly comparable to the block maxima one.

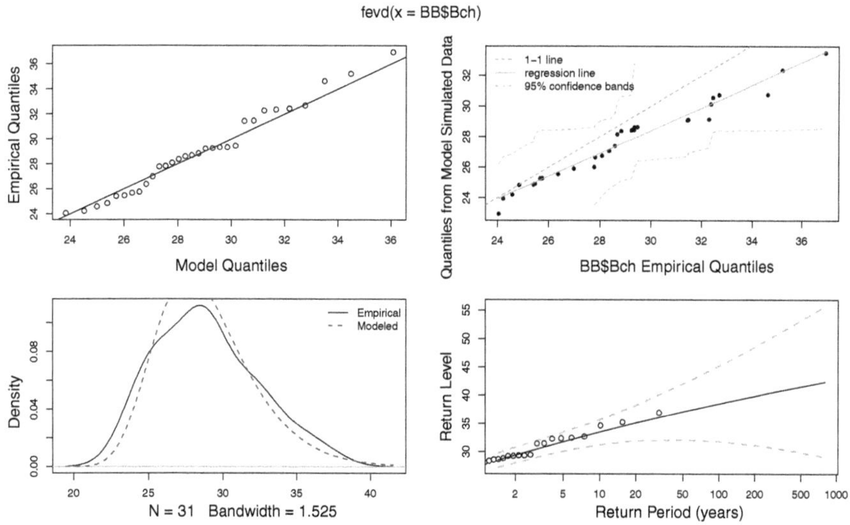

Fig. 15.7 Fitting a GEV distribution to the block maxima sample

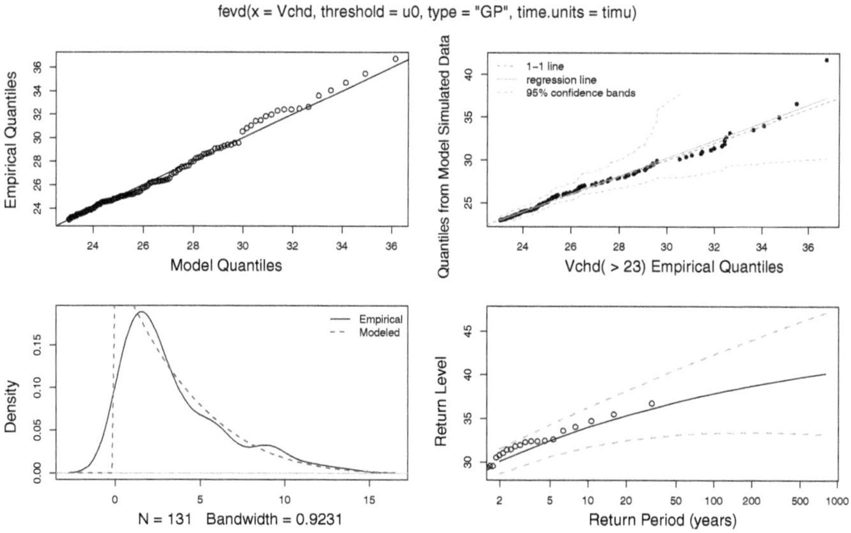

Fig. 15.8 Fitting a GPD distribution to the sample of values over the threshold of 23 m/s

15.5 Results

15.5.1 *Influence of Jittering*

In this investigation, the study variable is similar to the variable of interest, so calculations are straightforward. The results obtained for a 50-year return level and its 95% confidence interval are the following:

	MAXB	POT
Return level (m/s)	37.3	37.0
95% CI	[32.1 ; 42.4]	[33.3 ; 40.7]

The two methods give similar results. However, in order to better understand the impact of *jittering*, i.e., adding a small level of noise to the original discretized data, we use a *bootstrap* approach [244], performing jittering 1000 times and recalculating the return levels and confidence intervals, taking the mean at the end. We obtain the results placed in the following table:

	MAXB	POT
Return level (m/s)	37.0	36.9
95% CI	[32.1 ; 41.9]	[33.2 ; 40.7]

The results from this are very similar to what we originally found. Next, to get an idea of sensitivity to the choice of threshold, we run all of the POT calculations again with a threshold of 22.5 m/s, then 23.5 m/s. Averaging again over 1000 jitterings, we obtain

- for a threshold of 22.5 m/s (which leads to retaining 140 values): 36.7 [33,1 ; 40,3];
- for a threshold of 23.5 m/s (which leads to retaining 120 values): 37.1 [33.2 ; 41.1].

The results are thus little affected by small changes in the choice of threshold.

15.5.2 *Validation*

The highest value observed in this 32-year time series was 37 m/s on 3 February 1990. The maximum speed on 26 December 1999 was 35 m/s. In fact, 98% of the daily maximum speeds are below 25 m/s. Thus, it would appear reasonable that 37 m/s corresponds to a 50-year level.

We have also estimated the wind speeds associated with return levels of $1/p$. However, once the distribution of the maxima has been estimated, nothing stops us from estimating the probability associated with a given maximum, provided that this value is indeed in the domain of the series's non-stationary. For example, for the wind speed variable, we can estimate the return probability associated with the value

Table 15.2 Estimation of the probability that the wind speed exceeds two values (32 and 37.9 m/s) after GEV modeling

	$P(X > 32)$	$P(X > 37.9)$
Estimate	$1.6 \cdot 10^{-1}$	$2.0 \cdot 10^{-2}$
95% CI	$[0.8 \cdot 10^{-1} 2.8 \cdot 10^{-1}]$	$[1.10 \cdot 10^{-4}\ 9.0 \cdot 10^{-2}]$

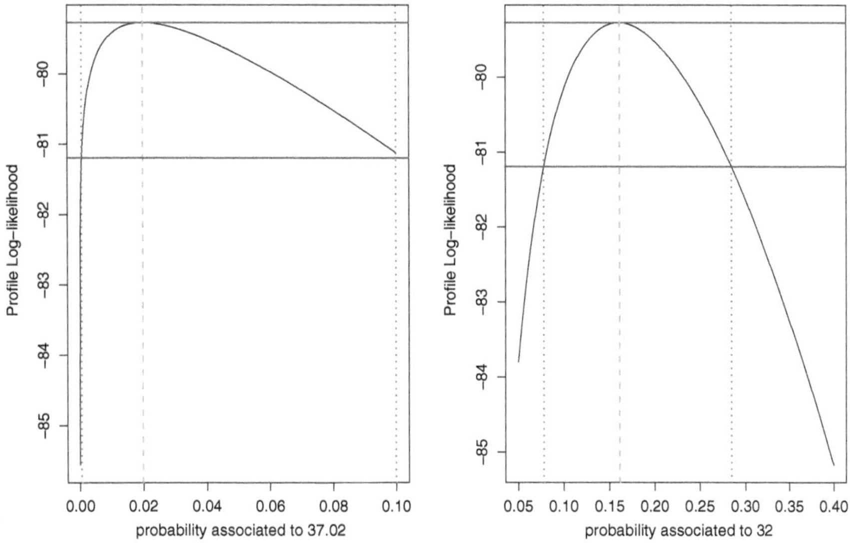

Fig. 15.9 Profile likelihoods obtained by varying the probability of exceeding a fixed value and optimizing for each of these probabilities the likelihood with respect to the distribution's scale parameter

32 m/s by noticing that $P(X > x) = 1 - P(X \leq x)$. In the GEV setting, we thus obtain

$$P(X > 32) = 1 - F_X(32) \simeq 0.16,$$

where here F_X is the cumulative distribution function of the GEV distribution estimated using the mean parameters obtained over the 1000 fits. In the POT setting, with the threshold of 23 m/s, we have

$$P(X > 32|X > 23) = 1 - F_X(32) \simeq 0.24,$$

where F_X is now the cumulative distribution function of the estimated GPD distribution associated with the threshold of 23 m/s. The estimated survival functions for the two approaches are shown in Fig. 15.10. Recall that the survival function is defined as $1 - F(x)$, where F is the cumulative distribution function of the random variable. These plots also show the estimation of the upper bound of extreme wind speeds for

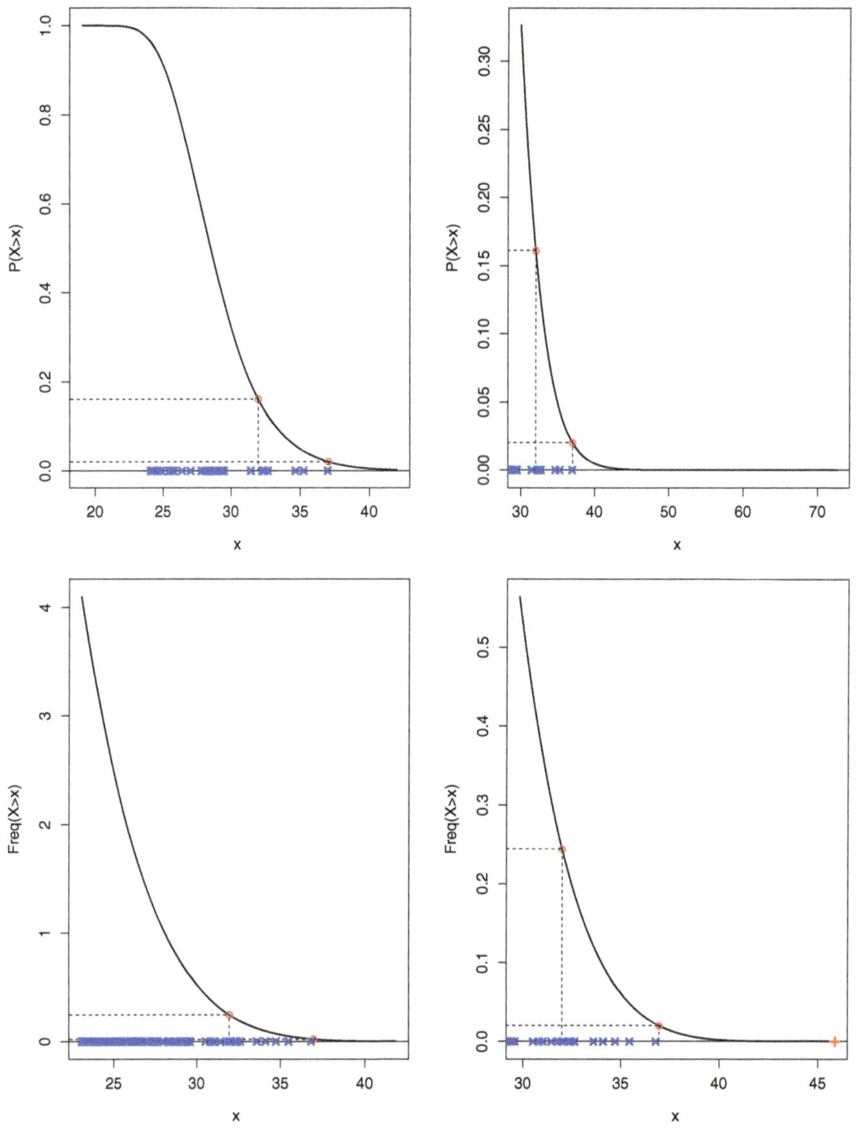

Fig. 15.10 Top: survival function for the GEV model estimated using 500 fits. Bottom: mean annual exceedance frequency for the POT model estimated for the threshold of 23 m/s for two distinct intervals. The upper bound for the maximal wind speed under this model is shown in the bottom-right plot (red cross)

the threshold exceedance approach. Under the MAXB approach, there is high uncertainty in the value of the shape parameter ξ, and in the average of the 1000 jitterings, its value is close to 0, leading to an extremely high and physically questionable upper bound. As for the POT approach, the shape parameter is on average -0.18 and the estimated upper bound of the distribution's support (finite when $\xi < 0$) is 45.7 m/s.

	GEV modeling	POT modeling
Upper bound (m/s)	73.7	45.7

Remark 15.29 Recall that when $\xi < 0$, the upper bound of the support of distribution of extreme values is finite. Its value does not correspond to a return level; in effect, above this upper bound, all exceedance probabilities are zero (conditional on the pertinence of the given distribution). If its value is quite dependent on that of ξ, the bound is unlikely to be physically realistic.

To conclude the study, the confidence intervals for the probabilities of exceeding a given value (that are part of the defined domain of the extreme values) have been obtained using profile likelihood. These are summarized in Table 15.2. The 50-year return level obtained with the GEV distribution, whose parameters are the mean ones, thus corresponds well to a 50-year level. Figure 15.9 illustrates the estimation of these probabilities along with their 95% confidence intervals obtained using profile likelihood.

Chapter 16
Conjunction of Rainfall in Neighboring Watersheds

Nicolas Roche and Anne Dutfoy

Abstract This chapter presents, in great detail, a statistical study of extreme values in a bivariate setting, where the hazards considered together are rains in nearby watersheds. This type of natural hazard can cause destructive flooding. The different steps of the treatment methodology are clarified, and particular care is paid to checking the conditions of the relevance of the theory.

16.1 Introduction

In France, the dimensioning of hydraulic works is based on the estimation of extreme quantiles of hydrometeorological events. The quantiles to examine, as well as the methods for estimating them, are not the subject of any rulebook—official or otherwise. Nevertheless, engineers can, for instance, base their work on the practices of CTPBOH[1] or the recommendations of CFBR[2] [5, 16].

In particular, EDF has developed the SCHADEX method, presented in Chap. 18, which aims to estimate extreme flow quantiles by combining extreme rainfall and the hydrological state of a watershed. The joint use of probabilistic rainfall, continuous climate chronicles on the watershed in question, and a rainfall-runoff model to evaluate hydrological outcomes for the latter, leads finally to determine the distribution of extreme flows.

In this type of approach, when studying neighboring watersheds, it becomes necessary to take into account dependencies between these two watersheds, in particular in view of anticipating the occurrence of extreme-valued events.

In the following, we are going to study the case of the watersheds of the Drac and Isère Rivers at Grenoble (Isère department of France, cf. Figure 0.1). After an analysis of the daily rainfall in the two watersheds, we will examine dependence in their

[1] Permanant Technical Committee for Dams and Hydraulic Works; see the glossary.
[2] The French Committee for Dams and Reservoirs; see the glossary.

N. Roche · A. Dutfoy (✉)
EDF R&D, Palaiseau, France
e-mail: anne.dutfoy@edf.fr

N. Bousquet and P. Bernardara (eds.), *Extreme Value Theory with Applications to Natural Hazards*, https://doi.org/10.1007/978-3-030-74942-2_16

extreme values. After modeling this, we will attempt to define relevant conditional probability distributions.

16.2 Data Analysis

The data used to characterize the Isère and Drac watersheds at Grenoble, recorded at the stations W1410010 and W2832020, respectively, can be obtained from the HYDRO databank*.

The Isère River is an important left-bank tributary of the Rhône River. Its source is in the Graian Alps of Savoie, and it merges with the Rhône a few kilometers north of Valence. At Grenoble, its watershed is 5720 km^2 and its flow around 180 m^3/s.

The Drac River is a tributary (affluent) of the Isère River in south-eastern France. With a length of around 130 km, its source is in the Hautes-Alpes department, and it flows into the Isère River a few kilometers downstream of Grenoble. Its watershed is 3550 km^2 and its flow around 100 m^3/s when it merges with the Isère River.

16.2.1 Rainfall Data in Watersheds

Daily rainfall at specific locations is measured with the help of rain gauges and given in mm (1 mm of rainfall corresponds to 1 L/m^2). The watershed rainfall data used in this example are the output of an original interpolation method based on EDF-defined weather patterns [348]. The latter makes it possible to deal with the problem of underestimation of precipitation at altitude where rainfall capture by measuring devices is sometimes deficient (the presence of strong winds and a preponderance of heavy rain are the two main factors aggravating precipitation capture at altitude). For this example, we have 53 years of daily rainfall data in the watersheds of the Rivers Drac (before it joins with the Isère River) and Isère (at Grenoble) from 1 January 1953 to 31 December 2005 (Fig. 16.1). This corresponds to 19,538 daily rainfall values for each watershed.

16.2.2 Marginal Data Analysis

For the statistical analysis that follows, it is necessary to render the daily rainfall continuous, i.e., as if coming from a continuous bivariate distribution. This is achieved here using *jittering* (cf. Sect. 5.3.4). It is customary to add a zero-mean noise to the data in such cases, but here the data has many 0 values corresponding to days without rain. So as not to introduce negative values, we add instead a uniform noise between 0.001 and 0.002 which does not have a mean of 0 but remains negligible with respect to the scale of the measurements.

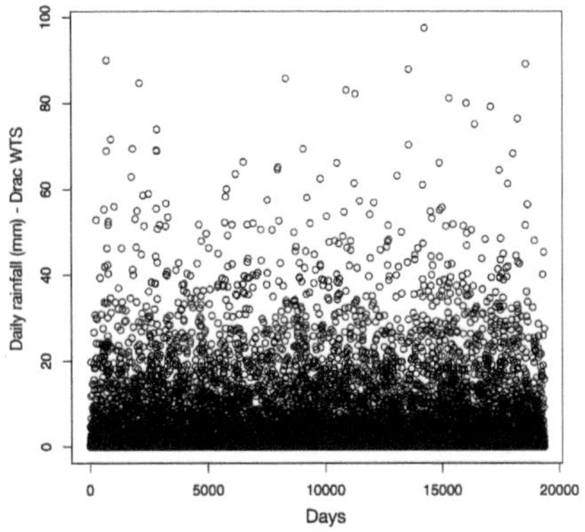

Fig. 16.1 Daily rainfalls from 1 January 1953 to 31 December 2005 in the watersheds (WTS) of the Drac (before it joins with the Isère River) and Isère Rivers (at Grenoble)

The Drac and Isère watershed rainfalls are thus considered random variables which we, respectively, denote X_1 and X_2, draws from which are written x_1 and x_2. The latter therefore correspond to observed rainfalls in each watershed.

The univariate analysis of each marginal distribution will be performed using the threshold exceedance approach (POT). We are going to look for the parameters of generalized Pareto distributions (GPD) that best characterize the data. Remember that the cumulative distribution function $F(x)$ of a random variable above a threshold follows a GPD given by

$$F_j(x) = 1 - \left[1 + \xi_j \left(\frac{x - u_j}{\sigma_j}\right)\right]^{-1/\xi_j}, \tag{16.1}$$

with $j \in \{1, 2\}$, where u is the chosen threshold, σ the scale parameter, and ξ the shape parameter.

As a first step, it is necessary to choose the thresholds beyond which events will be considered extreme in each of the watersheds independently. Typically, experts consider 3–4 such events per year to be the norm, so we can heuristically set the thresholds to be consistent with this: 39 mm and 34 mm for the Drac and Isère Rivers, respectively. These thresholds imply 188 exceedances for the Drac River (around 3.5 per year) and 200 for the Isère River (around 3.8 per year); see Figs. 16.2 and 16.3.

In order to ensure independence of the events thus selected, it is necessary to check that the distributions of the dates at which the thresholds are exceeded follow uniform distributions. For this, we show in Fig. 16.4 the empirical cumulative distribution functions of the exceedance dates, along with that of the uniform distribution.

We therefore decide to retain these thresholds, and now turn to estimating GPD distributions using maximum likelihood on the daily rainfall data. The parameters obtained are shown in Table 16.1. The QQ-plots in Fig. 16.5 help to judge the quality—relatively high in this case—of the fits.

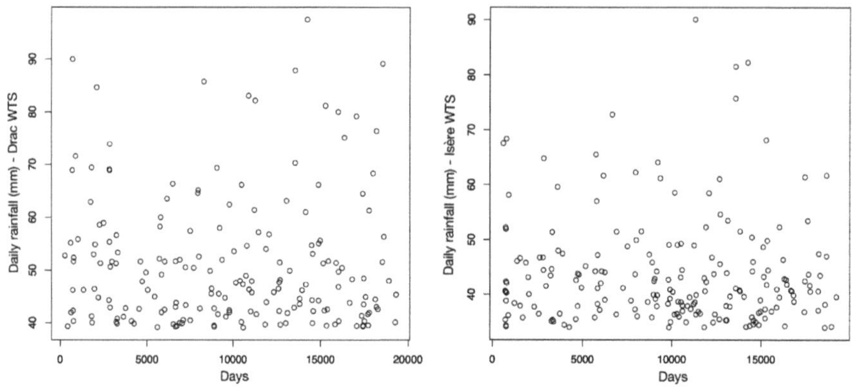

Fig. 16.2 Daily rainfalls above marginal thresholds for the watersheds (WTS) of the Drac and the Isère Rivers at Grenoble

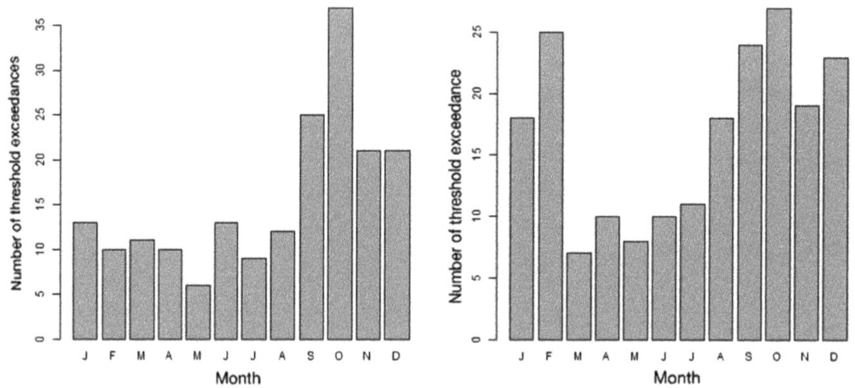

Fig. 16.3 Monthly distribution of the number of marginal threshold exceedances for the watersheds (WTS) of the Drac River and the Isère River at Grenoble

16.3 Asymptotic Dependence Analysis

The study of the conjunction of extreme rainfall events is of great importance when the intensity of such rainfalls increases simultaneously. To conduct such a study, we model the dependence structure between marginal distributions using a copula. Various tests can then be used to judge the usefulness of a multivariate statistical study of extreme rainfalls (see Chap. 9) including analyses of:

- bivariate measurements in the empirical rank space;
- the pseudo-polar coordinates associated with the measurements;
- asymptotic independence using Falk & Michel's test [262].

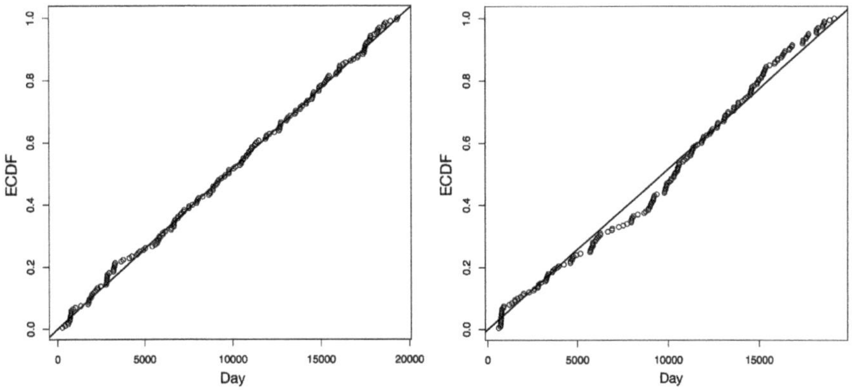

Fig. 16.4 Empirical cumulative distribution functions (CDF) of the dates at which the threshold is exceeded in the watersheds of the Drac and the Isère Rivers at Grenoble. The uniform CDF is shown as a solid line

Table 16.1 GPD parameters for the daily rainfall in the watersheds (WTS) of the Drac and Isère Rivers

Parameters	Drac River WTS	Isère River WTS
u	39 mm	34 mm
ξ	−0.0046	−0.015
σ	12.341	10.008

16.3.1 Analysis of Bivariate Measurements in the Rank Space

Figure 16.6 shows rainfall in the watersheds of the Drac and Isère Rivers in the physical space, as well as in the modified empirical rank space. The empirical cumulative distribution function for each is given by

$$\hat{F}_j(x) = \frac{1}{n+1} \sum_{i=1}^{n} \mathbf{1}_{\{x_j^i \le x\}}, \tag{16.2}$$

with $j \in \{1, 2\}$.

The point cloud in Fig. 16.6b is thus a realization from the empirical copula of the sample of measurements. In this plot, we see that the rainfalls seem to be independent in non-extreme situations (the data appears uniformly distributed over the square $[0; 0.5] \times [0; 0.5]$). On the other hand, there appears to be a certain dependence on higher-quantile rainfall events. To better illustrate this, Fig. 16.7 plots—again in the rank space—measured rainfalls (x_1, x_2) for high quantiles. We see that when there is high rainfall, it tends to occur in both watersheds simultaneously.

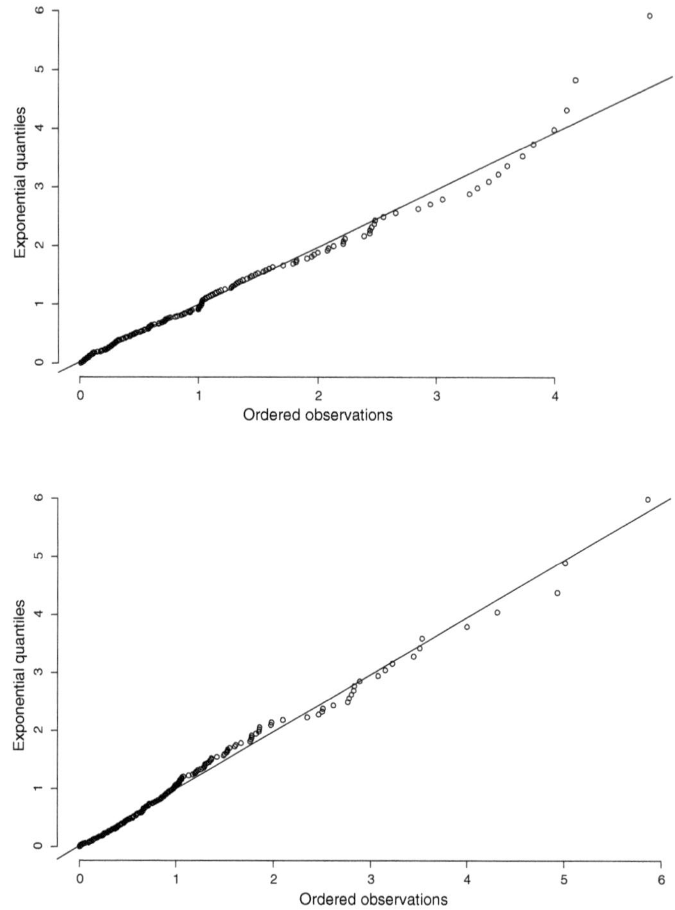

Fig. 16.5 QQ-plots of the GPD fits for the rainfall data from the watersheds of the Drac River (top) and the Isère River (bottom)

By limiting the data to events where both rainfall values exceed their respective thresholds (thus retaining 103 events), we obtain a realization of the extreme copula of our sample—see Fig. 16.7c.

16.3.2 Analysis of Pseudo-Polar Coordinates

For this analysis we first need to transform the random variables corresponding to watershed rainfalls using standard Fréchet distributions:

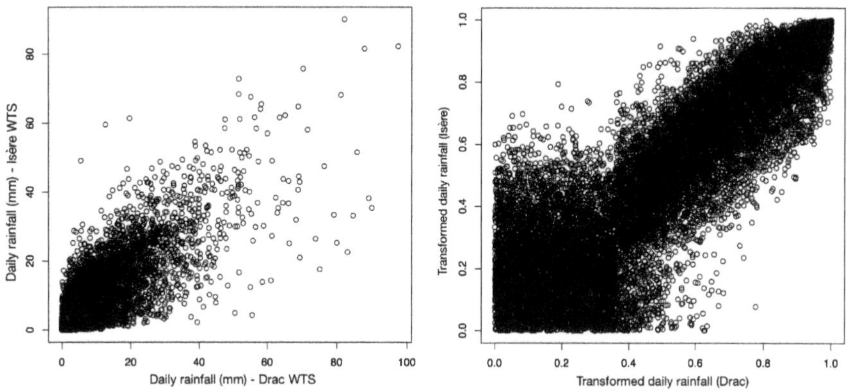

Fig. 16.6 Graphical analysis of the asymptotic dependence of daily rainfall in the watersheds of the Drac and Isère Rivers. Pairs (x_1, x_2) are shown in the physical space (left) and empirical rank space (right)

$$V_j = \frac{1}{\log F_j(X_j)}, \tag{16.3}$$

with $j \in \{1, 2\}$. It can be shown that the pair of random variables (V_1, V_2) follows a bivariate distribution with standard Fréchet distributions as its marginals, and its copula is that of the pair (X_1, X_2). The latter is ensured by the fact that the transformation using the standard Fréchet distribution is increasing and performed componentwise.

The standard Fréchet distribution has the cumulative distribution function

$$\Psi(x) = \exp(-1/x) \quad \forall x \geq 0. \tag{16.4}$$

As those for F_1 and F_2 are unknown, we can replace them by their modified empirical estimates $\hat{F}_1$ and $\hat{F}_2$. The rainfall data for the watersheds of the Rivers Isère and Drac are thus transformed using (16.3) into data (V_1, V_2). Next, we introduce pseudo-polar coordinates (W, R) defined by

$$W = \frac{V_1}{V_1 + V_2}, \tag{16.5}$$

$$R = V_1 + V_2. \tag{16.6}$$

Bivariate rainfall measurements, where one of the two values corresponds to a high quantile, lead to very large values of V_1 or V_2 and thus R. This variable R thus characterizes bivariate rainfall measurements in which at least one of the two values corresponds to a high quantile.

The variable W takes values in [0; 1]. When both V_1 and V_2 simultaneously correspond to high quantiles, W will be close to 0.5. However, the reverse is not true: the event $W = 0.5$ also contains all bivariate measurements (V_1, V_2) for which

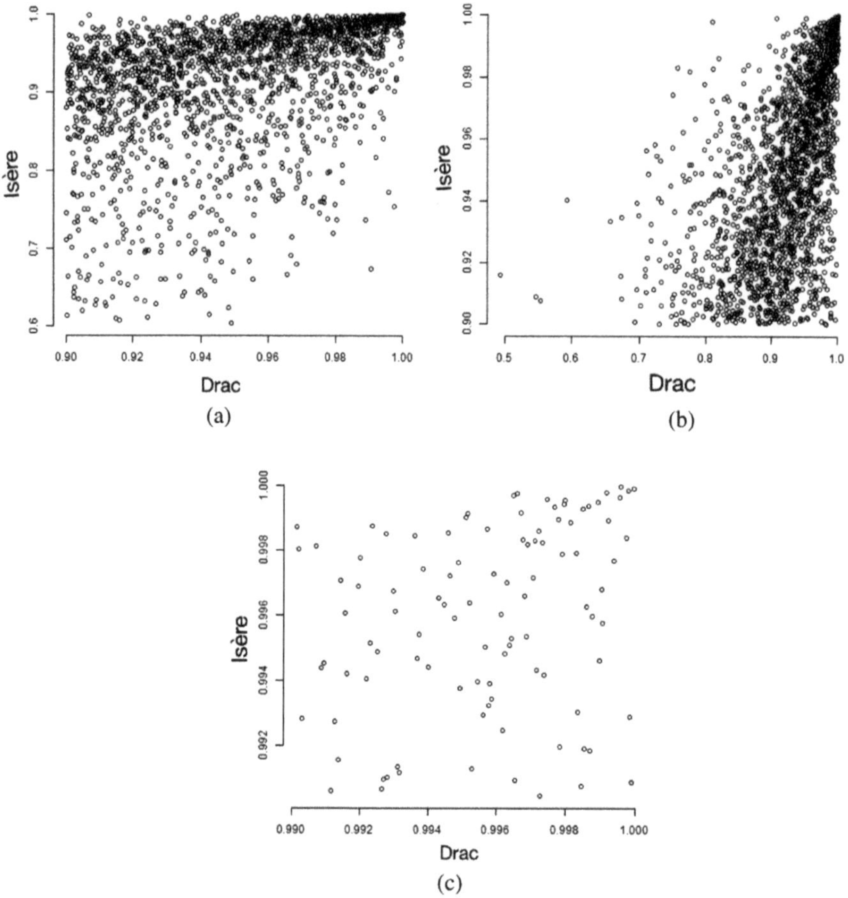

Fig. 16.7 High empirically ranked rainfalls in the watersheds of the (**a**) River Isère and (**b**) Drac River; (**c**) shows the empirical rank space where both rainfall values exceed their respective thresholds

the values of V_1 and V_2 correspond to the same quantile, whether it be extreme or not. Furthermore, when the value of V_1 corresponds to a high quantile and that of V_2 does not, W will be close to 1. Inversely, when the value of V_2 corresponds to a high quantile and that of V_1 does not, W will be close to 0 (Fig. 16.8).

The analysis of asymptotic dependence is performed on extreme-valued bivariate measurements, in the sense that at least one of V_1 and V_2 corresponds to a high quantile. These measurements are found in the top-right quarter of Fig. 16.8. The thresholds (μ_1, μ_2) were chosen using univariate analyses of the distributions exceeding the marginal thresholds (see Table 16.1).

Figure 16.9 shows a histogram of the values of W corresponding to the extreme-valued measurements associated with these thresholds. We see that there is no strong

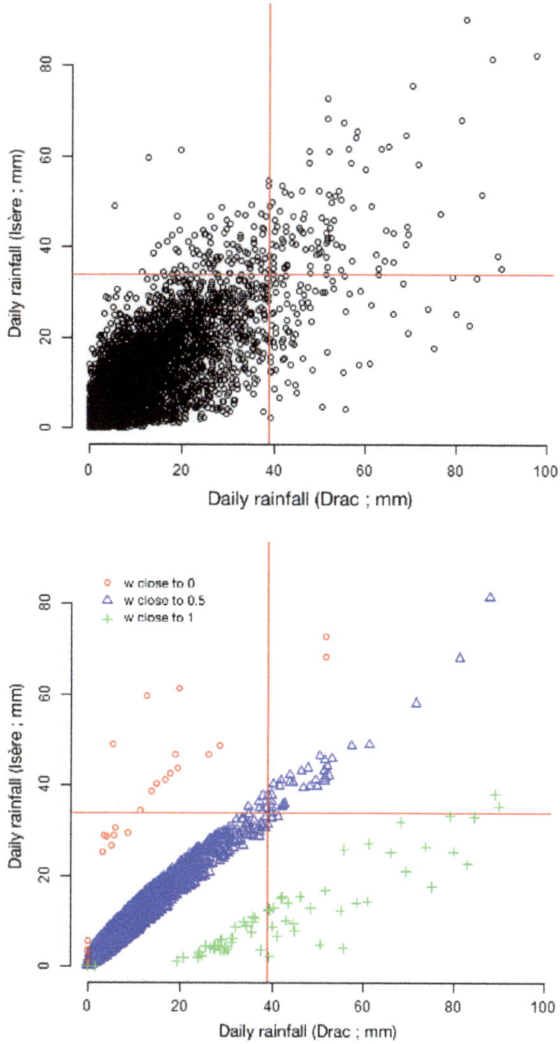

Fig. 16.8 (Top) All rainfall measurements (WTS_2, WTS_1), and (bottom) distinction made between measurements corresponding to values of W close to 0, 1, and 0.5 ($\in [0,4; 0,6]$). The rainfall thresholds $u_2 = 34$ mm and $u_1 = 39$ mm considered for the dependence analysis—one for each watershed—are shown

peak close to 0 or 1 and that the density around $W = 0.5$ is far from negligible. This suggests that there is indeed dependence between the marginal extreme values. This analysis of the histogram of values of W thus appears to confirm that the measurements are asymptotically dependent.

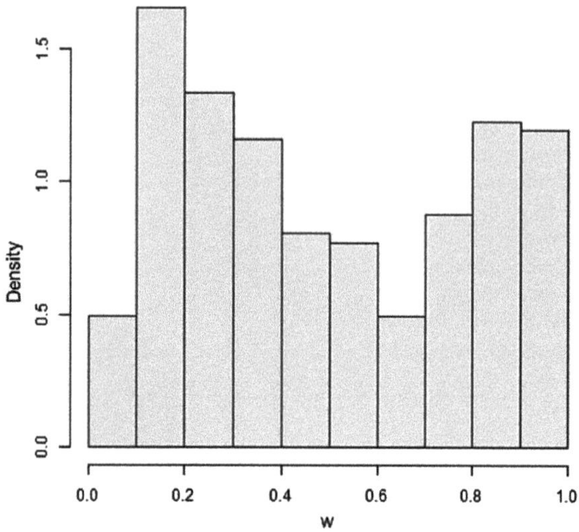

Fig. 16.9 Histogram of values of W corresponding to extreme measurements associated with the thresholds $u_2 = 34$ mm and $u_1 = 39$ mm

16.3.3 Falk and Michel's Independence Test

Falk and Michel's test [262] transforms the data into realizations of marginal Weibull distributions:

$$Z_j = \log F_j(X_j), \tag{16.7}$$

with $j \in \{1, 2\}$. These authors show that Z_1 and Z_2 are asymptotically independent if and only if the conditional distribution of $Z_1 + Z_2 | Z_1 + Z_2 > c$ has the cumulative distribution function $F(t) = t^2$ when c tends to zero from below. A statistical independence test can then be constructed based on a Kolmogorov–Smirnov metric (cf. Example 4.4). Here, the p-value of this test gives 10^{-15}, which allows us to strongly reject the hypothesis H_0 of asymptotic independence between Z_1 and Z_2.

16.3.4 First Conclusions

The above analyses confirm the absence of asymptotic independence between the rainfall measured in the watersheds of the Drac and Isère Rivers at Grenoble, which means that the theory of generalized extreme values can be used to model the tail of the joint rainfall distribution:

$$F(x_1, x_2) = \exp\left(\log\left\{F_1(x_1) \cdot A\left(\frac{\log F_1(x_1)}{\log\{F_1(x_1) \cdot F_2(x_2)\}}\right)\right\}\right) \tag{16.8}$$

for any $x_1 > u_1$ and $x_2 > u_2$, where Pickands' function $A(.)$, seen in Chap. 9 and discussed further in Sect. 16.4.1, characterizes the dependence structure. The distribution in (16.8) can be approximated, in practice, by

$$\hat{F}(x_1, x_2) = \exp\left(\log\left\{\hat{F}_1(x_1)\cdot\hat{A}\left(\frac{\log \hat{F}_1(x_1)}{\log\left\{\hat{F}_1(x_1)\cdot\hat{F}_2(x_2)\right\}}\right)\right\}\right), \quad (16.9)$$

where $\hat{F}_1$ and $\hat{F}_2$ are approximations to the tails of the marginal distributions, associated with the thresholds u_1 and u_2 and modeled by generalized Pareto distributions. In our approach, these are considered the best possible estimations:

$$\hat{F}_j(x) = 1 - \lambda_j\left[1 + \xi_j\left(\frac{x - u_j}{\sigma_j}\right)\right]^{-1/xi_j}, \quad (16.10)$$

with $j \in \{1, 2\}$ and

$$\lambda_1 = \mathbb{P}(X_1 > u_1) \simeq 1 - 0.9896,$$
$$\lambda_2 = \mathbb{P}(X_2 > u_2) \simeq 1 - 0.9903.$$

16.4 Dependence Structure

Two strategies for fitting the parameters of the bivariate distribution can be considered when studying the dependence between two dependent variables. The first consists of estimating together—in the same maximum likelihood procedure—the parameters of the marginal distributions and the dependence function. A model that results from this is called a *total parametric model.* The second strategy consists of estimating them separately, leading to a *partial parametric model.*

Here, we follow the second approach, which allows us to choose marginal distributions consistent with our own criteria and the aims of the study.

16.4.1 Pickands' Dependence Function

Pickands' function A is connected to the dependence function l via the equation:

$$l(V_2, V_1) = (V_2 + V_1)\, A\left(\frac{V_2}{V_2 + V_1}\right), \quad (16.11)$$

where (V_2, V_1) are given by (16.3). Formula (16.8) can then be rewritten as a function of the variables V_2 and V_1, and the dependence function l:

$$F(x_2, x_1) = \exp\left(\log\left\{-\frac{l(V_2, V_1)}{V_2 \cdot V_1}\right\}\right) \tag{16.12}$$

for any $x_2 > u_2$ and $x_1 > u_1$. Estimation of the dependence function is performed parametrically, starting by choosing an adequate model. Here, we propose to use the logistic model studied already in Chap. 9, defined as

$$l(V_2, V_1) = \left(V_2^{1/\alpha} + V_1^{1/\alpha}\right)^{\alpha}, \quad \text{with } 0 < \alpha \leq 1. \tag{16.13}$$

The parameter α represents the strength of the dependence between extreme rainfall events in the watersheds of the Isère and Drac Rivers at Grenoble. In particular, full dependence corresponds to $\alpha \to 0$, and independence to $\alpha = 1$.

16.4.2 Goodness-of-Fit Test

The test used here, developed by Genest et al. [327], allows us to reject (or not) the hypothesis that Pickands' function belongs to a given parametric family, e.g., Clayton, Gumbel–Hoougaard, Franck, etc. It uses the Cramér–von Mises criterion S_n, measuring the distance between a theoretical Pickands function $A(t)$ and the empirical estimate $A^*(t)$ of the Pickands function of a sample [270]. The latter, as well as the parameters of the Pickands estimator, are estimated using the sample of paired data from the two watersheds:

$$\left\{(x_{2,1}, x_{1,1}), \ldots, (x_{2,\mathrm{n}}, x_{1,\mathrm{n}})\right\}.$$

The Cramér–von Mises criterion is written as

$$S_n = \int_0^1 n\|A_n(t) - A_n^*(t)\|\, dt.$$

After an initial estimation of the empirical and parametric Pickands functions—respectively, denoted $A_n^0(t)$ and $A_n^{*0}(t)$ (conducted using the $(x_{2,\mathrm{i}}, x_{1,\mathrm{i}})_i$), and an initial calculation of the Cramér–von Mises criterion (denoted S_n^0), the statistical distribution of the latter can be evaluated using *parametric bootstrap* procedure [244], summarized as follows.

For sufficiently large N, and $k \in [1; N]$:

1. generate a new sample $(x_{2,i}^k, x_{1,i}^k)_i$ using the estimated parameters of the copula $A_n^{*(k-1)}(t)$,
2. calculate $A_n^k(t)$ and $A_n^{*k}(t)$ using this new sample,
3. calculate S_n^k.

The corresponding p-value can be estimated using the probability that the Cramér–von Mises criterion corresponding to the observed data is smaller than that of the bootstrap samples, i.e.,

$$p - \text{value} \simeq \frac{1}{N} \sum_{k=1}^{N} \mathbf{1}_{\{S_n < S_n^k\}}.$$

Here, we have used the logistic extreme copula (16.13). The p-value associated with this in the goodness-of-fit test is 0.0004995. As this value is very small, we consider that the logistic copula is a reasonable approximation to the dependence between extreme rainfall events in the Drac and Isère watersheds at Grenoble.

16.4.3 Choosing the Final Model

After having confirmed a lack of asymptotic independence between rainfall data pairs, we have chosen to model the tail of the joint rainfall distribution using formula (16.9). The latter is based on the marginal distributions of the observed data in each watershed (16.10) and the logistic dependence function (16.13). The parameters obtained for the marginal distributions are shown in Table 16.1, and the parameter α of the logistic distribution is estimated as 0.605 using maximum likelihood.

16.5 Model Use

16.5.1 The Tail of the Joint Distribution

Recall that the rainfalls in the watersheds of the Drac and Isère Rivers are represented as a bivariate random variable (X_1, X_2). Given the bivariate rainfall modeling justified in the preceding paragraphs, (X_1, X_2) has the cumulative distribution function $F(x_1, x_2)$ given in (16.9). This modeling is valid only in the extreme region:

$$A = \{(X_1, X_2) \mid X_1 > 39 \text{ mm}; \ X_2 > 34 \text{ mm}\}. \tag{16.14}$$

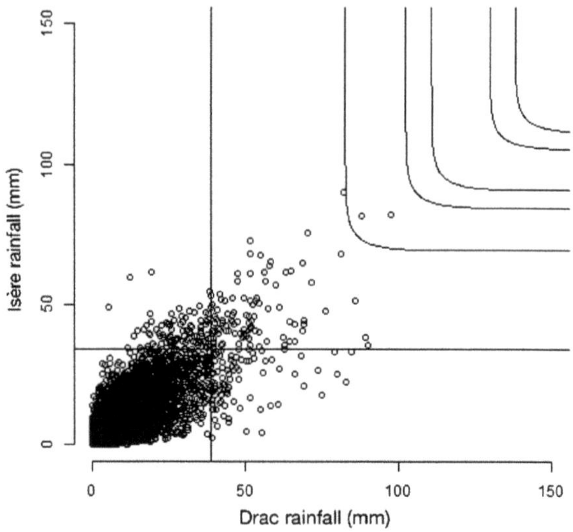

Fig. 16.10 Iso-quantile contours of the tail of the bivariate rainfall distribution in the Drac and Isère Rivers watersheds. Quantiles corresponding to return periods of 10, 50, 100, 500, and 1000 years are shown

Under $F(x_1, x_2)$, the extreme region A is attained with probability

$$\mathbb{P}(A) = \lambda_1 + \lambda_2 - [1 - F(u_1, u_2)] . \tag{16.15}$$

For the parameters in our study, this gives $\mathbb{P}(A) \simeq 3, 21.10^{-3}$.

Figure 16.10 plots the iso-quantile contours, i.e., the set of bivariate quantiles of the same order p.

16.5.2 Rainfall Distribution Conditional on the Conjunction of Extreme Values

We define the bivariate random variable $\mathbf{Y}$ as follows:

$$\mathbf{Y} = (Y_1, Y_2) = \mathbf{X} | \mathbf{X} \in A . \tag{16.16}$$

We would like to obtain the cumulative distribution function of this variable. That is, we wish to define the bivariate distribution of the rainfall in the two watersheds, given that both are above their univariate thresholds: (u_1, u_2) (Fig. 16.11). The cumulative distribution function of $\mathbf{Y}$ is given by

$$F_{\mathbf{Y}}(x_1, x_2) = \frac{F_{\mathbf{X}}(x_1, x_2) - F_{\mathbf{X}}(u_1, x_2) - F_{\mathbf{X}}(x_1, u_2) + F_{\mathbf{X}}(u_1, u_2)}{\mathbb{P}(A)}, \tag{16.17}$$

and its probability density by

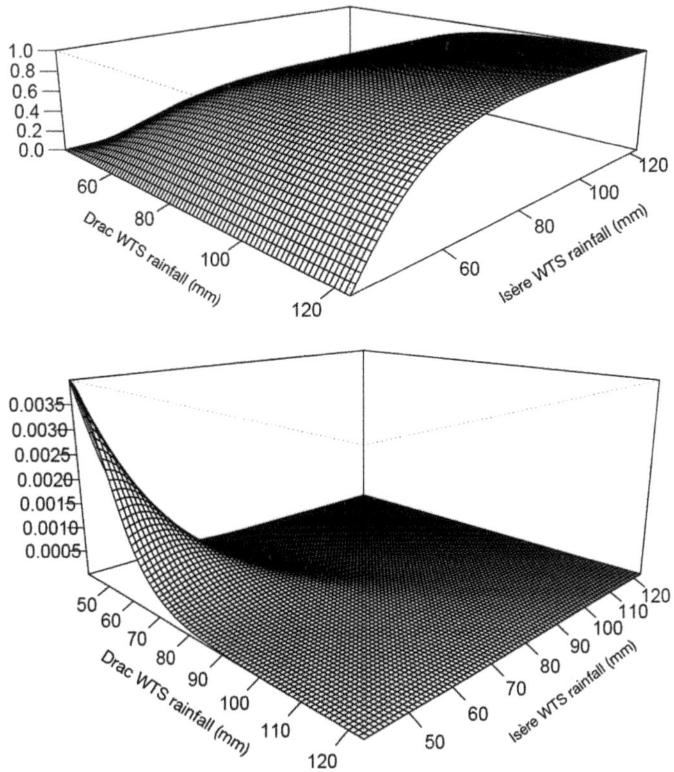

Fig. 16.11 Joint cumulative distribution function (top) and probability density (bottom) of Drac and Isère watersheds (WTS) rainfalls, at Grenoble in the region A, conditional on being in an extreme-valued situation

$$f_{\mathbf{Y}}(x_1, x_2) = \frac{\partial^2 F_{\mathbf{Y}}(x_1, x_2)}{\partial^2 x_1 \cdot x_2} = \frac{p_{\mathbf{X}}(x_1, x_2)}{\mathbb{P}(A)}. \tag{16.18}$$

16.5.3 Marginal Rainfall Distribution Conditional on the Conjunction of Extreme Rainfalls

Using Eq. (16.17), it is possible to calculate the distribution of extreme rainfall in one watershed, given the conjunction of extreme rainfalls in both (Figs. 16.12 and 16.13). In other words, we can calculate the distribution of extreme values Y_1 in the watershed of the Drac River (resp. Y_2 for the Isère River) given that there is strong rainfall in both (above the thresholds defined in the univariate analyses).

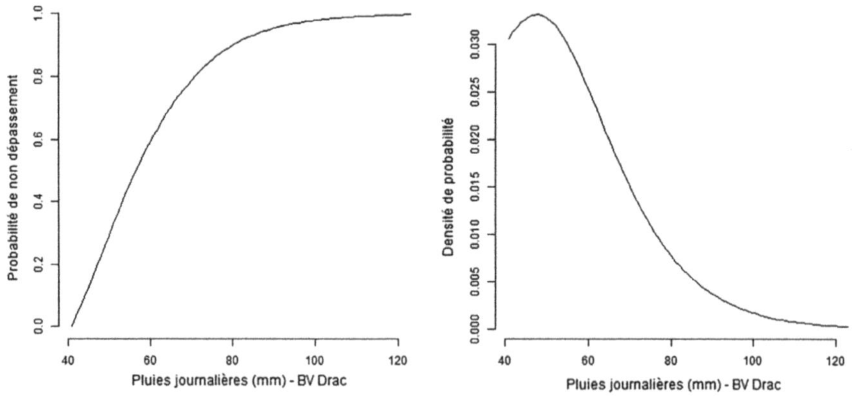

Fig. 16.12 Cumulative distribution function (CDF; left) and probability density function (PDF; right) of rainfall for the watershed of the Drac River, given the conjunction of extreme rainfalls in both watersheds

The corresponding cumulative distribution functions are given by

$$\begin{cases} F_{Y_2}(x_2) = \dfrac{\hat{F}_1(x_2) - \hat{F}_1(u_2) - F_{\mathbf{X}}(x_2, u_1) + F_{\mathbf{X}}(u_2, u_1)}{\mathbb{P}(A)}, \\ F_{Y_1}(x_1) = \dfrac{\hat{F}_2(x_1) - \hat{F}_2(u_1) - F_{\mathbf{X}}(u_2, x_1) + F_{\mathbf{X}}(u_2, u_1)}{\mathbb{P}(A)}. \end{cases} \tag{16.19}$$

As for the probability densities, they are given by

$$\begin{cases} p_{Y_2}(x_2) = \dfrac{p_{X_2}(x_2) - \dfrac{\partial F_{\mathbf{X}}}{\partial x_2}(x_2, u_1)}{\mathbb{P}(A)}, \\ p_{Y_1}(x_1) = \dfrac{p_{X_1}(x_1) - \dfrac{\partial F_{\mathbf{X}}}{\partial x_1}(u_2, x_1)}{\mathbb{P}(A)}. \end{cases}$$

16.5.4 Conditional Distribution of the Rainfall in One Watershed

It is of interest to know the rainfall distribution in one of the watersheds in the event of a conjunction of extreme rainfalls in both, given the value of the rainfall in the other (Figs. 16.14 and 16.15).

Thus, the rainfall distribution in the watershed of the Drac River (resp. Isère) in the event of a conjunction of extreme rainfalls in both, given the value of the rainfall

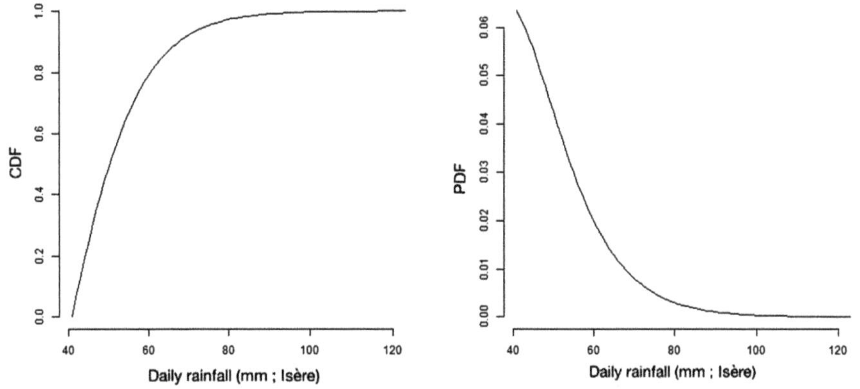

Fig. 16.13 Cumulative distribution function (CDF; left) and probability density function (PDF; right) of rainfall for the watershed of the Isère River, given the conjunction of extreme rainfalls in both watersheds

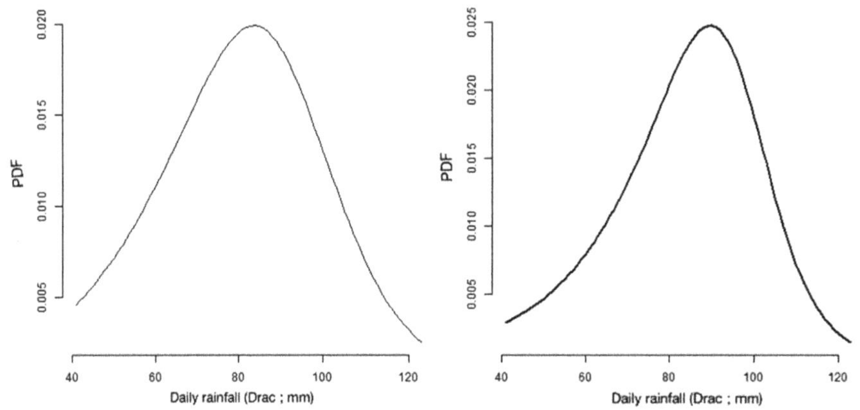

Fig. 16.14 Rainfall probability density function (PDF) in the watershed of the Drac River given a conjunction of extreme rainfalls in both watersheds and the value of the rainfall in the watershed of the Isère Rive (left) 50 mm, (right) 80 mm

in the watershed of the Isère River ($rain_{Is}$) (resp. Drac—$rain_{Dr}$), is given by the new random variable Z_2 (resp. Z_1). These variables can be formally written as

$$\begin{cases} Z_2 = Y_2|\{Y_1 = rain_{Dr}\} = X_1|\{X_2 > u_2,\ X_1 = rain_{Dr} > u_1\}, \\ Z_2 = Y_2|\{Y_2 = rain_{Is}\} = X_2|\{X_2 = rain_{Is} > u_2,\ X_1 > u_1\}. \end{cases} \tag{16.20}$$

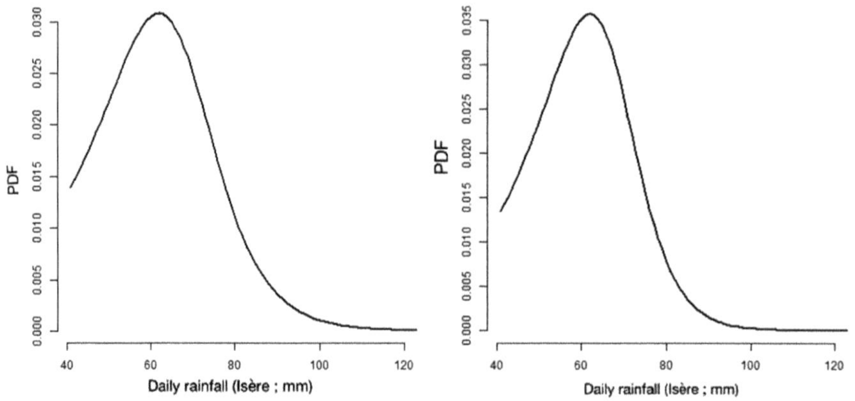

Fig. 16.15 Rainfall probability density function (PDF) in the watershed of the Isère River given a conjunction of extreme rainfalls in both watersheds and the value of the rainfall in the watershed of the Drac River: (left) 50 mm, (right) 80 mm

The probability densities p_{Z_2} of Z_2 and p_{Z_1} of Z_1 are then given by

$$\begin{cases} p_{Z_2}(x_2) = \dfrac{p_{\mathbf{Y}}(x_2, rain_{Dr})}{p_{Y_1}(rain_{Dr})}, \\ \\ p_Z(x_1) = \dfrac{p_{\mathbf{Y}}(rain_{Dr}, x_1)}{p_{Y_2}(rain_{Dr})}. \end{cases} \tag{16.21}$$

Chapter 17
Conjunction of a Flood and a Storm

Alain Sibler and Anne Dutfoy

Abstract The case study in this chapter involves the conjunction of two climate hazards: a flood (high river flow) and a storm (strong wind), which we label *Flow* and *Wind Speed* for simplicity. A combination of the two is likely to jeopardize industrial installations. The aim of the study is to calculate the probability of the conjunction of these extreme events and produce daily probability and annual frequency plots of such events.

17.1 Introduction

In this chapter, we focus on the bivariate analysis of the extreme values of the *Flow* and *Wind Speed* processes, after a brief initial summary of results from univariate marginal analyses. Data for the study comes from the Meuse department of the Grand-Est region of France. For more in-depth details on how the following univariate results were obtained, readers should consult the previous chapters. The analysis is conducted according to the following steps:

1. Univariate analysis of seasonality and trends in the flow and wind speed.
2. Bivariate analysis of the correlation in extreme values of the two processes.
3. Estimating the daily probability and annual frequency of the cumulative level exceedance for given return periods; annual frequency is either defined in terms of the mean number of co-occurring events per year or the mean number of co-occurring episodes per year (see Sect. 6.3.2).
4. Plotting daily iso-probability and annual iso-frequency contours in the space of marginal return periods.

The data consist of joint measurements of the daily mean flow and daily maximal instantaneous wind speed, and we analyze them using the `R` packages `evir`, `evd`, and `texmex`.

A. Sibler · A. Dutfoy (✉)
EDF R&D, Boulevard Gaspard Monge, 91120 Palaiseau, France
e-mail: anne.dutfoy@edf.fr

N. Bousquet and P. Bernardara (eds.), *Extreme Value Theory with Applications to Natural Hazards*, https://doi.org/10.1007/978-3-030-74942-2_17

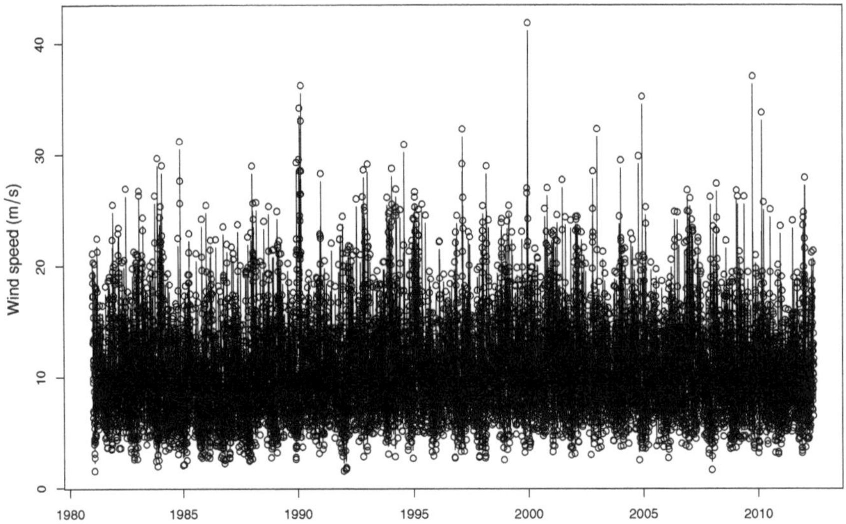

Fig. 17.1 Daily wind speeds (1981–2012) in the Meuse department of the Grand-Est region of France

17.2 Summary of Univariate Analyses

17.2.1 Available Data

In this case study, storms are represented in terms of daily maximal instantaneous wind speed. The data are from 1 January 1981 to 22 May 2012 and was recorded by Météo-France (Fig. 17.1). Though earlier data is available, we have not used it here since it is only considered homogeneous from 1981. Flooding is represented in terms of high daily water flow. These data were recorded for the Meuse river between 1 January 1953 to 31 December 2013 and are available in the HYDRO databank (Fig. 17.2).

17.2.2 Seasonality

It can be shown that, by restricting the study to the December–February period, the flow can be considered stationary. In addition, this period is the one which contains the extreme values we wish to analyze. Similarly, a study of the wind speed data leads to retaining the October–March period as the one which contains the highest wind speeds. The period in which both are stationary is therefore December–February.

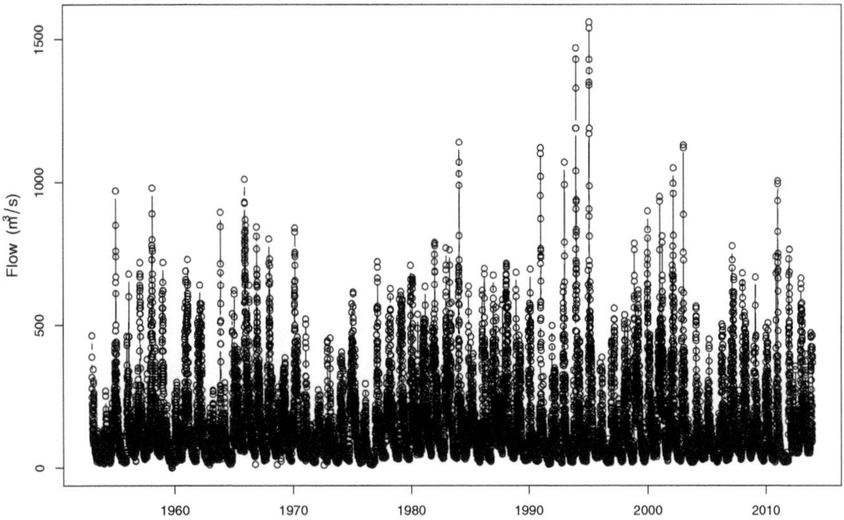

Fig. 17.2 Daily mean flow of the Meuse river (1953–2013)

17.2.3 POT Approach and GPD Models

We have analyzed and then modeled the behavior of both of the marginal processes in their extreme value regions, exactly as in previous chapters. For each process, a threshold exceedance approach (POT) has been retained for modeling the tails of the marginal distribution process.

To proceed in this way, we must extract from the data the sequence of independent cluster maxima, where each cluster contains several successive values above a high threshold. The latter must be chosen in such a way that the sequence of cluster maxima follows a homogeneous Poisson process, i.e., has the following properties:

1. the cluster maxima (one for each occurrence of a flood) are independent,
2. the distribution of the time between successive cluster maxima is exponential,
3. the annual number of cluster maxima follows a Poisson distribution.

Remember that once thresholds have been chosen, stationarity in the extreme values of the processes can be studied using the sequences of cluster maxima displayed graphically alongside statistical tests (Sect. 5.4.3.2), of which the most important are the Mann-Kendall test—which checks for the presence or absence of a trend—and the uniform distribution test on the set of dates at which cluster maxima occur. Other tools are also commonly used for validating chosen thresholds, using for instance the fact that the excesses of the cluster maxima over the thresholds should have GPD distributions. More details can be found in the previous chapters.

Table 17.1 gives the parameter estimates for the GPD model of the flow process, along with confidence intervals estimated using the delta method (Theorem 4.4),

Table 17.1 Parameters of a generalized Pareto distribution fitted to the cluster maxima exceedances over the threshold $q_s = 620\,\text{m}^3.\text{s}^{-1}$ of the flow of the Meuse river, along with 95% confidence intervals

	Value	Standard deviation	a_{inf}	b_{sup}
ξ	0.041	0.163	−0.278	0.360
σ	188	41	107	269

denoted by $[a_{inf}, b_{sup}]$. The threshold has been set at $q_s = 620\,\text{m}^3.\text{s}^{-1}$. As for the study of the wind speed time series, the threshold has been set at $v_s = 26.9\,\text{m.s}^{-1}$, which leads to the following parameters for the GPD model of the cluster maxima exceedances: $\xi = 0$ and $\sigma = 3.23$.

17.3 Bivariate Analysis of Flows and Wind Speeds

We now turn to study the dependence between the extreme values of the Meuse river's flow and the wind speed. This bivariate analysis takes place over the period in which both processes are stationary, i.e., daily measurements from December to February between 1 January 1981 and 22 May 2012.

17.3.1 Bivariate Data

Figure 17.3 shows paired measurements over the period of the study, first in the physical space (a), then in the rank space (b) using the empirical ranks. Figure 17.4 shows regions of these spaces in which at least one component corresponds to a 90% quantile or above. Figure 17.5 shows pairs for which both components are over their 90% quantiles. Lastly, Fig. 17.6 shows these measurements in the Fréchet space. These plots seem to indicate asymptotic independence of the flow and wind speed in extreme value settings, a result which will be confirmed by calculations in the following section.

17.3.2 Asymptotic Dependence Analysis

Figure 17.7 displays the functions $\chi(u)$ and $\bar{\chi}(u)$. Their analysis helps to estimate the asymptotic behavior of the vector of random flow and wind speed pairs (see Sect. 9.2.3). Recall that for two correlated hazards X_1 and X_2, with marginal distributions F_{X_1} and F_{X_2} and copula C_F,

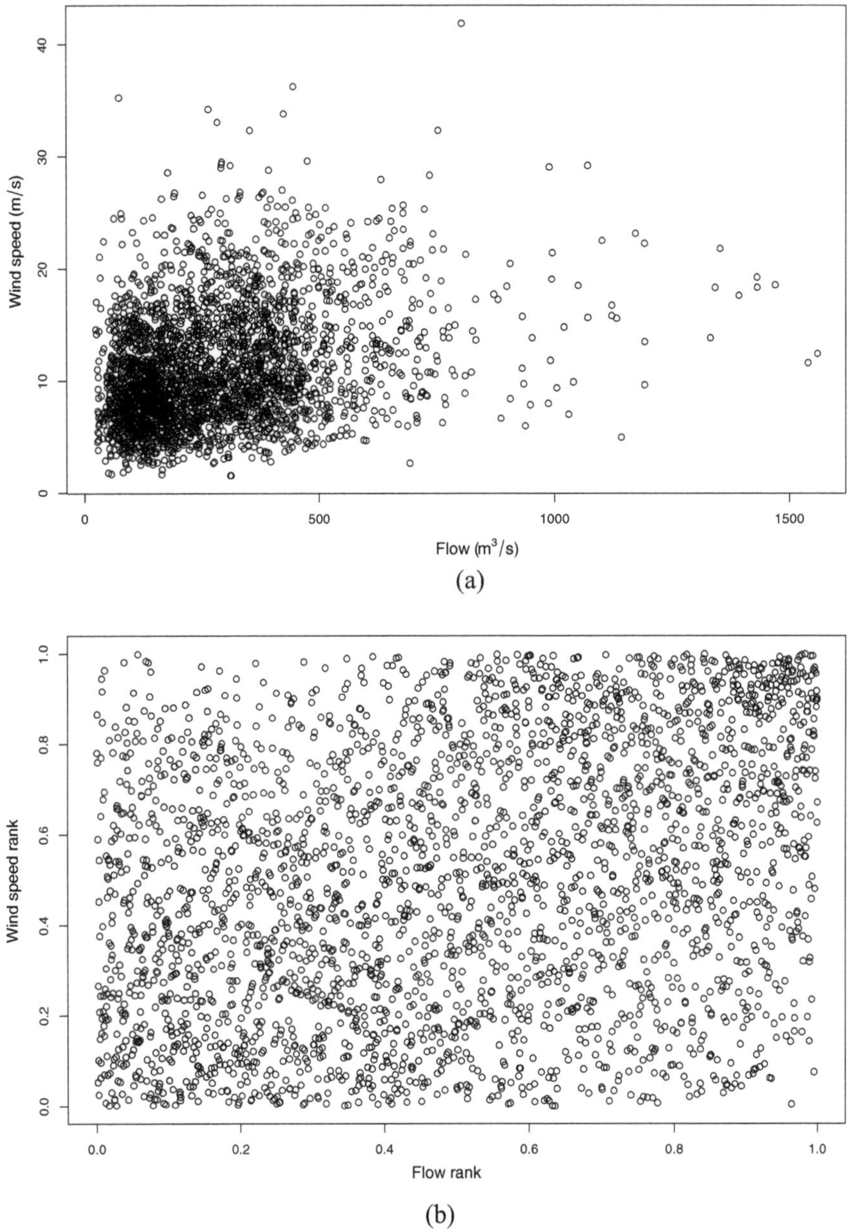

Fig. 17.3 Bivariate data points of the flow and wind speed between the months of December and February over the period 1981–2012; **a** physical space and **b** rank space

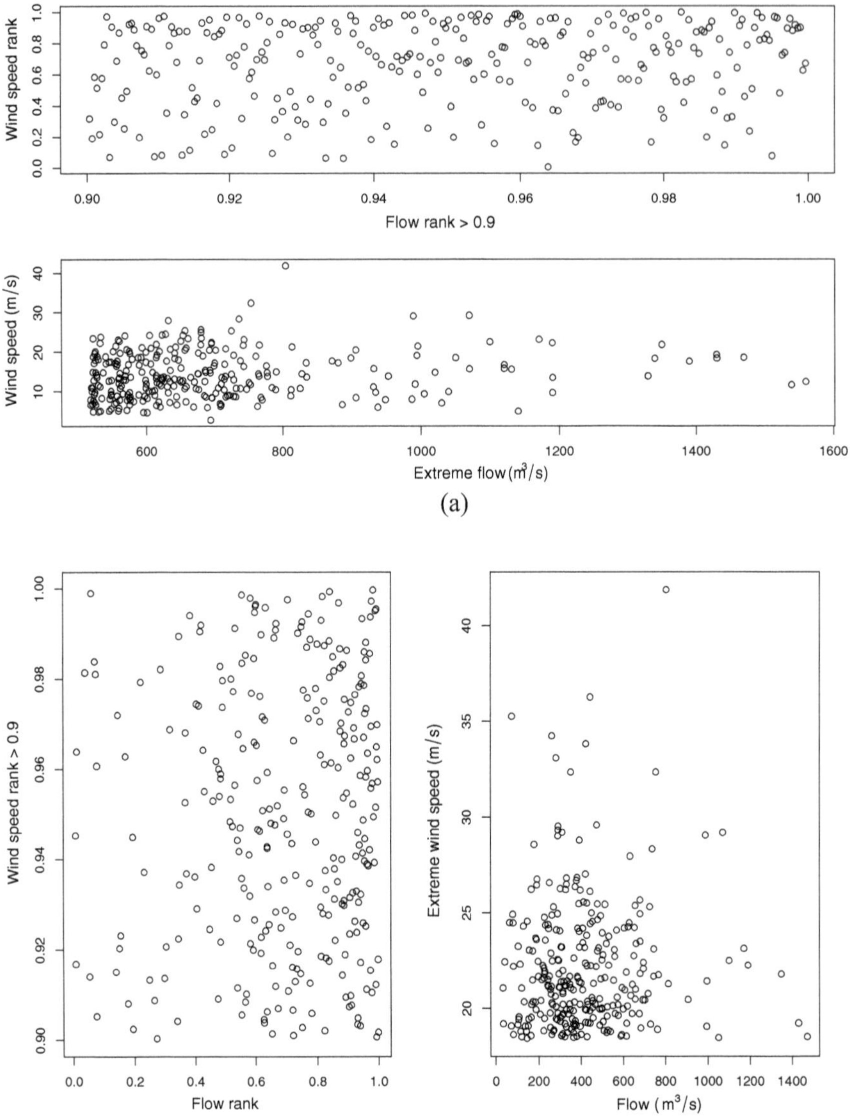

Fig. 17.4 Bivariate data points of the flow and wind speed between the months of December and February over the period 1981–2012; **a** high flow and **b** high wind speed

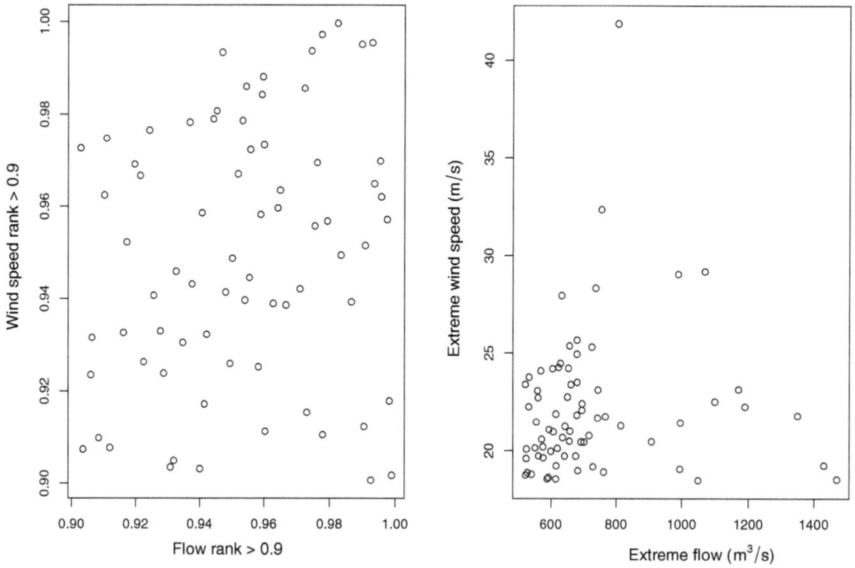

Fig. 17.5 Bivariate data points of the flow and wind speed between the months of December and February over the period 1981–2012; only data for which both components have high ranks are shown

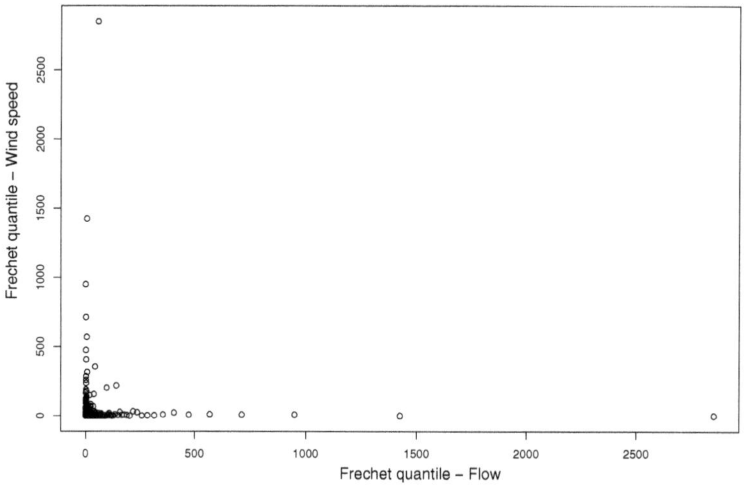

Fig. 17.6 Bivariate data points of the flow and wind speed between the months of December and February over the period 1981–2012; Fréchet space representation

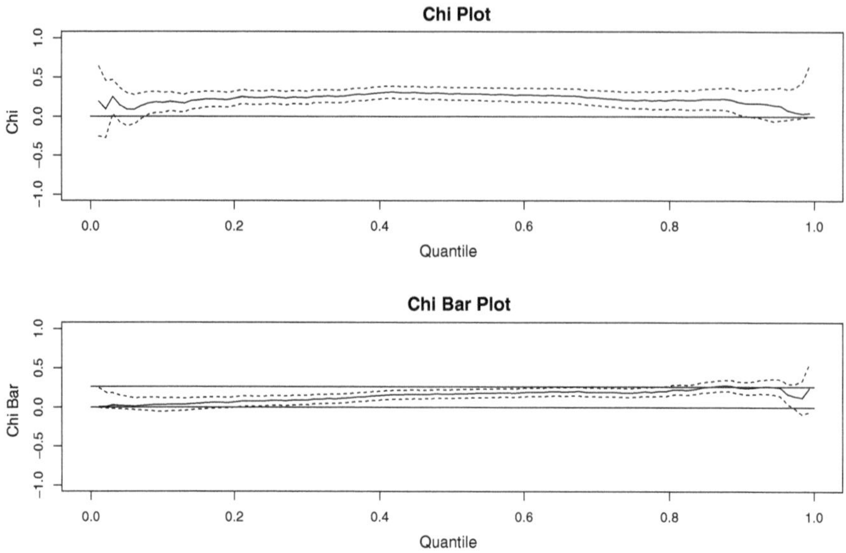

Fig. 17.7 Plots of the functions $u \mapsto \chi(u)$ and $u \mapsto \bar{\chi}(u)$ for the bivariate flow and wind speed measurements

$$\chi = \lim_{u \to 1^-} \chi(u),$$
$$\bar{\chi} = \lim_{u \to 1} \bar{\chi}(u),$$

where

$$\begin{aligned}\chi(u) &= 2 - \frac{\log \mathbb{P} F_{X_1}(X_1) < u,\, F_2(X_2) < u}{\log \mathbb{P} F_{X_1}(X_1) < u} \\ &= 2 - \frac{\log C_F(u, u)}{\log u},\end{aligned}$$

and

$$\begin{aligned}\bar{\chi}(u) &= \frac{2 \log \mathbb{P} F_1(X_1) > u}{\log \mathbb{P} F_1(X_1) > u,\, F_2(X_2) > u} - 1 \\ &= \frac{2 \log(1-u)}{\log \bar{C}_F(u, u)} - 1.\end{aligned}$$

We see that $\chi(u)$ tends to 0 when we tend toward the order 1 quantile, while $\bar{\chi}(u)$ remains above 0; these two together suggest positively associated asymptotic independence, i.e., random events exceeding each extreme threshold tend to favor each other, with this dependence decreasing as the thresholds become more and

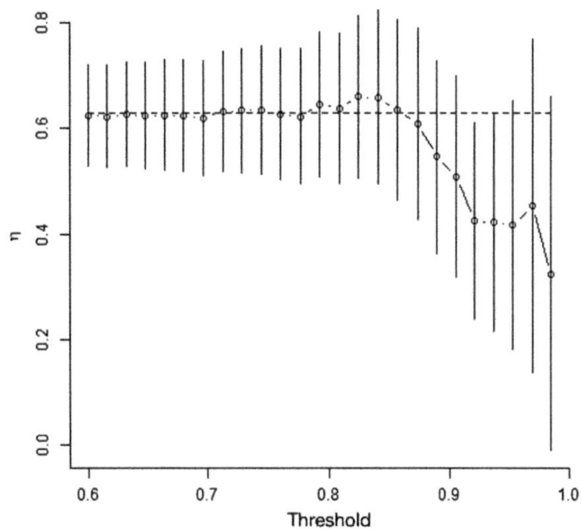

Fig. 17.8 The value of η in the Ledford and Tawn model as a function of the threshold

more extreme. The strength of the positive association is given by the limit value of $\bar{\chi}(u)$ as we tend toward the order 1 quantile; here we estimate this as $\bar{\chi} = 0.26$. Ledford and Tawn's test, run on the scalar variable $T = \min(Z_1, Z_2)$ (where $Z_i = -1/\log(\hat{F}_i(X_i))$), with X_1 corresponding to measures of the Meuse river's flow and X_2 to the wind speed, and F_i the cumulative distribution function of X_i, allows us to calculate the tail dependence coefficient η, which characterizes extreme values of the variable T (cf. Sect. 9.3.4).

Figure 17.8 plots the estimated value of η as a function of the threshold chosen for the GPD modeling of T. We obtain $\eta \simeq 0.63$. As we know that $\bar{\chi} = 2\eta - 1$, our two estimated values appear to be coherent with each other. Furthermore, the confidence intervals for η do not contain the value 1, meaning that the asymptotic dependence hypothesis can be rejected.

Finally, Fig. 17.9 shows the histogram of a nonparametric reconstruction of the spectral density of extreme-valued measurements. In the pseudo-polar space ($R = Z_1 + Z_2$, $W = V_1/R$), where (Z_1, Z_2) are the coordinates of the measurements in the Fréchet space (cf. Sect. 9.2.2.4), we plot the histogram of the W for the large values of R (those above their 95% quantiles). The "hollow belly" of this histogram is supporting evidence for asymptotic independence.

17.3.3 Physically Based Validation of the Statistical Results

In this section, we propose a physical process-based argument, provided by EDF hydrological experts, to validate the observed positively associated asymptotic independence that was statistically estimated in the present study, between the flow of

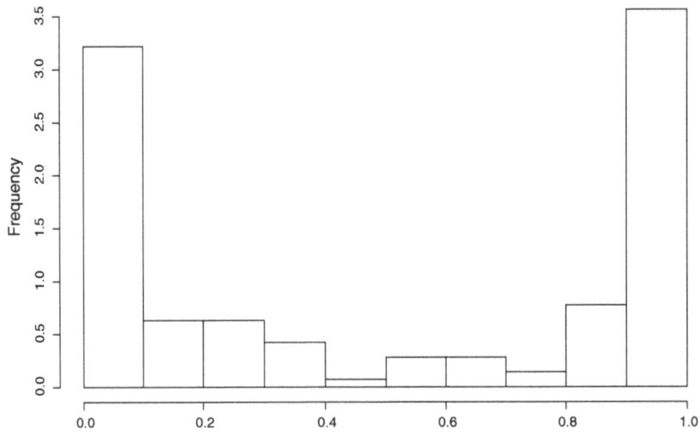

Fig. 17.9 Spectral density of W in the pseudo-polar space of extreme-valued bivariate measurements of the flow and wind speed

the Meuse river and the wind speed. Statistical links between floods and wind speed at a given site of interest are indeed plausible. In fact, rains which are at the origin of the most extreme flood events are a result—like the wind—of atmospheric dynamics, in particular the circulation behavior of forced air masses at the synoptic scale[1] induced by the configuration of the pressure field. Several factors nevertheless make it possible to qualify the quality of the expected statistical relationship and ultimately lead to a low probability of the conjunction of these two types of extreme event.

Spatial shifts. In the case of the wind speed, it is its value at the site of interest which is important to us, whereas for the rain, it is its cumulative value over the watershed above the site of interest. This introduces a shift proportional to the length of the watercourse. However, the wavelength of pressure anomalies is often greater than 2000 km, which limits the consequences of this spatial shift for typical watersheds.

Time lags. The flood peak value at the watershed outlet* is delayed (with respect to the rain that caused it) by the time it takes for the water to flow to the point of interest. Such delays depend—among other things—on the area of the watershed. In the case of large ones (which is more or less the case for the watersheds of France's largest rivers), the lag between the rainfall and maximal flow will get close to or exceed typical average autocorrelation time periods of the wind speed variable.

Determinants of extreme rainfall. Unlike typical wind storms, which behave relatively deterministically with respect to an atmospheric pressure field, and more precisely its horizontal gradient, the causes of rainfall are more complex. Even in large watersheds, it is not only the strength of the flow that counts but also its moisture load, along with all of the causes of ascending air masses (low-level horizontal convergence, forcings due to relief, advection, convection, upper troposphere divergence, absolute or potential vorticity advection, etc.). There exist statistical links of various strengths between these, but it must be remembered that even in the absence

[1] Meteorological and oceanographic term for phenomena occurring at the planetary scale.

of the spatial or temporal shifts mentioned above, the correlation between rainfall and wind speed is far from perfect and—in addition—differs from one watershed to the next.

Although the watershed of the Meuse river—in contrast with those of rivers originating in France's highest mountains—is often considered a floodplain, its *orography*[2] still plays an important role in rainfall analyses. Indeed, the mountainous reliefs of the Argonne (upstream of the Meuse river) and Ardennes (right bank tributaries: Chiers and Semoy rivers) are opposed to the very dominant westerly flows at this latitude, despite their moderate altitude. In the case of strong ocean winds, the orography increase in rainfall is exacerbated. However, ocean storms are often associated with a sustained strengthening of the *mean jet stream*[3] associated with a positive phase of the North Atlantic Oscillation (NAO). In the cold season in particular, repeated rainy episodes induced by high winds coming off the Atlantic lead to waterlogged land and subsequently to extreme high levels of the Meuse river, such as those which occurred at the end of January 1995. At the flood peak, an ongoing storm may still be generating strong or even extreme north to north-west winds. While nondeterministic, a statistical link between wind and flood seems to characterize the nature of these natural hazards.

17.4 Annual Probabilities and Frequencies

Our aim is to calculate the probability of the conjunction of the two extreme hazards, each given in terms of the exceedance of high thresholds. These marginal thresholds are defined in terms of return periods. For reasons of confidentiality, the return periods corresponding to the thresholds used in this chapter are not shown. However, Fig. 17.10 plots observed data pairs of the flow of the Meuse river and the wind speed from 1981–2012 during the season being studied. The shaded area corresponds to the zone for the thresholds used here.

Once the probabilities of such simultaneous events have been estimated, the annual frequencies can be calculated, and it is then possible to plot annual iso-probability and iso-frequency contours in the annual return period space (N_D, N_V) (corresponding to the *Flow* and *Wind Speed* processes).

[2] The domain of physical geography involving the formation and features of mountains.

[3] Rapid and concentrated air currents found between the troposphere (where the temperature decreases with altitude) and the stratosphere (where the temperature increases with altitude).

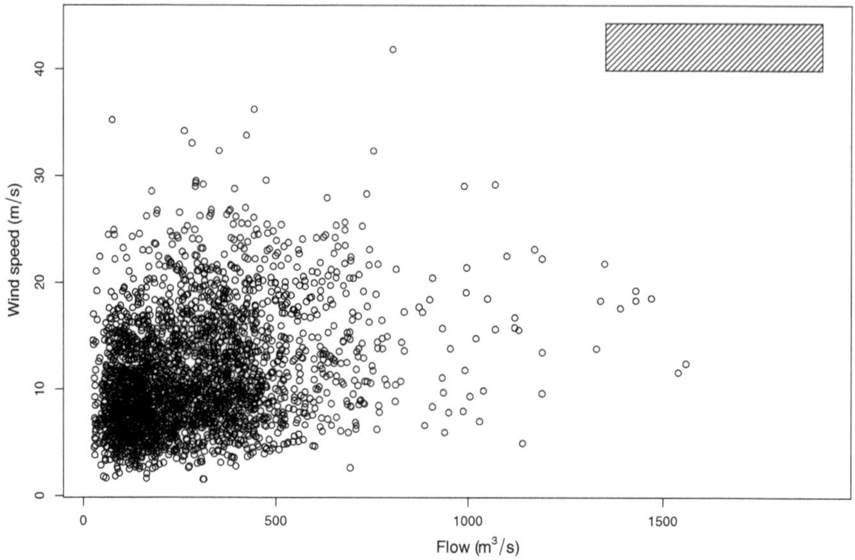

Fig. 17.10 Bivariate data for the flow and wind speed. The shaded zone corresponds to the area of interest for the study

17.4.1 *Probability of Simultaneous Extreme Events*

In Chap. 9, estimators for the probability of the conjunction of extreme events are presented in detail; we invite the reader to refer back to that chapter for the technical details of the methodology led in the following.

17.4.1.1 Estimation of the Threshold Probability p_η

We use four different estimators of the parameter η of the Ledford and Tawn model (cf. Sect. 9.3.4):

- Hill's estimator $\hat{\eta}_{he}$ directly determined using the data given as T;
- the maximum likelihood estimator $\hat{\eta}_{me}$ obtained by fitting a GPD distribution to T;
- Hill's estimator $\hat{\eta}_{hp}$ determined using the values of T obtained by sampling from the GPD distribution;
- the maximum likelihood estimator $\hat{\eta}_{mp}$ obtained by fitting a GPD distribution to the values of T obtained by sampling from the GPD distribution.

In Fig. 17.11, the four estimators of η are plotted as a function of the threshold probability p_η. The final value we select for the threshold probability is the one at which the distance between the four estimators is the smallest, over the stable part of the plot, i.e., $p_\eta = 0.85$ here.

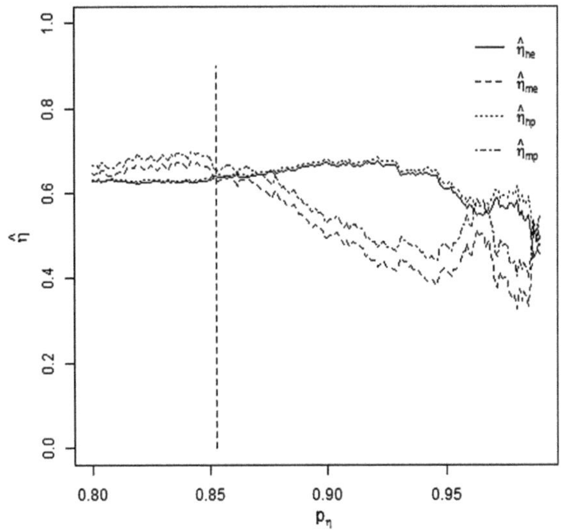

Fig. 17.11 Estimates of the parameter η of the Ledford and Tawn model as a function of the threshold probability p_η

17.4.1.2 Estimating *p*

We have available to us the four estimators for the probability of the conjunction of extreme hazards described in Sect. 9.4.3.4: $\hat{p}_e$, $\hat{p}_p$, $\hat{p}_T$, and $\hat{p}_r$. The first three of these are given in terms of the parameter η of the Ledford and Tawn model itself calculated with the help of the four estimators defined above. We have therefore a total of 13 useable statistical estimators.

For each conjunction of extreme events, defined by the respective thresholds for the flow of the Meuse river and the wind speed, with return periods (N_D, N_V), we plot the functions of the estimators in terms of a parameter which must be set: the thresholding parameter p_0, which is involved in all of the estimators. All of the estimates of the probability of both hazards exceeding their thresholds can thus be plotted on the same graph for comparison, as a function of the same threshold probability p_0, ranging across the extreme quantiles 0.8–0.95. The threshold p_0 corresponds to minimizing the spread between the various estimators across the considered range of values of p_0 in which all of the estimators are well-defined (i.e., non-zero).

Remark 17.1 For each return periods pair (N_D, N_V), the estimator $\hat{p}_r$ requires the calculation of the extremal index θ of the random variable T_r (not shown here).

We will not provide here all of the plots that have been used to compare estimators for each conjunction of extreme events, but only the conclusions drawn from analyzing them. We have chosen not to retain the estimator $\hat{p}_r$ when calculating the probability of simultaneous threshold exceedance, preferring to use the estimator $\hat{p}_T$, which is calculated with the help of the estimator $\hat{\eta}_{he}$ of η in the Ledford and Tawn model.

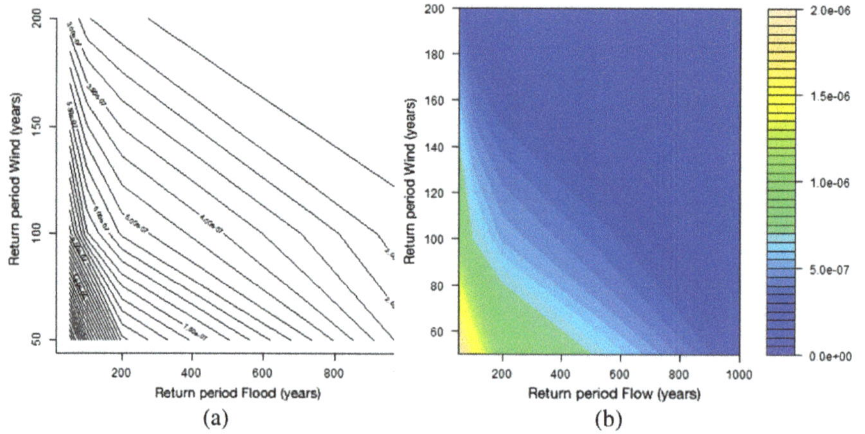

Fig. 17.12 Maps of the daily probability of the conjunction of extreme events: **a** iso-probability contours estimated using the estimators $\hat{p}_T - \hat{h}_e$, and **b** zones containing probabilities of the same order of magnitude

Remark 17.2 A bivariate extreme-valued event means that both marginal extreme events have occurred *on the same day*, during the overlapping seasonal period (December–February).

Probability maps, as in Fig. 17.12, help to delimit boundaries between zones containing probabilities with the same order of magnitude. Figure 17.12 shows two ways to visualize the probability of the conjunction of extreme events. Given the scarcity of available data, it is not possible to give a very detailed view of the most extreme levels.

17.4.2 Annual Frequencies

The period of stationarity for both high flow and high wind speed processes is from December–February, containing a total of $n_0 = 90$ days. Let us now consider events corresponding to high threshold exceedance, denoted by $\mathscr{E} = \{D > d, V > v\}$, whose probabilities of occurring—across a range of extreme thresholds (d, v)—were calculated in the previous section. The annual frequency of an event can be defined in several ways, as described in Sect. 6.3.2. We illustrate here the behavior of the two main definitions of it in our example.

The annual frequency, defined as the mean number of days per year when the event E occurs, is written as follows:

$$\mathbb{F}_a^{6.13}(\mathscr{E}) = n_0 \mathbb{P}(\mathscr{E}).$$

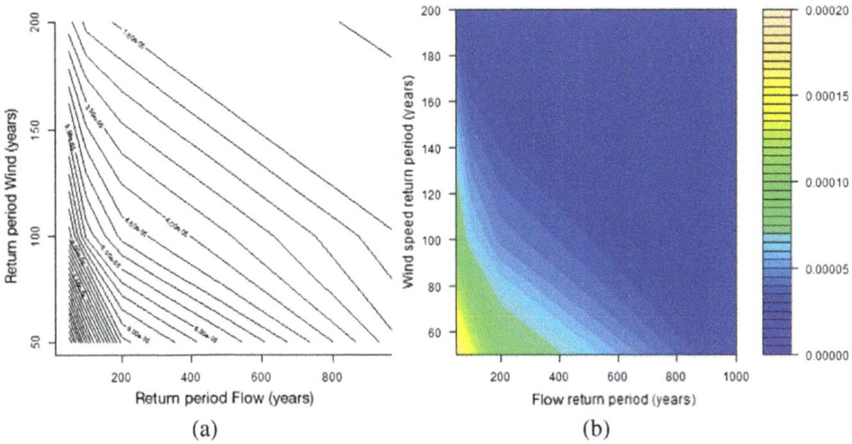

Fig. 17.13 Maps of the annual frequencies of the conjunction of extreme events (mean number of days per year): **a** annual iso-frequency contours estimated using the estimators $\hat{p}_T - \hat{h}_e$, and **b** zones containing annual frequencies of the same order of magnitude

The annual frequency, defined as the mean number of episodes per year of the event E, is given by

$$\mathbb{F}_a^{6.14}(\mathscr{E}) = n_0\, \theta_{(D,V)} \mathbb{P}(\mathscr{E}),$$

where $\theta_{(D,V)}$ is the multivariate extremal index of D and V, defined as the inverse of the mean duration (length) of independent clusters corresponding to simultaneous exceedance of the threshold q_s for the flow of the Meuse river and v_s for the wind speed, during the overlapping time period and season studied in the joint analysis. See Sect. 9.3.4 for more details.

The independence of simultaneous exceedance clusters is regulated by a redescent criterion defined as the maximum of the redescent criteria of the two variables: $max(r_D, r_V) = 5$ days. We then find

$$\theta_{(D,V)} = 0.36,$$

i.e., a mean length of clusters of simultaneous exceedances of $L_{(D,V)}$ of around 3 days.

These values make it possible to draw boundaries between zones of frequencies of a similar order of magnitude. Figures 17.13 and 17.14 plot the frequencies of the conjunction of extreme hazards in the ranges considered earlier. Given the scarcity of data, it is again not possible to give a very detailed view of the most extreme levels.

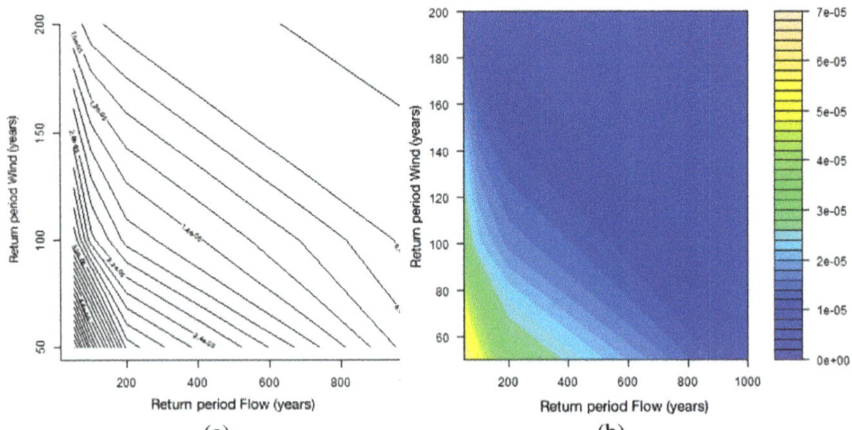

Fig. 17.14 Maps of the annual frequencies of the conjunction of extreme events (mean number of episodes per year): **a** annual iso-frequency contours estimated using the estimators $\hat{p}_T - \hat{h}_e$, and **b** zones containing annual frequencies of the same order of magnitude

17.5 Prospects

A study of wind speeds and river flows can be built upon with the help of other data and analyses. Clearly, as for any study of extreme values, the more data we have, the more precise the results can be; it would therefore be of interest to repeat the study in several years, based on longer time series. Two other ways to move forward are given below.

First, it would be possible to model the tail of the bivariate distribution of flows and wind speeds with the help of the Ledford and Tawn model. This bivariate distribution could then be used to model conditional scenarios and help answer the question: *When the Meuse river floods, what is the likely distribution of wind speeds?* The same question can be asked of the risk of flooding given a high wind speed.

Second, it is possible to define an iso-quantile curve corresponding to all combinations of flows and wind speeds having the same criticality. This criticality would then allow us to define the return periods proposed by Kendall [328, 329, 334]. From this iso-quantile curve, it would then be possible to choose a specific combination to then be used as a starting point for robustness analyses of installations potentially affected by the joint appearance of flooding and storm conditions. This choice could be based on probabilistic considerations (e.g., the most likely to occur combination) or practical ones (e.g., the combination most dangerous to the installation in question).

Chapter 18
SCHADEX: An Alternative to Extreme Value Statistics in Hydrology

Emmanuel Paquet

Abstract This chapter presents, in detail, a methodology used by EDF in many hydrological studies, which offers an alternative to the use of extreme value distributions. Based on multi-exponential weather patterns and hydrological models, this approach has proven to be robust and demonstrates that the response provided by statistical theory is not necessarily the most appropriate in practice. The existence of this approach demonstrates the need to continue the research of suitable statistical models, and the comparison of real data with models derived from theory.

18.1 Introduction

18.1.1 Objectives

The SCHADEX[1] (*Climate-Hydrological Simulation for the Assessment of Extreme Flows*) method [587] is a probabilistic method for extreme flood estimation at the outlet of a watershed. It was developed by EDF and has become since 2007 the reference method for dimensioning EDF's dam spillway*. During 2006–2016, it has been used on around 200 watersheds in France and the rest of the world (notably, Norway, Canada, and Central Europe).

SCHADEX is built around two models:

- a probabilistic *Multi-Exponential Weather Pattern* (MEWP) model which involves a combination of exponential distributions fitted to samples of rainfall measurements grouped by weather pattern. Its cumulative distribution function is of the following form:

---

[1] In French: *Simulation Climato-Hydrologique pour l'Appréciation des Débits Extrêmes*.

E. Paquet (✉)
EDF-DTG, Grenoble, France
e-mail: emmanuel.paquet@edf.fr

N. Bousquet and P. Bernardara (eds.), *Extreme Value Theory with Applications to Natural Hazards*, https://doi.org/10.1007/978-3-030-74942-2_18

$$F(x) = \sum_{i \in T_t} p_i \{1 - \exp(-\lambda_i x)\}, \tag{18.1}$$

where T_t is the set of weather patterns, λ_i a scale parameter, and p_i a weight reflecting the relative frequency of weather pattern i during a given period (typically, a season);

- a rainfall–runoff stochastic simulation process based on the MORDOR hydrological model [306] which allows rainfall episodes of any strength to be simulated under the whole range of possible water status of a watershed.

This method provides an estimate of the entire distribution (up to a return time of 10,000 years) of mean flows (for a relevant time step for the dynamics of the considered watershed), as well as that of peak flow*.

18.1.2 History

The estimation of extreme flows arriving at EDF's dams has been carried out since the 1960s using the GRADEX method, itself developed at EDF [2, 362]. The principle of the method is to deduce, using calculations based on the Gumbel-like behavior of the output of a simplified physical model (cf. Sect. 10.3), the asymptotics of the distribution of the water volume of rare floods with the help of a distribution modeling extreme rainfall.

While remaining faithful to its underlying principles, the GRADEX method has significantly evolved over the years, in particular, due to improved computing resources for dealing with rainfall measurements. It has become one of the go-to methods worldwide for dimensioning dams with regard to hydrological risks [2]. However, its industrial application has really been implemented by EDF and other major French engineering firms.

Starting in the mid-1990s, the method was regularly put into question by the hydrological community and reviewers of extreme flood studies, for the following reasons in particular.

- The return times associated with intense rainfall episodes in France by the probabilistic GRADEX rainfall model sometimes seemed very overestimated.
- The use of the exponential distribution for modeling extreme rainfall was quite controversial, in particular by statisticians, who thought it was better to use models with heavier tails coming from extreme value theory.
- When extrapolating flows, the assumption of a sudden switch from the distribution of annual maximal debits to that of extreme rainfall (above a return time of 10–50 years, depending on the watershed) appeared simplistic and difficult to justify from one study to the next.
- The difficulty of taking into account effects related to snow (rain-on-snow abatement, water sheet resulting from snowmelt).

These observations then led, in the early 2000s, to a detailed study of the methods used to fit exponential distributions, especially in the Pyrenees mountains for which the extreme rainfall quantiles were the most underestimated. Starting in 2005, subsampling with respect to weather patterns was introduced to try and form more homogeneous rainfall samples, leading to the MEWP model—a composite distribution of exponential distributions based on this sampling [303]. This model was then statistically analyzed at the French scale [304].

In parallel, a stochastic process for rainfall–runoff simulation was designed to cross-reference rainfall hazards and water status in a quasi-exhaustive way. It was based on the MORDOR hydrological model (which includes the effect of snow in mountainous watersheds), the use of which in studies and operational predictions became the norm in EDF's General Technical Division (DTG) in the early 2000s. The first simulations that allowed to get to the whole distributions of flows, including extreme quantiles, were run in 2005.

The SCHADEX method was presented to the public in 2006 at a conference organized by the France Hydrotechnical society, and the first real-world studies were published the same year. Since the end of 2007, it has been EDF's go-to method for dimensioning dams with respect to flood risk. Since that date, the majority of extreme flood studies related to EDF's dams, and those submitted to control authorities, have been carried out using this method.

Since its initial development, the method has been the subject of intensive scientific work, notable in three Ph.D. theses supervised by EDF: methods, applications, and sensitivity studies [302], an evaluation of SCHADEX in the non-stationary setting [123], and estimation at sites where measurements were not taken [606]. SCHADEX was tested and compared with other industry methods in the French project EXTRAFLO [453] and the European FloodFreq COST action [505]. It has been applied in numerous settings outside France across a range of collaborations. Important work has been carried out in Norway to evaluate the method [99, 457], with the aim of eventually merging it into reference methods used in dam safety analyses.

18.2 Methodology

18.2.1 General Information

SCHADEX is a probabilistic method that provides estimates of flow distributions (means and peaks) at an outlet* of a watershed. These distributions are constructed right out to high return times (10,000 years) using a stochastic process for rainfall–runoff simulation to simulate the response of a watershed to all types of rainfall (from normal to extreme), starting from a full range of initial hydrological conditions (soil saturation, snowpack, etc.). Stochastic simulation methods used to evaluate extreme flows are typically divided into two categories [105]:

- *Event simulation*: the hydrological model is fed using rainfall scenarios of reference, generally using return times expected for extreme flows. The watershed's initial conditions (as given in the model) are either constants or given in a statistical manner. A stochastic process is then used to model the combination of the two hazards.
- *Continuous simulation*: the hydrological model is fed with a continuous rainfall time series from the past or generated stochastically. A statistical analysis of the calculated flows induced by the model is then performed.

SCHADEX is a hybrid simulation method described as "semi-continuous". Synthetic rainfall events are independently added to historical rainfall–temperature time series (or *climate chronicle* or *record*).

Events are simulated independently, one by one, where the hydrological conditions of the watershed are updated based on the climate record up to where the simulated event was added. The use of the climate record makes it possible to take into account the distribution of the watershed's hydrological states well, given that they are involved in the simulation process with their true probabilities—obtained directly from the climate record—in particular, conditioned by rainfall occurring before the event.

18.2.2 Main Methodological Details

18.2.2.1 Time Step Used

The time step—typically a value between 4 and 24 h—defines the interval between times at which the data (flow, rainfall, and temperature) are combined, probabilistic and hydrological models are defined, and most calculations are performed. This time step needs to be close to the baseline time of typical floods observed within the watershed. It can be deduced by analyzing the hydrograph* of the largest floods observed in the watershed, or a regional analysis if such data is not available. Sometimes, only daily measurements are available and we are forced to work with these for a watershed that would be better-modeled using a shorter time step; this may lead to a loss of robustness, notably when calculating peak flows.

To simplify what follows, we will suppose that the time step of the study is a daily one.

18.2.2.2 The Shape Parameter of a Flood

The shape parameter of a flood is the ratio between its peak flow (the maximal pointwise or hourly flow recorded during the flood) and the average flow calculated over a given period. This value is different from one flood to another. This parameter, which can subsequently be introduced into the model used to simulate extreme floods, can be represented (and thus estimated) in two different ways.

1. *Mean shape parameter* K_c. This is the mean shape parameter over a collection of significant floods measured at the outlet of the watershed (or a neighboring one of a similar size with similar hydrological features). This collection of floods is made up of a set of peak or mean threshold-exceeding flows. If we have around n years of relatively continuous records, we generally select around $2n$ independent floods to work with. The nature (shape) of these floods is usually little affected by upstream developments and measurement problems. This choice supports the analysis on the upper floods occurring on the whole watershed, and allows the mean estimator of the shape parameter calculated using these floods to be used for extrapolation to extreme values.
2. *The* $K_v - K_c$ *model.* In some watersheds, a statistical link between the series of daily flood flows and the hydrograph's shape parameter for this same flood can be obtained. If j is the day where a highest daily flood flow is observed, the shape parameter of the series of daily flows is defined by

$$K_v = \frac{3Q(j)}{Q(j-1) + Q(j) + Q(j+1)}.$$

Over the set of floods chosen earlier, we can then establish the linear relationship

$$K_c - 1 = a(K_v - 1).$$

We consider that this model is *representative* (i.e., that there is a link between the daily flood flow and the shape parameter) if the coefficient a is above 0.25. If so, the model can then be applied to floods simulated by SCHADEX instead of using a constant shape parameter. Figure 18.1 shows a set of floods of the Arve river at Arthaz (see Sect. 18.5) and the corresponding $K_v - K_c$ model.

18.2.3 Areal Precipitation

Areal precipitation is the cumulative amount of rain that falls in a watershed or basin over a given time period (typically a day or year), assuming it is distributed uniformly across this whole watershed. This simplistic definition is strongly associated with the nature of the hydrological model used in SCHADEX, which does not include hydrological processes occurring at a smaller scale than a watershed.

The value of this areal precipitation is calculated using a linear combination of rainfall records within or near the watershed. The weight accorded to each measurement station could be calculated, for example, using the Thiessen polygon method (also known as Voronoi tesselation [574]) which calculates the proportion of the watershed's surface collected by each rain gauge, given its proximity to the others.

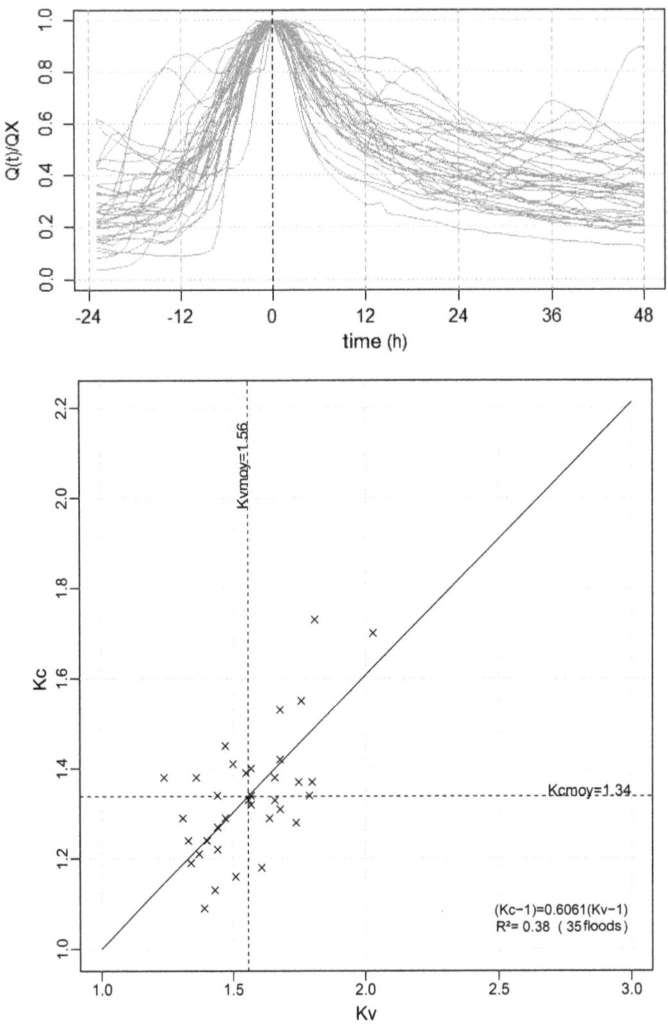

Fig. 18.1 Set of flood hydrographs for the Arve river at Arthaz, and the associated $K_v - K_c$ model

18.2.3.1 Central Rainfall/Centered Precipitation Events

Flood simulation and its probabilistic implementation is based on a simple "triangular" rainfall event, over 3 days, made up of a *central rainfall* (of at least 1 mm) and two *adjacent rainfalls* (one before and one after the central one) with smaller values than the central one. This set of three is called a *centered precipitation event* (Fig. 18.2).

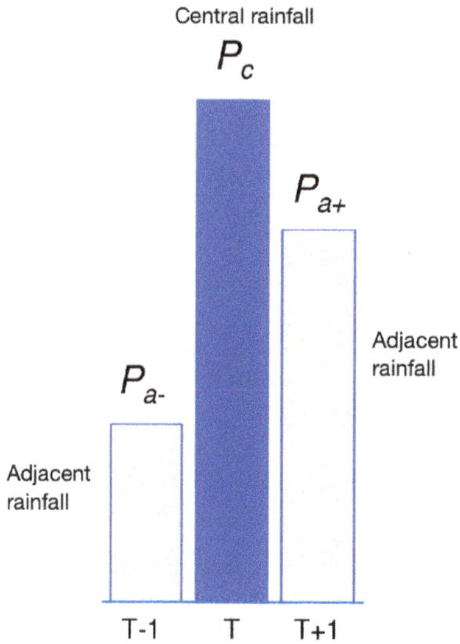

Fig. 18.2 A centered precipitation event

The MEWP model (18.1) can be applied to central rainfalls extracted from the precipitation time series of a watershed. According to this probabilistic model, working with central rainfall amounts to working on independent threshold exceedance data points.

18.2.3.2 Flow Rates After a Centered Event

We can associate with each observed centered precipitation event the maximal observed flow (over a day or at the peak) a short time (0–2 days) after the central rainfall. In some sense, this corresponds to sampling the flow related to the rainfall that produced it. In watersheds where rainfall is the cause of the biggest floods, this distribution converges asymptotically to the distribution of annual maxima, which are essentially maximal flows following intense rainfall events. Figure 18.3 shows this distribution (and the corresponding annual maxima) for flow rates of the Tarn river at Millau.

This distribution of flows is similar to the distribution of flows simulated by SCHADEX, and therefore provides an empirical distribution with which SCHADEX estimates can be compared over the range of observable return times.

18.2.3.3 Climate Record

This is a continuous watershed rainfall/air temperature time series at least 15 years long on which the stochastic rainfall/runoff simulation will be based. Provided with

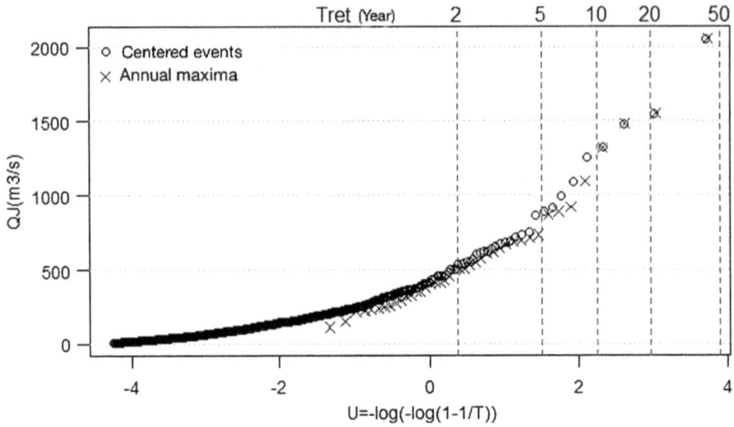

Fig. 18.3 The flow rate distribution following centered precipitation events

input data from the hydrological model, it allows for the generation of a sequence of the hydrological states found in the model (in particular: soil saturation and snowpacks), where all "observable" states are given with their true probabilities extracted from alternating dry/humid and cold/hot periods in the record.

18.2.4 The MORDOR Hydrological Model

MORDOR is a hydrological model, developed since the 1990s by EDF-DTG, that has been applied to several hundred watersheds for a range of different studies (general hydrology, inflow reconstruction, extreme floods, and climate change), as well as for operational hydrological prediction. Its main features are the following.

- *Full representation of the water cycle* in mountain watersheds: evapotranspiration, runoff, drainage flow, storage, snowmelt, etc.
- *Reservoir model*: the different hydrological "floors" of the watershed (subsurface layer, vegetation, slopes, and water table) are represented by interconnected reservoirs (see Fig. 18.4).
- *Conceptual model*: hydrological processes are given in terms of simple mathematical terms which are not intended to give a rigorous physical characterization of the phenomena (like Darcy's law or laws of hydraulics theory would do [418]).
- *While model*: the watershed is considered to be a homogeneous entity. Only its altitude distribution is taken into account, in order to deal with snow.

The input data for MORDOR is, in addition to hypsometry (the land elevation above sea level) and the area of the watershed, the mean rainfall over the latter and the air temperature (averaged or occasional). These data are continuous series of values, considered at the study's time step.

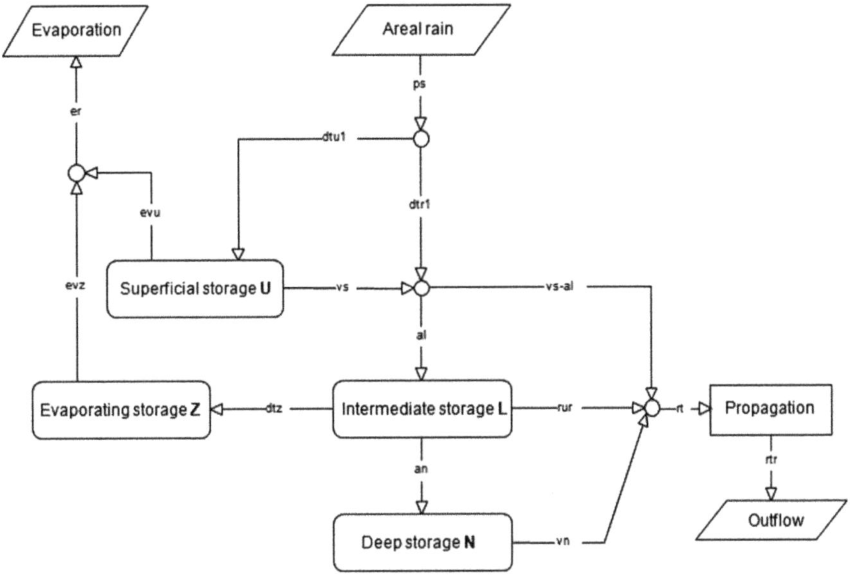

Fig. 18.4 Schematic diagram of the MORDOR model (*source* [587])

In its more complete version, the model has more than 25 free parameters to optimize. This can be done with the help of a modified genetic algorithm which generates a large number of sets of parameters and selects the best ones based on their ability to generate typical sequences of observed flows at the watershed's outlet. This ability is underlyingly defined by an objective function that combines statistical indices (Nash–Sutcliffe coefficient [547], Kling–Gupta efficiency [365, 440]) quantifying the bias between observed and modeled flows, simultaneously on the whole time series of flows, the inter-annual hydrological regimes, and the classified flow magnitudes.

18.2.4.1 Stochastic Rainfall–Runoff Simulation

This step consists in simulating the response of the watershed to simulated centered precipitation events of all magnitudes, starting from the whole range of hydrological conditions modeled with the help of the climate record. This process is summarized in Fig. 18.5. Several hundred simulated events are independently generated for each day where a centered precipitation event occurred, leading to around two million simulated events in all. Generated events are drawn randomly from the model for centered precipitation events shown in Fig. 18.2.

The maximal daily mean flows obtained for each simulated flood are stored with the probability of the associated simulated precipitation event. The latter is equal to the product of the probability of the central rainfall P_c (coming from the MEWP

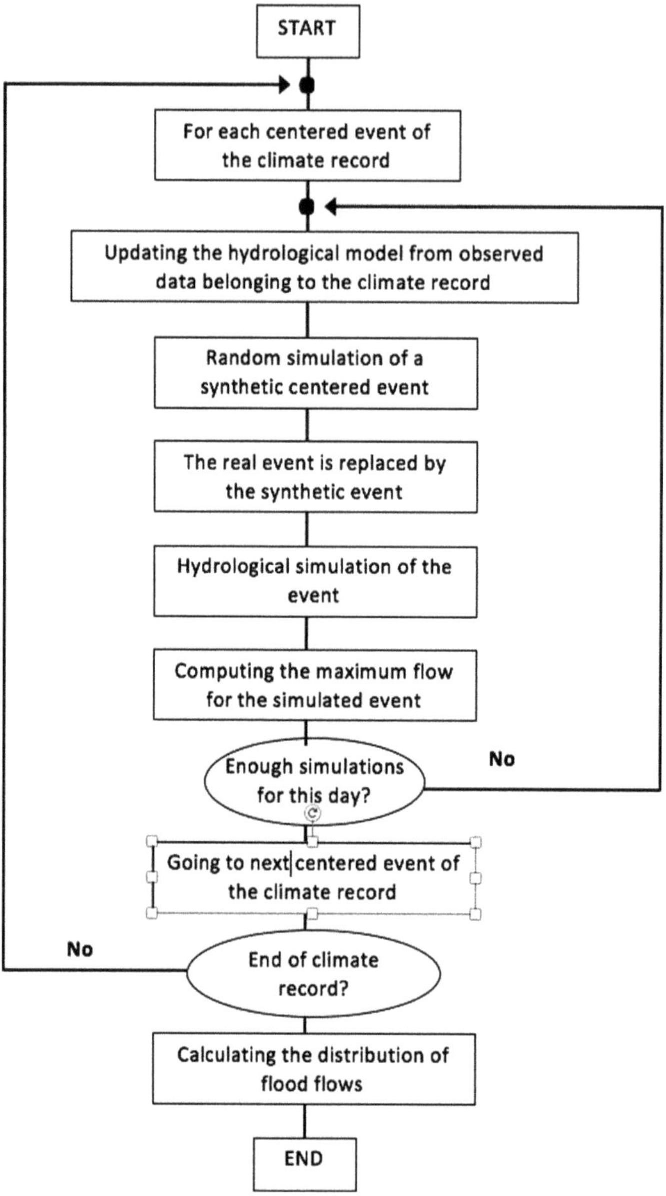

Fig. 18.5 SCHADEX simulation flowchart

model (18.1) for the season containing the day of simulation) with the probabilities of the adjacent rainfalls (preceding and following the central rainfall): P_{a-} and P_{a+}. They are possibly supplemented by a formal description of the probabilistic link between the simulated precipitations and the true rainfalls observed in the previous days. For more details on the hypotheses involved in introducing these probabilities, we suggest consulting [302, 587].

18.2.4.2 Assessing Mean Flows and Peak Flows

The cumulative distribution function of mean daily flows is then built using the set of flood flows stored for each event and their associated probabilities (after normalizing by the sum of all of the probabilities). This cumulative distribution function (CDF) is plotted using the Gumbel distribution (cf. Fig. 18.12 for the Tarn river) and compared with the CDF of the observed flows consecutive to a centered event.

The distribution of peak flows can then be established in two ways:

- using a mean shape coefficient and scaling the mean flow distribution;
- using a variable shape coefficient ($K_v - K_c$ model), by calculating it for each simulated event then forming the corresponding CDF.

18.3 Running an Analysis

SCHADEX studies should be conducted by researchers competent in statistical hydrology (able to criticize data and perform usual frequency analysis) and hydrological modeling. A good hydro-climatological understanding of the watershed is also a plus when it comes to choosing how to model and criticizing results. A successful application of the method requires an in-depth hydrological analysis of the watershed in question and the data available. To obtain credible and robust extrapolations, all assumptions and data must be carefully examined before moving to the actual computational stage. The main steps involved in applying SCHADEX are the following.

1. *Data collection.* Rainfall and air temperature measurements (in and near the watershed), flow rates (at its outlet or nearby—upstream and/or downstream), surface area and hypsometry (after a precise delineation of the watershed on a geographic information system or GIS), and historical flood data are brought together. Continuous (unbroken) series of at least 15 years of data are needed to robustly parameterize the probabilistic models and the hydrological model involved.
2. *Critical appraisal of the data.* In particular, we focus on eliminating outliers, detecting significant absences of stationarity, and identifying break points related to changing a measurement device or its location. For flow rates, we focus

on checking the height–flow rating (calibration) curves, on identifying major changes of gauging sections. It is also necessary to correct for the influence of installations on the measured flow in order to get back to an unaltered hydrological signal.

3. *Establish the water balance.* Coherently with the rest of the region, and taking into account specific features of the watershed in question, we can establish the water balance of the watershed, i.e., evaluate the mean annual precipitation that falls on it. In mountain watersheds, this volume of precipitations can be very different from the volume measured in a downstream rainfall station.
4. *Choosing the time step.* Calculations and analyses will be made at a time step close to the typical duration of floods in the watershed. This duration can be estimated using hydrographs of significant floods that occurred in the watershed. In practice, we often work with a daily time step for watersheds from one hundred to several thousand square kilometers, especially since available data (rainfall and flow rate) is often no more frequent than daily.
5. *Defining watershed rainfall.* A watershed is modeled as a single object; we therefore work with a variable corresponding to overall watershed rainfall (also known as *spatial rainfall*), corresponding to the average amount of precipitation that falls on the watershed at each time step. This value is calculated using a linear combination of the observed value at each rain gauge in the watershed. The rainfall stations are chosen in the function of their representativity and the quality of their measurements. Their corresponding weights are optimized with the help of an initial "guess" for the hydrological model, or calculated using the Thiessen polygon method.
6. *Probabilistic rainfall models.* The following probabilistic models are fitted to the watershed's rainfall time series.

   - *The MEWP model* (18.1) *for extreme rainfall quantiles* at the study's time step: the year is divided into three or four appropriate consecutive periods. It is required to group together months in which there is homogeneity in the statistical nature of central rainfalls, while hopefully forming periods with contrasting rainfall risk factors. An algorithm inserted within the fitting tool suggests a division into seasons which can then be modified by trial and error. The threshold quantile above which an exponential model is proposed for each set of months (and thus weather pattern) is chosen by a hydrologist in such a way that the model characterizes the tail of each sub-population's distribution well. The threshold will be found somewhere between the median and the 90% quantile of each sub-population, which is segmented by type of weather and season.
   - *Contingency table for adjacent rainfall events.*
   - *Correcting for previous rainfall events*: a statistical test is run to look at the duration of probabilistic dependence between observed centered rainfall events and earlier rain, and see in which months of the year the dependence is significant.

7. *Hydrological model parameter calibration.* The hydrologist needs to choose the least biased and most homogeneous periods of the flow rate's time series

data to calibrate the model, and in some cases put constraints on certain model parameters which are poorly determined by the calibration. This step generally requires a good level of experience in interpreting hydrological models, good knowledge of specific features of the watershed, and a capacity to judge the quality of hydrometric data.

8. *Calculating the shape coefficient.* A selection of flood hydrographs is conducted using a specific software tool that can access the flow data. It then performs an over-threshold selection with independence criteria between selected hydrographs. This threshold is chosen to give on average two floods per year over the whole record. Floods with strong measurement anomalies or very atypical hydrographs are manually removed. The mean shape parameter and the model for the varying one ($K_v - K_c$) are then calculated for the remaining sample.
9. *SCHADEX simulation.* The rainfall–runoff stochastic simulation process produces the distribution of extreme flows, based on the various models described earlier. A large number of simulated distributions are generated and compared with empirical distributions in order to evaluate extrapolations. For a given study defined with a daily time step, comparisons involve daily flow rate distributions, mean and peak flow distributions over 3 days, then distributions defined by months and seasons. The hydrologist needs to identify any biases that may weaken the proposed extrapolations, and correct the options and parameters related to these biases. A range of model and option combinations should be tested. In general, most attention tends to be on fitting the MEWP model (18.1) and parameterizing the MORDOR hydrological model.
10. *Validation.* Results are validated based on a number of criteria:

    - Coherence with the empirical distributions and possibly available historical data from in or near the watershed.
    - Coherence with regional values obtained during earlier studies.
    - The plausibility of the model parameters used (in particular in the hydrological model).
    - Robustness of the proposed solution with respect to the various options tested.

The time required for a typical study is generally from a few days to a couple of months. It strongly depends on the quality of the data and on the watershed being studied. Small high-altitude watersheds are usually the most complex as they tend to be poorly served by measurement stations and are subject to more variable hydrometeorological phenomena.

Numerical calculations and data processing are performed with the help of a range of tools and dedicated software packages (FORTRAN executables and `R` packages) developed by EDF for applying the method in industrial settings. A great number of graphical outputs are provided to the hydrologist to facilitate their analysis of results.

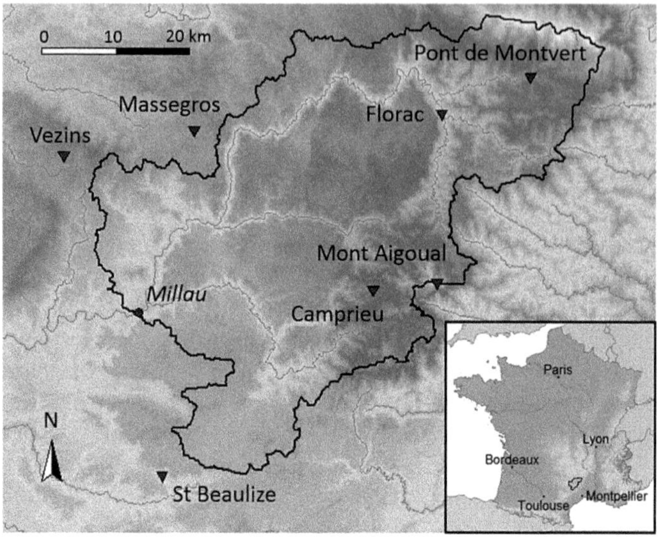

Fig. 18.6 The Tarn river's watershed at Millau (triangles are the rainfall stations used to calculate watershed rainfall)

18.4 Case Study I: Tarn River's Watershed at Millau

The Tarn river's watershed at Millau (2170 km^2) is found on the southwest face of the Massif Central, about 85 km northwest of Montpellier (Fig. 18.6). Its median altitude is 889 m, and its highest point is Mont Aigoual (1699 m), close to the sources of the Tarn. The eastern part of the watershed is on the western face of the Cévennes mountain range, characterized by granitic soil and steep slopes. In the rest of the watershed, the Tarn and its main tributaries (the rivers Jonte and Dourbie) flow through gorges that cut through the Causses limestone plateaus.

The hydrological regime of the Tarn river is Mediterranean pluvial with clear seasonal differences (cf. Fig. 18.7) including a pronounced baseflow during June–August, and significant floods due to intense rainfall during September–May, particularly in the October–January period.

The average annual flow calculated over 1969–2010 is 47 m^3/s (i.e., corresponding to a height of 684 mm when averaged out across the watershed's area), which can be contrasted with an estimated mean annual rainfall of 1262 mm falling on the watershed during that period. The largest flood during this observation period was on November 5, 1994, with a peak flow of 2600 m^3/s following a total of 270 mm rainfall over the preceding days.

The daily flow rates of the Tarn river at Millau are available from 1965, and hourly ones from 1988. The latter is of good quality with not many missing values. The watershed's rainfall is calculated using a linear combination of daily measures at seven rain gauges in or near the watershed (cf. Fig. 18.6). The weights used in

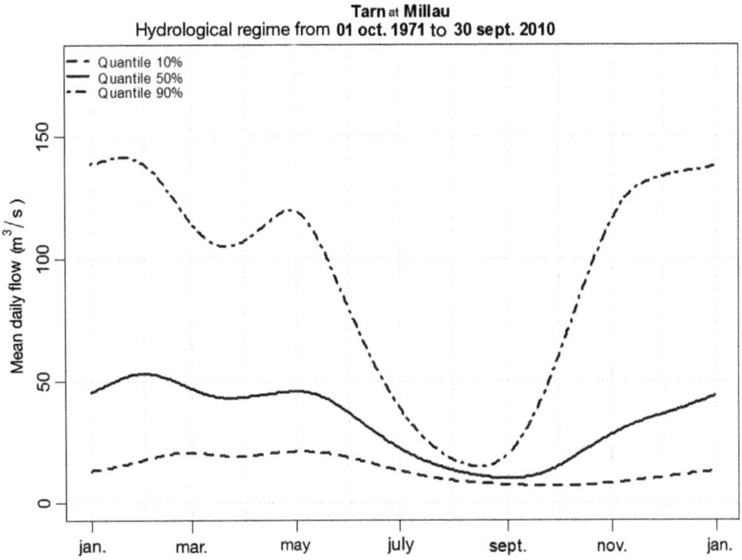

Fig. 18.7 The hydrological regime of the Tarn river at Millau

the linear combination have been tuned by minimizing the discrepancy between simulated flows from MORDOR and observed ones.

The distributions for extreme rainfall in the watershed were computed using the MEWP model (18.1), divided into four seasons: December–February, March–May, June–September, and October–November. The MEWP distribution for the October–November season of higher rainfall risk and its related marginal distributions by type of weather are shown in Figs. 18.8 and 18.9.

The risk of heavy rainfall events is quite characteristic of the Cevennes region, with more pronounced heavy rainfall in autumn; the millennial watershed rainfall for October–November is 309 mm/day, whereas it ranges 155–220 mm/day in the other seasons. During the risky season, note also that—logically for the specific climate of this region—the most hazardous weather pattern is of type 4 (Class 4 on graphics of Fig. 18.9). In terms of annual risk, the 1000-year return period for this watershed's rainfall is estimated at 311 mm/day.

The parameters of the MORDOR hydrological model have been calibrated using observed flow rates during 1971–2010. Due to the minimal role played by snow in this watershed, that part of MORDOR was not activated here, leading to a much smaller number of parameters to deal with. The statistical behavior of the obtained model is quite satisfactory. The Nash–Sutcliffe coefficient [547] between the observed and modeled flows is 0.89, and the coefficient determining the flow variations is 0.80. The hydrological regimes and distribution of the observed and modeled flows are shown in Figs. 18.10 and 18.11, highlighting that the hydrological model is well able to exhibit observed seasonal effects in the watershed's flows and the distribution of its flood flows.

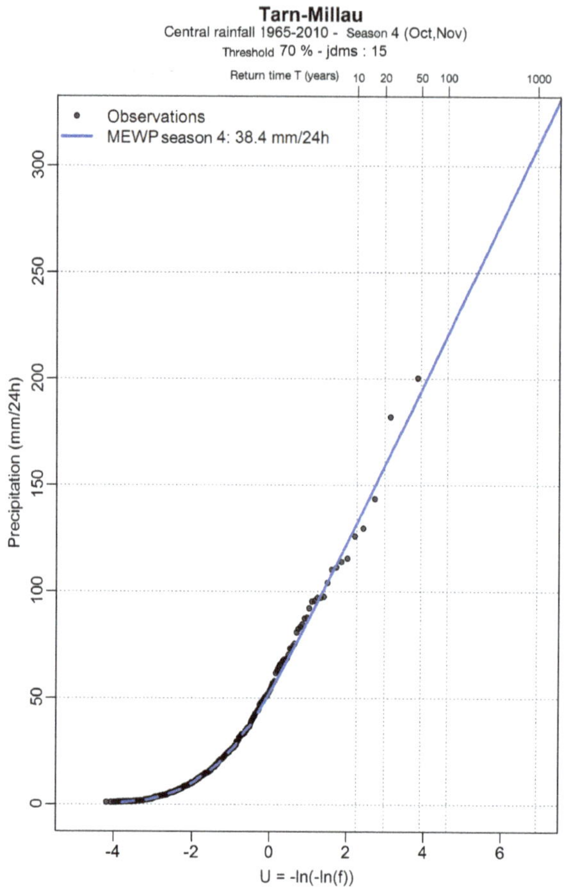

Fig. 18.8 MEWP distribution with respect to weather patterns for the rainfall in the watershed of the Tarn river at Millau (during October–November)

The SCHADEX simulation was performed over the climate record during 1971–2010. A total of 2106 events were simulated with central rainfalls located between 1 and 490 mm: more precisely, 785 events were simulated for each day in the record where a central rainfall was observed. The CDF of the daily flow rate of the Tarn river at Millau can then be constructed from the simulated flood events, and compared with the observed one for events following episodes of rainfall over the period 1969–2010; see Fig. 18.12. It should be emphasized that the proposed CDF, even if it appears as a continuous line in the graph provided, does not come from an analytical formula but is actually constructed using the 2106 simulated floods. It is extremely coherent with the empirical distribution of true flow rates, which is reassuring about the robustness of the extrapolation proposed. The daily flow with a return time of 1000 years is estimated at 4795 m^3/s.

One feature of SCHADEX is that it covers almost entirely the range of extreme rainfall × hydrological state (of the watershed) pairs with its flood simulations at all times of the year. Thus, for each month we can construct a cumulative distribution

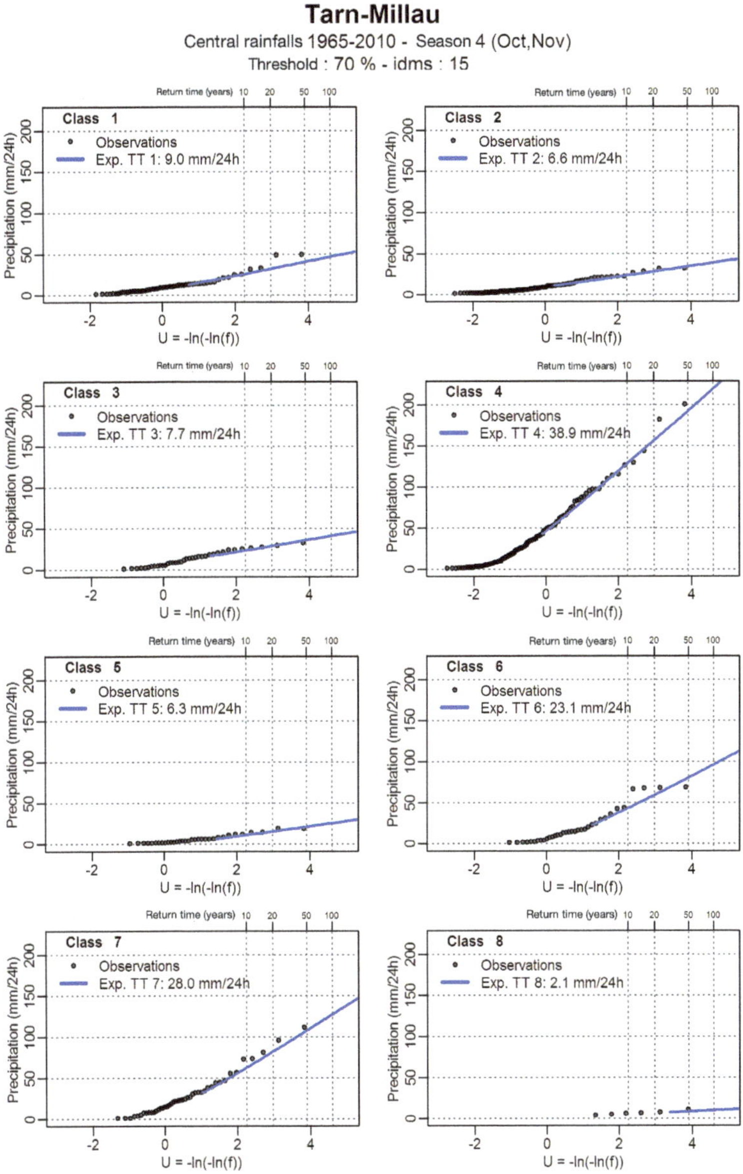

Fig. 18.9 Marginal distributions with respect to weather patterns for the rainfall in the watershed of the Tarn river at Millau (during October–November)

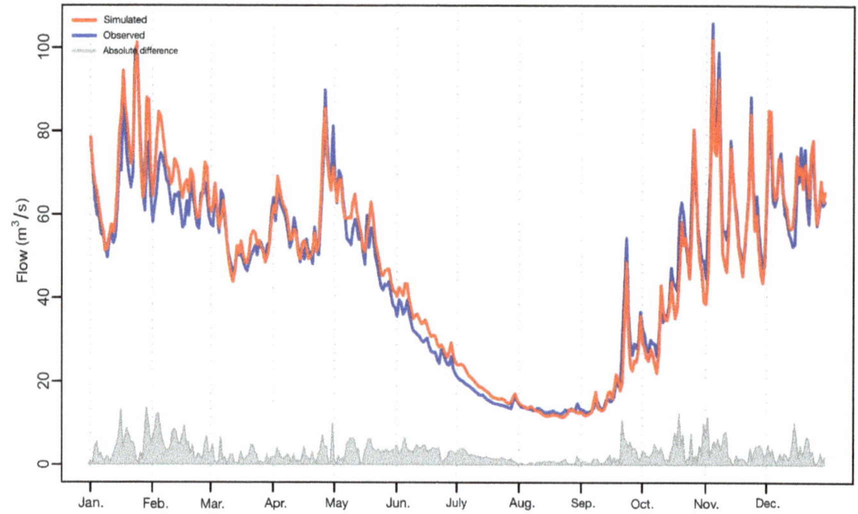

Fig. 18.10 Hydrological regime of observed and modeled flow rates of the Tarn river at Millau

Fig. 18.11 Distribution of observed and modeled flows of the Tarn river at Millau

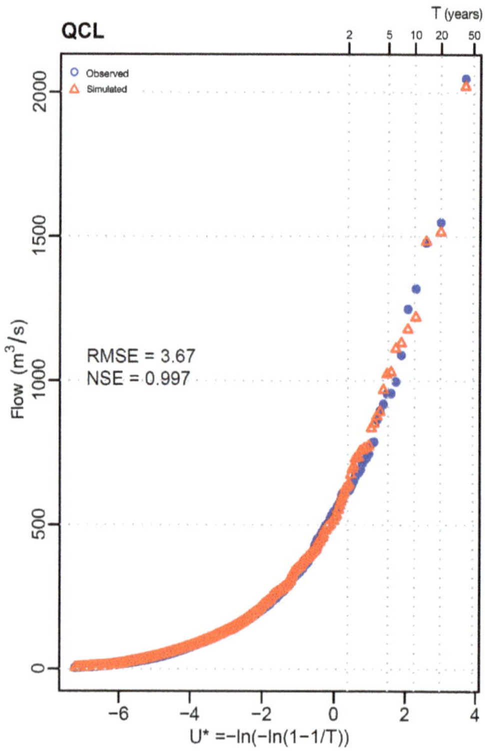

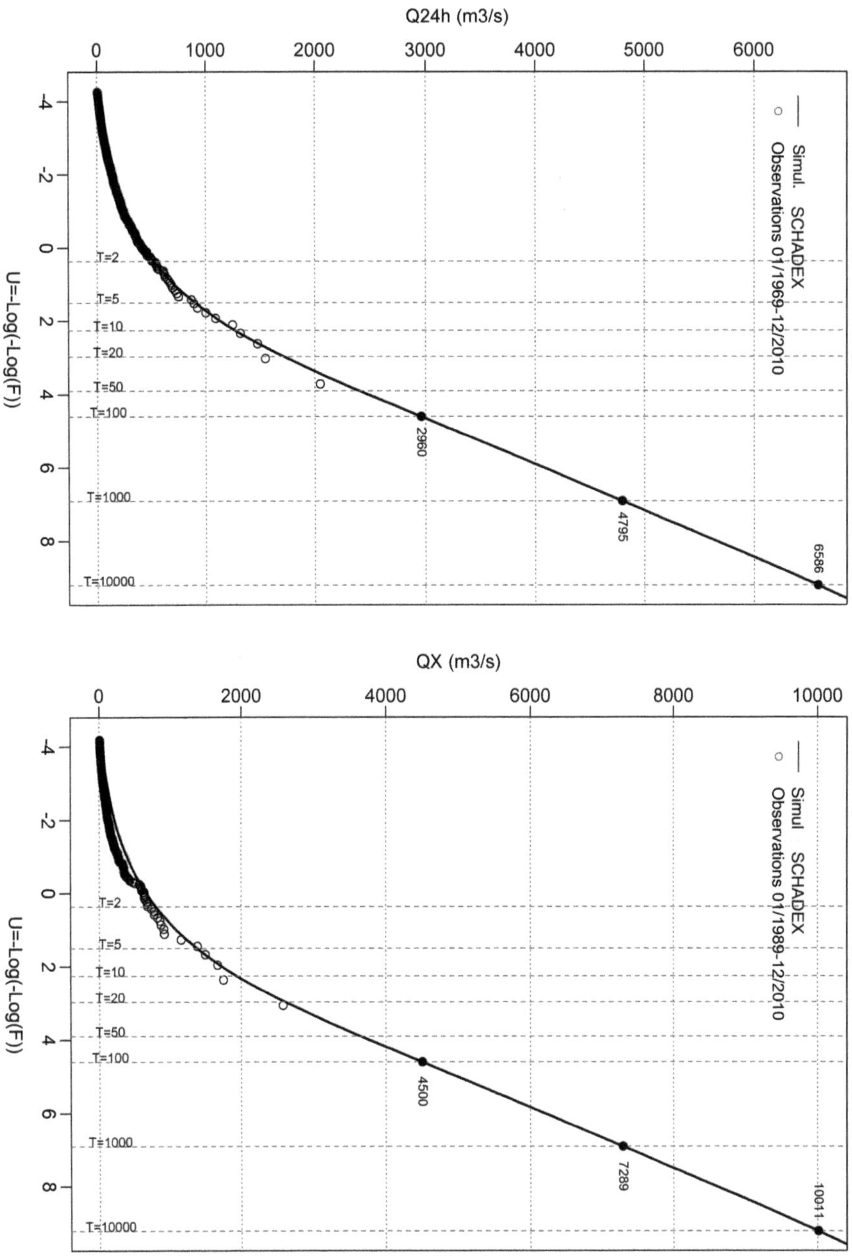

Fig. 18.12 Distributions of daily flow rates (Q24h) and peak flows (QX) estimated by SCHADEX for the Tarn river at Millau

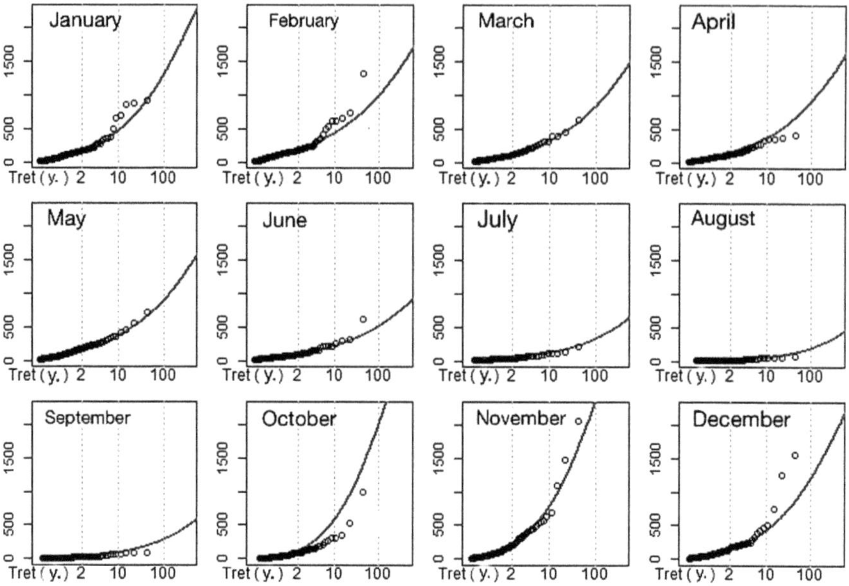

Fig. 18.13 Monthly distributions of empirical and simulated daily flows

function of simulated flood events and compare it with the corresponding empirical one (see Fig. 18.13). These monthly distributions clearly show differences in the risk of flooding in the Tarn river's watershed depending on the time of year, with a clear risk identified during October–May, peaking strongly during October–November. We also remark that even at the monthly scale, there is a tight fit with the empirical distributions (perhaps slightly less so for October and December), which reassures us that seasonality in the hydrological processes is reproduced well by the set of models encompassed within SCHADEX (especially MEWP and MORDOR).

The mean shape parameter of flood events was calculated using a set of 31 hydrographs recorded at Millau between 1989 and 2010 involving floods with peak flows of 334–2582 m^3/s. These hydrographs, synchronized at the peak and mapped to 1-d, are shown in Fig. 18.14 along with the mean hydrograph. The mean shape parameter of these 31 flood events is 1.52.

The cumulative distribution function of peak flows of the Tarn river at Millau can be formed using that of daily flows simply by multiplying by the mean shape parameter 1.52 from above. This distribution can be compared with the empirical one of maximal peak flows following centered rainfall events during 1989–2010 (Fig. 18.12). We see that the SCHADEX distribution is still very coherent with respect to the empirical one, suggesting that the extrapolation involved can be used with confidence. The peak flow with a return time of 1000 years is estimated at 7289 m^3/s, corresponding to a specific flow of 3.4 $m^3/s/km^2$.

The study of extreme flows of the Tarn river at Millau is helped by the quantity and quality of precipitation and hydrometric data available for this particular watershed,

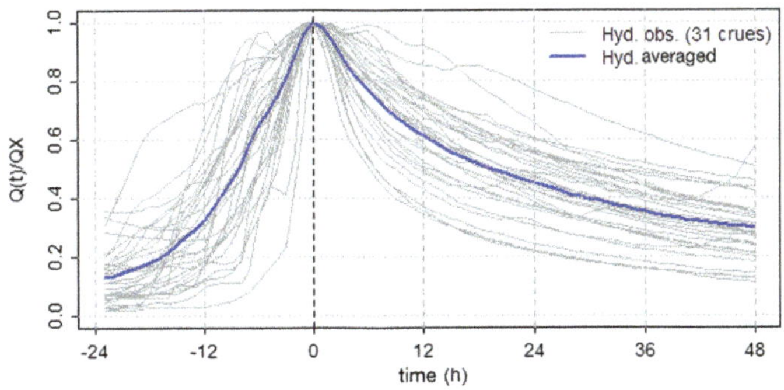

Fig. 18.14 1-d mappings of hydrographs of 31 floods of the Tarn river at Millau, synchronized at their peaks

as well as the clarity of the hydrological processes at the origin of extreme flood events in this watershed: the risk of extreme rainfall is highly contrasted, very concentrated on the beginning of autumn, and a large amount of flood data is available for that period, helping the search for parameters of the hydrological model that lead to a good fit to the data.

18.5 Case Study II: Arve River's Watershed at Arthaz

The watershed of the Arve river at Arthaz (1650 km^2) is located in Haute-Savoie (Fig. 18.15). It contains in particular the Mont Blanc and Aiguilles Rouges massifs in its eastern part, as well as part of the Aravis mountain range. It stretches over a large range of altitudes, culminating at the peak of Mont Blanc (4809 m) with a median altitude of 1402 m, and its outlet is at an altitude of around 400 m. Glaciers cover about 7% of its surface, especially toward the east. Geologically, there is quite a difference between the eastern part—essentially crystalline, and the west, made up of the limestone Aravis and Bornes massifs. The main tributaries of the Arve river are, from upstream down, the Diosaz, Bon-Nant, Giffre, and Borne rivers.

The Arve has a snow-glacier hydrological regime (Fig. 18.16), with a maximum in June entirely due to snowmelt and flows that usually remain steady over the year due to well spread out rainfalls. A moderate median altitude in the western part of the watershed (the most exposed to precipitations) is also in favor of this steady flow.

The mean annual flow calculated over 1991–2012 (the period in which available hydrometric data are the most reliable) is around 65 m^3/s (i.e., corresponding to a height of 1246 mm when averaged out over the watershed's area), with the watershed's mean annual precipitation estimated at 1737 mm. In terms of daily flows, the largest one over the period of available data (1961 and after) was 535 m^3/s on September 22, 1968.

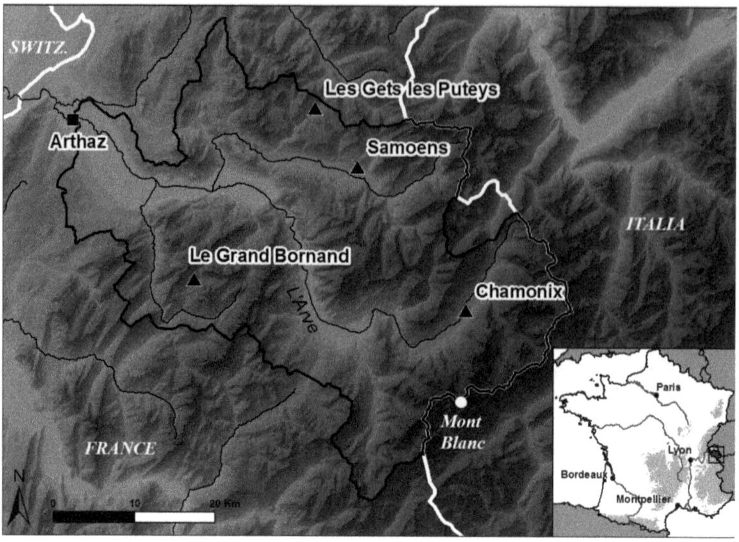

Fig. 18.15 Arve river's watershed at Arthaz (triangles are the rainfall stations used to calculate watershed rainfall)

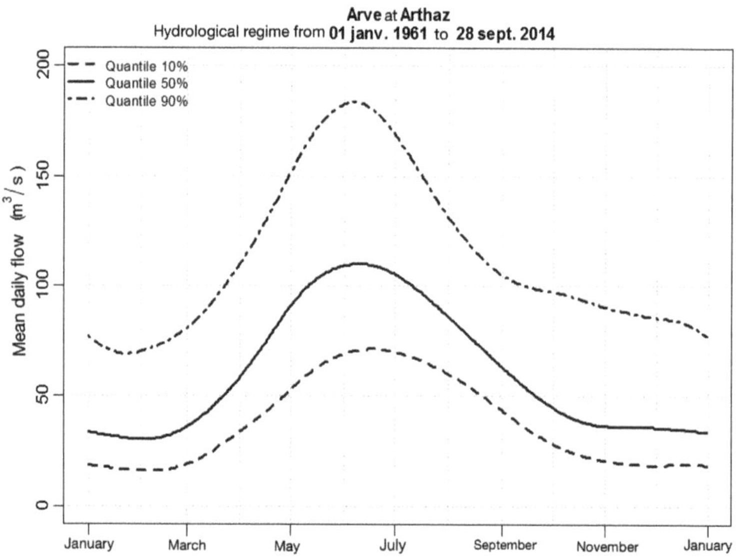

Fig. 18.16 Snow-glacier hydrological regime of the Arve river at Arthaz

Daily flow rates for the Arve river at Arthaz are available since 1961, and hourly ones since 1994. This hydrometric time series is relatively complete but suffers from significant measurement biases linked to the upper part of the calibration curve (which converts water heights to flows). The watershed's rainfall can be calculated over the period 1966–2013 using a linear combination of the daily measures taken by the four hydrometric (or rainfall) stations (see Fig. 18.15). As for the Tarn river example, the four weights are optimized with help from the hydrological model.

Modeling extreme rainfall in the watershed using MEWP distinguishes four different seasons: January–March, April–May, June–August, and September–December. The distributions for the latter and April–May, as well as the corresponding data, are shown in Fig. 18.17. The risk of severe rainfall is maximal during September–December, with a value of 157 mm for a 1000-year return time. The value ranges 96–127 mm in the other periods. Seasonal differences are therefore less pronounced than those in the Tarn example. In the September–December and January–March periods, type 2 weather patterns (known as *stationary fronts*) bring the highest risk of severe rainfall. In terms of annual risk, the 1000-year return period is estimated at 159 mm/day.

The MORDOR parameters were calibrated using the observed flows over the years 1991–2012, because the hydrometric time series seemed to have the highest quality for this period. Obviously, snow (and ice) accumulation and melting processes were taken into account in this model. The statistical performance of the model is acceptable: the Nash–Sutcliffe coefficient [547] between the observed and modeled flows was 0.85, similar to the coefficient determining flow variations. The regimes and distributions of the observed and modeled flows are shown in Figs. 18.18 and 18.19 and they highlight that the hydrological model is capable of reproducing seasonal flows and the distribution of the most extreme ones (due to high rainfall and the peak period for snowmelt). The largest values appear to be slightly underestimated by the model here, but this parameterization of the model is preferred over others based on a range of criteria (plausibility of parameters, robustness of the extrapolation, etc.).

The SCHADEX simulation was then run on the climate record for the years 1962–2012. Again, 2016 events were simulated with central rainfalls taking values between 1 and 240 mm.

The CDF of the daily flow rate of the Arve river at Arthaz was then constructed from the simulated flood events, and compared with the observed one for events following episodes of rainfall over the period 1961–2013; see Fig. 18.20. At this level of the study, all of the available hydrometric information was used (while only the period 1991–2012 was used for model calibration) in order to better evaluate the robustness of the proposed extrapolations.

The results are coherent with the empirical distribution of observed flows, though the extrapolation moves away from it for 10-year return periods and longer. To clarify this, we can take a closer look at the simulated seasonal distributions and the corresponding recorded data (Fig. 18.21). We see that the hydrological risk does not appear to vary much across seasons up to a 20-year return time; however, extrapolating to higher return times does confirm that the September–December period is the high-risk one.

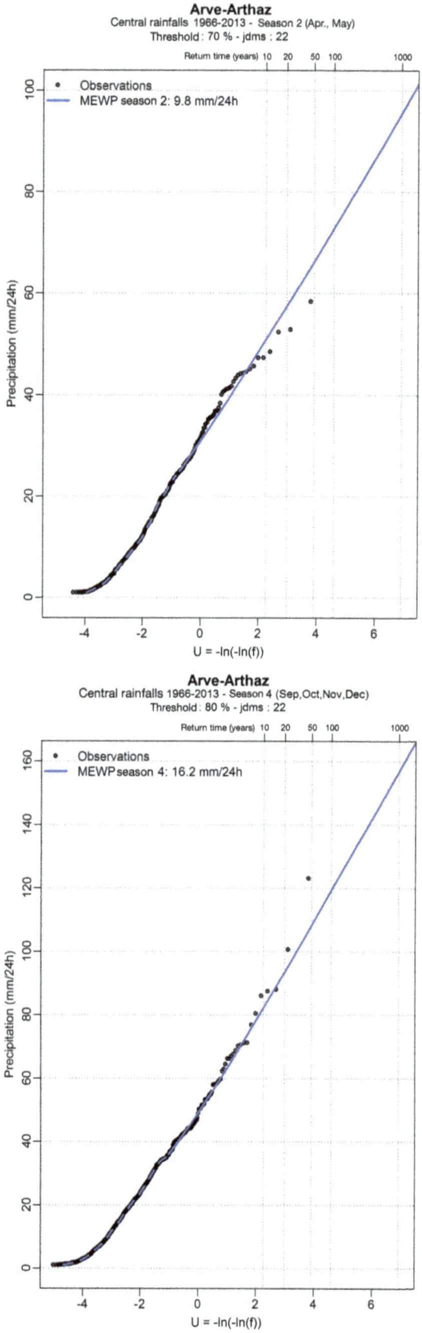

Fig. 18.17 MEWP distributions of rainfall in the Arve river's watershed at Arthaz (April–May and September–December)

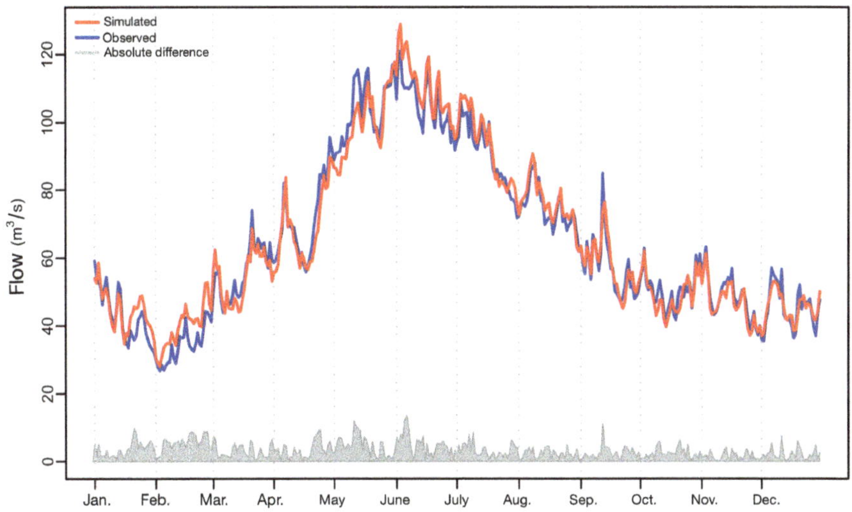

Fig. 18.18 Hydrological regime of the Arve river at Arthaz

Fig. 18.19 Distribution of observed and modeled flow rates of the Arve river at Arthaz

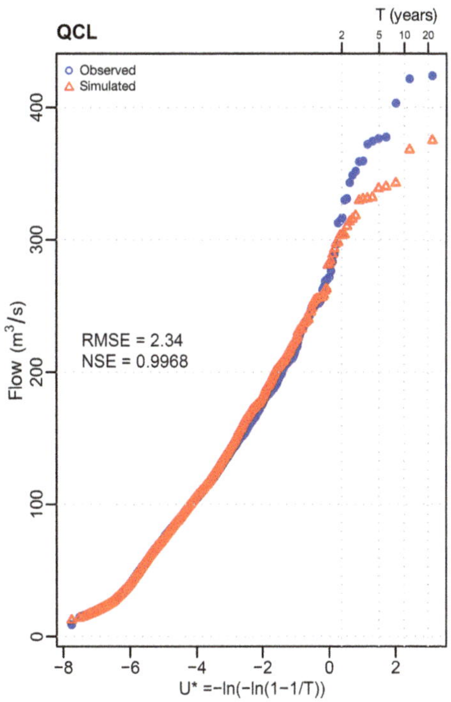

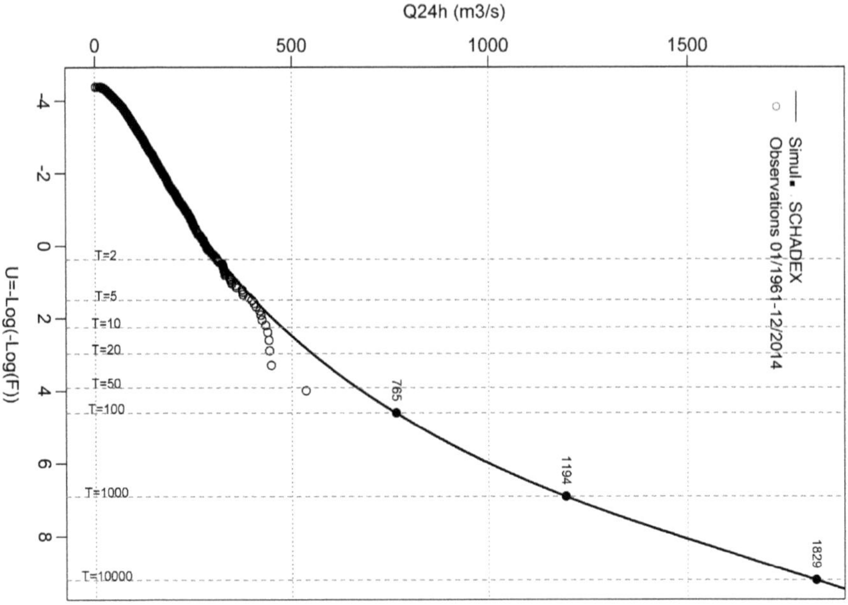

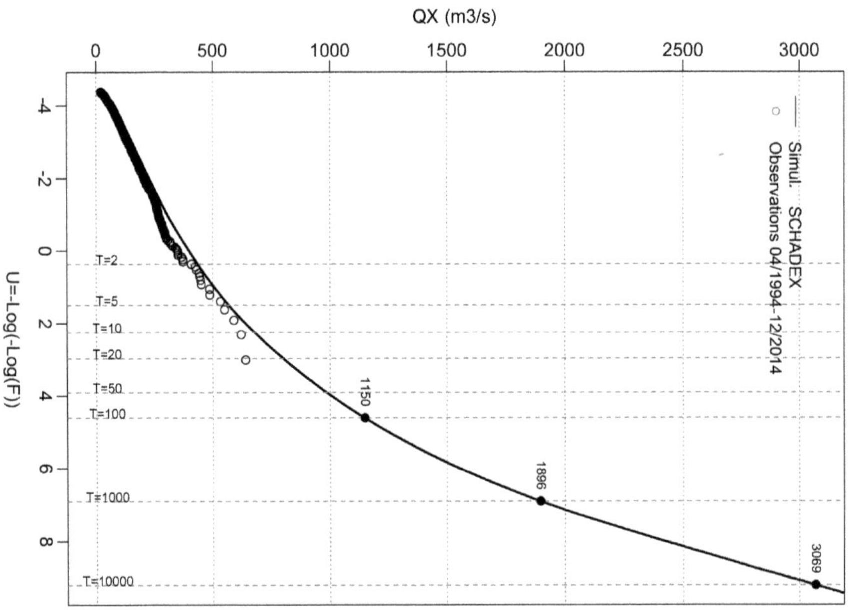

Fig. 18.20 Distributions of daily flow rates (Q24h) and peak flows (QX) estimated by SCHADEX for the Arve river at Arthaz

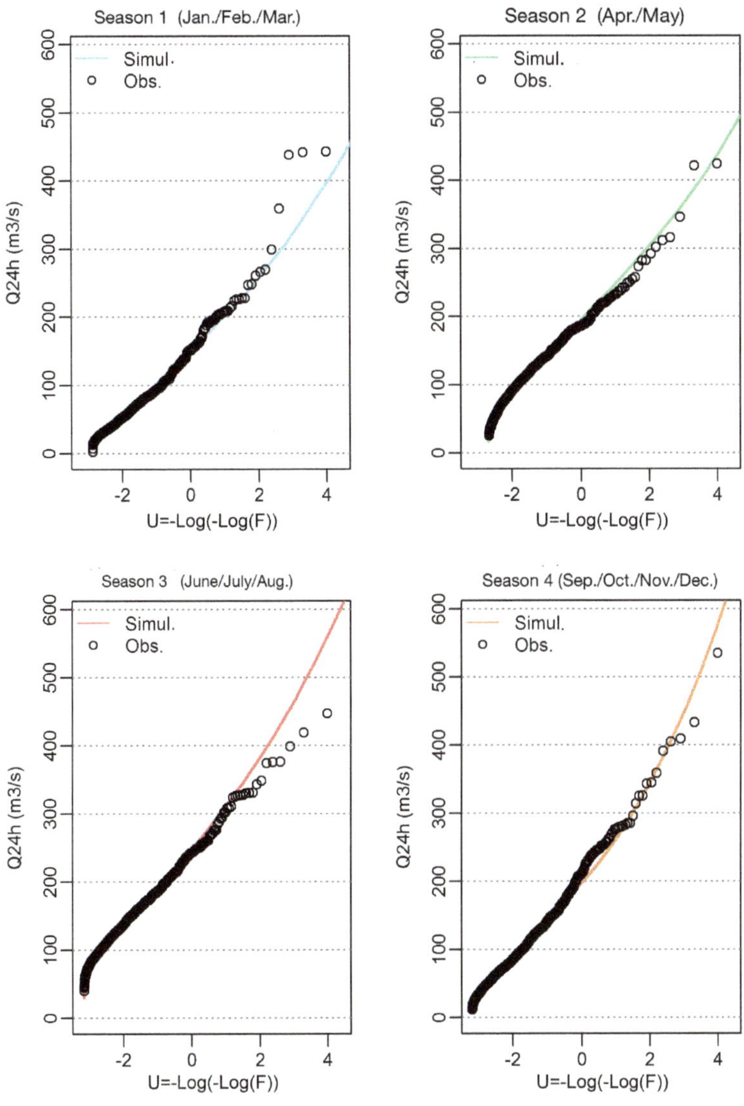

Fig. 18.21 Distributions of daily flow rates for each season estimated by SCHADEX for the Arve river at Arthaz

The concordance with the empirical seasonal distributions is good, in particular for the at-risk season. For January–March, four events have empirical return times which deviate from the distribution simulated by SCHADEX. The daily flow corresponding to a 1000-year return time is estimated at 1194 m^3/s.

In this study, the calculation of peak flows is performed using the variable shape parameter ($K_v - K_c$) model based on a set of 35 hydrographs recorded at Arthaz between 1994 and 2014, with peak flows in the range 272–640 m^3/s. This model is written as

$$K_c - 1 = 0.61(K_v - 1).$$

The ratio of the peak flow over the mean daily flow is thus calculated for each simulated flood event, and depends on the dynamics of daily flows. The CDF of peak flows of the Arve river at Arthaz has thus been produced by simulation and subsequently compared to the empirical distribution of maximal peak flows consecutive to centered rainfall events over the period 1994–2013 (Fig. 18.20). The coherence between the SCHADEX distribution and the empirical one is acceptable. The peak flow with a 1000-year return time is estimated at 1888 m^3/s (i.e., a specific flow of 1.14 $m^3/s/km^2$). The ratio of the peak flow over the daily flow quantile increases as the return time T does; when $T = 10$ years, it is equal to 1.42, rising to 1.58 for $T = 1000$ years.

An estimation of the risk of extreme floods of the Arve river is made complicated by several things:

- varying quality in the bhydrometric time series, depending on time and the magnitude of observed values;
- a risk of extreme rainfall that does not change much during the year;
- high recorded floods, caused simultaneously by strong rainfall and snowmelt in varying proportions from one occurrence to the next; treating such data is a real challenge for the hydrological model and for selecting its parameters robustly.

Nevertheless, SCHADEX does allow us to provide a detailed view of hydrological risks by attempting to characterize seasonal effects in the statistical and hydrological processes involved as well as possible. This ability to bring together a large quantity of hydro-climatological information for making robust extrapolations is without doubt one of the most important features of the method.

Appendix A
Glossary

Accretion. Accretion refers to the formation and growth of a body, structure or object by the addition and/or agglomeration of matter.

Albedo. Albedo is the reflective power of a surface, defined as the ratio of the reflective light energy to the incident light energy.

Anchor variable. An anchor variable is a variable about which expert knowledge can be added using a decision heuristic. The nature and order of magnitude of this variable is known to the expert, who can use its properties to state: "the variable X takes values between x_a and x_b". An observable variable, i.e., one that is directly accessible to the expert, often has good anchoring properties.

ANEMOC. The *Atlas Numérique d'États de mer Océanique et Côtier* (ANEMOC: http://anemoc.cetmef.developpement-durable.gouv.fr) is a database for oceanic data (climate, bathymetry, wave energy, etc.) developed by EDF-LNHE with the help of CETMEF, built using retrospective simulation (with the TOMAWAC software) over a period of around 24 years (1979–2002) for the Atlantic, Channel, and North Sea coasts, and 30 years (1979–2008) for the Mediterranean Sea.

Asymptotic behavior. The asymptotic behavior of a mathematic object $\Sigma(t)$ depending on a value t means a description of what happens to this object as t tends to a certain limit. If $t = n$, where n the number of observed values and $\Sigma(t) = \hat{\Sigma}_n$ a statistical estimator of the quantity Σ, then its *asymptotic properties* characterize the nature of its convergence or divergence to a value Σ_0 when $n \to \infty$; the function $n \mapsto \hat{\Sigma}_n$ tends to a (straight line) asymptote $n \mapsto \Sigma_0$.

ASN. The French *Autorité de Sûreté Nucléaire* (ASN) is the independent administrative authority which has the role, on behalf of the state, to overlook civil nuclear safety and radiation protection, and to disseminate information on risks associated with nuclear activity.

Baseflow. The baseflow is the minimal annual flow of a watercourse, corresponding to a dry period due to drought or water extraction (e.g., for agricultural use).

CFBR. The *Comité Français des Barrages et Réservoirs* (CFBR; http://www.barrages-cfbr.eu) is an association composed of members representing the administration, national societies, public establishments, local authorities, businesses,

N. Bousquet and P. Bernardara (eds.), *Extreme Value Theory with Applications to Natural Hazards*, https://doi.org/10.1007/978-3-030-74942-2

and engineering companies, which is coordinated by EDF and whose role it is, through various internal working groups, to establish technical criteria and recommendations for the design, implementation, monitoring (etc.) of dams and tanks/reservoirs (e.g., retention basins) in France.

Chop. Waves generated by local wind along an expanse of water (sea, estuary, lake, etc.).

Cluster. A cluster is a group of consecutive and correlated values from the same process, sharing a certain feature. For example, a cluster of exceedances can be seen as a wave of values above a given threshold.

CTPBOH. The *Comité Technique Permanent des Barrages et Ouvrages Hydrauliques* (CTPBOH) is a consultative body of the French Ministry of Enviroment, created in 1996 and composed of experts chosen according to their particular technical skills in the field of hydraulic structures. This body is consulted in advance about all construction or modification projects related to large dams and dikes.

Decision heuristic. A decision heuristic is a cognitive shortcut used by an individual to make a quick judgement (or decision) about a phenomenon known at least partially to them, based on their knowledge of the situation. Untangling the nature of the heuristic used (for instance, whether it is universally valid or not, or linked to personal experience, etc.) can help to protect against potential biases contained within it. See [253, 335] for more details and examples.

Declustering. A *cluster* is, in the context of extreme value statistics, a sequence of consecutive exceedances of a threshold that can be considered as part of the same *wave*. *Declustering* is a procedure to replace such dependent data with a single data point (for example, the maximum exceedance) in order to produce an independent sample.

Entropy. The entropy $H(X)$ of a random variable X with density $f(x)$ is a measure of the information quantity contained in the distribution of X. In the discrete case, Shannon's entropy formula defines it as the minimum number of *bits* required to reconstruct a message coded in the data. The concept can be extended to the continuous case with the formula $H(X) = -\mathbb{E}_f[\log f(X)]$. Every integrable density has finite entropy, and the greater the entropy, the less informative the distribution f is considered to be. This concept is one of the most important in Shannon's (probabilistic) information theory, and also corresponds to the notion of entropy in the physics (thermodynamics) sense. See [185] for more details.

Eustastic change. Eustastic change is a measure of variation in the mean sea level, with respect to the supposed fixed continents. This cyclic phenomenon of raising/lowering of the sea level, in the order of mm/year, has several causes: melting glaciers, changes in the tectonic plates, rate of activity in oceanic ridges, etc.

Extreme tide. An extreme high tide is an abnormally high or low tide, induced by unusual weather conditions on top of the usual tides induced by the moon and sun.

Flood plain. A watercourse's flood plain is the zone over which water flows if it breaks its usual banks; this can be a vast area if flooding occurs

Gauging. In hydrology, gauging is the process of measuring water flow at a gauging station at a given time. This can be performed using kinematics, dynamics (hydraulics), or physics (using dilution).

Heteroskedasticity. Heteroskedasticity describes the nature of a collection of random variables if some subcollections of these variables have different values of variability (e.g., variances).

High tide exceedance. A high tideexceedance is a positive *high tide residual*, or *surge*.

Hydro Bank. Hydro Bank (www.hydro.eaufrance.fr) is a department of the French Ministry of Ecology, Sustainable Development and Energy, storing the water level values recorded at stations in service on French rivers by the State services (Regional Environmental Directorates, of Development and Housing (DREAL), Territorial Departmental Directorates (DDT), water agencies, etc.), companies such as EDF, development companies such as the Compagnie Nationale du Rhône and various research organizations (IRSTEA, universities...).

Hydrograph. A hydrograph is a plot of the temporal variation in water flow, either at a specific place in a watershed (e.g., the outlet) or at some point along a watercourse. A flood, shown on a hydrograph, is characterized by a *rising limb*, a peak, following by a *recession limb*.

Industrial risk. Industrial risk is generally defined as the product of a probability of occurrence of a hazard by a measure of the cost incurred by it. Note that according to the ISO 73:2009 standard [7], a less formal and more general definition of risk is *the effect of uncertainty on the achievement of objectives*.

Kurtosis. The *kurtosis* or Pearson flattening coefficient of a probability distribution of a real random variable is defined as the third order moment of this reduced centered variable.

Non-inductive logic. Non-inductive logic is a type of logic that is not based on the observation of specific events in order to deduce the general behavior of a phenomenon.

Outlet (of a watershed or bassin). The outlet of a watershed or bassin is the point of convergence of all waters belonging to it. It can be a lake, watercourse, sea, ocean, etc.

Outlier. An outlier is a data point whose value is significantly different to the other ones in a statistical sample [619].

Peak flow. Peak flow is the maximum instantaneous flow recorded during a high water event.

Phenology. Phenology is the study of the appearance of periodic (generally annual) events that depend on seasonal variation in the climate, usually in terms of living things (trees, vines, etc.). It can help us to distinguish markers of catastrophic natural events from the past, and to reconstruct historical extreme values.

Possibility theory. Possibility theory is a mathematical theory, alternative to probability theory, for dealing with certain types of uncertainty.

Reinsurance. Reinsurance means insuring insurers themselves. It works by transferring to a company (the reinsurer) a random risk (typically, the consequences of a extreme hazard event) in return for payment corresponding to the transferred risk and its characteristics. This mechanism originates in the industrial revolution in Germany, where insurance companies were unable to cover the totality of risk potentially affecting rapidly developing industry.

Risk aversion. Risk aversion is a behavior that pushes an arbitrator (real or virtual, such as a questioned expert) to avoid that the (possible) consequences of their arbitration could be turned against him or her. This leads to bias corresponding to excessive caution, or even a refusal to provide expertise. Thus, an expert questioned about the imminence of an extreme hazard may tend to over-estimate its likelihood, because a posteriori, an under-estimation could be much more costly to them professionally.

Risk neutrality. A risk neutral arbitrator is one whose behavior is not affected by personal biases.

River bed. The river bed of a stream, river or creek, is the physical space in which water flows normally, held in by the watercourse's banks.

SHOM. The *Service Hydrographique et Océanographique de la Marine* (SHOM) is a French public administrative body, under the supervision of the Ministry of Defense, whose mission is to understand and describe the oceanic physical environment in its relationship with the atmosphere, seabed and littoral zones, to predict the evolution of these interactions, and to disseminate this information (www.shom.fr). Its flagship activity is to build reference databases characterizing the geophysical, maritime, and coastal environments.

Skewness. The skewness or asymmetry of a probability distribution of a real random variable is defined as the third order moment of this reduced centered variable.

Spillway. A spillway is a structure attached to dams which helps to control the maximum water level behind them. It does so by diverting water down itself if more water arrives at the dam than can be safely stored behind it, thus protecting the structure.

Swell. Swell is a relatively regular wave movement of the sea surface. Unlike chop, swell is generated by the wind blowing away from the coast. It is more regular than chop in terms of period and height.

Tidal range. The tidal range is the difference in level between high and low tide of a tide. Specifically, it is generally defined, for a given day and within a high water—low water interval, as the difference in water level between the high water level and the low water level immediately following or preceding it. It is sometimes referred to as *tide height* (a term sometimes also used for height of water) or *tide amplitude* (referring to half tide, i.e., the difference in height of water at high or low water with that at mid-tide).

Watershed. A watershed is a territory, surrounded by natural boundaries, whose waters all converge toward the same exit point—called an *outlet*.

Water status. Water status refers to the quantity of water retained in soil (qualitatively: from very dry to very wet soil).

Appendix B
Software Tools

Tools Used by EDF

The calculations of most of the examples reported in this book have been carried out using the packages available in free access on the Internet and dedicated to extreme statistics. See http://www.r-project.org for more information on the environment `R`.

Other software tools are used at EDF. The ASTEX software has been developed by the LNHE to merge, restructure, and improve three existing tools for statistical adjustment of extreme laws to time series of flows (CRUE), swells (EVENAL) and excess tides (SWELL). Today ASTEX is used in the following laboratories (see [40, 57] for case studies):

- the *National Laboratory of Hydraulics and Environment* (LNHE), belonging to EDF Research & Development;
- the *General Technical Division* (DTG), an entity of EDF, in particular in charge of the measurement and expertise concerning the hydrometric and meteorological forecasts necessary for the proper functioning of EDF's hydroelectric and nuclear fleet;
- le *Centre for Studies and Expertise on Risks, the Environment, Mobility and Development* (CEREMA): https://www.cerema.fr/.

We can also mention the *EXTREMES* software tool, running in a MATLAB[1] environment and which gathers different tools dedicated to the study of extreme values. It is the result of a collaboration between the IS2 team of INRIA Rhône-Alpes and EDF R&D. A toolbox running in a SCILAB environment is currently being developed at the LNHE.

Useful R Packages

List of useful `R` packages used for the treatment of case studies:

- `evd` (functions for Extreme Value Distributions)
- `evir` (Extreme Values in `R`)

[1] http://mistis.inrialpes.fr/software/EXTREMES/accueil.html

N. Bousquet and P. Bernardara (eds.), *Extreme Value Theory with Applications to Natural Hazards*, https://doi.org/10.1007/978-3-030-74942-2

- `extRemes` (*Extreme value toolkit*): its graphical interface allows it to be of simple use; various methods used in [173]
- `ismev` (An Introduction to Statistical Modeling of Extreme Values)
- `Renext` (*Renewal method for extreme values extrapolation*): developed by IRSN with user's graphical interface on `R` (`RenextGUI`) ;
- `POT` (Peaks Over Threshold)
- `RFA` (Regional Frequency Analysis)
- `nsRFA` (Non-supervised Regional Frequency Analysis)
- `chron` (Chronological objects which can handle dates and times)
- `Kendall` (Kendall rank correlation and Mann-Kendall trend test)
- `texmex` (Statistical Modelling of Extreme Values)

It should be noted that some examples of studies under `R`, using the graphical interface of the `extRemes` package on an example of wind data, are detailed in depth in [576].

Other Software Tools

The environment *S-plus*, competitor of `R`, also presents a chargeable set of libraries for the use of extreme statistics (rather in finance): `The EVIS library`, `The Stuart Cole's Routine` and `The S+Finmetrics Module`.

Software tools comparable to *EXTREMES* are also available from RISKTEC (*Xtremes*: http://www.risktec.de/software.htm) and WRPLLC (*HYFRAN*: http://www.wrpllc.com). They present a wide range of tools and methods to evaluate extremes, as well as an easy-to-use graphical interface.

References

1. Règles Fondamentales de Sûreté des Installations Nucléaires (French Fundamental Safety Rules applicables to nuclear installations). *Série REP, Règle No.2.e, 12 april* (1984)
2. Les crues de projet des barrages: méthode du gradex - Design flood determination by the GRADEX method. *Bulletin du Comité Français des Grands Barrages (1994), 18ième Congrés CIGB/ICOLD*, 2 (1994)
3. *Flood Estimation Handbook* (Institute of Hydrology, Wallingford, 1999)
4. *Measuring River Eischarge in High Flow (Flood) or High Sediment Concentration Conditions*. Application Note: R&E Instruments – Acoustic Eoppler Current Profilers. Communication Technology Technical Report (1999)
5. *Petits barrages – Recommandations pour la conception, la justification et le suivi (Small dams – Rrecommendations for design, justification and monitoring)*. Comité Français des Barrages et Réservoirs (2002)
6. *Risk management - Vocabulary. ISO/CEI Guide 73:2002* (2002)
7. *Risk management - Vocabulary. ISO/CEI Guide 73:2009* (2009)
8. *Guide Mesures marégraphiques*. http://agora.shom.fr/docQual/2008/GU/GU2008-056. Service Hydrographique et Océanographique de la Marine (SHOM) (2010)
9. Blackett Review of High Impact Low Probability Risks, *Report of the UK Government Office for Science*, pp. 589–612 (2011)
10. *Détermination des volumes prélevables sur le périmètre du SAGE Cher Amont*. Rapport Daucéa pour la Commission Locale de l'Eau du SAGE Cher amont, Office International de l'Eau (2011)
11. *ISO-772: Hydrométrie - Vocabulaire et symboles*. The International Organization for Standardization (ISO) (2011)
12. *Meteorological and Hydrological Hazards in Site Evaluation for Nuclear Installations*. International Atomic Energy Agency Safety Standards. Specific Safety Guide No. SSG-18 (2011)
13. Assessing the Reliability of Complex Models, *Mathematical and Statistical Foundations of Verification, Validation and Uncertainty Quantification* (National Research Council of the National Academies, Washington D.C., 2012)
14. L'aléa Inondation. État de l'art préalable à l'élaboration du guide Inondation pour les installations nucléaires. *Rapport de l'Institut de Radioprotection et de Sûreté Nucléaire* (2013)
15. *Protection des installations nucléaires de base contre les inondations externes*. Guide No. 13 INB de l'Autorité de Sûreté Nucléaire (2013)
16. *Recommandations pour le dimensionnement des évacuateurs de crues de barrages*. Comité Français des Barrages et Réservoirs (2013)

N. Bousquet and P. Bernardara (eds.), *Extreme Value Theory with Applications to Natural Hazards*, https://doi.org/10.1007/978-3-030-74942-2

17. *External Hazards*. Nuclear Safety Assessment Guide NS-TAST-GE-013. Report of the Office for Nuclear Regulation (2014)
18. *Working Group II Contribution to AR5: "Climate Change 2014: Impacts, Adaptation and Vulnerability". Chapter 9.5: Extreme Events and Weather Disasters*. Intergovernmental Panel on Climate Change. Fifth Assessment Report (2014)
19. *Guidance Document. Issue T: Natural Hazards*. Guidance for the WENRA Safety Reference Levels for Natural Hazards introduced as lesson learned from TEPCO Fukushima Daiichi accident, 22 April (2015)
20. *OpenTURNS 1.7 Reference Guide*. http://www.openturns.org (2017)
21. K. Abbas, Y. Tang, Analysis of Fréchet distribution using reference priors. Commun. Stat. - Theory Methods **44**, 2945–2956 (2015)
22. F.J. Acero, S. Parey, T.T.H. Hoang, D. Dacunha Castelle, J.A. Garcia, M.C. Gallego, Non stationary future return levels for extreme rainfall over Extremadura (South-west Iberian Peninsula). *Hydrol. Sci. J.***62**, 1394–1411 (2017)
23. A. AgaKouchak, D. Easterling, K. Hsu, S. Schubert (eds.), *Sorooshian* (Extremes in a Changing Climate. Detection, Analysis and Uncertainty (Springer, Berlin, 2013)
24. P. Ailliot, C. Thompson, P. Thomson, Space-time modelling of precipitation by using a hidden Markov model and censored Gaussian distributions. J. Roy. Stat. Soc.: Ser. C (Appl. Stat.) **58**(3), 405–426 (2009)
25. H. Akaike, A new look at the statistical model identification. IEEE Trans. Autom. Control **19**, 716–723 (1974)
26. H. Akaike, *On Entropy Maximization Principle, Applications of Statistics* (North-Holland, Amsterdam, 1977), pp. 27–41
27. A.O. Akan, *Open Channels Hydraulics* (Elsevier, Amsterdam, 2006)
28. B.G. Akinoglu, A. Ecevit, Construction of a quadratic model using modified Angstrom coefficients to estimate global solar radiation. Sol. Energy **45**(2), 85–92 (1990)
29. C. Albert, A. Dutfoy, S. Girard, On the extrapolation limits of extreme-value theory for risk management. Spatial Stat. **21**, 187–208 (2017)
30. I. Albert, S. Donnet, C. Guihenneuc-Jouyaux, S. Low-Choy, K. Mengersen, J. Rousseau, Combining expert opinions in prior elicitation. Bayesian Anal. **7**, 503–532 (2012)
31. H. Alexandersson, A statistical analysis of wind, wind profiles and gust ratios at Granby. *Uppsala Report No 55* (1979)
32. H. Alexandersson, A Homogeneity test applied to precipitation data. J. Climatol. **6**, 661–675 (1986)
33. M. Allais, Le Comportement de l'Homme Rationnel devant le Risque: Critique des Postulats et Axiomes de l'école Américaine. Econometrica **21**, 503–546 (1953)
34. C. Alston, P. Kunhert, S. Low Choy, R. McVinish, K. Mengersen, Bayesian model comparison: review and discussion, in *International Statistical Insitute, 55th session* (2005)
35. E. Alvarado, E. Sandberg, S. Pickford, Modeling large forest fires as extreme events. Northwest Sci. **72**, 66–75 (1998)
36. A. Alvarez, P. Olivares, Méthodes d'estimation pour des lois stables, avec des applications en finance (estimation methods for stable laws with financial applications). Journal de la Société Française de Statistique **146**, 23–54 (2005)
37. D. Ambrosi, Approximation of shallow water equations by Roe's Riemann solver. Int. J. Numer. Methods **20**, 157–168 (1995)
38. The American Institute of Aeronautics and Astronautics, *Guide for the verification and validation of computational fluid dynamics simulations*. AIAA Standard Series G-077-2002 (2002)
39. The American Society of Mechanical Engineers. *Guide for Verification and Validation in Computational Solid Mechanics*. ASME V&V (10) (2006)
40. M. Andreewsky, X. Kergadallan, P. Bernardara, M. Benoit, P. Gaufres, C. Trmal, Comparaison de différentes méthodes d'estimation des niveaux extrêmes en site à fort et à faible marnage. La Houille Blanche **4**, 26–36 (2014)
41. C. Andrieu, J. Thoms, A tutorial on adaptive MCMC. Stat. Comput. **18**, 343–373 (2008)
42. M. Arjovsky, S. Chintala, L. Bottou, Wasserstein GAN (2017). arXiv:1701.07875v2

43. J. Arnau, R. Bono, Short time series analysis: C statistic vs Edgington model. Qual. Quant. **32**, 63–75 (1998)
44. B.C. Arnold, N. Balakrishnan, H.N. Nagajara, *A First Course in Order Statistics*. Society for Industrial and Applied Mathematics (2008)
45. P. Asadi, S. Engelke, A.C. Davison, Statistical regionalization for estimation of extreme river discharge (2016). arXiv:1611.03219
46. Y.F. Atchadé, J.S. Rosenthal, On adaptive Markov chain Monte Carlo algorithms. Bernoulli **11**, 815–828 (2005)
47. A.B. Atkinson, T. Piketty, E. Saez, Top incomes in the long run of history. J. Econ. Liter. **49**, 3–71 (2011)
48. S.K. Au, J.L. Beck, Estimation of small failure probabilities in high dimensions by subset simulation. Probab. Eng. Mech. **16**, 263–277 (2001)
49. Y. Auffray, P. Barbillon, J.-M. Marin, Maximin design on non hypercube domain and kernel interpolation. Stat. Comput. **22**, 703–712 (2012)
50. T. Aven, On the need for restricting the probabilistic analysis in risk assessments to variability. Risk Anal. **30**, 354–360 (2010)
51. M. Balesdent, Kriging-based adaptive importance sampling algorithms for rare event estimation. Struct. Saf. **44**, 1–10 (2013)
52. Y. Balkanski, L. Menut, E. Garnier, R. Wang, N. Evangeliou, S. Jourdain, C. Eschstruth, M. Vrac, P. Yiou. Mortality induced by pm 2.5 exposure following the 1783 laki eruption using reconstructed meteorological fields. Sci. Rep. **8**, 15896 (2018)
53. G. Baloutsos, D. Koutsoyiannis, Analysis of a long record of annual maximum rainfall in Athens, Greece, and design rainfall inferences. Nat. Hazards **22**, 31–51 (2000)
54. J.L. Bamber, W.P. Aspinall, An expert judgement assessment of future sea levels, in *Nature Climate Change* (2013)
55. J.L. Bamber, M. Oppenheimer, R.E. Kopp, W.P. Aspinall, R.M. Cooke, Ice sheet contributions to future sea-level rise from structured expert judgment. Proc. US Natl. Acad. Sci. **116**, 11195–11200 (2019)
56. P. Barbillon, *Méthodes d'interpolations à noyaux pour l'approximation de fonctions type boîte noire coûteuses*. Ph.D. thesis, Université Paris XI (2010)
57. L. Bardet, C.M. Euluc, V. Rebour, J. L'Her, Regional frequency analysis of extreme storm surges along the french coast. Nat. Hazard. **11**, 1627–1639 (2011)
58. O. Barndorff-Nielsen, On a formula for the distribution of the maximum likelihood estimator. Biometrika **70**, 343–365 (1983)
59. D. Barriopedro, E.M. Fischer, J. Luterbacher, R.M. Trigo, R. Garcia-Herrera, The hot summer of 2010: redrawing the temperature record map of Europe. Science **332**, 220–224 (2011)
60. D. Basu, On the elimination of nuisance parameters. J. Am. Stat. Assoc. **77**, 355–366 (1977)
61. C. Baudrit, D. Dubois, D. Guyonnet, Joint propagation and exploitation of probabilistic and possibilistic information in risk assessment. IEEE Trans. Fuzzy Syst. **14**, 593–608 (2006)
62. M. Baum, U.D. Hanebeck, Tracking a minimum bounding rectangle based on extreme value theory, in *Multisensor Fusion and Integration for Intelligent Systems (MFI), 2010 IEEE Conference* (2010), pp. 56–61
63. J. Bect, D. Ginsbourger, L. Li, V. Picheny, E. Vazquez, Sequential design of computer experiments for the estimation of a probability of failure. Stat. Comput. **22**, 773–793 (2012)
64. J. Bect, L. Li, E. Vazquez, Bayesian subset simulation. SIAM J. Uncertainty Quantif. **5**, 762–786 (2017)
65. J. Beirlant, *Statistics of Extremes, Theory and Application* (Wiley-Probability and Statistics, New York, 2005)
66. J. Beirlant, Y. Goegebeur, J. Teugels, J. Segers, D. De Waal, C. Ferro, *Statistics of Extremes* (Wiley, New York, 2004)
67. J. Beirlant, J.L. Teugels, *Asymptotic normality of Hill's estimator, in Extreme Value Theory*, vol. 51 (Springer, Berlin, 1987), pp. 148–155
68. J. Beirlant, B. Vandewalle, Some comments on the estimation of a dependence index in bivariate extreme value statistics. Stat. Probab. Lett. **60**, 265–278 (2002)

69. J. Beirlant, P. Vynckier, J.L. Teugels, Tail index estimation, Pareto quantile plots, and regression diagnostics. J. Am. Stat. Assoc. **91**, 1659–1667 (1996)
70. R. Bell, M. Danscheid, T. Glad, Challenges in defining acceptable risk levels, in *Proceedings of the RISK21 Workshop, Monte Verità, Ascona, Switzerland*, vol. 61 (2006), pp. 77–87
71. Y. Ben-Haim, *Information-Gap Theory: Decisions Under Severe Uncertainty* (Academic, London, 2001)
72. G. Benito, M. Lang, M. Barriendos, M. Llasat, F. Frances, T. Ouarda, V. Thorndycraft, Y. Enzel, A. Bardossy, D. Couer, B. Bobbe, Use of systematic, paleoflood and historical data for the improvement of flood risk estimation: review of scientific methods. Nat. Hazards **31**, 623–643 (2004)
73. D.J. Benjamin, J.O. Berger, V.E. Johnson, Redefine statistical significance. Nat. Human Behav. **2**, 6–10 (2018)
74. M. Benoit, F. Lafon, A nearshore wave atlas along the coasts of france based on the numerical modelling of wave climate over 25 Years, in *29th International Conference on Coastal Engineering (ICCE'2004)* (Lisboa, Portugal, 2004)
75. M. Benoit, F. Marcos, F. Becq, Development of a third generation shallow water wave model with unstructured spatial meshing, in *25th International Conference on Coastal Engineering (ICCE'1996)* (Orlando, Floride, USA, 1996)
76. N. Benoumechiara, B. Michel, P. Saint-Pierre, N. Bousquet, Detecting and modeling worst-case dependence structures between random inputs of computational reliability models. HAL-01779484 (2018)
77. P.G. Benson, M.L. Nichols, An investigation of motivational bias in subjective predictive probability distribution. Dec. Sci. **13**, 10–59 (1982)
78. J.O. Berger, *Statistical Decision Theory and Bayesian Analysis*, 2nd edn. (Springer, New York, 1985)
79. J.O. Berger, The case for objective Bayesian analysis. Bayesian Anal. **1**, 385–402 (2006)
80. J.O. Berger, J.M. Bernardo, D. Sun, The formal definition of reference priors. Ann. Stat. **37**, 905–938 (2009)
81. J.O. Berger, L.R. Pericchi, The intrinsic bayes factor for model selection and prediction. J. Am. Stat. Assoc. **91**(433), 109–122 (1996)
82. J.O. Berger, D. Rios Insua, F. Ruggeri, *Bayesian Robustness* (Springer, New York, 2000)
83. H. Bergstrom, Distribution of wind speed. Wind Energy Report WE **92**, 2 (1992)
84. S.M. Berman, Convergence to bivariate limiting extreme value. Ann. Inst. Stat. Math. **13**, 217–223 (1961)
85. S.M. Berman, Maxima and large excursions of stationary Gaussian processes. Trans. Am. Math. Soc. **160**, 67–85 (1973)
86. P. Bernardara, M. Andreewsky, M. Benoit, Application of the regional frequency analysis to the estimation of extreme storm surges. J. Geophys. Res. **116**, 1–11 (2011)
87. P. Bernardara, F. Mazas, X. Kergdallan, L. Hamm, A two-step framework for over-threshold modelling of environmental extremes. Nat. Hazard. **14**, 635–647 (2014)
88. P. Bernardara, F. Mazas, J. Weiss, M. Andreewsky, X. Kergadallan, M. Benoît, L. Hamm, On the two-step threshold selection for over-threshold modelling, in *Coastal Engineering Proceedings*, vol. 33 (2008)
89. J.M. Bernardo, A.F.M. Smith, *Bayesian Theory* (Wiley, New York, 1994)
90. V.J. Berrocal, A.E. Raftery, T. Gneiting, Combining spatial statistical and ensemble information in probabilistic weather forecasts. Mon. Weather Rev. **135**, 1386–1402 (2007)
91. M. Berveiller, *Éléments finis stochastiques: approches intrusives et non-intrusives pour des analyses de fiabilité*. Ph.D. thesis, Université Clermont II (2005)
92. M. Betancourt, A conceptual introduction to Hamiltonian Monte Carlo (2017), arXiv:1701.02434
93. H.G. Beyer, C. Costanzo, D. Heinemann, Modifications of the Heliosat procedure for irradiance estimates from satellite images. Sol. Energy **56**, 207–212 (1996)
94. A. Bhattacharya, D. Pati, Y. Yang, Bayesian fractional posteriors. Ann. Stat. **47**, 39–56 (2016)

95. C. Bioche, *Approximation de lois impropres et applications (Approximation of improper laws and applications)*. Ph.D thesis, Université Blaise Pascal, Clermont-Ferrand (2015)
96. E.K. Bird, The use of questionnaires for acquiring information on public perception of natural hazards and risk mitigation - a review of current knowledge and practice. Nat. Hazard. **9**, 1307–1325 (2009)
97. L. Birgé, P. Massart, Gaussian model selection. J. Eur. Math. Soc. **3**, 203–268 (2001)
98. J. Blanchet, K.R.A. Murthy, On distributionally robust extreme value analysis (2016), arXiv:1601.06858
99. J. Blanchet, J. Touati, D. Lawrence, F. Garavaglia, E. Paquet, Evaluation of a compound distribution based on weather pattern subsampling for extreme rainfall in Norway. Nat. Hazard. **15**(12), 2653–2667 (2015)
100. T. Bollerslev, Generalized autoregressive conditional heteroskedasticity. J. Econom. **31**, 309–328 (1986)
101. M. Borga, C. Vezzani, G. Ealla Fontana, Regional rainfall depth-duration-frequency equations for an alpine region. Nat. Hazards **36**, 221–235 (2005)
102. L. Bottolo, G. Consonni, P. Dellaportas, A. Lijoi, Bayesian analysis of extreme values by mixture modeling. Extremes **6**, 25–47 (2003)
103. J. Boucard, F. (coord.) Surville, E. Garnier (rapp.), *La crise Xynthia à l'aune de l'histoire. Enseignements et enjeux contemporains d'une histoire des submersions*. Contribution aux missions d'enquête parlementaire et sénatoriale sur Xynthia par le groupe de recherche SUBMERSIONS (1984)
104. S. Boucheron, M. Thomas, Tail index estimation, concentration and adaptivity. Electron. J. Stat. **9**, 2751–2792 (2015)
105. W. Boughton, O. Droop, Continuous simulation for design flood estimation - a review. Environ. Model. Softw. **18**(4), 309–318 (2003)
106. N. Bouleau, *Probabilités de l'ingénieur* (Hermann, 1986)
107. N. Bouleau, Splendeurs et misères des lois de valeurs extrêmes. Risques **3**, 85–92 (1991)
108. N. Bouleau, *Mathématiques et risques financiers* (Odile Jacob, 2009)
109. N. Bouleau, *La modélisation critique* (Quae, 2014)
110. M. Bourotte, *Générateur stochastique de temps multisite basé sur un champs gaussien multivarié*. Ph.D. thesis, Université d'Avignon et des Pays de Vaucluse (2016)
111. P.-H. Bourrelier, J. Dunglas, Des évènements naturels extrêmes aux figures de la catastrophe. Responsabilité and Environnement **56**, 41–47 (2009)
112. N. Bousquet, A Bayesian analysis of industrial lifetime data with Weibull distributions. *INRIA Research Report n.6025* (2006)
113. N. Bousquet, Diagnostics of prior-data agreement in applied Bayesian analysis. J. Appl. Stat. **35**, 1011–1029 (2008)
114. N. Bousquet, Vague but proper maximal entropy priors in Bayesian experiments. Stat. Pap. **51**, 613–628 (2010)
115. N. Bousquet, Accelerated Monte Carlo estimation of exceedance probabilities under monotonicity constraints. Annales de la Faculté des Sciences de Toulouse **21**, 557–591 (2012)
116. N. Bousquet, Extreme events modelling in energy companies, in *Encyclopedia of Statistics (Wiley StatRef)* (2018)
117. N. Bousquet, F. Gamboa, T. Klein, V. Moutoussamy, Fast and conservative quantile estimators in monotonic Monte Carlo settings. Submitted (2019)
118. N. Bousquet, M. Keller, Bayesian prior elicitation and selection for extreme values (2018), arXiv:1712.00685
119. N. Bousquet, T. Klein, V. Moutoussamy, Approximation of limit state surfaces in monotonic Monte Carlo settings, with applications to classification. SIAM J. Uncertainty Quantif. **6**, 1–33 (2018)
120. S. Bozzi, G. Passoni, P. Bernardara, N. Goutal, A. Arnaud, Roughness and discharge uncertainty in 1D water level calculations. Environ. Model. Assess. **20**, 343–353 (2015)
121. J.-F. Breihl, *Les surcotes et les submersions marines dans la partie centrale du golfe de Gascogne: les enseignements de la tempête Xynthia*. Ph.D. thesis, Université de la Rochelle (2014)

122. L. Breiman, J.H. Friedman, R.A. Olshen, C.J. Stone, *Classification and Regression Trees* (Monterey, Wadsworth, 1984)
123. P. Brigode, *Changement climatique et risque hydrologique: évaluation de la méthode SCHADEX en contexte non-stationnaire*. Ph.D. thesis, Université Pierre and Marie Curie - Paris VI (2013)
124. E. Brodin, H. Rootzén, Univariate and bivariate GPE methods for predicting extreme wind storm losses. Insurance Math. Econom. **44**, 345–356 (2009)
125. A. Bücher, H. Dette, S. Volgushev, New estimators of the Pickands dependence function and a test for extreme-value dependence. Ann. Stat. **39**, 1963–2006 (2011)
126. M. Buchinsky, Recent advances in quantile regression models: a practical guideline for empirical research. J. Hum. Resour. **33**, 88–126 (1998)
127. TA. Buishand, Rainfall depth-duration-frequency curves; a problem of dependent extremes, in *Statistics for the Environment* (Wiley, New York, 1993)
128. T. Bulteau, D. Idier, J. Lambert, M. Garcin, How historical information can improve extreme coastal water levels probability prediction: application to the Xynthia event at La Rochelle (France). Natl. Hazards Earth Syst. Sci. Discuss. **2**, 7061–7088 (2014)
129. J. Burbea, C.R. Rao, Entropy differential metric, distance and divergence measures in probability space: a unified approach. J. Multivar. Anal. **12**, 575–596 (1982)
130. K.P. Burnham, D. Anderson, *Model Selection and Multi-Model Inference* (Springer, New York, 2002)
131. K.P. Burnham, D.R. Anderson, Multimodel inference: understanding AIC and BIC in model selection. Sociol. Methods Res. **33**, 261–304 (2004)
132. P. Burridge, A.M. Robert Taylor, Additive outlier detection via extreme value theory. J. Time Ser. Anal. **27**, 685–701 (2006)
133. J.J. Cai, A.L. Fougères, C. Mercadier, Environmental data: multivariate extreme value theory in practice. Journal de la Société Française de Statistique **154**, 178–199 (2013)
134. Y. Cai, D. Reeve, Extreme value prediction via a quantile function model. Coast. Eng. **77**, 91–98 (2013)
135. K.W. Campbell, Bayesian analysis of extreme earthquake occurrences. Bull. Seismol. Soc. Am. **73** (1983)
136. S.D. Campell, F.X. Diebold, Weather forecasting for weather derivatives. J. Am. Stat. Assoc. **100**, 6–16 (2005)
137. C. Cannamela, J. Garnier, B. Iooss, Controlled stratification for quantile estimation. Ann. Appl. Stat. **2**, 1554–1580 (2008)
138. A.J. Cannon, A flexible nonlinear modelling framework for nonstationary generalized extreme value analysis in hydroclimatology. Hydrol. Process. **24**, 673–685 (2010)
139. O. Cappe, R. Douc, A. Guillin, J.M. Marin, C.P. Robert, Adaptive importance sampling in general mixture class. Stat. Comput. **18**, 447–459 (2008)
140. B.P. Carlin, S. Chib, Bayesian model choice via Markov Chain Monte Carlo. J. R. Stat. Soc., Ser. B (Methodol.) **57**, 473–484 (1995)
141. B. Carpenter, A. Gelman, M.D. Hoffman, D. Lee, B. Goodrich, M. Betancourt, M. Brubaker, J. Guo, P. Li, A. Riddell, Stan: A probabilistic programming language. J. Stat. Softw. **76** (2017). https://doi.org/10.18637/jss.v076.i01
142. J.A. Carta, P. Ramírez, S. Velázquez, A review of wind speed probability distributions used in wind energy analysis: case studies in the Canary Islands. Renew. Sustain. Energy **13**, 933–955 (2009)
143. M. Carvalho, A. Ramos, Bivariate extreme statistics. Ii. REVSTAT **10**, 83–107 (2012)
144. G. Casella, An introduction to empirical Bayes analysis. Am. Stat. **39**, 83–87 (1985)
145. E. Castillo, A.S. Hadi, N. Balakrishnan, J.M. Sarabia, *Extreme Value and Related Models with Applications in Engineering and Science* (Wiley, New York, 2005)
146. G. Celeux, J. Diebolt, The SEM algorithm: A probabilistic teacher algorithm derived from the EM algorithm for the mixture problem. Comput. Stat. Q. **2**, 73–82 (1986)
147. F. Cérou, P. Del Moral, T. Furon, A. Guyader, Rare event simulation for a static distribution, in *Proceedings of the 7th International Workshop on Rare Event Simulation* (France, Rennes, 2008), pp. 107–115

148. F. Cérou, P. Del Moral, T. Furon, A. Guyader, Sequential Monte Carlo for rare event estimation. Stat. Comput. **22**, 795–808 (2012)
149. F. Cérou, A. Guyader, Adaptive multilevel splitting for rare event analysis. Stoch. Anal. Appl. **25**, 417–443 (2007)
150. G.J. Chaitin, *Information Randomness and Incompleteness*, 2nd edn. (World Scientific, Singapore, 1990)
151. S.C.Y. Chan, Y. Niv, K.A. Norman, A probability distribution over latent causes, in the orbitofrontal cortex. J. Neurosci. **36**, 7817–7828 (2016)
152. J.C. Chang, S.R. Hanna, Air quality model performance evaluation. Meteorol. Atmos. Phys. **87**, 167–196 (2004)
153. J.C. Chang, S.R. Hanna, Technical Descriptions and User's Guide for the BOOT Statistical Model Evaluation Software Package, Version 2.0 (2005), http://www.harmo.org/kit/Download/BOOT_UG.pdf
154. A. Charpentier, Copules et risque multiples. Technip 139–212 (2014)
155. V. Chavez-Demoulin, *Two problems in environmental statistics: capture- recapture analysis and smooth extremal models*. Ph.D. thesis, École Polytechnique Fédérale de Lausanne (1999)
156. V. Chavez-Demoulin, A.C. Davison, Generalized additive models for sample extremes. Appl. Stat. **54**, 207–222 (2005)
157. V. Chavez-Demoulin, P. Embrechts, M. Hofert, An extreme value approach for modeling operational risk losses depending on covariates. J. Risk Insur. **83**, 735–776 (2015)
158. F. Chebana, T.B.M.J. Ouarda, Index flood-based multivariate regional frequency analysis. Water Resour. Res. **45** (2012)
159. R.A. Chechile, Bayesian analysis of Gumbel distributed data. Commun. Stat. - Theory Methods **30**(3), 485–496 (2001)
160. M.H. Chen, J.G. Ibrahim, Power prior distributions for regression models. Stat. Sci. **15**, 46–60 (2000)
161. L. Cheng, A. AghaKouchak, E. Gilleland, R.W. Katz, Non-stationarity extreme value analysis in a changing climate. Clim. Change **127**, 353–369 (2014)
162. L. Cheng, E. Gilleland, M.J. Heaton, A. AghaKouchak, Empirical Bayes estimation for the conditional extreme value model. Stat: The ISI's J. Rapid Dissemin. Stat. Res. (2014). https://doi.org/10.1002/sta4.71
163. C.K. Chu, J.S. Marron, Comparison of two bandwidth selectors with dependent errors. Ann. Stat. **19**, 1906–1918 (1991)
164. B.S. Clarke, A.R. Barron, Jeffreys' prior is asymptotically least favorable under entropy risk. J. Stat. Plan. Infer. **41**, 37–60 (1994)
165. R.T. Clarke, Uncertainty in the estimation of mean annual flood due to rating-curve indefinition. J. Hydrol. **222**, 185–190 (1999)
166. R.T. Clemen, R.L. Winkler, Combining probability distribution from experts in risk analysis. Risk Anal. **19**, 187–203 (1999)
167. R.B. Cleveland, W.S. Cleveland, J.E. Mcrae, I. Terpenning, STL: a seasonal-trend decomposition procedure based on Loess (with discussion). J. Offic. Stat. **6**, 3–73 (1990)
168. W.S. Cleveland, Robust locally weighted regression and smoothing scatterplots. J. Am. Stat. Assoc. **74**, 829–836 (1979)
169. W.S. Cleveland, S.J. Devlin, Locally-weighted regression: an approach to regression analysis by local fitting. J. Am. Stat. Assoc. **83**, 596–610 (1988)
170. F.C. Coehlo, C.T. Codeço, M.G.M. Gomes, A Bayesian framework for parameter estimation in dynamical models. PLOS One (2011), https://doi.org/10.1371/journal.pone.0019616
171. E. Cœur, M. Lang, L'information historique des inondations: l'histoire ne donne-t-elle que des leçons? La Houille Blanche **2**, 79–84 (2000)
172. E. Cœur, M. Lang, R. Naulet, R. Burnet, D. Strazzeri, Histoire et connaissance des phénomènes naturels extrêmes. Ingénieries - EAT - Risques naturels, 15–26 (1998)
173. S. Coles, *An Introduction to Statistical Modelling of Extreme Values* (Springer, Berlin, 2001)
174. S. Coles, A. Davison, *Statistical Modelling of Extreme Values, Techical Report of the Competence Center Environment and Sustainability* (ETH Zurich, 2008)

175. S. Coles, L. Pericchi, Anticipating catastrophes through extreme value modelling. J. Roy. Stat. Soc.: Ser. C (Appl. Stat.) **52**, 405–416 (2003)
176. S.G. Coles, L.R. Pericchi, S.A. Sisson, A fully probabilistic approach to extreme rainfall modeling. J. Hydrol. **273**, 35–50 (2003)
177. S.G. Coles, E.A. Powell, Bayesian methods in extreme value modelling: a review and new developments. Int. Stat. Rev. **64**, 119–136 (1996)
178. S.G. Coles, J.A. Tawn, Modeling extreme multivariate events. J. R. Stat. Soc. **53**, 377–392 (1991)
179. S.G. Coles, J.A. Tawn, A Bayesian analysis of extreme rainfall data. J. R. Stat. Soc. Ser. C (Appl. Stat.)) **43**, 1–48 (1996)
180. R.M. Cooke, *Experts in Uncertainty: Opinion and Subjective Probability in Science* (Oxford University Press, Oxford, 1991)
181. D. Cooley, Extreme value analysis and the study of climate change. Clim. Change **97**, 77–83 (2009)
182. D. Cooley, S.R. Sain, Spatial hierarchical modeling of precipitation extremes from a regional climate model. J. Agric. Biol. Environ. Stat. **15**, 381–402 (2010)
183. J.M. Cornuet, J.M. Marin, A. Mira, C.P. Robert, Adaptive multiple importance sampling. Scand. J. Stat. **39**, 798–812 (2012)
184. J. Cotton, Bayesian priors and estimation of extreme rainfall in Camborne, in *Report of the University of Exeter* (2008)
185. T.M. Cover, J.A. Thomas, *Elements of Information Theory*, 2nd edn. (Wiley, New York, 2006)
186. R.T. Cox, Probability, frequency and reasonable expectation. Am. J. Phys. **14**, 1–10 (1946)
187. E. Cristiano, M.-C. ten Veldhuis, N. van de Giesen, Spatial and temporal variability of rainfall and their effects on hydrological response in urban areas - a review. Hydrol. Earth Syst. Sci. **21**, 3859–3878 (2017)
188. S. Csorgo, P. Deheuvels, D.M. Mason, Kernel estimates of the tail index of a distribution. Ann. Stat. **13**, 1050–1077 (1985)
189. C. Cunnane, Methods and merits of regional flood frequency analysis. J. Hydrol. **100**, 453–459 (1988)
190. D. Kwiatkowski, P.C.B. Phillips, P. Schmidt, Y. Shin, Testing the null hypothesis of stationarity against the alternative of a unit root. J. Econ. **54**, 159–178 (1992)
191. D. Dacunha-Castelle, T.T.H. Hoang, S. Parey, Modeling of air temperatures: preprocessing and trends, reduced stationary process, extremes, simulation. Journal de la Société Française de Statistique **156**(1) (2015)
192. G. Damblin, *Contributions statistiques au calage et à la validation des codes de calcul*. Ph.D. thesis, Université Paris Saclay & AgroParisTech (2015)
193. G. Damblin, M. Couplet, B. Iooss, Numerical studies of space filling designs. J. Simul. **7**, 276–289 (2013)
194. E. Damsleth, Conjugate classes of Gamma distributions. Scandinavian J. Stat.: Theory Appl. **2**, 80–84 (1975)
195. R.M. Darbra, E. Eljarrat, D. Barceló, How to measure uncertainties in environmental risk assessment. Trends Anal. Chem. **27**, 377–385 (2008)
196. T. Darlymple, Flood frequency analysis, in *U.S. Geol. Surv. Water Supply Pap. 1543-A* (1960)
197. C. Daubord, G. Jan, G. André, V. Goirand, *Amélioration de la connaissance des tempêtes passées sur le littoral Atlantique-Manche (1850-2010). Inventaire des niveaux extrêmes observés au niveau des marégraphes. Analyses et réanalyses des enregistrements*. Rapport du projet NIVEXT du Service Hydrographique et Océanographique de la Marine (SHOM) (2016)
198. A.G. Davenport, The dependence of wind loads on meteorological parameters, in *Pro Seminar, Wind Effects on Buildings and Structures* (1967)
199. H.A. David, H.N. Nagaraja, *Order Statistics*, 3rd edn. (Wiley, New York, 2003)
200. R.A. Davis, Maximum and minimum of one-dimensional diffusions. Stoch. Process. Appl. **13**, 1–9 (1982)

201. R.A. Davis, T. Mikosch, The extremogram: A correlogram for extreme events. Bernoulli **15**, 977–1009 (2009)
202. A.C. Davison, N.L. Ramesh, Local likelihood smoothing of sample extremes. J. Roy. Stat. Soc. B **62**, 191–208 (2000)
203. C.A. de Fomento. Efectos de la lluvias caídas en Venezuela en Diciembre de 1999. *CAF-Program of the United Nations for the Development* (2000)
204. Agence Internationale de l'Énergie Atomique. Phénomènes météorologiques extrêmes (cyclones tropicaux exceptés) et choix des sites de centrales nucléaires. *Guide de Sûreté No. 50-SG-S1A* (1983)
205. C. De Michele, R. Rosso, Uncertainty assessment of regionalized flood frequency estimates. J. Hydrol. Eng. **6**, 453–459 (2001)
206. F. De Santis, J. Mortera, A. Nardi, Jeffreys priors for survival models with censored data. J. Stat. Plan. Infer. **99**, 193–209 (2001)
207. M.J. de Smith, *Statistical Analysis Handbook* (The Winchelsea Press, London, 2018)
208. L.T. DeCarlo, W.W. Tryon, Estimating and testing autocorrelation with small samples: a comparison of the C-statistic to a modified estimator. Behav. Res. Ther. **31**, 781–788 (1993)
209. S. Dehaene, *Consciousness and the Brain: Deciphering How the Brain Codes our Thoughts* (Viking Press, New York, 2014)
210. P. Deheuvels, E. Haeusler, D.M. Mason, Almost sure convergence of the Hill estimator. Math. Proc. Cambridge Philos. Soc. **104**(2), 371–381 (1988)
211. A.L.M. Dekkers, J.H.J. Einmahl, L. de Haan, A moment estimator for the index of an extreme value distribution. Ann. Stat. **17**, 1833–1855 (1989)
212. J.-P. Delahaye, Randomness, unpredictability and absence of order, in *Philosophy of Probability*, ed. by J.-P. Dubucs (Dordrecht, Kluwer, 1993)
213. A. Delis, T. Katsaounis, Numerical solution of the two-dimensional shallow water equations by the application of relaxation methods. Appl. Math. Model. **29**, 754–783 (2005)
214. V. Deneau, P. Guillot, La neige collante. Essai de cartographie du risque sur l'ensemble de la France (avec discussion). La Houille Blanche **6**, 465–474 (1984)
215. A. Der Kiureghian, Y. Zhang, C. Li, Inverse reliability problem. J. Eng. Mech. **120**, 1154–1159 (1994)
216. A. Der Kiuriegan, Aleatory or epistemic? Does it matter? in *Special Workshop on Risk Acceptance and Risk Communication. Stanford University, USA, March 26–27* (2007)
217. L. Dezileau, B. Terrier, J.F. Berger, P. Blanchemanche, A. Latapie, R. Freydier, A. Paquier, M. Lang, J.L. Delgado, Reconstitution des crues extrêmes du Gardon à partir d'une analyse paléo-hydrologique. La Houille Blanche **4**, 44–52 (2014)
218. G. Di Baldassare, A. Montanari, Uncertainty in river discharge observations: a quantitative analysis. Hydrol. Earth Syst. Sci. **13**, 913–921 (2009)
219. F.X. Diebold, T. Schuermann, J.D. Stroughair, Pitfalls and opportunities in the use of extreme value theory in risk management. J. Risk Financ. **1**, 30–35 (2000)
220. J. Diebolt, M.A. El Aroui, M. Garrido, S. Girard, Quasi-conjugate Bayes estimates for GPE parameters and application to heavy tails modelling. Extremes **8**, 57–78 (2005)
221. D. Dipak, J. Yan (eds.), *Extreme Value Modeling and Risk Analysis. Methods and Applications* (Chapman & Hall, Boca Raton, 2016)
222. C. Dombry, M. Ribatet, Functional regular variations, pareto processes and peaks over thresholds. Stat. Interface **8**, 9–17 (2015)
223. A. Domeneghetti, A. Castellarin, A. Brath, Assessing rating-curve uncertainty and its effects on hydraulic model calibration. Hydrol. Earth Syst. Sci. Discuss. **8**, 10501–10533 (2011)
224. H. Dong, M.K. Nakayama, Constructing confidence intervals for a quantile using batching and sectioning when applying Latin Hypercube sampling, in *Proceedings of the 2014 Winter Simulation Conference* (2014)
225. H. Douglas, *Science Policy and the Value-Free Ideal* (University of Pittsburgh Press, Pittsburgh, 2009)
226. G. Draisma, H. Grees, A. Ferreira, and L. Ee Haan, Bivariate tail estimation dependence in asymptotic independence. J. Bernoulli Soc. Math. Stat. Probab. **10**, 251–280 (2004)

227. H. Drees, Weighted approximations of tail processes under mixing conditions. Ann. Appl. Probab. **10**, 1274–1301 (2000)
228. M. Drescher, A.H. Perera, C.J. Johson, L.J. Buse, C.A. Drew, M.A. Burgman, Toward rigorous use of expert knowledge in ecological research. Ecosphere **4**, 1–26 (2013)
229. T. Du, L. Xiong, C.-Y. Xu, C.J. Gippel, S. Guo, P. Liu, Return period and risk analysis of nonstationary low-flow series under climate change. J. Hydrol. **527**, 234–250 (2015)
230. D. Dubois, H. Prade, *Représentations formelles de l'incertain et de l'imprécis*. In: Concepts et méthodes pour l'aide à la décision (volume 1), D. Bouyssou, D. Dubois, M. Pirlot et H. Prade (eds.) (Lavoisier, 2006)
231. D. Dubois, H. Prade, *Possibility Theory* (Springer, Berlin, 2012)
232. V. Dubourg, B. Sudret, F. Deheeger, Meta-model based importance sampling for structural reliability analysis. Probab. Eng. Mech. **33**, 47–57 (2013)
233. D. Dumas, The two memorable floods on the Isère in Grenoble (1651 and 1859): An analysis of estimates by M. Pardé. Revue de Géographie Alpine **92**, 39–49 (2004)
234. J. Dunning-Davies, Some comments on extreme values statistics. *Report of the University of Hull* (2005), arXiv:physics/0503134
235. T. Dupont-Courtade, Insurance demand under ambiguity and conflict for extreme risks: an experimental analysis. *Document de travail du Centre d'Économie de la Sorbonne* (2012). ISSN:1955-611X.2012
236. J. Durbin, G. Watson, Testing for serial correlation in least squares regression. Biometrika **37**, 159–179 & 409–428 (1950)
237. K. Durga Rao, H. Kushwaha, A. Verma, A. Srividya, Dynamic fault tree analysis using Monte Carlo simulation in probabilistic safety assessment. Reliabil. Eng. Syst. Safety **54**, 822–833 (2008)
238. K. Durga Rao, H.S. Kushwaha, A.K. Verma, A. Srividya, Epistemic uncertainty propagation in reliability assessment of complex systems. Int. J. Perform. Eng. **3**, 71–84 (2007)
239. A. Dutfoy, S. Parey, N. Roche, Multivariate extreme value theory - a tutorial with applications to hydrology and meteorology. Depend. Model. **2**, 30–48 (2014)
240. A. Eagle, Randomness is unpredictability. British J. Philos. Sci. **56**, 749–790 (2005)
241. A. Eagle, *Philosophy of Probability: Contemporary Readings* (Routledge, New York, 2011)
242. R.G. Easterling, J.O. Berger, Statistical foundations for the validation of computer models, in *Proceedings of the Workshop on Foundations for V&V in the 21st Century*, ed. by D. Pace, S. Stevenson (Society for Modeling and Simulation International, 2002)
243. E.F. Eastoe, J.E. Hefferman, J.A. Tawn, Nonparametric estimation of the spectral measure, and associated dependence measures, for multivariate extreme values using a limiting conditional representation. Extremes **17**, 25–43 (2014)
244. B. Efron, The Jackknife, the bootstrap and other resampling plans, in *CBMS-NSF Regional Conference Series in Applied Mathematics* (SIAM, Philadelphia, 1982)
245. B. Efron, Better bootstrap confidence intervals. J. Am. Stat. Assoc. **82**, 171–185 (1987)
246. D. Egloff, M. Leippold, Quantile estimation with adaptive importance sampling. Ann. Stat. **38**, 1244–1278 (2010)
247. A. Ehlert, M. Schlather, Capturing the multivariate extremal index: bounds and interconnections. Extremes **11**, 353–377 (2008)
248. P. Elancheran, Gauss-Galerkin approximation of diffusion processes with boundary conditions. Commun. Stat. Stoch. Models **15**, 503–515 (1996)
249. P. Embrechts, C. Klüppelberg, T. Mikosch, *Modelling Extremal Events for Insurance and Finance* (Chapman & Hall, Boca Raton, 1997)
250. S. Engelke, J. Ivanovs, Robust bounds in multivariate extremes. Ann. Appl. Probab. **27**, 3706–3734 (2017)
251. S. Engelund, R. Rackwitz, On predictive distribution functions for the three asymptotic extreme value distributions. Struct. Saf. **11**, 255–258 (1992)
252. R.F. Engle, Autoregressive conditional heteroskedasticity with estimates of the variance of UK inflation. Econometrica **50**, 987–1008 (1982)

253. B. Englich, K. Soder, Moody experts - how mood and expertise influence judgmental anchoring. Judgmen. Decis. Mak. **4**, 41–50 (2009)
254. S.L. Engwell, R.S.J. Sparks, W.P. Aspinall, Quantifying uncertainties in the measurement of tephra fall thickness. J. Appl. Volcanol. **2**(1), 1–12 (2013)
255. I. Epifani, L. Ladelli, A. Pievatolo, Bayesian estimation for a parametric Markov Renewal model applied to seismic data. Electron. J. Stat. **8**, 2264–2295 (2014)
256. C. Escalante Sandoval, Bivariate distribution with two component extreme value marginals to model extreme wind speed. Atmosfera **21**, 373–387 (2008)
257. M. Eugenia Castellanos, S. Cabras, A default Bayesian procedure for the generalized Pareto distribution. J. Stat. Plan. Infer. **137**, 473–483 (2007)
258. M. Evans, Measuring statistical evidence using relative belief. Comput. Struct. Biotechnol. J. **14**, 91–96 (2016)
259. M. Evans, H. Moshonov, Checking for prior-data conflict. Bayesian Anal. **1**, 893–914 (2006)
260. M. Faber, J. Koehler, K. Nishijima, *Applications of Statistics and Probability in Civil Engineering* (CRC Press, New York, 2011)
261. M. Falk, J. Hüsler, R.D. Reiss, *Law of Small Numbers: Extreme and Rare Events*, 3ed edn. (Birkhäuser, 2011)
262. M. Falk, R. Michel, Testing for tail independence in extreme value models. Ann. Inst. Stat. Math. **58**, 261–290 (2006)
263. J. Fan, I. Gijbels, *Local Polynomial Modelling and Its Applications* (Chapman & Hall, London, 1996)
264. J. Fan, W. Zhang, Statistical methods with varying coefficient models. Stat. Interface **1**, 179–195 (2008)
265. D. Faranda, J.M. Freitas, V. Lucarini, G. Turchetti, S. Vaienti, Extreme value statistics for dynamical systems with noise. Nonlinearity **26**, 2597 (2013)
266. D. Faranda, S. Vaienti, A recurrence-based technique for detecting genuine extremes in instrumental temperature records. Geophys. Res. Lett. **40**, 5782–5786 (2013)
267. D. Faranda, S. Vaienti, Extreme value statistics for dynamical systems under observational noise. Phys. D **280**, 86–94 (2014)
268. H. Faure, Discrépance des suites associées à un système de numération. Acta Arith **41**, 337–351 (1982)
269. L. Fawcett, D. Walshaw, Markov chain models for extreme wind speeds. Environmetrics **17**, 795–809 (2006)
270. J.-D. Fermanian, *An Overview of the Goodness-of-Fit Test Problem for Copulas*. Lecture Notes in Statistics. (Springer, Berlin, 2013), pp. 61–89
271. L.T. Fernholz, Robustness issues regarding content-corrected tolerance limits. Metrika **55**, 53–66 (2002)
272. O. Ferreira, V. Vousdoukas, P. Ciavola (eds.), *Review of climate change impacts on storm occurence*. Report of the European Project 202798 MICORE: Morphological Impacts and Coastal Risks induced by Extreme Storm Events (1984)
273. C.A.T. Ferro, J. Segers, Inference for clusters of extreme values. J. R. Stat. Soc. B **65**, 545–556 (2009)
274. S. Ferson, W.L. Oberkampf, L. Ginzburg, Model validation and predictive capability for the thermal challenge problem. Comput. Methods Appl. Mech. Eng. **197**, 2408–2430 (2008)
275. S. Few, *Solutions to the Problem of Over-Plotting in Graphs, Perceptual Edge - Visual Business Intelligence Newsletter, September/October* (2008)
276. S.E. Fienberg, Bayesian models and methods in public policy and government settings. Stat. Sci. **26**, 212–226 (2011)
277. H.D. Fill, J.R. Stedinger, Using regional regression within index flood procedures and an empirical Bayesian estimator. J. Hydrol. **210**, 128–145 (1998)
278. N. Fillion, R.M. Corless, On the epistemological analysis of modeling and computational error in the mathematical sciences. Synthese **191**, 1451–1467 (2014)
279. D. Fink, *A Compendium of Conjugate Priors, Computing Science Institute Nijmegen* (Elsevier, Amsterdam, 1997)

280. B. Fischhoff, D. MacGregor, Subjective confidence in forecasts. J. Forecast. **1**, 155–172 (1982)
281. P.C. Fishburn, The axioms of subjective probability. Stat. Sci. **1**, 335–345 (1986)
282. R.A. Fisher, *Statistical Methods for Research Workers* (Oliver and Boyd, Edinburgh, 1926)
283. R.A. Fisher, L.H.C. Tippett, Limiting forms of the frequency distribution of the largest and smallest member of a sample. Proc. Camb. Philos. Soc. **24**, 180–190 (1928)
284. R. de Fondeville, A.C. Davison, Functional peaks-over-threshold analysis and generalized r-pareto process (2019). arXiv:2002.02711
285. A.-L. Fougères, Valeurs extrêmes multivariées: aspects divers et modèles. Technip 213–230 (2014)
286. A.L. Fougères, J.P. Nolan, H. Rootzén, Models for dependent extremes using stables mixtures. Scand. J. Stat. **36**, 42–59 (2009)
287. K. Fraedrich, H. Jansen, U. Luksch, F. Lunkeit, The Planet Simulator: towards a user friendly model. Meterologische Zeitschrift **14**, 299–304 (2005)
288. F. Alves, A location invariant Hill-type estimator. Extremes **4**, 199–213 (2001)
289. M. Francisco-Fernández, J.M. Vilar-Fernández, Bandwidth selection for the local polynomial estimator under dependence: a simulation study. Comput. Stat. **20**, 539–558 (2005)
290. J. Franklin, S. Sisson, Assessments of strategies for evaluating extreme risks. Technical report, Report of the ACERA Project No 0602, University of New South Wales, Australian Centre of Excellence for Risk Analysis (2007)
291. M. Fréchet, Sur la loi de probabilité de l'écart maximum. Annales de la Société Polonaise de Mathématiques **6**, 93–116 (1927)
292. P. Friederichs, Statistical downscaling of extreme precipitation events using extreme value theory. Extremes **13**, 109–132 (2010)
293. P. Friederichs, T.L. Thorarinsdottir, Forecast verification for extreme value distributions with an application to probabilistic peak wind prediction. Environmetrics **23**(7), 579–594 (2012)
294. S. Fu, M. Couplet, N. Bousquet, An adaptive kriging method for solving nonlinear inverse statistical problems. Environmetrics **28**, e2439 (2017)
295. M. Fujita, R. Rackwitz, Updating first- and second-order reliability estimates by importance sampling. Proc. Jpn. Soc. Civil Eng. **5**, 53–60 (1988)
296. E.M. Furrer, R.W. Katz, Improving the simulation of extreme precipitation events by stochastic weather generators. Water Resour. Res. **44** (2008)
297. G. Elliott, T.J. Rothenberg, J.H. Stock, Efficient tests for an autoregressive unit root. Econometrica **64**, 813–836 (1996)
298. E. Gaioni, D. Dey, F. Ruggeri, Bayesian modeling of flash floods using generalized extreme value distribution with prior elicitation. Chilean J. Stat. **1**, 75–90 (2010)
299. M. Galevski, La corrélation entre les pluies torrentielles et l'intensité de l'érosion (avant-propos de P. Reneuve). *Annales de l'École Nationale des Eaux e Forêts and de la station de recherches et expériences* **14**, 379–428 (1955)
300. P. Galiatsatou, P. Prinos, Outliers and trend detection tests in rainfall extremes, in *Proceedings of the 32nd IAHR Congress, SS10-15-O* (Venice, Italy, 2011)
301. J.C. Galland, N. Goutal, J.M. Hervouet, TELEMAC: a new numerical model for solving shallow water equations. Adv. Water Resour. **14**, 138–148 (1991)
302. F. Garavaglia, *Méthode SCHADEX de prédétermination des crues extrêmes. Méthodologie, applications, études de sensibilité*. Ph.D. thesis, ENSE Grenoble (2011)
303. F. Garavaglia, J. Gailhard, E. Paquet, M. Lang, R. Garçon, P. Bernardara, Introducing a rainfall compound distribution model based on weather patterns sub-sampling. Hydrol. Earth Syst. Sci. **14**, 951–964 (1995)
304. F. Garavaglia, M. Lang, E. Paquet, J. Gailhard, R. Garçon, B. Renard, Reliability and robustness of rainfall compound distribution model based on weather pattern sub-sampling. Hydrol. Earth Syst. Sci. **15**, 519–532 (2011)
305. J.A. García, J. Martin, L. Naranjo, F.J. Acero, A Bayesian hierarchical spatio-temporal model for extreme rainfall in Extremadura (Spain). Hydrol. Sci. J. (2018). https://doi.org/10.1080/0262667.2018.1457219

306. R. Garcon, Prévision opérationnelle des apports de la Durance à Serre-Ponçon à l'aide du modèle MORDOR. La Houille Blanche **5** (1996)
307. E. Garnier, De la mémoire des catastrophes dans nos sociétés modernes: Lothar-Martin (déc., et les tempêtes des siècles. Cités (Philosophie, Politique, Histoire). Hors-série 1Oième anniversaire. Presses Universitaires de France **2010**, 381–390 (1999)
308. E. Garnier, *Les dérangements du temps. 500 ans de chaud et de froid en Europe (The changes in the weather. 500 years of hot and cold in Europe)* (Plon, 2010)
309. E. Garnier, The Laki fog of 1783. Volcanic activity and health crises in europe. *Bulletin de l'Académie Nationale de Médecine* **195**, 1043–1055 (2011)
310. E. Garnier, Histoire des tempêtes et assurance (storm history and insurance). Les cahiers de l'assurance **91**, 22–30 (2012)
311. E. Garnier, Comment l'historien peut-il contribuer à reconstituer les climats du passé? *Le climat en question, 10 septembre* (2013)
312. E. Garnier, *A Historic Experience for a Strengthened Resilience. European Societies in Front of Hydro-meteors 16th–20th Centuries* (Wiley, Chichester, 2015), pp. 3–26
313. E. Garnier, *Genève face à la catastrophe 1350–1950* (Slatkine, Un retour d'expérience pour une meilleure résilience urbaine. Genève, 2016)
314. E. Garnier, *Historic Droughts From Archives: Beyond the Instrumental Records* (Wiley, Chichester, 2017), pp. 45–67
315. E. Garnier, *Xynthia, February 2010: Autopsy of a Forseable Catastrophe* (Wiley, Chichester, 2017), pp. 111–148
316. E. Garnier, Lessons learned from the past for a better resilience to contemporary risks. Disaster Prev Manag **28**(6), 778–795 (2019)
317. E. Garnier, P. Ciavola, C. Armaroli, T. Spencer, O. Ferreira, *Historical analysis of storms events: case studies in France* (Portugal and Italy, in Coastal Engineering, England, 2017)
318. E. Garnier, V. Daux, P. Yiou, I. Garcia de Cortazar, Grapevine harvest dates in Besançon between 1525 and 1847: Social outcome or climatic evidence? Clim. Change **104**, 703–727 (2010)
319. E. Garnier, F. Surville, *La tempête Xynthia face à l'histoire; submersions et tsunamis sur les littoraux français du Moyen Âge à nos jours* (Le Croît Vif, Saintes, 2010)
320. P.H. Garthwaite, J.B. Kadane, A. O'Hagan, Statistical methods for eliciting probability distributions. J. Am. Stat. Assoc. **100**, 680–701 (2005)
321. P. Gaufrès, Extreme storm surge distributions and practical application at Marseille (france), in *7th International Conference on Hydroinformatics, Advanced Application of Modelling, Vol III, Nice, France* (2006), pp. 1601–1608
322. E. Gaume, L. Gaál, A. Viglione, J. Szolgay, S. Kohnová, G. Blöschl, Bayesian MCMC approach to regional flood frequency analyses involving extraordinary flood events at ungauged sites. J. Hydrol. **394**, 101–117 (2010)
323. E. Gaume, J.P. Villeneuve, M. Desbordes, Uncertainty assessment and analysis of the calibrated parameter values of an urban storm water quality model. J. Hydrol. **210**, 39–50 (1998)
324. E.L. Geist, T. Parsons, T. Brink, H.J. Lee, *Tsunami Probability* (Harvard University Press, U.S., 2013), pp. 93–135
325. A. Gelman, Prior distributions for variance parameters in hierarchical models. Bayesian Anal. **1**, 515–534 (2006)
326. A. Gelman, C. Hennig, Beyond subjective and objective in statistics. J. R. Stat. Soc. Ser. A **180**, 1–31 (2017)
327. C. Genest, I. Kojadinovic, J. Neslehov, J. Yan, A goodness-of-fit test for bivariate extreme-value copulas. Bernoulli **17**, 253–275 (2011)
328. C. Genest, L.P. Rivest, Statistical inference procedures for bivariate Archimedean copulas. J. Am. Stat. Assoc. **88**(423), 1034–1043 (1993)
329. C. Genest, L.P. Rivest, On the multivariate probability integral transformation. Stat. Probab. Lett. **53**(4), 391–399 (2001)
330. C. Genest, J.V. Zidek, Combining probability distributions: a critique and an annotated bibliography. Stat. Sci. **1**, 114–148 (1986)

331. M. Ghil, P. Yiou, S. Hallegatte, B.D. Malamud, P. Naveau, A. Soloviev, P. Friederichs, V. Keilis-Bork, D. Kondrashov, V. Kossobokov, O. Mestre, C. Nicolis, H.W. Rust, P. Shebalin, M. Vrac, A. Witt, I. Zaliapin, Extreme events: dynamics, statistics and prediction. Nonlinear Process. Geophys. **18**, 295–350 (2011)
332. S. Ghosh, B.K. Mallick, A hierarchical Bayesian spatio-temporal model for extreme precipitation events. Environmetrics **22**, 192–204 (2011)
333. S. Ghosh, S. Resnick, A discussion on mean excess plots. Stoch. Process Appl. **120**, 1492–1517 (2010)
334. K. Ghoudi, A. Khoudraji, L.P. Rivest, Propriétés statistiques des couples de valeurs extrêmes bidimensionnelles. Canadian J. Stat. **26**, 187–197 (1998)
335. G. Gigerenzer, P.M. Todd, and the ABC Research Group, in *Simple Heuristics that Make Us Smart* (Oxford University Press, Oxford, 1999)
336. A. Gille-Genest, *Utilisation des méthodes numériques probabilistes dans les applications au domaine de la fiabilité des structures*. Ph.D. thesis, Université Paris 6 (1999)
337. P. Girard, E. Parent, The deductive phase of statistical analysis via predictive simulations: test, validation and control of a linear model with autocorrelated errors representing a food process. J. Stat. Plan. Infer. **124**, 99–120 (2004)
338. N. Gissibi, C. Klüppelberg, J. Mager, *Big Data: Progress in Automatic Extreme Risk Analysis* (Springer, Berlin, 2017)
339. G.J. Glasser, R.F. Winter, Critical values of the coefficient of rank correlation for testing the hypothesis of independence. Biometrika **48**, 444–448 (1961)
340. P.W. Glynn, Importance sampling for Monte Carlo estimation of quantiles, in *Proceedings of Second International Workshop on Mathematical Methods in Stochastic Simulation and Experimental Design* (Publishing House of Saint Petersburg University, 1996), pp. 180–185
341. P.W. Glynn, G. Rubino, B. Tuffin, Robustness properties and confidence interval reliability, in *Rare Event Simulation* (Wiley, New York, 2009)
342. Y. Goda, Plotting-position estimator for the L-moment method and quantile confidence interval for the GEV, GPa, and Weibull distributions applied for extreme wave analysis. Coast. Eng. **53**, 111–149 (2011)
343. Y. Goda, M. Judaka, H. Kawai, Incorporating of Weibull distribution in L-moments method for regional frequency analysis of peak over threshold wave heights, in *Proceedings of the 32th International Conference on Coastal Engineering (ICCE 2010), Shangai* (2010)
344. N.K. Goel, E.H. Burn, M.E. Pandey, Y. An, Wind quantile estimation using a pooled frequency analysis approach. J. Wind Eng. Ind. Aerodyn. **92**, 509–528 (2004)
345. B. Goertzel, *Probability Theory Ensues from Assumptions of Approximate Consistency: A Simple Derivation and Its Implications for AGI, Proceedings of AGI-13* (Springer, 2013)
346. J.I. Gold, H.R. Heekeren, Neural mechanisms for perceptual decision making, in *Neuroeconomics (Chapter 19)*, 2nd edn. ed. by P.W. Glimcher, R. Fehr (2013)
347. I. Goodfellow, J. Pouget-Abadie, M. Mirza, B. Xu, D. Warde-Farley, S. Ozair, A. Courville, Y. Bengio, Generative adversarial nets, in *Advances in Neural Information Processing Systems 27 (NIPS 2014)* (2014)
348. F. Gottardi, *Estimation statistique et réanalyse des précipitations en montagne - Utilisation d'ébauches par types de temps et assimilation de données d'enneigement – Application aux grands massifs montagneux français*. Ph.D. thesis, Institut National Polytechnique de Grenoble (2009)
349. C. Gourerioux, A. Monfort, *Statistics and Econometric Models* (Cambridge University Press, Cambridge, 2008)
350. T. Gouriou, *Evolution des composantes du niveau marin à partir d'observations de marégraphie effectuées depuis la fin du 18ième siècle en Charente-Maritime*. Ph.D. thesis, Université de la Rochelle (2012)
351. N. Goutal, J.-M. Lacombe, F. Zaoui, K. El Kadi-Abderrezzak, MASCARET: a 1-D open-source software for flow hydrodynamic and water quality in open channel networks, in *River Flow* (2012), pp. 1169–1174

352. D. Grabaskas, M.K. Nakayama, R. Denning, T. Aldemir, Advantages of variance reduction techniques in establishing confidence intervals for quantiles. Reliab. Eng. Syst. Saf. **149**, 187–203 (2016)
353. R.B. Gramacy, H.K.H. Lee, Bayesian treed Gaussian process models with an application to computer modeling. J. Am. Stat. Assoc. **103**, 1119–1130 (2007)
354. B. Grechuk, M. Zabarankin, Risk averse decision making under catastrophic risk. Eur. J. Oper. Res. **239**, 166–176 (2014)
355. P.J. Green, B.W. Silverman, *Nonparametric Regression and Generalized Linear Models - A roughness penalty approach* (Chapman and Hall/CRC, London, 2000)
356. S. Greenland, S.J. Senn, K.J. Rothman, J.B. Carlin, C. Poole, S.N. Goodman, D. Altman, Statistical tests, P-values, confidence interval, and power: a guide to misinterpretations. Eur. J. Epidemiol. **31**, 227–350 (2016)
357. J.A. Greenwood, J.M. Landwehr, N.C. Matalas, J.R. Wallis, Probability weighted moments: definition and relation to parameters of several distributions expressable in inverse form. Water Resour. Res. **15**, 1049–1054 (1979)
358. Q.F. Gronaud, A. Sarafoglou, D. Matzke, A. Ly, U. Boehm, M. Marsman, D.S. Leslie, J.J. Forster, E.-J. Wagenmakers, H. Steingroever, A tutorial on bridge sampling. J. Math. Psychol. **81**, 80–97 (2017)
359. P. Grünwald, The safe Bayesian: learning the learning rate via the mixability gap, in *ALT* (2012), pp. 169–183
360. C. Gu, *Smoothing Spline ANOVA Models* (Springer, Berlin, 2002)
361. S.D. Guikema, Natural disaster risk analysis for critical infrastructure systems: an approach based on statistical learning theory. Reliab. Eng. Syst. Saf. **94**, 855–860 (2009)
362. P. Guillot, E. Euband, La méthode du Gradex pour le calcul des probabilités des crues à partir des pluies, 84 (IAHS Publications, 1967)
363. S. Guillotte, F. Perron, Polynomial Pickands functions. Bernoulli **22**, 213–241 (2016)
364. E.J. Gumbel, *Statistics of Extreme* (Columbia University Press, Columbia, 1958)
365. H.V. Gupta, H. Kling, K.K. Yilmaz, G.F. Martinez, Decomposition of the mean squared error and NSE performance criteria: implications for improving hydrological modelling. J. Hydrol. **377**, 80–91 (2009)
366. R.D. Gupta, D. Kundu, Generalized exponential distribution. Aust. N. Z. J. Stat. **41**, 173–188 (1999)
367. A. Guyader, N. Hengartner, E. Matzner-Lober, Simulation and estimation of extreme quantiles and extreme probabilities. Appl. Math. Optim. **64**, 171–196 (2011)
368. D. Guyonnet, B. Bourgine, D. Dubois, H. Fargier, Hybrid approach for addressing uncertainty in risk assessment. J. Environ. Eng. **129** (2003)
369. L. Haan, S.I. Resnick, On asymptotic normality of the Hill estimator. Stoch. Model. **14**, 849–867 (1998)
370. P. Hall, S.N. Lahiri, J. Polzehl, On bandwidth choice in nonparametric regression with both short and long-range dependent errors. Ann. Stat. **23**, 1921–1936 (1995)
371. P. Hall, N. Tajvidi, Nonparametric analysis of temporal trend when fitting parametric models to extreme value data. Stat. Sci. **15**, 153–167 (2000)
372. J.Y. Halpern, A counterexample to theorems of Cox and Fine. J. Artif. Intell. Res. **10**, 67–85 (1999)
373. J.H. Halton, On the efficiency of certain quasi-random sequences of points in evaluating multidimensional integrals. Numer. Math. **2**, 84–90 (1960)
374. Y. Hamdi, L. Bardet, C.-M. Duluc, V. Rebour, Use of historical information in extreme-surge frequency estimation: the case of marine flooding on the La Rochelle site in France. Nat. Hazard. **15**, 1515–1531 (2015)
375. Y. Hamdi, E. Garnier, N. Giloy, M.C. Duluc, V. Rebours, Analysis of the risk associated with coastal flooding hazards: a new historical extreme storm dataset for dunkirk, france. Nat. Hazard. **18**, 3383–3402 (2018)
376. K.H. Hamed, A.R. Rao, A modified mann-kendall trend test for autocorrelated data. J. Hydrol. **204**, 182–196 (1998)

377. I.S. Hanaish, K. Ibrahim, A.A. Jemain, Stochastic modeling of rainfall in Peninsular Malaysia using Barlette Lewis rectangular pulses models. Model. Simul. Eng. **37** (2011)
378. M. Hanel, T.A. Buishand, C.A.T. Ferro, A non-stationary index-flood model for precipitation extremes in transient regional climate model simulations. J. Geophy. Res. **114** (2009)
379. S.O. Hansson, Ethical criteria of risk acceptance. Erkenntnis **59**, 291–309 (2003)
380. T.J. Hastie, R.J. Tibshirani, *Generalized Additive Models* (Chapman and Hall, London, 1990)
381. A.S. Hedayat, N.J.A. Sloane, J. Stufken, *Orthogonal Arrays, Theory and Applications* (Springer, Berlin, 1999)
382. J.E. Hefferman, A directory of coefficients of tail dependence extremes. Biometrika **3**, 279–290 (2000)
383. J.E. Hefferman, J.A. Tawn, A conditional approach for multivariate extreme values (with discussion). J. R. Stat. Soc., Ser. B (Methodol.) **66**, 497–546 (2004)
384. L. Held, M. Ott, On $p-$values and Bayes factors. Ann. Rev. Stat. Appl. 5 (2018). https://doi.org/10.1146/annurev--statistics--031017--100307
385. J.C. Helton, F.J. Davis, Latin hypercube sampling and the propagation of uncertainty in analyses of complex systems. Reliab. Eng. Syst. Saf. **81**, 23–69 (2003)
386. J.C. Helton, J.E. Johnson, W.L. Oberkampf, C.J. Sallabery, Representation of analysis results involving aleatory and epistemic uncertainty, in *SANEIA Report SANE2008-4379* (2008)
387. R.W. Herschy (ed.), *Hydrometry - Principles and Practices* (Wiley, Chichester, 1999)
388. T.C. Hesterberg, B.L. Nelson, Control variates for probability and quantile estimation. Manag. Sci. **44**, 1295–1312 (1998)
389. B. Hill, A simple general approach to inference about the tail of a distribution. Ann. Stat. **3**, 1163–1174 (1975)
390. Y.-C. Ho, E.L. Pepyne, Simple explanation of the no free lunch theorem of optimization, in *Proceedings of the 40th IEEE Conference on Decision and Control, Orlando, Florida, USA, Eecember*, 1 (2001), pp. 4409–4414
391. T.T.H. Hoang, *Modeling of nonstationary, nonlinear time series. Application to the definition of trends on mean, variability and extremes of air temperatures in Europe*. Ph.D. thesis, Université Paris-XI (Orsay) (2010)
392. E. Hofer, When to separate uncertainties and when not to separate. Reliab. Eng. Syst. Saf. **54**, 113–118 (1996)
393. F.O. Hoffman, J.S. Hammonds, Propagation of uncertainty in risk assessments: the need to distinguish between uncertainty due to lack of knowledge and uncertainty due to variability. Risk Anal. **14**, 707–712 (1994)
394. L.H. Holtuijsen, *Waves in Oceanic and Coastal Waters* (Cambridge University Press, Cambridge, 2010)
395. J.R.M. Hosking, Towards statistical modeling of tsunami occurrence with regional frequency analysis. J. Math. Ind. **4**, 41–48 (2012)
396. J.R.M. Hosking, J.R. Wallis, Some statistics useful in regional frequency analysis. Water Resour. Res. **29**, 271–281 (1993)
397. J.R.M. Hosking, J.R. Wallis, *Regional Frequency Analysis, An Approach Based on L-Moments* (Cambridge University Press, Cambridge, 1997)
398. F. Hu, J. Zidek, Incorporating relevant sample information using the likelihood, in *Technical Report No 161, Department of Statistics* (The University of British Columbia, Vancouver, 1995), p. 6
399. C.M. Hurvich, C.L. Tsai, Regression and time series model selection in small samples. Biometrika **76**, 297–307 (1989)
400. C.M. Hurwich, C.-L. Tsai, Regression and time series model selection in small samples. Biometrika **76**, 297–307 (1989)
401. J. Husler, E. Li, Testing asymptotic independence in bivariate extremes. J. Stat. Plann. Inference **139**, 990–998 (2009)
402. J. Hüsler, A note on exceedances and rare events of non-stationary sequences. J. Appl. Probab. **30**, 877–888 (1993)

403. J. Hüsler, M. Schmidt, A note on the point processes of rare events. J. Appl. Probab. **33**, 654–663 (1996)
404. R.L. Iman, W.J. Conover, A distribution-free approach to inducing rank correlation among input variables. Commun. Stat. **B11**, 311–334 (1982)
405. B. Iooss, *Contributions au traitement des incertitudes en modélisation numérique*. Habilitation Thesis, Université Toulouse-III (2009)
406. B. Iooss, Traitement des incertitudes en simulation numérique pour la maîtrise de procédés industriels. *Séminaire Aristote, École Polytechnique, 19 décembre* (2014)
407. B. Iooss, P. Lemaître, A review on global sensitivity analysis. Uncertainty Manag. Simul. Optim. Complex Syst. **59**, 101–122 (2015)
408. T.Z. Irony, N. Singpurwalla, J.M. Bernardo, Non-informative priors do not exist: a dialogue with José M. Bernardo (with discussion with A.P. Dawid). J. Stat. Plan. Infer. **65**, 159–189 (1997)
409. E. Jahani, M. Shayanfar, M. Barkhordari, A new adaptive importance sampling Monte Carlo method for structural reliability. KSCE J. Civ. Eng. **17**, 210–215 (2013)
410. P. Javelle, T.B.M.J. Ouarda, M. Lang, B. Bobee, G. Galea, J.M. Gresillon, Development of regional flood-duration-frequency curves based on the index-flood method. J. Hydrol. **258**, 249–259 (2002)
411. E.T. Jaynes, *Probability Theory: The Logic of Science* (Cambridge University Press, Cambridge, 2003)
412. H. Jeffreys, An invariant form for the prior probability in estimation problems. Proc. R. Soc. Lond. Ser. A **186**, 453–461 (1946)
413. A.F. Jenkinson, The frequency distribution of the annual maximum (or minimum) values of meteorological elements. Q. J. R. Meteorol. Soc. **81**, 158–171 (1955)
414. L. Jing, *Hastings-within-Gibbs Algorithm: Introduction and Application on Hierarchical Models* (University of Texas at San Antonio, Lecture Notes, 2010)
415. J.M. Hammersley, Monte Carlo methods for solving multivariate problems. Ann. N. Y. Acad. Sci. **86**, 844–874 (1960)
416. M.E. Johnson, L.M. Moore, D. Ylvisaker, Minimax and maximin distance design. J. Stat. Plan. Infer. **26**, 131–148 (1990)
417. V.E. Johnson, Revised standards for statistical evidence. Proc. Natl. Acad. Sci. **110**, 19313–19317 (2013)
418. J.B. Jones, P.J. Mulholland, *Streams and Ground Waters* (Academic, San Diego, 2000)
419. J.B. Kadane, *Predictive and Structural Methods for Eliciting Prior Distributions* (North Holland Publishing Company, Amsterdam, 1980), pp. 89–93
420. J.B. Kadane, J.A. Wolfson, Experiences in elicitation. The Statistician **47**, 3–19 (1998)
421. D. Kahneman, P. Slovic, A. Tversky (eds.), *Judgment Under Uncertainty: Heuristics and Biases* (Cambridge University Press, Cambridge, 1982)
422. M. Kamary, K. Mengersen, C.P. Robert, J. Rousseau, Testing hypotheses via a mixture estimation model (2018), arXiv:1412.2044v3
423. M.P. Kaminskiy, V.V. Krivtsov, A simple procedure for Bayesian estimation of the Weibull distribution. IEEE Trans. Reliab. **54**, 612–616 (2005)
424. J. Kanda, Application of an empirical extreme value distribution to load models. J. Res. Nat. Inst. Stand. Technol. **99**(4), 413–420 (1994)
425. R.E. Kass, A.E. Raftery, Bayes factors. J. Am. Stat. Assoc. **90**, 773–795 (1995)
426. R.E. Kass, L. Wasserman, The selection of prior distributions by formal rules. J. Am. Stat. Assoc. **91**, 1343–1370 (1996)
427. R.W. Katz, Overview of extreme value analysis under climate change, in *US CLIVAR/NCAR ASP Researcher Colloquium on Statistical Assessment of Extreme Weather Phenomena under Climate Change* (NCAR, Boulder, 2011)
428. R.W. Katz, G.S. Brush, M.B. Parlange, Statistics of extremes: modeling ecological disturbances. Ecology **86**, 1124–1134 (2005)
429. R.W. Katz, M.B. Parlange, P. Naveau, Statistics of extremes in hydrology. Adv. Water Resour. **25**, 1287–1304 (2002)

430. C. Keef, J. Tawn, R. Lamb, Estimating the probability of widespread flood events. Environmetrics **24**, 13–21 (2013)
431. O. Kempthorne, L. Folks, *Probability, Statistics and Data Analysis* (Iowa State University Press, Ames, Iowa, 1971)
432. M.G. Kendall, *Rank Correlation Methods* (Griffin, London, 1975)
433. X. Kergadallan, *Analyse statistique des niveaux extrêmes. Environnements maritime et estuarien*. État de l'art CETMEF (2013)
434. X. Kergadallan, C. Trmal, C. Perherin, Étude des surcotes météorologiques extrêmes le long des côtes métropolitaines. *Actes du congrès SHF "Événements extrêmes d'inondation", Lyon* (2013)
435. C. Keribin, Les méthodes bayésiennes variationnelles et leur application en neuroimagerie: une étude de l'existant. *NRIA Research Report RR-7091* (2009)
436. P. Kersaudy, B. Sudret, N. Varsier, O. Picon, J. Wiart, A new surrogate modeling technique combining kriging and polynomial chaos expansions - Application to uncertainty analysis in computational dosimetry. J. Comput. Phys. **286**, 103–117 (2015)
437. J. Kiefer, J. Wolfowitz, Stochastic estimation of the maximum of a regression function. Ann. Math. Stat. **23**, 462–466 (1952)
438. M.Y. Kim, X. Xue, The analysis of multivariate interval-censored survival data. Stat. Med. **21**, 3715–3726 (2002)
439. J.P.C. Kleijnen, *Design and Analysis of Simulation Experiments*, 2nd edn. (Springer, Berlin, 2015)
440. H. Kling, M. Fuchs, M. Paulin, Runoff conditions in the upper Danube basin under an ensemble of climate change scenarios. J. Hydrol. **425**, 2564–277 (2012)
441. R. Koenker, G. Bassett, Regression quantiles. Econometrica **46**, 33–50 (1978)
442. A.N. Kolmogorov, *Foundations of the Theory of Probability* (Chelsea Publishing Co., Oxford, 1950)
443. S. Kotz, S. Nadarajah, *Extreme Value Distributions: Theory and Applications* (Imperial College Press, London, 2000)
444. D. Koutsoyiannis, Statistics of extremes and estimation of extreme rainfall I: Theoretical investigations. Hydrol. Sci. J. **49**, 575–590 (2004)
445. D. Koutsoyiannis, Statistics of extremes and estimation of extreme rainfall II: Empirical investigations of long rainfall records. Hydrol. Sci. J. **49**, 591–610 (2004)
446. J.K. Kruschke, *Doing Bayesian Analysis, Second Edition: A Tutorial with R JAGS and Stan* (Academic, New York, 2015)
447. R. Krzysztofowicz, Bayesian theory of probabilistic forecasting via deterministic hydrologic model. Water Resour. Res. **35**, 2739–2750 (1999)
448. J. Kysely, A cautionary note on the use of nonparametric bootstrap for estimating uncertainties in extreme-value models. J. Appl. Meteorol. Climatol. **47**, 3236–3251 (2008)
449. T. Labopin-Richard, F. Gamboa, A. Garivier, B. Iooss, Bregman superquantiles. Estimation methods and applications. Depend. Model. **4**, 76–108 (2016)
450. I. Lakatos, Falsification and the methodology of scientific research programmes, in *Criticism and the Growth of Knowledge*, ed. by I. Lakatos, A. Musgrave (Cambridge University Press, Cambridge, 1970)
451. H.H. Lamb, K. Frydendahl, *Historic Storms of the North Sea, British Isles and Northwest Europe* (Cambridge University Press, Cambridge, 1991)
452. J.H. Lambert, N.C. Matalas, C.W. Ling, Y.Y. Haimes, D. Li, Selection of probability distributions in characterizing risk of extreme events. Risk Anal. **14**, 731–742 (1993)
453. M. Lang, P. Arnaud, J. Carreau et al., Résultats du projet ExtraFlo (ANR 2009–2013) sur l'estimation des pluies et crues extrêmes. La Houille Blanche - Revue Internationale de l'eau **2**, 5–13 (2014)
454. M. Lang, K. Pobanz, B. Renard, E. Renouf, E. Sauquet, Extrapolation of rating curves by hydraulic modelling, with application to flood frequency analysis. Hydrol. Sci. J. **55**, 883–898 (2010)

455. N. Laptev, J. Yosinski, L.E. Li, S. Smyl, Time-series extreme event forecasting with neural networks at uber, in *Proceedings of ICML 2017 Time Series Workshop* (2017)
456. A. LaRocque, *Probable Maximum Flood* (Springer, Berlin, 2016), pp. 777–779
457. D. Lawrence, E. Paquet, J. Gailhard, A.K. Fleig, Stochastic semi-continuous simulation for extreme flood estimation in catchments with combined rainfall-snowmelt flood regimes. Nat. Hazard. **14**(5), 1283–1298 (2014)
458. G. Le Cozannet, J.-C. Manceau, J. Rohmer, Bounding probabilistic sea-level projections within the framework of the possibility theory. Environ. Res. Lett. **12** (2017). https://doi.org/10.1088/1748-9326/aa5528
459. J.F. Le Gall, *Intégration, Probabilités et Processus Aléatoires*. Cours de l'École Normale Supérieure (2006)
460. E. Le Roy Ladurie, *Histoire du climat depuis l'an mil* (Flammarion, Paris, 1967)
461. M.R. Leadbetter, On extreme values in stationary sequences. Zeitschrift für Wahrscheinlichkeitstheorie und Verwandte Gebiete **28**, 289–303 (1974)
462. M.R. Leadbetter, On a basis for 'Peaks over Threshold' Modelling. Stat. Probab. Lett. **12**, 357–362 (1991)
463. M.R. Leadbetter, G. Lindgren, H. Rootzén, *Extremes and Related Properties of Random Sequences and series* (Springer, New York, 1983)
464. J. Leal, S. Wordsworth, R. Legood, E. Blair, Eliciting expert opinion for economic models: an applied example. Value Health **10**, 195–203 (2007)
465. C. Leary, Systematic errors in operational National Meteorological Center primitive-equation surface prognoses. Mon. Weather Rev. **99**, 409–413 (1971)
466. E. Lebarbier, T. Mary-Huard, Le critère BIC: fondements théoriques et interprétation. Technical report, INRIA Research Report RR-5315 (2004)
467. L. Lebart, A. Morineau, M. Piron, *Statistique exploratoire multidimensionnelle* (Dunod, Paris, 1995)
468. H. Lebesgue, *Oeuvres scientifiques (en cinq volumes)*. Institut de Mathématiques de l'Université de Genève (1972)
469. G.C. Leckebusch, D. Renggli, U. Ulbrich, Development and application of an objective storm severity measure for the northeast Atlantic region. Meteorol. Z. **17**, 575–587 (2008)
470. O. Leclerc, A. Gallucio, Expert's Liability and Risk Management - Lessons from L'Aquila Trial. Cahiers Droit, Sci. Technol. **5**, 243–250 (2015)
471. B. Lecoutre, *Pourquoi les méthodes bayésiennes sont-elles si peu utilisées dans les publications expérimentales? Les pratiques actuelles réexaminées à partir des conceptions des fondateurs de l'inférence statistique*. I. Drouet (ed.). Éditions Matériologiques, Paris (2016)
472. P. L'Ecuyer, V. Demers, B. Tuffin, Splitting for rare event simulation, in *Proceedings of the Winter Simulation Conference* (2006), pp. 137–148
473. A. Ledford, J.A. Tawn, Statistics for near independence in multivariate extreme values. Biometrika **83**, 169–187 (1996)
474. E.L. Lehman, *Fisher, Neyman, and the Creation of Classical Statistics* (Springer, New York, 2011)
475. M. Lemaire, A. Chateauneuf, J.-C. Mitteau, *Structural Reliability* (Wiley, New York, 2010)
476. C. Lemieux, *Monte Carlo and quasi-Monte Carlo sampling* (Springer, Berlin, 2009)
477. F. Leone, F. Winet (coord.), *La vulnérabilité des sociétés and des territoires face aux menaces naturelles*. Géorisques. Publications de l'Université Montpellier III (2011)
478. J. Leslie, In an era of extreme weather, concerns grow over dam safety. Yale Environment **360** (2019)
479. Z. Liang, Y. Hu, H. Huang, J. Wang, B. Li, Study on the estimation of design value under non-stationary environment. South-to-North Water Trans. Water Sci. Technol. **14**, 50–53 (2016)
480. X. Lin, J. Pittman, B. Clarke, Information conversion, effective samples, and parameter size. IEEE Trans. Inf. Theory **53**, 4438–4456 (2007)
481. D.V. Lindley, *Bayesian Statistics: A Review* (SIAM, Philadelphia, 1972)
482. K. Linnerud, T.K. Mideksa, G.S. Eskeland, The impact of climate change on nuclear power supply. Energy J. **32**, 149–168 (2011)

483. J.S. Liu, *Monte Carlo Strategies in Scientific Computing* (Springer, Berlin, 2001)
484. S. Liu. Ph.D. thesis (Iowa State University, 2013)
485. F. Longin, *Extreme Events in Finance: A Handbook of Extreme Value Theory and its Applications* (Wiley, New York, 2016)
486. Y. Lu, X. Qin, Comparison of stochastic point process models of rainfall in Singapore, in *Proceedings of World Academy of Science, Engineering and Technology* (2012)
487. V. Lucarini, D. Faranda, J. Wouters, Universal behavior of extreme value statistics for selected observables of dynamical systems. J. Stat. Phys. **147**, 63–73 (2012)
488. L.J. Lucas, H. Owhadi, M. Ortiz, Rigorous verification, validation, uncertainty quantification and certification through concentration-of-measure inequalities. Comput. Methods Appl. Mech. Eng. **57**, 4591–4609 (2009)
489. A. MacDonald, C.J. Scarrott, D. Lee, B. Darlow, M. Reale, G. Russell, A flexible extreme value mixture model. Comput. Stat. Data Anal. **55**, 2137–2157 (2011)
490. M.T. Madi, M.Z. Raqab, Bayesian prediction of temperature records using the Pareto model. Environmetrics **15**, 701–710 (2004)
491. M.T. Madi, M.Z. Raqab, Bayesian prediction of rainfall records using the generalized exponential distribution. Environmetrics **18**, 541–549 (2007)
492. H. Madsen, C.P. Pearson, E. Rosbjerg, Comparison of annual maximum series and partial duration series methods for modeling extreme hydrologic events: 2. Regional modeling. Water Resour. Res. **33**, 759–769 (1997)
493. H. Madsen, E. Rosbjerg, Generalized least squares and empirical Bayes estimation in regional index-flood modeling. Water Resour. Res. **33**, 771–781 (1997)
494. H. Madsen, E. Rosbjerg, The partial duration series method in regional index-flood modeling. Water Resour. Res. **33**, 737–746 (1997)
495. S. Maity, K. Sasmal, S. Bonthu, H.V. Warrior, Role of parallel computing in numerical weather forecasting models. Int. J. Comput. Appl. **4**, 22–27 (2003)
496. Y. Malevergne, V. Pisarenko, D. Sornette, On the power of generalized extreme value (gev) and generalized Pareto distribution (GPD) estimators for empirical distributions of stock returns. Appl. Financ. Econ. **16**, 271–289 (2006)
497. I. Malinowska, D. Szynal, On a family of Bayesian estimators and predictors for a Gumbel model based on the kth lower records. Appl. Math. **31**, 107–115 (2004)
498. B. Mandelbrot, R. Hudson, *Une approche fractale des marchés: risque, perdre et gagner (A fractal approach to markets: risk, loss and gain)* (Odile Jacob, 2005)
499. H.B. Mann, Nonparametric tests against trend. Econometrica **13**, 245–259 (1945)
500. N.R. Mann, R.E. Schafer, N.D. Singpurwalla, *Methods for Statistical Analysis of Reliability and Life Data*, Wiley Series in Probability and Statistics (Wiley, New York, 1974)
501. J.-M. Marin, C.P. Robert, *Bayesian Core: A Practical Approach to Computational Bayesian Statistics* (Springer, Berlin, 2007)
502. F. Marohn, Local asymptotic normality in extreme value index estimation. Ann. Inst. Stat. Math. **49**, 645–666 (1997)
503. J.S. Marron, Partitioned cross-validation. Economet. Rev. **6**, 271–284 (1987)
504. D. Marsan, C.J. Bean, S. Steacy, J. McClosey, Observation of diffusion processes in earthquake populations and implications for the predictability of seismicity systems. J. Geophys. Res. **105**, 28081–28094 (2000)
505. M. Martinkova, A review of applied methods in Europe for flood-frequency analysis in a changing environment, in *Floodfreq COST action ES0901: European procedures for flood frequency estimation, Centre for Ecology & Hydrology on behalf of COST* (2013)
506. A.P. Martins, H. Ferreira, The multivariate extremal index and the dependence structure of a multivariate extreme value distribution. TEST **14**, 433–448 (2005)
507. D.M. Mason, Laws of large numbers for sums of extreme values. Ann. Probab. **10**(3), 754–764 (1982)
508. P. Massart, *Concentration Inequalities and Model Selection*, Lecture Notes in Mathematics (Springer, Berlin, 2007)

509. M.D. McKay, R.J. Beckman, W.J. Conover, A comparison of three methods for selecting values of input variables in the analysis of output from a computer code. Technometrics **21**, 239–245 (1979)
510. G. McLachlan, T. Khrisnan, *The EM Algorithm and Extensions*, 2nd edn. (Wiley, New York, 2008)
511. M.D. Morris, T.J. Mitchell, Exploratory designs for computational experiments. J. Stat. Plan. Infer. **43**, 381–402 (1995)
512. B. Merz, G. Bloschl, A process typology of regional floods. Water Resour. Res. **39**, 1340 (2003)
513. M. Mhanna, W. Bauwens, A stochastic space-time model for the generation of daily rainfall in the gaza strip. Int. J. Climatol. **32**(7), 1098–1112 (2012)
514. J. Michalakes, M. Vachharajani, GPU acceleration of numerical weather prediction, in *IEEE International Symposium on Parallel and Distributed Processing*, 129 (2008)
515. B. Miladinovic, C.P. Tsokos, Ordinary, Bayes, empirical Bayes, and non-parametric reliability analysis for the modified Gumbel failure model. Nonlinear Anal. **71**, 1426–1436 (2009)
516. C.A. Miller, P.N. Edwards (eds.), *Changing the Atmosphere. Expert Knowledge and Environmental Governance* (MIT Press, New York, 2001)
517. P.C.D. Milly, J. Betancourt, M. Falkenmark, R.M. Hirsch, Z.W. Kundzewicz, D.P. Lettenmaier, R.J. Stouffer, Stationarity is dead: whither water management? Science **319**, 573–574 (2014)
518. J. Miquel, *Guide pratique d'estimation des probabilités des crues* (Collection de la Direction des Études et Recherches d'EDF, Eyrolles, 1984)
519. Y. Miserey, L'histoire du climat de la France sort de l'oubli. *Le Figaro – Sciences et Environnement* (2007)
520. J.F.B. Mitchell, J. Lowe, R.A. Wood, M. Vellinga, Extreme events due to human-induced climate change. Phil. Trans. R. Soc. A **364**, 2117–2133 (2009)
521. A. Montanari, D. Koutsoyiannis, Modeling and mitigating natural hazards: stationarity is immortal!. Water Resour. Res. **50**, 9748–9756 (2014)
522. M.G. Morgan, Use (and abuse) of expert elicitation in support of decision making for public policy. Proc. US Natl. Acad. Sci. **111**, 7176–7184 (2014)
523. J. Morio, Extreme quantile estimation with nonparametric adaptive importance sampling. Simul. Model. Pract. Theory **27**, 76–89 (2012)
524. J. Morio, *Contributions à la simulation des événements rares dans les systèmes complexes*. Habilitation Thesis, Université Rennes I (2013)
525. J. Morio, M. Balesdent, D. Jacquemart, C. Vergé, A survey of rare event simulation methods for static input-output models. Simul. Model. Pract. Theory **49**, 287–304 (2014)
526. J. Morio, R. Pastel, F. Le Gland, An overview of importance splitting for rare event simulation. Eur. J. Phys. **31**, 1295–1303 (2010)
527. J. Morio, R. Pastel, F. Le Gland, Estimation de probabilités et de quantiles rares pour la caractérisation d'une zone de retombée d'un engin. Journal de la Société Française de Statistique **152**, 1–29 (2011)
528. S. Morita, P.F. Thall, P. Mueller, Determining the effective sample size of a parametric prior, in *Anderson Cancer Center Department of Biostatistics, Working Paper Series 2007. Working Paper 36* (2007)
529. T. Morlot, C. Perret, A.C. Favre, J. Jalbert, Dynamic rating curve assessment for hydrometric stations and computation of the associated uncertainties: Quality and station management indicators. J. Hydrol. **517**, 173–186 (2014)
530. R.D. Moser, T.A. Oliver, Validation of physical models in the presence of uncertainty, in *Handbook of Uncertainty Quantification*, ed. by R. Ghanem et al. (Springer International Publishing, Switzerland, 2015)
531. V.A. Mousseau, B.J. Williams, Uncertainty quantification in a regulatory environment, in *Handbook of Uncertainty Quantification*, ed. by R. Ghanem et al. (Springer International Publishing, Switzerland, 2015)
532. M. Moustapha, B. Sudret, J.-M. Bourinet, M. Guillaume, Quantile-based optimization uncer uncertainties using adaptive kriging surrogate models. J. Struct. Multidiscip. Optim. **54**, 1403–1421 (2016)

533. V. Moutoussamy, *Contributions to structural reliability analysis: accounting for monotonicity constraints in numerical models*. Ph.D. thesis, Université Toulouse III (2015)
534. R. Moyeed, R. Clarke, The use of Bayesian methods for fitting rating curves, with case studies. Adv. Water Resour. **28**, 807–818 (2005)
535. R. Mueller, K. Dagestad, P. Ineichen, M. Schroedter, S. Cros, D. Dumortier, R. Kuhlemann, J. Olseth, G. Piernavieja, C. Reise, L. Wald, D. Heinneman, Rethinking satellite based solar irradiance modelling: The Solis clear sky module. Remote Sens. Environ. **91**, 160–174 (2004)
536. A. Muller, P. Arnaud, M. Lang, J. Lavabre, Uncertainties of extreme rainfall quantiles estimated by a stochastic rainfall model and by a generalized Pareto distribution. Hydrol. Sci. **54**, 417–429 (2009)
537. M. Munoz Zuniga, *Méthodes stochastiques pour l'estimation contrôlée de faibles probabilités sur des modèles physiques complexes. Application au domaine nucléaire*. Ph.D. thesis, Université Paris VII (2011)
538. A. Murangira, M. Munoz Zuniga, T. Perdrizet, Structural reliability assessment through metamodel-based importance sampling with dimension reduction, in *HAL-01230454* (2015)
539. A.H. Murphy, What is a good forecast? An essay on the nature of goodness in weather forecasting. Weather Forecast. **8**, 281–293 (1993)
540. A.H. Murphy, R.L. Winkler, A general framework for forecast verification. Mon. Weather Rev. **115**, 1330–1338 (1987)
541. C. Murphy, P. Gardoni, The acceptability and the tolerability of societal risks: a capabilities-based approach. Sci. Eng. Ethics **14**, 77–92 (2008)
542. S. Nadarajah, Multivariate declustering techniques. Environmetrics **12**, 357–365 (2001)
543. E.A. Nadaraya, On estimating regression. Theory Probab. Appl. **9**, 141–142 (1954)
544. M.K. Nakayama, Asymptotically valid confidence intervals for quantiles and values-at-risk when applying Latin Hypercube Sampling. Int. J. Advan. Syst. Measur. **4**, 86–94 (2011)
545. S. Nandagopalan, *Multivariate extremes and estimation of the extremal index*. Ph.D. thesis, University of North Carolina at Chapel Hill (1990)
546. S. Nandagopalan, On the multivariate extremal index. J. Res. Nat. Inst. Stand. Technol. **99**, 543–550 (1994)
547. J.E. Nash, J.V. Sutcliffe, River flow forecasting through conceptual models part i - A discussion of principles. J. Hydrol. **10**, 282–290 (1970)
548. J.C. Neal, N.A. Odoni, M.A. Trigg, J.E. Freer, J. Garcia-Pintado, D.C. Mason, M. Wood, P.D. Bastes, Efficient incorporation of channel cross-section geometry uncertainty into regional and global scale flood models. J. Hydrol. **529**, 169–183 (2015)
549. R.M. Neal, Transferring prior information between models using imaginary data, *Technical Report 0108, Dept. Statistics, Univ. Toronto, 2001* (2001)
550. R.M. Neal, *MCMC Using Hamiltonian Dynamics* (Chapman & Hall, Boca Raton, 2011)
551. R.E. Neapolitan, A limiting frequency approach to probability based on the weak law of large numbers. Philos. Sci. **59**, 389–407 (1992)
552. C. Nemeth, P. Fearnhead, Stochastic gradient markov monte carlo (2019), arXiv:1907.06986
553. L. Neppel, B. Renard, M. Lang, P.-A. Ayral, D. Coeur, E. Gaume, N. Jacob, O. Payrastre, K. Pobanz, F. Vinet, Flood frequency analysis using historical data: accounting for random and systematic errors. Hydrol. Sci. J. **55**, 192–208 (2010)
554. R. Neukom, N. Steiger, J.J. Gomez-Navarro, J. Wang, J.P. Werner, No evidence for globally coherent warm and cold periods over the preindustrial common era. Nature **571**, 550–573 (2019)
555. C. Neves, M.I.F. Alves, Testing extreme value conditions - an overview and recent approaches. REVSTAT **6**, 83–100 (2008)
556. S.E. Neville, M.J. Palmer, M.P. Wand, Generalized extreme value additive model analysis via mean field variational Bayes. Aust. N. Z. J. Stat. **53**, 305–330 (2011)
557. C.C. Nguyen, E. Gaume, O. Payrastre, Regional flood frequency analyses involving extraordinary flood events at ungauged sites: further developments and validations. J. Hydrol. **508**, 385–396 (2014)
558. H. Niederreiter, *Random Number Generation and Quasi-monte Carlo Methods* (SIAM, 1992)

559. T. Nilsen, T. Aven, Models and model uncertainty in the context of risk analysis. Reliab. Eng. Syst. Saf. **79**, 309–317 (2003)
560. B. Ning, P. Bloomfield, Bayesian inference for generalized extreme value distribution with Gaussian copula dependence (2017), arXiv:1703.00968v1
561. K.M. Nissen, G.C. Leckebusch, J.G. Pinto, D. Renggli, S. Ulbrich, U. Ulbrich, Cyclones causing wind storms in the mediterranean: characteristics, trends and links to largescale patterns. Nat. Hazard. **10**, 1379–1391 (2010)
562. P.J. Northrop, Likelihood-based approaches to flood frequency estimation. J. Hydrol. **292**, 96–113 (2004)
563. P.J. Northrop, N. Attalides, Posterior propriety in Bayesian extreme value analyses using reference priors. Stat. Sin. **26**, 721–743 (2015)
564. W.T. Nutt, G.B. Wallis, Evaluation of nuclear safety from the outputs of computer codes in the presence of uncertainties. Reliab. Eng. Syst. Saf. **83**, 57–77 (2004)
565. R. Nuzzo, Scientific method: statistical errors. Nature **506**, 150–152 (2014)
566. J.O. Oakley, Estimating percentiles of uncertain computer code outputs. Appl. Stat. **58**, 83–93 (2004)
567. W.L. Oberkampf, M.F. Barone, Measures of agreement between computation and experiment: Validation metrics. J. Comput. Phys. **217**, 5–36 (2006)
568. W.L. Oberkampf, S.M. EeLand, B.M. Rutherford, K.V. Eiegertet, K.F. Alvin, Error and uncertainty in modelling and simulation. Reliab. Eng. Syst. Saf. **75**, 333–357 (2002)
569. W.L. Oberkampf, C.J. Roy, *Verification and Validation in Scientific Computing* (Cambridge University Press, Cambridge, 2010)
570. G.W. Oehlert, A note on the delta method. Am. Stat. **46**, 27–29 (1992)
571. H. Ogelmen, A. Ecevit, E. Tasmiroglu, A new method for estimating solar radiation from bright sunshine data. Sol. Energy **33**(6), 619–625 (1984)
572. A. O'Hagan, C.E. Buck, A. Daneshkhah, J.R. Eiser, P. Garthwaite, D.J. Jenkinson, J.E. Oakley, T. Rakow, *Uncertain Judgments: Eliciting Expert's Probabilities* (Wiley, New York, 2006)
573. T. O'Hagan, Fractional Bayes factor for model comparison (with discussion). J. R. Stat. Soc. Ser. B (Methodol.) **57**, 99–138 (1995)
574. A. Okabe, B. Boots, K. Sugihara, S.N. Chiu, *Spatial Tessellations - Concepts and Applications of Voronoi Diagrams*, 2nd edn. (Wiley, New York, 2000)
575. R. Olsen, J.H. Lambert, Y.Y. Haimes, Risk of extreme events under nonstationary conditions. Risk Anal. **18**, 497–510 (1998)
576. E. Omey, F. Mallor, E. Nualart, *An introduction to statistical modelling of extreme values. Application to calculate extreme wind speeds*. Hogeschool-Universiteit Brussel Research Paper 200/36, November 2009 (2009)
577. B. O'Neill, Exchangeability, correlation and Bayes' Effect. Int. Stat. Rev. **77**, 241–250 (2011)
578. N. Oreskes, K. Shrader-Frechette, K. Belitz, Verification, validation, and confirmation of numerical models in the earth sciences. Science **263**, 641–646 (1994)
579. F. O'Sullivan, B.S. Yandell, W.J. Raynor, Automatic smoothing of regression functions in generalized linear models. J. Am. Stat. Assoc. **81**, 96–103 (1986)
580. C.E.G. Otiniano, C.R. Gonçalves, C.C.Y. Dorea, Mixture of extreme-value distributions: identifiability and estimation, in *Communications in Statistics: Theory and Methods*, 46 (2017)
581. T. Ouarda, P.F. Rasmussen, B. Bobee, J. Bernier, Utilisation de l'information historique en analyse hydrologique fréquentielle. Revue des Sciences de l'Eau 41–49 (1998)
582. A.B. Owen, Variance and discrepancy with alternative scramblings. ACM Trans. Model. Comput. Simul. **13**, 363–378 (1993)
583. M.P. Paine, Asteroid impacts: the extra hazard due to tsunami. Sci. Tsunami Haz. **17**, 155–166 (1999)
584. E. Panagoulia, P. Economou, C. Caroni, Stationary and nonstationary generalized extreme value modelling of extreme precipitation over a mountainous area under climate change. Environmetrics **25**, 29–43 (2014)
585. E. Paquet, Critique et homogénéisation de séries climatologiques, Outil KRITIkOR, Notice Méthodologique. *EDF DTG, D4163/NT/2005-00174-A* (2005)

586. E. Paquet, J. Gailhard, R. Garçon, Évolution de la méthode du Gradex: approche par type de temps and modélisation hydrologique. La Houille Blanche **5**, 37–50 (2006)
587. E. Paquet, F. Garavaglia, R. Garçon, J. Gailhard, The SCHADEX method: a semi-continuous rainfall-runoff simulation for extreme flood estimation. J. Hydrol. **495**, 23–37 (2013)
588. E. Parent, J. Bernier, Bayesian POT modeling for historical data. J. Hydrol. **274**, 95–108 (2003)
589. E. Parent, J. Bernier, Encoding prior experts judgments to improve risk analysis of extreme hydrological events via POT modeling. J. Hydrol. **283**, 1–18 (2003)
590. E. Parent, J. Bernier, Une procédure bayésienne de sélection/validation différentielle pour déterminer le domaine d'attraction des valeurs extrêmes. Revue de Statistique Appliquée **52**, 5–31 (2004)
591. E. Parent, J. Bernier, *Le raisonnement bayésien Modélisation and inférence* (Springer, Berlin, 2007)
592. E. Parent, A. Pasanisi, N. Bousquet, M. Keller, J. Bernier, Considérations décisionnelles pour la construction d'un ouvrage de protection contre les crues. Technip 299–333 (2014)
593. S Parey, T.T.H. Hoang, D. Dacunha Castelle, Different ways to compute temperature return levels in the climate change context. Environmentrics **21**, 698–718 (2010)
594. S Parey, T.T.H. Hoang, D. Dacunha Castelle, The importance of mean and variance in predicting changes in temperature extremes. J. Geophys. Res. **118**, 8285–8296 (2013)
595. S. Parey, T.T.T Hoang, D. Dacunha-Castelle, The role of variance in the evolution of observed temperature extremes in Eurasia and in the United States (2013)
596. S. Parey, F. Malek, C. Laurent, E. Eacunha Castelle, Statistical approach for very high temperatures in France, Trends and climatic evolution. Clim. Change **81**, 331–352 (2007)
597. J.B. Paris, *The Uncertain Reasoner's Companion: a Mathematical Perspective* (Cambridge University Press, Cambridge, 1994)
598. A. Pasanisi, M. Keller, E. Parent, Estimation of a quantity of interest in uncertainty analysis: some help from Bayesian Decision Theory. Reliab. Eng. Syst. Saf. **100**, 93–101 (2012)
599. R. Pastel, *Estimation de probabilités d'événements rares et de quantiles extrêmes - Applications dans le domaine aérospatial*. Ph.D. thesis, Université Rennes I (2013)
600. F. Pauli, S.G. Coles, Penalized likelihood inference in extreme value analyses. J. Appl. Stat. **28**, 547–560 (2001)
601. O. Payrastre, Utilité de l'information historique pour l'étude du risque de crues. *14ième Journées Scientifiques de l'Environnement: l'Eau, la Ville, la Vie, 12-13 mai* (2003)
602. O. Payrastre, E. Gaume, H. Andrieu, Usefulness of historical information for flood frequency analyses: developments based on a case study. Water Resour. Res. **47**, 8 (2011)
603. O. Payrastre, E. Gaume, H. Andrieu, Information historique et étude statistique des crues extrêmes: quelles caractéristiques souhaitables pour les inventaires de crues historiques? La Houille Blanche **3**, 5–11 (2013)
604. N. Pedroni, E. Zio, E. Ferrario, A. Pasanisi, M. Couplet, *Propagation of aleatory and epistemic uncertainties in the model for the design of a flood protection dike, in Proceedings of the PSAM and ESREL Conferences* (Helsinki, Finland, 2012)
605. L. Peng, Estimation of the coefficient of tail dependence in bivariate extremes. Stat. Probab. Lett. **43**, 399–409 (1999)
606. D. Penot, *Cartographie des événements hydrologiques extrêmes et estimation SCHADEX en sites non jaugés*. Ph.D. thesis, Université de Grenoble (2014)
607. J.P. Penot, Topex-Poséidon and Jason 1, mesureurs des océans. *BT Espace No. 1103, Centre National d'Études Spatiales* (2001)
608. D.F. Percy, *Subjective Reliability Analysis Using Predictive Elicitation*, vol. 7 (World Scientific, Singapore, 2003), pp. 57–72
609. N. Pérot, N. Bousquet, Functional Weibull-based models of steel fracture toughness for structural risk analysis: estimation and selection. Reliab. Eng. Syst. Saf. **165**, 335–367 (2017)
610. F. Perrin, *Prise en compte des données expérimentales dans les modèles probabilistes pour la prévision de la durée de vie des structures*. Ph.D. thesis, Université Clermont II (2008)

611. M. Petelet, B. Iooss, O. Asserin, A. Loredo, Latin hypercube sampling with inequality constraint. Adv. Stat. Anal. **94**, 325–339 (2010)
612. A.N. Pettitt, A non-parametric approach to the change-point problem. Appl. Stat. **28**, 126–135 (1979)
613. C. Pfister, E. Garnier, M.-J. Alcoforado, D. Wheeler, J. Luterbacher, M.-F. Nunes, J.-P. Taborda, The Meteorological framework and the cultural memory of three severe winter-storms in early eighteenh century Europe. Clim. Change **101**, 281–310 (2010)
614. P.C.B. Phillips, P. Perron, Testing for a unit root in time series regression. Biometrika **75**, 335–346 (1988)
615. J. Pickands, Statistical inference using extreme order statistics. Ann. Stat. **3**, 119–131 (1975)
616. J. Pickands, Multivariate extreme value distributions. Bull. Inst. Int. Stat. **49**, 859–878 (1981)
617. J. Pickands, Bayes quantile estimation and threshold selection for the generalized Pareto family, in *Extreme Value Theory and Applications*, ed. by J. Galambos, J. Lechner, E. Simiu (Springer, US, 1994), pp. 123–138
618. R. Piessens, E. deDoncker Kapenga, C. Uberhuber, D. Kahaner, *Quadpack: a Subroutine Package for Automatic Integration* (Springer, New York, 1983)
619. L. Pineau Guillou, C. Lathuiliere, R. Magne, S. Louazel, E. Corman, C. Perherin, *Caractérisation des niveaux marins et modélisation des surcotes pendant la tempête Xynthia*. Éditions Paralia, 11-1/11-8 (2010)
620. J. Planzalg, R. Hamböker, *Parametric Statistical Theory* (Walter de Gruyter, Berlin, 1994)
621. M. Plummer, JAGS: a program for analysis of Bayesian graphical models using Gibbs sampling, in *Proceedings of the 3rd International Workshop on Distributed Statistical Computing* (2003)
622. K. Popper, *Objective Knowledge* (Clarendon Press, Oxford, 1972)
623. C.P. Porter, Kolmogorov on the role of randomness in probability theory. Math. Struct. Comput. Sci. **24** (2014)
624. A. Pouget, J.M. Beck, W.J. Ma, P.E. Latham, Probabilistic brains: knowns and unknowns. Nat. Neurosci. **16**, 1170–1178 (2016)
625. N. Pouvreau, *Trois cents ans de mesures marégraphiques en France: outils, méthodes et tendances des composantes du niveau de la mer au port de Brest*. Ph.D. thesis, Université de la Rochelle (2008)
626. S. James Press, *Subjective and Objective Bayesian Statistics: Principles, Models, and Applications*, 2nd edn., Revised and expanded (Wiley, New York, 2003)
627. M. Prevosto, Extremes and long term extrapolation, in *Report of the European Project EquiMar (Equitable Testing and Evaluation of Marine Energy Extraction Eevices in terms of Performance, Cost and Environmental Impact)* (2011)
628. L. Pronzato, Minimax and maximin space-filling designs: some properties and methods for construction. Journal de la Société Française de Statistique **158**, 7–36 (2017)
629. L. Pronzato, W. Müller, Design of computer experiments: space filling and beyond. Stat. Comput. **22**, 681–701 (2012)
630. Y. Qi, Bootstrap and empirical likelihood methods in extremes. Extremes **11**, 81–97 (2008)
631. Z. Qu, *The new Heliosat-4 method for the assessment of solar irradiance at ground level. La nouvelle méthode Heliosat-4 pour l'évaluation du rayonnement solaire au sol*. Ph.D. thesis, Mines de Paris (2013)
632. A.E. Raftery, T. Gneiting, F. Balabdaoui, M. Polakowski, Using Bayesian model averaging to calibrate forecast ensembles. Mon. Weather Rev. **133**, 1155–1174 (2005)
633. F. Ragone, J. Wouter, F. Bouchet, Computation of extreme heat waves in climate models using a large deviation algorithm. Proc. US Natl. Acad. Sci. **115**, 24–29 (2017)
634. G. Ragulina, T. Reitan, Generalized extreme value shape parameter and its nature for extreme precipitation using long time series and the Bayesian approach. Hydrol. Sci. J. (2017). https://doi.org/10.1080/02626667.2016.1260134
635. N.I. Ramesh, A.C. Davison, Local models for exploratory analysis of hydrological extremes. J. Hydrol. **256**, 106–119 (2002)

636. M.P.M. Reddy, M. Affholder, *Descriptive Physical Oceanography: State of the Art* (Taylor and Francis, Milton Park, 2002)
637. B.J. Reich, E. Cooley, K.M. Foley, S.L. Napelenok, B.A. Shaby, Extreme value analysis for evaluating ozone control strategies. Ann. Appl. Stat. **7**, 739–762 (2013)
638. D.S. Reis, J.R. Stedinger, Bayesian MCMC flood frequency analysis with historical information. J. Hydrol. **313**, 97–116 (2005)
639. R. Reiss, M. Thomas, *Statistical Analysis of Extreme Values: with Applications to Insurance Finance, Hydrology and Other Fields* (Birkhauser, 2001)
640. G. Réméniéras, *L'Hydrologie de l'Ingénieur* (Eyrolles, 1999)
641. B. Renard, Détection and prise en compte d'éventuels impacts du changement climatique sur les extrêmes hydrologiques en France. La Houille Blanche **1**, 109–117 (2008)
642. B. Renard, M. Lang, P. Bois, Statistical analysis of extreme events in a non-stationary context via a Bayesian framework: case-study with peak-over-threshold data. Stoch. Env. Res. Risk Assess. **21**, 97–112 (2006)
643. B. Renard, X. Sun, M. Lang, *Bayesian Methods for Non-Stationary Extreme Value Analysis* (2013)
644. S. Resnick, Discussion of the Danish data on large fire insurance losses. ASTIN Bull. **27**, 139–151 (1997)
645. S.I. Resnick, *Heavy-Tail Phenomena: Probabilistic and Statistical Modeling* (Springer, New York, 2007)
646. M. Ribatet, POT: modelling peaks over a threshold. R News **7**, 34–36 (2007)
647. M. Ribatet, T.B.M.J. Ouarda, E. Sauquet, J.-M. Gresillon, Modeling all exceedances above a threshold using an extremal dependence structure: Inferences on several flood characteristics. Water Resour. Res. **45** (2009)
648. M. Ribatet, E. Sauquet, J.M. Grésillon, T.B.M.J. Ouarda, A regional Bayesian POT model for flood frequency analysis. Stoch. Environ. Res. Risk Assess. **21**, 327–339 (2007)
649. C.W. Richardson, Stochastic simulation of daily precipitation, temperature, and solar radiation. Water Resour. Res. **17**, 182–190 (1981)
650. H. Rinne, *The Weibull Distribution: A Handbook* (CRC Press, Boca Raton, 2008)
651. C.P. Robert, *The Bayesian Choice: From Decision-Theoretic Foundations to Computational Implementation*, 2nd edn. (Springer, Berlin, 2007)
652. C.P. Robert, G. Casella, *Monte Carlo Statistical Methods*, 2nd edn. (Springer, Berlin, 2004)
653. C.Y. Robert, Estimating the multivariate extremal index function. Bernoulli **14**, 1027–1064 (2008)
654. G.O. Roberts, J.S. Rosenthal, Examples of adaptive MCMC. J. Comput. Graph. Stat. **18**, 349–367 (2009)
655. D.G. Robinson, *A Research Perspective* (Taylor and Francis, Milton Park, 2005)
656. R.T. Rockafellar, *Coherent Approaches to Risk in Optimization Under Uncertainty* (Maryland, INFORMS, 2007), pp. 38–61
657. R.T. Rockafellar, *Return Periods and Return Levels Under Climate Change* (Springer, Berlin, 2013)
658. E. de Rocquigny, Quantifying uncertainty in an industrial approach: an emerging consensus in an old epistemological debate. SAPIENS: Surv. Perspect. Integrating Environ. Soc. **2**, 1–18 (2009)
659. E. de Rocquigny, Structural reliability under monotony: A review of property of FORM and associated simulation methods and a new class of Monotonous Reliability Methods (MRM). Struct. Safety **31**, 363–374 (2009)
660. E. de Rocquigny, N. Devictor, S. Tarantola (eds.), *Uncertainty in Industrial Practice - A Guide to Quantitative Uncertainty Management* (Wiley, New York, 2008)
661. H.J.E. Rodda, M.A. Little, *Understanding Mathematical and Statistical Techniques in Hydrology, An Examples-Based Approach* (Wiley Blackwell, New York, 2015)
662. I. Rodriguez-Iturbe, D. Cox, V. Isham, Some models for rainfall based on stochastic point processes. Proc. R. Soc. Lond. A **410**(1839), 269–288 (1987)

663. J. Rohmer, C. Baudrit, The use of the possibility theory to investigate the epistemic uncertainties within scenario-based earthquake risk assessments. Nat. Hazards **56**, 613–632 (2011)
664. R. Romanowicz, K.J. Beven, J. Tawn, *Bayesian Calibration of Flood Inundation Models* (Wiley, Chichester, 1996), pp. 333–360
665. H. Rootzén, R.W. Katz, Design life level: quantifying risk in a changing climate. Water Resour. Res. **49**, 5964–5972 (2013)
666. H. Rootzén, N. Tajvidi, Extreme value statistics and wind storm losses: a case study. Scand. Actuar. J. **1**, 70–94 (1997)
667. E. Rosbjerg, Estimation in partial duration series with independent and dependent peak values. J. Hydrol. **76**, 183–195 (1985)
668. G. Rosen, A. Cohen, Extreme percentile regression, in *Theory and Computational Aspects of Smoothing: Proceedings of the COMPSTAT '94 Satellite Meeting* (1994), pp. 200–214
669. M. Roth, T.A. Buishand, G. Jongbloed, A.M.G. Klein Tank, J.H. van Zanten, A regional peaks-over-threshold model in a nonstationary climate. Water Resour. Res. 48 (2012)
670. O. Roustant, D. Ginsbourger, Y. Deville, DiceKriging, DiceOptim: Two R packages for the analysis of computer experiments by Kriging-based metamodeling and optimization. J. Stat. Softw. **51** (2013), https://doi.org/10.18637/jss.v051.i01
671. W.E. Rowe, Understanding uncertainty. Risk Anal. **14**, 743–750 (1994)
672. G. Rubino, B. Tuffin, *Rare Event Simulation Using Monte Carlo Methods* (Wiley Online Library, New York, 2009)
673. R.Y. Rubinstein, D.P. Kroese, *Simulation and the Monte Carlo Method* (Wiley, New York, 2011)
674. F. Ruggeri, N. Bousquet, Robust Bayesian analysis for generalized extreme value models, *unpublished preprint* (2017)
675. D. Ruppert, M.P. Wand, Multivariate locally weighted least squares regression. Ann. Stat. **22**, 1346–1370 (1994)
676. A. Sabourin, Semi-parametric modeling of excesses above high multivariate thresholds with censored data. J. Multivar. Anal. **136**, 126–146 (2015)
677. A. Sabourin, P. Naveau, Bayesian Eirichlet mixture model for multivariate extremes: a re-parametrization. Comput. Stat. Data Anal. **71**, 542–567 (2014)
678. A. Sabourin, P. Naveau, A.L. Fougères, Bayesian model averaging for multivariate extremes. Extremes **16**, 325–350 (2013)
679. A. Sabourin, B. Renard, Combining regional estimation and historical floods: a multivariate semi-parametric peaks-over-thresold model with censored data. Water Resour. Res. **51**, 9646–9664 (2015)
680. J. Sacks, W.J. Welch, T.J. Mitchell, H.P. Wynn, Design and analysis of computer experiments. Stat. Sci. **4**, 409–435 (1989)
681. H. Saidi, M. Ciampittiello, C. Dresti, L. Turconi, Extreme rainfall events: evaluation with different instruments and measurement reliability. Environ. Earth Sci. **72**, 4607–4616 (2014)
682. J.D. Salas, J. Obeysekera, Revisiting the concepts of return period and risk for nonstationary hydrologic extreme events. J. Hydrol. Eng. **19**, 554–568 (2014)
683. E. Salinas, Prior and prejudice. Nat. Neurosci. **14**, 943–945 (2011)
684. L. Sanders, The Probabilistic Mind. Sci. News **180**, 18 (2011)
685. N.E. Savin, K.J. White, The Durbin-Watson test for serial correlation with extreme sample sizes or many regressors. Econometrica **45**, 1989–1996 (1977)
686. C. Scarrott, A. MacDonald, A review of extreme value threshold estimation and uncertainty quantification. REVSTAT **10**, 33–60 (2012)
687. M.G. Schaefer, Regional analysis of precipitation annual maxima in Washington state. Water Resour. Res. **26**, 119–131 (1990)
688. M. Schlather, Examples for the coefficient of tail dependence and the domain of attraction of a bivariate extreme value distribution. Stat. Probab. Lett. **53**, 325–329 (2001)
689. G. Schmidt, The uncertainty in climate modeling. Bull. Atomic Sci. (2007)
690. W. Schnedler, Likelihood estimation for censored random vectors. Econ. Rev. **24**, 195–217 (2005)

691. R. Schöbi, B. Sudret, S. Marelli, Rare event estimation using polynomial-chaos kriging. ASCE-ASME J. Risk Uncertainty Eng. Syst., Part A: Civil Eng. 3 (2016)
692. A.Y. Schumann, N.R. Moloney, D. Jörn, *Extreme Value and Record Statistics in Heavy-Tailed Processes with Long-Range Memory* (American Geophysical Union, 2013), pp. 315–334
693. G.E. Schwarz, Estimating the dimension of a model. Ann. Stat. **6**, 461–464 (1978)
694. E. Gideon, H. Schwarz, Estimating the dimension of a model. Ann. Stat. **6**, 461–464 (1978)
695. Q. Sebille, A.-L. Fougères, C. Mercadier, Modeling extreme rainfall. A comparative study of spatial extreme value models. Spatial Stat. **21**, 187–208 (2017)
696. A. Sen, M. Srivastava, *Regression Analysis - Theory, Methods, and Applications*, 4th edn. (Springer, Berlin, 2011)
697. S. Seo, *A review and comparison of methods for detecting outliers in univariate data sets*. Ph.D. thesis, University of Pittsburgh (2006)
698. F. Serinaldi, C.G. Kilsby, Stationarity is undead: uncertainty dominates the distribution of extremes. Adv. Water Resour. **77**, 17–36 (2015)
699. G. Shafer, *A Mathematical Theory of Evidence* (Princeton University Press, Princeton, 1976)
700. G. Shafer, V. Vovk, The sources of Kolmogorov's grundbegriffe. Stat. Sci. **21**, 70–98 (2006)
701. S. Shenoy, D. Gorinevsky, *Predictive Analysis for Extreme Events in Big Data* (2015), pp. 184–193
702. P. Silberzahn, *Nate Silver, prévisionniste extraordinaire et ... inutile* (2017). Internet blog: https://philippesilberzahn.com/
703. D. Singh, A. Ganju, A. Singh, Weather prediction using nearest-neighbour model. Curr. Sci. **88**, 1283–1289 (2005)
704. A. Sklar, Fonctions de répartition à n dimensions et leurs marges. Publications de l'Institut de Statistique de l'Université de Paris **8**, 229–231 (1959)
705. P. Smets, *The Transferable Belief Model and Other Interpretations of Dempster-Schafer's Model* (Elsevier Science Publishers, Amsterdam, 1991), pp. 375–383
706. E. Smith, Bayesian modelling of extreme rainfall data. Ph.D. Thesis, University of Newcastle (2005)
707. R.L. Smith, Extreme value theory based on the r largest annual events. J. Hydrol. **86**, 27–43 (1986)
708. R.L. Smith, Bayesian and frequentist approaches to parametric predictive inference, in *Bayesian Statistics*, vol. 6, ed. by J.M. Bernardo, J.O. Berger, A.P. Dawid, A.F.M. Smith (Oxford University Press, Oxford, 1999), pp. 589–612
709. R.L. Smith, I. Weissman, Characterization and estimation of the multivariate extremal index. Technical Report, University of North Carolina at Chapel Hill, USA (1996)
710. P. Snow, On the correctness and reasonableness of Cox's theorem for finite domains. Comput. Intell. **17**, 178–192 (1998)
711. I.M. Sobol, On the distributions of points in a cube and the approximate evaluation of integrals. USSR Comput. Math. Math. Phys. **7**, 86–112 (1967)
712. L. Soler, *Introduction à l'épistémologie* (Ellipses, 2000)
713. P.X.K Song, *Correlated Data Analysis: Modeling, Analytics, and Applications*. Series in Statistics. (Springer, Berlin, 2007)
714. M.G. Sotillo, R. Aznar, F. Valero, Mediterranean offshore extreme wind analysis from the 44-year HIPOCAS database: different approaches towards the estimation of return periods and levels of extreme values. Adv. Geosci. **7**, 275–278 (2006)
715. J. Sprenger, *Bayésianisme versus fréquentisme en inférence statistique*. I. Drouet (ed.). Éditions Matériologiques, Paris (2017)
716. J. Sprenger, Discussion: beyond subjective and objective in bayesian statistics. J. R. Stat. Soc., Ser. A **180** (2017)
717. J. Sprenger, The objectivity of subjective bayesian inference. Eur. J. Philos. **8**, 539–558 (2018)
718. A.-M. Staicu, N.M. Reid, On probability matching priors. Can. J. Stat. **36**, 613–622 (2008)
719. J.R. Stedinger, Estimating a regional flood frequency distribution. Water Resour. Res. **19**, 503–510 (1983)

720. J.R. Stedinger, L.H. Lu, Appraisal of regional and index flood quantile estimators. Stoch. Hydrol. Hydraul. **9**, 49–75 (1995)
721. M.L. Stein, *Interpolation of Spatial Data* (Springer, Berlin, 1999)
722. S.M. Steinberg, Confidence intervals for functions of quantiles using linear combinations of order statistics. Institute of Statistics Mimeo Series No. **1433** (1983)
723. M.A. Stephens, EDF Statistics for goodness of fit and some comparisons. J. Am. Stat. Assoc. **69**, 730–737 (1974)
724. A.G. Stephenson, B.A. Shavy, B.J. Reich, A.L. Sullivan, Estimating spatially varying severity thresholds of the forest fire danger rating system using max-stable extreme events modelling. J. Appl. Meteorol. Climatol. **54**, 395–407 (2015)
725. D. Sun, J.O. Berger, Bayesian sequential reliability for Weibull and related distributions. Ann. Inst. Stat. Math. **46**, 221–249 (1994)
726. L.M. Surhone, M.T. Tennoe, S.F. Henssonow, *OpenBUGS*. Digital Library (2010)
727. R.E. Swain, E. Bowles, E. Ostenaa, A framework for characterization of extreme floods for dam safety risk assessments, in *Proceedings of the 1998 USCOLE Annual Lecture* (Buffalo, New York, 1998)
728. R. Swinbanks, P. Friederichs, S. Wahl, *Forecasting High-Impact Weather Using Ensemble Prediction Systems* (Cambridge University Press, Cambridge, 2016)
729. G.J. Székely, E-statistics: energy of statistical samples. Technical Report 03-05, Department of Mathematics and Statistics, Bowling Green State University (2004)
730. G.J. Székely, M.L. Rizzo, Testing for equal distribution in high dimension. InterStat **5** (2004)
731. G.J. Székely, M.L. Rizzo, Hierarchical clustering via joint between-within distances: extending Ward's minimum variance method. J. Classif. **22**, 151–183 (2005)
732. G.J. Székely, M.L. Rizzo, The energy of data. Ann. Rev. Stat. Appl. **4**, 447–449 (2017)
733. N. Taleb, Black swans and the domains of statistics. Am. Stat. **61**, 198–200 (2007)
734. N.N. Taleb, *The Black Swan: The Impact of the Highly Improbable* (Random House, New York, 2009)
735. N.N. Taleb, *Antifragile: Things That Gain from Disorder* (Random House, New York, 2012)
736. N.N. Taleb, D.G. Goldstein, M.W. Spitznagel, The six mistakes executives make in risk management. Harvard Busin. Rev. (2009)
737. S. Tanaka, K. Takara, A study on threshold selection in POT analysis of extreme floods, in *The Extremes of the Extremes: Extraordinary Floods*, vol. 271 (IAHS Publishing, Wallingford, 2002)
738. P. Tankov, Tails of weakly dependent random vectors. J. Multivar. Anal. **145**, 73–86 (2016)
739. A. Terenin, D. Draper, Rigorizing and extending the Cox-Jaynes derivation of probability: implications for statistical practice (2015). arXiv:1507.06597
740. D. Tetzlaff, S. Uhlenbrook, Significance of spatial variability in precipitation for process-oriented modelling: results from two nested catchments using radar and ground station data. Hydrol. Earth Syst. Sci. **9**, 29–41 (2005)
741. G. Tetzlaff, J. Zimmer, *Probable Maximum Precipitation* (Springer, Berlin, 2016), pp. 778–779
742. M. Thomas, *Concentration results on extreme value theory*. Ph.D. thesis, Université Paris Diderot (2015)
743. C.S. Thompson, P.J. Thomson, X. Zheng, Fitting a multisite daily rainfall model to New Zealand data. J. Hydrol. **340**(1), 25–39 (2007)
744. A.-L. Tiberi Wadier, A. Laugel, M. Benoit, Construction of the numerical wave databases Anemoc-2 on the Mediterranean Sea and the Atlantic Ocean through hindcast simulations over the period 1979–2010, in *Advances in Hydroinformatics*, ed. by P. Gourbesville, J.A. Cunge, G. Caignaert. Part of the Series Springer Water (2015)
745. H. Touchette, The large deviation approach to statistical mechanics. Phys. Rep. **478**, 1–69 (2009)
746. G. Toulemonde, *Estimation et tests en théorie des valeurs extrêmes*. Ph.D. thesis, Université Paris VI (2008)

747. V.L. Tran, *Stochastic models of solar radiation processes*. Ph.D. thesis, Université d'Orléans, 2013
748. E.R.W. Tredger, J.T.H. Lo, S. Haria, H.H.K. Lay, N. Bonello, B. Hlavka, C. Scullion, Bias, guess and expert judgement in actuarial work. Br. Actuar. J. **21**, 545–578 (2016)
749. W.W. Tryon, A simplified time-series analysis for evaluating treatment interventions. J. Appl. Behav. Anal. **15**, 423–429 (1982)
750. Y.L. Tsai, D.J. Murdoch, E.J. Dupuis, Influence measures and robust estimators of dependence in multivariate extremes. Extremes **14**(4), 343–363 (2011)
751. Y.L. Tsai, D.J. Murdoch, E.J. Dupuis, A robust test for asymptotic independence of bivariate extremes. Extremes **47**(1), 172–183 (2011)
752. C.P. Tsokos, B. Miladinovic, Bayesian reliability analysis for the Gumbel failure model. Neural Parallel Sci. Comput. **16**, 93–104 (2008)
753. Y. Tsompanakis, N.D. Lagaros, M. Papadrakakis, *Structural Design Optimization Considering Uncertainties* (Taylor and Francis, Milton Park, 2008)
754. S. Tuhkanen, Climatic parameters and indices in plant geography. Folia Geobotanica et Phytotaxonomica **16**, 292 (1981)
755. A. Tversky, D. Kahneman, Judgment under Uncertainty: Heuristics and Biases. Science **185**, 1124–1131 (1974)
756. H. Uehara, K. Ando, Studies for prevention and mitigation of natural disasters using the K computer. Fujitsu Sci. Tech. J. **50**, 1–6 (2014)
757. H. Van de Vyver, A.W. Delcloo, Stable estimations for extreme wind speeds. an application to Belgium. Theor. Appl. Climatol. **105**, 417–429 (2011)
758. A.W. van der Vaart, *Asymptotic Statistics* (Cambridge University Press, Cambridge, 1998)
759. P.H.A.J.M. Van Gelder, A new statistical model for extreme water levels along the Dutch coast, in *Stochastic Hydraulics'96, Proceedings of the 7th IAHR International Symposium*, ed. by K.S. Tickle, I.C. Goulter, C.C. Xu, S.A. Wasimi, F. Bouchart (Rotterdam, Balkema, 1996), pp. 243–249
760. P.H.A.J.M. Van Gelder, J. De Ronde, N.M. Neykov, P. Neytchev, Regional frequency analysis of extreme wave heights: Trading space for time, in *Proceedings of the 27th International Conference on Coastal Engineerong (ICCE 2000), Sydney* (2000)
761. P.H.A.J.M. Van Gelder, N.M. Neykov, Regional frequency analysis of extreme water level along the Dutch coast using l-moments: a preliminary study, in *International Scientific Conference on Stochastic Models of Hydrological Processes and their Application to Problems of Environmental Preservation, Moscow* (1998)
762. K.S. Van Horn, Constructing a logic of plausible inference: a guide to Cox's theorem. Int. J. Approx. Reason. **34**, 3–24 (2003)
763. J.M. van Noortwijk, H.J. Kalk, E.H. Chbab, Bayesian estimation of design loads. HERON **49**, 189–203 (2004)
764. M.T.H. van Vliet, F. Ludwig, J.J.G. Zwolsman, G.P. Weedon, P. Kabat, Global river temperatures and sensitivity to atmospheric warming and changes in river flow. Water Resour. Res. **47** (2011). https://doi.org/10.1029/2010WR009198
765. C.J. Van Westen, Remote sensing and GIS for natural hazards assessment and disaster risk management. J. Time Ser. Anal. **27**, 685–701 (2006)
766. S.B. Vardeman, C.S. Lee, Likelihood-based statistical estimation from quantized data. IEEE Trans. Instrum. Meas. **54**, 409–414 (2005)
767. F. Varenne, Épistémologie des modèles et des simulations: tour d'horizon et tendances, in *Les Modèles, possibilités et limites. Jusqu'où va le réel ?, J.M. Levy (dir.), Matériologiques*, avril:13–46 (2014)
768. A. Vehtari, A. Gelman, J. Gabry, Practical Bayesian evaluation using leave-one-out cross-validation and WAIC. Stat. Comput. **27**, 1413–1432 (2017)
769. A.K. Verma, S. Ajit, D.R. Karanki, *Reliability and Safety Engineering*. Springer Series in Reliability Engineering, 2nd edn. (2015)
770. J.L. Vilar Zanón, C. Lozano Colomer, On Pareto conjugate priors and their application to large claims reinsurance premium calculation. ASTIN Bull. **37**, 405–428 (2007)

771. E.L. Vischer, *Histoire de la protection contre les crues en Suisse. Des origines jusqu'au 19ième siècle*. Rapport de l'Office Fédéral de l'Environnement (OFEV) (2003)
772. F.J. Walchan, G. Deodatis, Getting the most form your extreme wind data: a step-by-step guide. J. Res. Natl. Inst. Stand. Technol. **99**, 399–412 (1994)
773. A. Wald, An extension of Wilks's method for setting tolerance limits. Ann. Math. Stat. **29**, 45–55 (1943)
774. H.M. Walker, The contributions of Karl Pearson. J. Am. Stat. Assoc. **58**, 11–22 (1958)
775. S.N. Ward, E. Asphaug, Asteroid impact tsunami: a probabilistic hazard assessment. Icarus (1999)
776. L. Wasserman, Bayesian model comparison: review and discussion. J. Math. Psychol. **44**, 92–107 (2000)
777. G.S. Watson, Smooth regression analysis. Sankhyā Indian J. Stat. Ser. A **26**, 359–372 (1954)
778. M.D. Weinberg, Computing the bayes factor from a Markov Chain Monte Carlo simulation of the posterior distribution. Bayesian Anal. **7**, 737–770 (2012)
779. A. Weisheimer, T.N. Palmer, On the reliability of seasonal climate forecasts. J. R. Soc. Interface **11**, 20131162 (2014)
780. J. Weiss, *Analyse régionale des aléas maritimes extrêmes*. Ph.D. thesis, Université Paris Est (2012)
781. J. Weiss, P. Bernardara, Comparison of local indices for regional frequency analysis with an application to extreme skew surges. Water Resour. Res. **49**, 2940–2951 (2013)
782. J. Weiss, P. Bernardara, M. Benoit, Formation of homogeneous regions for regional frequency analysis of extreme significant wave heights. J. Geophys. Res. **119**, 2906–2922 (2014)
783. J. Weiss, P. Bernardara, M. Benoit, Modeling intersite dependence for regional frequency analysis of extreme marine events. Water Resour. Res. **50**, 5926–5940 (2014)
784. W.G. Cochran, The χ^2 test of goodness of fit. Ann. Math. Stat. **23**, 315–345 (1952)
785. P.A. Wilderer (ed.), *Treatise on Water Science* (Elsevier, Amsterdam, 2010)
786. D. Wilks, Multisite generalization of a daily stochastic precipitation generation model. J. Hydrol. **210**, 178–191 (1998)
787. S.S. Wilks, Determination of sample sizes for setting tolerance limits. Ann. Math. Stat. **12**, 91–96 (1941)
788. P. Willems, J. Olsson, K. Arnbjerg-Nielsen, S. Beecham, A. Pathirana, I. Bulow Gregersen, H. Madsen, V.T.V Nguyen, *Impacts of Climate Change on Rainfall Extremes and Urban Drainage Systems* (IWA Publishing, London, 2012)
789. L.J. Wolfson, N. Bousquet, *Elicitation Encyclopedia of Statistics (StatRef)* (Wiley, New York, 2016)
790. S.A. Wood, J. Wang, Priors for a Bayesian analysis of extreme values, in *Report of Melbourne Business School* (2009), http://works.bepress.com/sally_wood/5/
791. W. Xie, B.L. Nelson, R.B. Barton, Multivariate input uncertainty in output analysis for stochastic simulation. ACM Trans. Model. Comput. Simul. **27** (2016)
792. L. Yan, L. Xiong, S. Guo, C.-H. Xu, J. Xia, T. Du, Comparison of four nonstationary hydrologic design methods for changing environment. J. Hydrol. **551**, 132–150 (2017)
793. L.C. Young, On randomness in ordered sequences. Ann. Math. Stat. **12**, 293–300 (1941)
794. K. Yu, Z. Lu, J. Stander, Quantile regression: applications and current research areas. The Statistician **52**, 331–350 (2003)
795. A. Zellner, Maximal data information prior distributions, in *New Developments in the Applications of Bayesian Methods*, eds. by A. Aykac, C. Brunat (Amsterdam, North-Holland, 1977), pp. 201–215
796. A. Zellner, *On Assessing Prior Distributions and Bayesian Analysis with G-Prior Distributions* (Elsevier North-Holland, Amsterdam, 1986), pp. 233–243
797. A. Zellner, Models, prior information and Bayesian analysis. J. Econ. **75**, 51–68 (1996)
798. A. Zellner, Past and recent results on maximal data information priors. J. Stat. Res. **32**, 1–22 (1998)

799. Z. Zeng, A.D. Ziegler, T. Searchinger, L. Yang, A. Chen, K. Ju, S. Piao, L.Z.X. Li, P. Ciais, J. Chen, D. an Liu, C. Azorin-Molina, A. Chappell, D. Medvigy, E.F. Wood, A reversal in global terrestrial stilling and its implication for wind energy production, in *Nature Climate Change* (2019)
800. X. Zhang, F.W. Zwiers, G. Li, Monte Carlo experiments on the detection of trends in extreme values. J. Clim. **17**, 1945–1952 (2004)
801. Z. Zhang, Y. Qi, X. Ma, Asymptotic independence of correlation cofficients with application to testing hypothesis of independence. Electron. J. Stat. **5**, 342–372 (2011)
802. D. Zholud, *Extreme value analysis of huge datasets*. Ph.D. thesis, University of Gothenburg (2011)
803. R.D. Zhou, P. Eng, C. Richard Donnelly, D.G. Judge, On the relationship between the 10,000 year flood and the maximum probable maximum flood, in *Proceedings of the HydroVision, Conference* (Denver, CO, USA, 2008), p. 2008
804. L. Zhu, X. Liu, R. Lund, A likelihood for correlated extreme series. Environmetrics **30** (2018)
805. R. Zieliński, Estimating quantiles with linex loss function applications to VaR estimation. Appl. Math. **32**, 367–373 (2005)
806. E. Zio, F. Di Maio, Bootstrap and order statistics for quantifying thermal-hydraulic code uncertainties in the estimation of safety margins. Sci. Technol. Nuclear Install., ID 340164 (2008). https://doi.org/10.1155/2008/340164
807. E. Zio, N. Pedroni, *Methods for representing uncertainty. A literature review*. Les Cahiers de la Sécurité Industrielle, no. 2013-03 (2013)

Index

N. Bousquet and P. Bernardara (eds.), *Extreme Value Theory with Applications to Natural Hazards*, https://doi.org/10.1007/978-3-030-74942-2

Milton Keynes UK
Ingram Content Group UK Ltd.
UKHW052121200324
439639UK00001B/6

9 783030 749415